Soil Biology

Volume 51

Series Editor
Ajit Varma, Amity Institute of Microbial Technology,
Amity University Uttar Pradesh, Noida, UP, India

More information about this series at http://www.springer.com/series/5138

Muhammad Zaffar Hashmi • Vladimir Strezov •
Ajit Varma
Editors

Antibiotics and Antibiotics Resistance Genes in Soils

Monitoring, Toxicity, Risk Assessment and Management

Springer

Editors
Muhammad Zaffar Hashmi
Department of Meteorology
COMSATS Institute of Information Technology
Islamabad, Pakistan

Vladimir Strezov
Faculty of Science and Engineering,
Department of Environmental Sciences
Macquarie University
Sydney, New South Wales
Australia

Ajit Varma
Amity Institute of Microbial Technology
Amity University Uttar Pradesh
Noida, UP
India

ISSN 1613-3382 ISSN 2196-4831 (electronic)
Soil Biology
ISBN 978-3-030-09765-3 ISBN 978-3-319-66260-2 (eBook)
DOI 10.1007/978-3-319-66260-2

Printed on acid-free paper

This Springer imprint is published by Springer Nature
The registered company is Springer International Publishing AG
The registered company address is: Gewerbestrasse 11, 6330 Cham, Switzerland

Preface

Worldwide consumption of antibiotics has increased drastically in the past few decades. The application of antibiotics has led to the production of antibiotic resistance genes (ARGs) which represent a growing and serious human health threat worldwide. Recent research highlighted that the main sources, reservoirs, and recipients of antibiotics and ARGs are water, air, and soil. Antibiotics and ARGs are increasingly being recognized as emerging contaminants, threatening effective treatment of infections and carrying a great risk to public health. Anthropogenic activities such as the rise in antibiotic use for medical and agricultural purposes are considered a major cause for escalating the threat. The environmental risks of pharmaceuticals, in general, were first identified in the 1990s followed by a series of monitoring and effect studies.

The overarching theme of this book is to summarize the current state of knowledge of antibiotics and ARGs in the soil environment. The book covers a wide range of topics for understanding the antibiotics and ARGs in soils, their risk to the environment, and options for effective control. It presents some very important tools and methodologies that can be used to address antibiotics and ARGs in a consistent, efficient, and cost-effective manner. Furthermore, this book includes antibiotic producing microorganisms, the routes of entry and fate of antibiotics and resistance genes, biomonitoring approaches, dissemination of ARGs in soils, risk assessment, the impact of antibiotics and ARGs on the soil microbial community and other biota, bioremediation and biodegradation approaches, as well as soil management strategies for antibiotics and ARG-contaminated soils. Special emphasis was given to dissemination mechanisms of ARGs in soil, as soil plays a crucial role in the development of antibiotic resistance traits in bacteria and the distribution of antibiotic-resistant microbial species, resistant genetic material, and antibiotic compounds. Resistance genes appear to be everywhere in nature—in pathogens, commensals, and environmental microorganisms. We invite you to gain a broader insight into the role of the soils in the mechanisms of resistance development, the dissemination of antibiotic-resistant genetic elements, and the transport of ARGs or antibiotics as environmental contaminants through the presentations of our contributing authors in this book.

Antibiotics are produced by several groups of microbes such as bacteria, fungi, and actinomycetes as their natural defense system against other microbes living in their vicinity. Soil microorganism had always been the primary source for the production of antibiotics and still continues to maintain its significance. But indiscriminate use of antibiotics and disinfectants in medicine, agriculture, and fish culture and their release in environment have given birth to another critical problem of multidrug-resistant pathogenic microbes, and hence, we are still in need of effective metabolites that can be used as antibiotics to combat these resistant strains. The first chapters of book include the history, consumption, physicochemical properties, and sources of antibiotics in soil, monitoring, mode of action, and applications of antibiotics. Recent reports showed that dosing livestock animal with antibiotics increases greenhouse gas emissions especially from the cow dung. The use of antibiotics also disrupts the microbes, and microbes may not perform vital functions to combat against the climate change. These antibiotics change the microbes present in the digestive system of dung beetles, which are considered important in carbon cycling and improving soil. Some chapters provide an overview of antibiotics in soil–plant system including the accumulation of antibiotics in different plants. As antibiotics are used in escalating quantities, there is a growing concern over the presence, toxicity, and fate of antibiotics in soil which may pose adverse effects on plants, soil biology, crop yield, and quality of production. Keeping in view latest advances in the instrument to detect antibiotics and ARGs such as application of next-generation sequencing, other metagenomic techniques, antimicrobial resistance dashboard application, and point of care (POC), i.e., lab on a chip, and gas chromatography techniques, their strengths and limitations were also discussed. But successful assessment of antibiotics and ARGs requires appropriate research questions.

This book provides a brief overview of recent research in determining the impact of antibiotics, their bioactive metabolites, and ARGs that enter the soil on the structure, diversity, and function of soil microbial communities and human health. It has been established that the release of these drugs, their residues, and/or metabolites disturbs the environment and threatens soil inhabitants such as earthworms. In vitro and in vivo assays have largely focused on the acute genotoxicity and biochemical toxicity of these compounds in earthworms. These drugs can be poisonous to earthworms and other inhabitants of soil ecosystems; the majority of them have been identified as genotoxic and many as causing biochemical toxicity. In the long term, genomc disturbances due to genotoxicity as well as biochemical toxicity may impair growth, reproduction, and population dynamics in these organisms. The book also describes the risk assessment process developed by various government bodies in order to determine the risks of releasing antibiotics and ARGs into the soil and also identifies the existing challenges.

Indeed, regional management regimes for agricultural and clinical use of antibiotics, together with good hygiene, have in many cases proved successful in minimizing resistance on a national basis. The last chapter aims to identify management options for reducing the spread of antibiotics and antibiotic resistance determinants via environmental pathways. Management options with respect to bioremediation as a sustainable soil management; limiting agricultural sources; and treatment of hospital, domestic, and industrial wastewater were also discussed.

Most chapters in the book are written to a fairly advanced level and should be of interest to the graduate student and practicing scientist. We also hope that the subject matter treated will be of interest to people outside medicine, soil, biology, and chemistry and to scientists in industry as well as government and regulatory bodies.

Islamabad, Pakistan — Muhammad Zaffar Hashmi
Sydney, Australia — Vladimir Strezov
Noida, India — Ajit Varma

Contents

1 **Antibiotics Producing Soil Microorganisms** 1
Niharika Chandra and Sunil Kumar

2 **Antibiotics Resistance Genes** 19
Ahmad Hasnain, Wajid Nasim, Hussani Mubarak, Nosheen Mirza, Sohaib Khan, Xiaomei Su, Sarfraz Ahmed, and Muhammad Zaffar Hashmi

3 **A Review on Antibiotics Consumption, Physico-Chemical Properties and Their Sources in Asian Soil** 39
Sija Arun, Moitraiyee Mukhopadhyay, and Paromita Chakraborty

4 **Entry Routes of Veterinary Antibiotics in the Environment** 55
Reep P. Tasho and Jae Young Cho

5 **Monitoring of Antibiotics and Antibiotic Resistance Genes in Agroecosystems** 71
Sarfraz Ahmed, Muhammad Ibrahim, Rabia Khushi Muhammad, Iqra Naseem, Rubbca Idrees, Muhammad Zaffar Hashmi, and Ateeq-ur- Rehman

6 **Role of Antibiotics in Climate Change** 97
Rida Akram, Muhammad Zaffar Hashmi, and Wajid Nasim

7 **Potential Dissemination of ARB and ARGs into Soil Through the Use of Treated Wastewater for Agricultural Irrigation: Is It a True Cause for Concern?** 105
Nada Al-Jassim and Pei-Ying Hong

8 **Antibiotic Resistance Gene Due to Manure Application** 141
Srujana Kathi

9 **Antibiotics and Antibiotics Resistance Genes Dissemination in Soils** 151
Eddie Cytryn, Zdzisław Markiewicz, and Magdalena Popowska

10 **Dissemination Mechanism of Antibiotic Resistance Genes in Environment** 191
Manoj Kumar, Ram Prasad, Shivesh Sharma, Ajit Varma, and Vivek Kumar

11 **Fate of Antibiotics in Soil** 207
Rida Akram, Asad Amin, Muhammad Zaffar Hashmi, Abdul Wahid, Muhammad Mubeen, Hafiz Mohkum Hammad, Shah Fahad, and Wajid Nasim

12 **Uptake of Antibiotics by Plants** 221
Sandra A. Youssef and Isam I. Bashour

13 **Recent Advances in Methods for the Detection of Antibiotics and Antibiotics Resistance Genes in Soil** 239
Hussani Mubarak, Nosheen Mirza, and Muhammad Zaffar Hashmi

14 **Elucidation of Emerging Nanomaterials Impacts on Antibiotic Resistance Against Soil and Aquatic Microflora** 259
Toqeer Ahmed

15 **The Effects of Antibiotics on the Structure, Diversity, and Function of a Soil Microbial Community** 283
Anna Piotrowska-Długosz

16 **Soil Antibiotics and Transfer of Antibiotic Resistance Genes Affecting Wildlife** 313
Vanessa Silva, Isabel Carvalho, Gilberto Igrejas, and Patrícia Poeta

17 **Genotoxicity and Biochemical Toxicity of Soil Antibiotics to Earthworms** 327
Sarfraz Ahmed, Muhammad Ibrahim, Rashem Waheed, Abu Bakar Hassan Azdee, Muhammad Zaffar Hashmi, and Shabir Ahmed

18 **Potential Environmental, Ecological and Health Effects of Soil Antibiotics and ARGs** 341
Biljana Balabanova

19 **Risk Assessment of Antibiotics and Antibiotic-Resistant Genes in Soil** 367
Khushbu Salian and Vladimir Strezov

20 **Antibiotics in the Soil: Sources, Environmental Issues, and Bioremediation** 387
Umesh B. Jagtap

21 Management and Regulation of Antibiotics and Antibiotics Resistance Genes in Soils 397
Muhammad Zaffar Hashmi, Wajid Nasim, Hussani Mubarak, Nosheen Mirza, Sohaib Khan, Xiaomei Su, Sarfraz Ahmed, and Ahmad Hasnain

Erratum to: Antibiotics and Antibiotics Resistance Genes in Soils E1

Index 411

List of Contributors

Sarfraz Ahmed Department of Biochemistry, Bahauddin Zakariya University, Multan, Pakistan

Shabir Ahmed Department of Biochemistry, Bahauddin Zakariya University, Multan, Pakistan

Toqeer Ahmed Centre for Climate Research and Development (CCRD), COMSATS Institute of Information Technology, Islamabad, Pakistan

Rida Akram Department of Environmental Sciences, COMSATS Institute of Information Technology (CIIT), Vehari, Pakistan

Nada Al-Jassim Biological and Environmental Science & Engineering Division (BESE), King Abdullah University of Science and Technology (KAUST), Thuwal, Saudi Arabia

Water Desalination and Reuse Center (WDRC), King Abdullah University of Science and Technology (KAUST), Thuwal, Saudi Arabia

Asad Amin Department of Environmental Sciences, COMSATS Institute of Information Technology (CIIT), Vehari, Pakistan

Sija Arun Department of Civil Engineering, SRM University, Kattankulathur, Kancheepuram District, Tamil Nadu, India

Abu Bakar Hassan Azdee Department of Biochemistry, Bahauddin Zakariya University, Multan, Pakistan

Biljana Balabanova Faculty of Agriculture, University "Goce Delčev", Štip, Republic of Macedonia

Isam I. Bashour Faculty of Agricultural and Food Sciences, American University of Beirut, Beirut, Lebanon

Isabel Carvalho Veterinary Sciences Department, University of Trás-os-Montes and Alto Douro, Vila Real, Portugal

Paromita Chakraborty Department of Civil Engineering, SRM University, Kattankulathur, Kancheepuram District, Tamil Nadu, India

SRM Research Institute, SRM University, Kattankulathur, Kancheepuram District, Tamil Nadu, India

Niharika Chandra Faculty of Biotechnology, Institute of Bioscience and Technology, Shri Ramswaroop Memorial University, Barabanki, Uttar Pradesh, India

Jae Young Cho Department of Agriculture Chemistry, Chonbuk National University, Jeonju-si, Jeollabuk-do, Republic of Korea

Eddie Cytryn Institute of Soil, Water and Environmental Sciences, Volcani Center, Agricultural Research Organization, Bet Dagan, Israel

Shah Fahad College of Plant Science and Technology, Huazhong Agricultural University, Wuhan, China

Hafiz Mohkum Hammad Department of Environmental Sciences, COMSATS Institute of Information Technology (CIIT), Vehari, Pakistan

Muhammad Zaffar Hashmi Department of Meteorology, COMSATS Institute of Information Technology (CIIT), Islamabad, Pakistan

Pei-Ying Hong Water Desalination and Reuse Center (WDRC), King Abdullah University of Science and Technology (KAUST), Thuwal, Saudi Arabia

Biological and Environmental Science & Engineering Division (BESE), King Abdullah University of Science and Technology (KAUST), Thuwal, Saudi Arabia

Muhammad Ibrahim Department of Biochemistry, Bahauddin Zakariya University, Multan, Pakistan

Rubbea Idrees Department of Biochemistry, Bahauddin Zakariya University, Multan, Pakistan

Gilberto Igrejas Department of Genetics and Biotechnology, Functional Genomics and Proteomics' Unit, University of Trás-os-Montes and Alto Douro, Vila Real, Portugal

Functional Genomics and Proteomics Unit, University of Tras-os-Montes and Alto Douro (UTAD), Vila Real, Portugal

Research Unit on Applied Molecular Biosciences (UCIBIO-REQUIMTE), University NOVA of Lisboa, Lisboa, Caparica, Portugal

Veterinary Sciences Department, University of Trás-os-Montes and Alto Douro, Vila Real, Portugal

Umesh B. Jagtap Department of Botany, Government Vidarbha Institute of Science and Humanities, Amravati, Maharashtra, India

Srujana Kathi Pondicherry University, Kalapet, Puducherry, India

Manoj Kumar Amity Institute of Microbial Technology, Amity University, Noida, Uttar Pradesh, India

Vivek Kumar Amity Institute of Microbial Technology, Amity University, Noida, Uttar Pradesh, India

Sunil Kumar Faculty of Biotechnology, Institute of Bioscience and Technology, Shri Ramswaroop Memorial University, Barabanki, Uttar Pradesh, India

Zdzisław Markiewicz Department of Applied Microbiology, Institute of Microbiology, University of Warsaw, Warsaw, Poland

Nosheen Mirza School of Metallurgy and Environment, Central South University, Changsha, China

Chinese National Engineering Research Center for Control and Treatment of Heavy Metal Pollution, Changsha, China

Department of Environmental Sciences, COMSATS Institute of Information Technology, Abbottabad, Pakistan

Hussani Mubarak Department of Soil and Environmental Sciences, Ghazi University, Dera Ghazi Khan, Pakistan

School of Metallurgy and Environment, Central South University, Changsha, China

Chinese National Engineering Research Center for Control and Treatment of Heavy Metal Pollution, Changsha, China

Muhammad Mubeen Department of Environmental Sciences, COMSATS Institute of Information Technology (CIIT), Vehari, Pakistan

Rabia Khushi Muhammad Department of Biochemistry, Bahauddin Zakariya University, Multan, Pakistan

Moitraiyee Mukhopadhyay SRM Research Institute, SRM University, Kattankulathur, Kancheepuram District, Tamil Nadu, India

Iqra Naseem Department of Biochemistry, Bahauddin Zakariya University, Multan, Pakistan

Wajid Nasim Department of Environmental Sciences, COMSATS Institute of Information Technology (CIIT), Vehari, Pakistan

CIHEAM-Institut Agronomique Mediterraneen de Montpellier (IAMM), Montpellier, France

National Research Flagship, CSIRO Sustainable Ecosystems, Toowoomba, QLD, Australia

Anna Piotrowska-Długosz Faculty of Agriculture and Biotechnology, Division of Biochemistry, Department of Soil Science and Soil Protection, UTP University of Science and Technology, Bydgoszcz, Poland

Patrícia Poeta Veterinary Sciences Department, University of Trás-os-Montes and Alto Douro, Vila Real, Portugal

Research Unit on Applied Molecular Biosciences (UCIBIO-REQUIMTE), University NOVA of Lisboa, Lisboa, Caparica, Portugal

Magdalena Popowska Department of Applied Microbiology, Institute of Microbiology, Faculty of Biology, University of Warsaw, Warsaw, Poland

Ram Prasad Amity Institute of Microbial Technology, Amity University, Noida, Uttar Pradesh, India

Ateeq-ur- Rehman Department of Biochemistry, Bahauddin Zakariya University, Multan, Pakistan

Department of Plant Pathology, Bahauddin Zakariya University, Multan, Pakistan

Khushbu Salian Faculty of Science and Engineering, Department of Environmental Sciences, Macquarie University, North Ryde, NSW, Australia

Shivesh Sharma Department of Biotechnology, MLN National Institute of Technology, Allahabad, Uttar Pradesh, India

Vanessa Silva Veterinary Sciences Department, University of Trás-os-Montes and Alto Douro, Vila Real, Portugal

Vladimir Strezov Faculty of Science and Engineering, Department of Environmental Sciences, Macquarie University, North Ryde, NSW, Australia

Reep P. Tasho Department of Agriculture Chemistry, Chonbuk National University, Jeonju-si, Jeollabuk-do, Republic of Korea

Ajit Varma Amity Institute of Microbial Technology, Amity University, Noida, Uttar Pradesh, India

Abdul Wahid Department of Environmental Sciences, Bahauddin Zakariya University, Multan, Pakistan

Rashem Waheed Department of Biochemistry, Bahauddin Zakariya University, Multan, Pakistan

Sandra A. Youssef Faculty of Agricultural and Food Sciences, American University of Beirut, Beirut, Lebanon

List of Abbreviations

AAC(6′)	6′-N-aminoglycoside acetyltransferase
ARGs	Antibiotic resistance genes
ARB	Antibiotic-resistant bacteria
ASP	Activated sludge process
ATP	Adenosine triphosphate
AWCD	Average well color development
Bp	Base-pairs
CLPP	Community-level physiological profiling
CTC	Chlortetracycline
CH_4	Methane
CO_2	Carbon dioxide
CFU	Colony-forming units
CRISPR	Clustered regularly interspaced short palindromic repeats
CRISPR-Cas	CRISPR and CRISPR-associated
DGGE	Denaturing gradient gel electrophoresis
DHA	Dehydrogenase
DOM	Dissolved organic matter
DNA	Deoxyribonucleic acid
DOX	Doxycycline
DT50	Dissipation half-lives
EC10	Effective dose with 10% inhibition
ED50	Effective dose with 50% inhibition
FAMEs	Fatty acid methyl esters
GC-MS	Gas chromatography-mass spectrometry
GSTs	Glutathione S-transferases
GHG	Greenhouse gases
GC content	Guanine and cytosine content
HGT	Horizontal gene transfer
IPCC	Intergovernmental Panel on Climate Change
Incgroup	Incompatibility group
ISCR	Insertion sequence common region

K*d*	Distribution coefficient
Kb	Kilo base-pairs
MBL	Metallo-beta-lactamase
MOI	Multiplicity of infection
mRNA	Messenger ribonucleic acid
NADH	Nicotinamide adenine dinucleotide
NADPH	Nicotinamide adenine dinucleotide phosphate
NDM	New Delhi metallo-beta-lactamase
OTC	Oxytetracycline
PBPs	Penicillin binding proteins
PCR	Polymerase chain reaction
pKa	Acid dissociation constant
PLFAs	Phospholipid fatty acids
$PLFA_{tot}$	Total phospholipid fatty acids
RNA	Ribonucleic acid
rRNA	Ribosomal ribonucleic acid
ROS	Reactive oxygen species
SDT	Sulfadimethoxine
SDZ	Sulfadiazine
SIR	Substrate-induced respiration
SMX	Sulfamethoxazole
SMZ	Sulfamethazine
SOM	Soil organic matter
STZ	Sulfathiazole
TC	Tetracycline
TYL	Tylosin
URE	Urease
WWTP	Wastewater treatment plant
WHO	World Health Organization

Chapter 1
Antibiotics Producing Soil Microorganisms

Niharika Chandra and Sunil Kumar

1.1 Introduction

Antibiotics are secondary metabolites produced by microorganism, which have antimicrobial properties and have been used as a chemotherapeutic agent against infectious and disease causing microbes for decades. The use of antibiotics along with proper sanitation and vaccination has led to drastic decrease in mortality due to infectious diseases that were primarily lethal (Procopio et al. 2012). Isolation of antibiotics from microorganism is much easier than chemical synthesis of these compounds and hence has resulted in discovery of countless novel antibiotics till date (Shlaes 2010). These antibiotics are being used to prevent and cure microbial infections in various spheres of human development such as human medicine, veterinary science, animal husbandry and maintenance of livestock, agriculture, and aquaculture (Kummerer 2009).

Soil is home to a large and diverse population of microorganisms due to its heterogeneous nature. Large variation in biotic and abiotic conditions of soil makes its microbial inhabitants to adapt and develop strategies for survival and successful reproduction. Production of antimicrobials is one of the most potent strategies for this adaptation (Davies 1990). Antibiotics such as β-lactams, aminoglycosides, streptomycins, and tetracyclines and others are being produced by soil bacteria and fungi. Fungal antibiotics such as penicillins, cephalosporin, fusidic acid griseofulvin, and fumagillin have been obtained by fungal species *Penicillium*, *Cephalosporium*, and *Aspergillus*. Several *Pseudomonas* species and *Bacillus* species are among the soil bacteria which have been exploited for the production

N. Chandra • S. Kumar (✉)
Faculty of Biotechnology, Institute of Biosciences and Technology, Shri Ramswaroop Memorial University, Hadauri, Tindola, Lucknow-Deva Road, Barabanki 225003, Uttar Pradesh, India
e-mail: sunilsbt@gmail.com; sunil.bio@srmu.ac.in

M.Z. Hashmi et al. (eds.), *Antibiotics and Antibiotics Resistance Genes in Soils*,
Soil Biology 51, DOI 10.1007/978-3-319-66260-2_1

of antibiotics like gramicidin, bacitracin, tyrothricin, pyocyanin, and pyrrolnitrin (Berdy 1974). *Streptomyces* species is one of the soil actinomycetes which have provided the highest number of commercial antibiotics such as tetracyclines, streptomycin, viomycin, and kanamycin. Several other commonly used antibiotics, gentamicin and rifamycin, have been isolated from actinomycetes like *Micromonospora*, *Actinomadura*, and *Nocardia* species (Berdy 1980).

Indiscriminate use of antibiotics and disinfectants in medicine, agriculture, and fish culture and their release in environment has given birth to another critical problem of multidrug-resistant pathogenic bacteria. Pathogenic microbes have come across a huge range of naturally occurring and synthetic antibiotics since long and have developed several defense mechanisms against them (Kummerer 2004; Demain and Sanchez 2009). The development of resistance is quick for those antimicrobials which target a single mechanism or enzyme to cause cell death. Whereas, those antibiotics which work against several targets or mechanisms to kill a microbe get resisted at a slower pace (Demain and Sanchez 2009). Therefore, new effective antibiotics are immediately required to combat these evolving resistant strains. Natural microbial metabolites still remain one of the most potent options to discover new chemotherapeutic agents (Rath et al. 2011). Soil from different ecosystems and biogeographical areas can be analyzed for discovery of new antimicrobial producing strains with novel mode of actions and secondary metabolites against the pathogenic resistant microbes.

1.2 Antibiotics

The word antibiotic which literally means "against life," has been defined in several ways by different researchers. Gottlieb (1967) defines antibiotics as low molecular weight organic compounds produced by microorganisms which inhibit the growth of other organisms in its vicinity at a very low concentration (Gottlieb 1967). Similarly, Thomashow et al. (1997) refer to antibiotics as chemically heterogeneous organic compounds of low molecular weight which are produced by some microorganisms to prevent growth and arrest the metabolic activity of other microorganisms (Thomashow et al. 1997). These compounds are produced by microbes as secondary metabolites, which do not play any part in the growth and reproduction. Instead, these secondary metabolites are produced in stationary phase and are released in the surrounding environment (Koberl et al. 2013). It was later realized that higher forms of life such as algae, plants, and animals also produced low molecular weight substances as secondary metabolites which possessed antimicrobial activity (Berdy 1980). Thus, Okafor (1987) defined antibiotics as compounds produced by any living form (microbes, algae, plants, or animals) which had the potential to inhibit the growth of other living form, at a low concentration (Okafor 1987). In a wider sense, antibiotics are chemotherapeutic agents which prevent the growth of life forms such as bacteria, fungi, virus, and protozoa.

1.2.1 History

The discovery of first antibiotic occurred by chance when a *staphylococci* agar plate inoculated by Sir Alexander Fleming got contaminated by a mold. The mold colony displayed a clear zone of bacterial inhibition around itself (Fleming 1980). The compound which was suppressing bacterial growth was identified as penicillin, which was then used as antibiotics to cure many infections and diseases caused by bacteria (Sköld 2006). Discovery of sulfonamides and β-lactam antibiotics in 1930s leads to an immense improvement in health and medicine services as diseases and bacterial infections which were fatal earlier became curable. Introduction of streptomycin in 1944 was another achievement for the cure of tuberculosis. The golden era for antibiotic discovery was marked till 1970, up to where maximum classes of antibiotics were discovered (Shlaes 2010). Table 1.1 lists the antibiotics discovered along with the source and year of discovery.

1.2.2 Classification

Antibiotics have been classified on the basis of several criteria such as mode of action, producing organism, and route of biosynthesis (Berdy 1974; Queener et al. 1978). However, some microbes may produce many antibiotics or more than one mode of action may operate simultaneously, making these criteria unmanageable. Antibiotics were then classified into 13 groups on the basis of their chemical structure. This classification is the most accepted one because it can easily accommodate the newly discovered antibiotics. The groups are aminoglycosides (e.g., kanamycin, neomycin, streptomycin), ansamacrolides (e.g., rifamycin), beta-lactams (e.g., ampicillin, meropenem, penicillin G, ceftiofur, cefotiam), chloramphenicol and analogues, lincosaminides (e.g., lincomycin), macrolides (e.g., erythromycin, oleandomycin, tylosin), nucleosides (e.g., puromycin), oligosaccharides (e.g., curamycin), peptides (e.g., neomycin, bacitracin, avermectin), phenazymes (e.g., myxin), polyenes (e.g., amphotericin B), polyethers (e.g., nigericin, monensin, salinomycin), and tetracyclines (e.g., tetracycline, chlortetracycline, oxytetracycline) (Queener et al. 1978) (Fig. 1.1).

1.2.3 Mechanism of Action

An ideal antibiotic should not interfere with the normal functioning of the host cell but should bring about inhibition of target microbe by disturbing its metabolism. This disturbance can be interference or inhibition of any biomolecule such as enzyme, nucleic acid, polysaccharides, or metabolites critical for survival (Shlaes 2010) (Fig. 1.2).

Table 1.1 Antibiotics produced by various microorganisms with discovery date

Antibiotic	Microorganism/source	Discovery year
Penicillin	*Penicillium notatum*	1929
Griseofulvin	*Penicillium griseojitlvum*	1939
Tyrothricin	*Bacillus brevis*	1939
Streptomycin	*Streptomyces griseus*	1944
Bacitracin	*Bacillus subtilis*	1945
Cephalosporins	*Streptomyces clavuligerus*	1945
Chloramphenicol	*Streptomyces venezualae*	1947
Neomycin	*Streptomyces fradiae*	1949
Tetracyclines	*Streptomyces aureofaciens*	1952
Viomycin	*Streptomyces vinaceus*	1951
Vancomycin	*Streptomyces orientalis*	1956
Kanamycin	*Streptomyces kanamyceticus*	1957
Rifamycin	*Amycolatopsis mediterrane*	1957
Fusidic acid	*Fusidium coccineum*	1963
Gentamicin	*Micromonospora purpurea*	1963
Fosfomycin	*Streptomyces fradiae*	1969
Ribostamycin	*Streptomyces ribosidificus*	1970
Mupirocin	*Pseudomonas fluorescens*	1985
Biapenem	Actinomycete	2002
Doripenem	Actinomycete	2005
Retapamulin	Fungus	2007
Ceftobiprole Medocaril	Fungus	2008
Ceftaroline fosamil	Fungus	2010

Adapted from: Handbook of Antibiotics, edited by Baron (1950) (Butler and Cooper 2011; Procopio et al. 2012)

1.2.3.1 Replication (of Genetic Material) and Transcription Inhibitors

DNA (deoxyribonucleic acid) replication is one of the essential functions of growing cells and hence inhibition of this process results in inhibition of cell division. Antibiotics which bind to form complex with important components of DNA replication such as topoisomerase can cause bacterial death (Franklin and Snow 1981; Chopra et al. 2002). Similarly, all kinds of RNA (ribonucleic acid) are synthesized by the action of DNA-dependent RNA polymerase through transcription, which is again essential for growth. Rifamycins bind with the β subunit of DNA-dependent RNA polymerase and inhibit the initiation and ribonucleotide chain growth in transcription (Hmmond and Lambert 1978; Nikaido 2009). Acridines disturb the phosphate backbone of DNA helix and actinomycin D binds with DNA double helix to hamper the movement of RNA polymerase across the DNA chain to inhibit transcription and hence cause bactericidal effect (Hmmond and Lambert 1978; Franklin and Snow 1981).

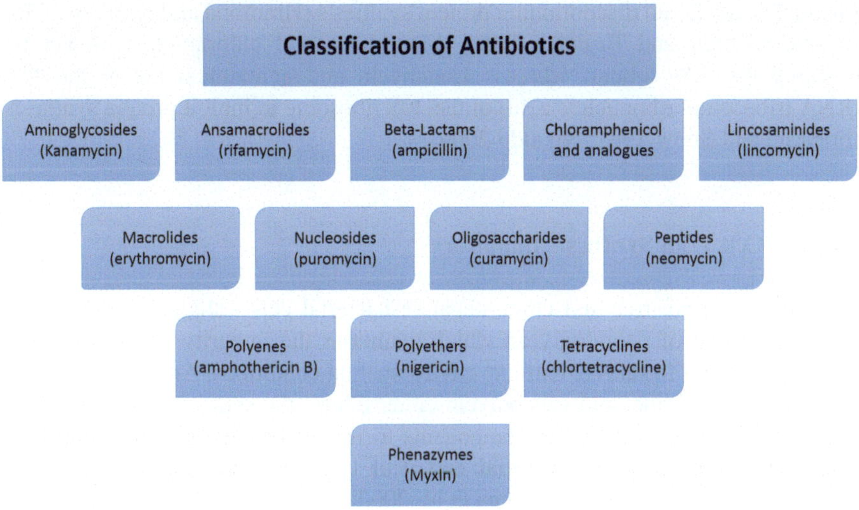

Fig. 1.1 Classification of antibiotics

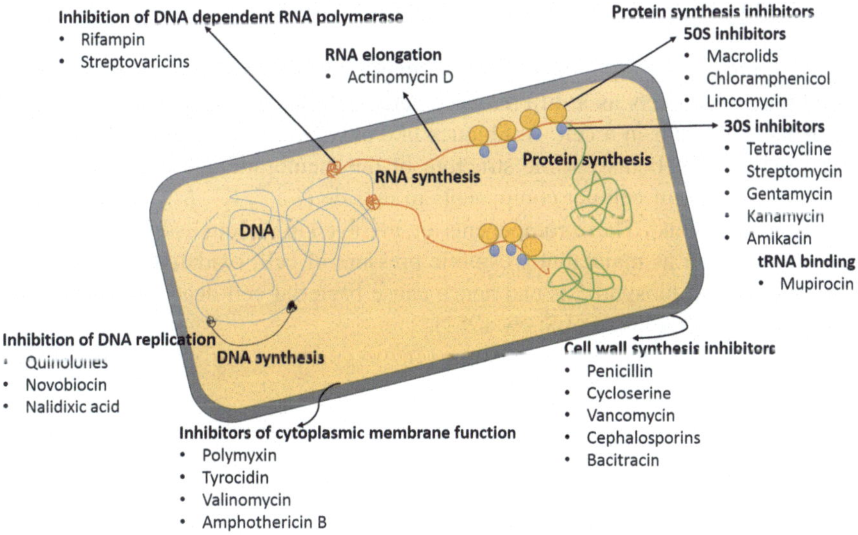

Fig. 1.2 Mechanism of action of antibiotics

1.2.3.2 Protein Synthesis Inhibitors

Synthesis of mRNA (messenger ribonucleic acid) involves the role of ribosomes (30S and 50S) and its interaction with other components, which is the target of antibiotics inhibiting translation. Streptomycin and tetracycline disrupt the 30S subunit of ribosome to block its interaction with amino acyl tRNA (transfer ribonucleic acid) and discontinuing protein synthesis (Hmmond and Lambert 1978; Brotz-Oesterhelt and Brunner 2008). Macrolides and chloramphenicol act by blocking the 50S subunit whereas kanamycin and gentamicin act at the 16S rRNA (ribosomal ribonucleic acid) of the 30S ribosome to inhibit protein synthesis (Franklin and Snow 1981; Nikaido 2009).

1.2.3.3 Cell Wall Synthesis Inhibitors

Both the Gram-positive and Gram-negative bacterial cell walls are composed of variable amount of peptidoglycan which maintains the integrity of cell wall and helps the bacteria to survive extreme environmental conditions. Antibiotics such as penicillin, cycloserine, and vancomycin either inhibit the synthesis of peptidoglycan or its assembly with other components to form an intact and functional cell wall, which then leads to bacterial cell wall lysis (Lancini and Parenti 1982; Hmmond and Lambert 1978; Chopra et al. 2002).

1.2.3.4 Inhibitors of Cytoplasmic Membrane Function

Lipids, proteins, and lipoproteins are the major constituents to form cytoplasmic membrane which acts as a differentially permeable barrier for exchange of ions, nutrients, and water in a cell. Antibiotics like polymyxin, tyrocidin, valinomycin, and amphotericin B disrupt the structure of the membrane and cause unwanted exchange or leak of cellular components to outside (Ca^{2+}, K^{+}, Mg^{2+} ions, metabolites, and nutrients). These components are involved in various essential processes for survival such as maintaining osmotic pressure of cell, oxidative phosphorylation, and protein biosynthesis and hence cause bacterial cell death (Hmmond and Lambert 1978; Franklin and Snow 1981).

1.2.4 Applications of Antibiotics

Antibiotics find their applications in medicine (human and animal), aquaculture, and agriculture (Kummerer 2009). β-lactam antibiotics including penicillins and cephalosporins are the most widely used group of antibiotics for human medicine all around the world. In case of animals, antibiotics are also used for animal breeding and promoting growth, other than veterinary practices (Gaskins et al. 2002). Antibiotics such as streptomycin with oxytetracycline are extensively used to prevent infections in crops, vegetables, and fruit yielding plants. Oxytetracycline, sulphonamides, premix, sarafloxacin, erythromycin, florfenicol, and several other antibiotics have been applied in farming of aquatic animals like fish and molluscs (Cabello 2004; Wolff 2004).

1.3 Soil as a Source of Microorganisms

The surface layer of Earth's crust is called soil which was formed by the weathering of geological rocks, and the process still continues. This outer, loose part of lithosphere is the region which supports the growth of plants, microbes, and other soil inhabitants by supplying nutrients and providing anchorage. Soil is the live and dynamic site for countless biological reactions and interactions which are life supporting (Aminov 2009; Alvarez Lerma et al. 2010; Allen et al. 2010). Soils also play significant roles in protecting and purifying freshwater ecosystems. Various natural services such as land for construction, food production, deriving raw material for several manufacturing processes, and other natural resources are all supported by soil. Hence, soil is of significant importance to humans. Soil consists of solid, liquid, and gaseous phases, and each type of soil varies in its characteristics depending upon the parent rock material, geomorphological and land use history, climate, and soil flora and fauna (Butler and Buss 2006).

1.3.1 Components of Soil

Soil is composed of several components such as minerals, organic matter, water, air, and soil inhabitants (Fig. 1.3). The mineral or inorganic content of soil (which consists of iron, aluminum, calcium, manganese, potassium, silicon, sulfur, phosphorus, and many other trace elements) is derived from the withering and corrosion of parent rock material. Therefore, the mineral content and compounds vary in different soil according to the parent rock. The organic matter in the soil which primarily consists of carbon, nitrogen, and phosphate sources is derived from living, fecal, and dead residues of plants and animals. Organic matter serves as a source of energy for microorganisms and other soil dwellers. Soil's water content depends upon the available rain, snowfall, and irrigation, and hence, varies according to the topographical position of the region. Soil gases consist of the atmospheric air that fill the pores present between soil particles. Both soil water and air are required for nutrient absorption, growth, and proper metabolism of plants and microbial population (Baron 1950; Martinez et al. 2009).

Another very important component of soil is the soil organisms. These soil dwellers can further be divided into soil flora and soil fauna. Soil microflora includes bacteria, actinomycetes, algae, and fungi. Soil fauna is composed of animals like protozoa, earthworms, nematodes, ants, rodents, etc. These soil organisms decompose the organic matter present in the soil and facilitate nutrient availability and absorption by higher plants and other organisms. Hence, these organisms have a significant role to play in any soil system, but still some soil organisms (bacteria, fungi, insects, or nematodes) can be parasitic and disease causing in plants and animals (Martinez et al. 2009).

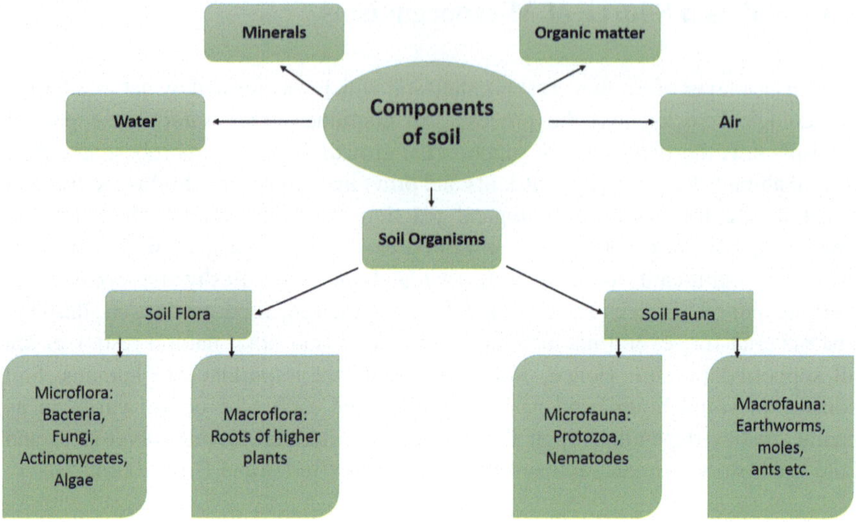

Fig. 1.3 Components of soil

1.3.2 Types of Microorganisms in Soil

Microorganisms are microscopic creatures that either exists as single cells or from colonies of cells. They can be grouped as bacteria, fungi, protozoa, micro-algae, and viruses. These organisms can be found in water, food, animal intestines, and several other different and extreme environments including soils (Lancini and Parenti 1982; Fischbach and Walsh 2009).

1.3.2.1 Bacteria

Bacteria can thrive in habitat and are the most abundant microorganisms present in soil. Bacteria are prokaryotic and unicellular microorganisms without any chlorophyll. They have been divided in three groups: *Cocci* (round-shaped), *Bacilli* (rod-shaped), and *Spirilla* (long wavy chains of cells). Out of these, *Bacilli* predominate in soil, followed by *Cocci* and *Spirilla*. *Pseudomonas*, *Arthrobacter*, *Achromobacter*, *Clostridium*, *Enterobacter*, *Sarcina*, *Micrococcus*, *Cyptophaga*, *and Chondrococcus* are some of the bacterial genera that are most commonly found in soil. Soil bacteria are very beneficial to plants and other soil organisms as they are the one to start the decomposition process and increase nutrient availability. Certain nitrogen fixing, nitrifying, denitrifying, and ammonifying bacteria are very helpful in making the atmospheric nitrogen available to plants (Ruan 2013).

1.3.2.2 Fungi

A fungus is another microorganism present in soil which has filamentous mycelium. Most common soil fungi belong to genera *Alternaria, Aspergillus, Botrytis, Cladosporium, Cephalosporium Chaetomium, Mucor, Monilia, Penicillium, Fusarium, Rhizopus, Gliocladium, Trichoderma, Pythium, Verticillium*, etc. Fungi also help to initiate decomposition of fresh organic matter and are resistant to acidic soil conditions. Fungi are the prime decomposers of lignin and also important in decay of cellulose, hemicellulose, and pectin (Ruan 2013).

1.3.2.3 Actinomycetes

These are microbes which have characteristics of both bacteria and fungi and have been classified into a separate category. Actinomycetes are unicellular similar to bacteria and have hyphae and conidia like fungi. *Streptomyces*, *Nocardia*, *Micromonospora*, and *Actinoplanes* are the most common and predominant genera present in soil. They also contribute in decomposition of organic residues and humus in soil (Ruan 2013; Qin et al. 2009; Koberl et al. 2013).

1.3.2.4 Algae

Algae grow in swamps, paddy fields, depressions, or flooded land where plenty of water and sunlight are available. Similar to plants, algae use sunlight to convert atmospheric carbon into sugar molecules, which are then utilized to derive energy. They help to maintain soil fertility and form mutualistic relationship with other organisms (Martinez et al. 2009).

1.3.2.5 Protozoa

Protozoa are single celled animals which move with the help of several organs like flagella, cilia, or pseudopodia. They are secondary consumers which feed on organic material of soil, soil bacteria and fungi, and other protozoans. They help in releasing nutrients from organic matter and maintain desirable microbial population in soil (Martinez et al. 2009).

1.3.3 Importance of Soil Microorganisms

Soil microbes play a vital role in almost all reactions and interactions occurring in soil, which support all the plants and consequently animals. As already discussed, soil microbes are primarily involved in breakdown and decomposition of organic matter. Soil microorganisms are extremely important for recycling of life-supporting nutrients such as carbon, nitrogen, and phosphorus which are critical for the formation of biomolecules. Soil bacteria can fix atmospheric nitrogen by the process of biological nitrogen fixation and ammonification. Several other nitrifying and denitrifying bacteria are also involved in cycling of this fixed nitrogen. Interaction of plants with soil microbes is known to promote plant growth by increasing nutrient availability and production of growth hormones and antibiotics. Beneficial microbes prevent plant diseases by shielding the roots and suppressing the growth of pathogenic bacteria and pests. Soil microbes also maintain soil structure by increasing its porosity, formation of soil aggregates, and increasing water infiltration. Soil microbes are also important for recycling of waste material and its detoxification. Several naturally occurring microbes are involved in biodegradation of oil, pesticides, insecticides, heavy metals, xenobiotic compounds, and other toxic contaminants. Many soil microbes also have potential uses in biofertilizers, bioremediation, biocomposting, and bioleaching. Another huge application of soil microbes is in production of secondary metabolites such as antibiotic which has a great impact on plants, animals, and human development, which will further be discussed in detail (Falconer and Brown 2009; Aminov 2009; Baron 1950; Baltz 2008; Demain and Sanchez 2009) (Fig. 1.4).

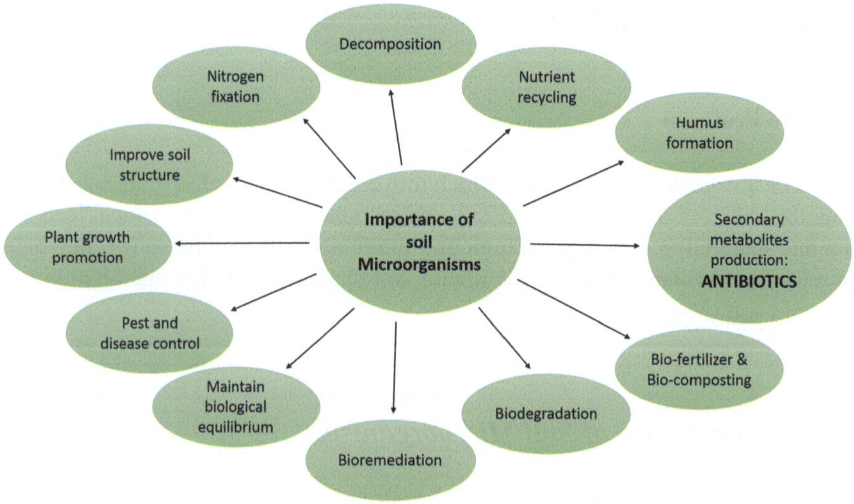

Fig. 1.4 Various aspects of benefit from soil microorganisms

1.3.4 *Isolation, Screening, and Identification of Antibiotic Producing Microorganisms from Soil*

The most common methods for isolation and primary screening of antibiotic producing microbes from soil include the crowded plate method and dilution plate method. In the crowded plate method, an aqueous dilution of soil sample is plated on agar plates such that it develops a lawn of microbial growth. Microbial colonies which show clear zone of inhibition around themselves are isolated for further studies. Dilution plate method is applied to isolate antibiotic producing microbes against a known microorganism. The diluted soil sample is mixed with the melted agar medium, poured, solidified, and incubated till microbial colonies appear. The plate is then inoculated or flooded with a growing culture of test organism and incubated again. The microbial colonies which show a clear zone of inhibition against the confluent growth of test microbe are selected for antibiotic production. The media and growth conditions can be altered to isolate specific groups of antibiotic producing microorganisms, i.e., bacteria, fungi, or actinomycetes (Okafor 1987). Once a pure culture of antibiotic producing microbe has been successfully isolated it is further tested for antibiotic production by several methods such as cross streak, agar plug, replica plating, ditch plating, and gradient plating methods (Okafor 1987).

After successful isolation of pure cultures, reliable identification of antibiotic producing microbes is accomplished by the analysis of several physiological, biochemical, and molecular characteristics of the microbe. Such identification techniques comprise studying cell morphology, differential staining, protein analysis, serological studies, flow cytometry studies, similarity of nucleotide (DNA and RNA) sequences, PCR (polymerase chain reaction), and RT PCR (reverse transcription polymerase chain reaction) techniques (Belgrader et al. 1999; Bentley et al. 2002; Omura et al. 2001). Several techniques being used, which are less time consuming and tedious, include mass spectrometry, fluorescence spectrometry, and capillary electrophoresis (Jarman et al. 2000, Saenz et al. 1999; Desai and Armstrong 2003).

1.3.5 *Antibiotics Producing Soil Microorganisms*

Antibiotics producing microorganisms can be isolated from various sources like soil, marine sources, endophytes, lichens, and even animals. According to Rinehart (1992), some scientists have been regularly searching for novel antimicrobial microorganisms from different places such as deep sea mud and seaweeds (Gonzalez del Val et al. 2001; Rinehart 1992; Perez et al. 2016). In addition, endophytes which reside in higher plants also comprise one of the important sources of antibiotics which are effective against different types of pathogens (Strobel and Daisy 2003).

A previous study conducted by Burkholder et al. (1944) had shown that lichens have high potential to produce useful antibacterial metabolites (Burkholder et al. 1944). In their study of 42 lichens species, 64.29% of the species did show active antimicrobial activity against *S. aureus* and *B. subtilis* (Burkholder et al. 1944). Besides, animals also become target for scientists to look for novel antibiotics. According to Moore et al. (1993), a broad-spectrum antibiotic, squalamine, has been successfully isolated from the stomach tissues of *Squalus acanthias* (dogfish shark) (Moore et al. 1993).

As discussed above, there are several sources where antibiotics can be discovered but soil still remains the most important target for a large group of researchers in their efforts to discover novel antimicrobial metabolite that have pharmaceutical values. This is because many microbes especially bacteria that inhabit in soil have the ability to produce biologically active secondary metabolites such as useful antibiotics.

Soil is a rich source where most antimicrobial producing microorganisms and their secondary metabolites can be found. Gram-positive bacteria, actinomycetes which form spore and filament are the most important group of antimicrobial producing soil microbes since they contribute to 75% of the identified products which are widely used in clinical applications (Oskay et al. 2004; Ceylan et al. 2008). According to Demain and Fang (2000), there will be about 500 antibiotics from actinomycetes continually being discovered each year (Demain and Fang 2000).

As reported by Oskay et al. (2004), of the 50 isolates that they obtained from actinomycetes, 34% of the isolates did produce antibacterial compounds (Oskay et al. 2004). Besides, from the recent research work conducted by Ceylan et al. (2008), they also reported that 15 isolates obtained from the genus of Streptomycetes showed the capability of producing antibacterial compounds towards both Gram-positive and Gram-negative bacteria which are resistance to several antibiotics (Ceylan et al. 2008). Streptomycetes are categorized in the family of Streptomycetaceae (Anderson and Wellington 2001) and different species of bacteria classified under this genus contribute mostly to the useful biologically active substances such as antibiotics that have been authorized (Anderson and Wellington 2001).

Besides, the ability of *Streptomyces* to act as useful biological control agents in retarding the growth of pathogenic fungi which infected plants also has been reported by many researchers. These pathogenic fungi may either arise from soil or air (Oskay 2009). In addition, the current research conducted by Oskay (2009) has discovered that there was a novel strain of *Streptomyces* assigned as *Streptomyces* sp. KEH23 that has a high potential to produce useful antibiotics which can actively act against pathogens that infected plants and human being (Oskay 2009). Streptomycetes can be commonly found in both terrestrial and aquatic environments especially where nutrients are highly abundant such as in the soil, hay, and composts (Locci 1989). Besides, there are several factors which can influence the distribution of streptomycetes which includes the temperature, moisture, pH, and climate (Williams et al. 1983; Locci 1989).

Besides Streptomycetes, other Gram-positive soil bacterium such as *Rhodococcus* has also been identified to have high potential of producing useful antimicrobial compounds when they are in the stress condition. Kurosawa et al. (2008) has isolated the amino glyceride antibiotic, rhodostreptomycin produced by Rhodococcus (Kurosawa et al. 2008). This antibiotic is effective against many types of test microorganism which including the hardy strain Streptomycetes. Table 1.2 shows some clinical important antimicrobial compounds which have been isolated from soil microorganisms.

1.4 Current Status

Despite the success of the discovery of antibiotics and advances in the process of their production, infectious diseases still remain the second leading cause of death worldwide, and bacterial infections cause approximately 17 million deaths annually, affecting mainly children and the elderly. The history of antibiotics derived from *Streptomyces* began with the discovery of streptothricin in 1942 (Baron 1950), and with the discovery of streptomycin 2 years later, scientists intensified the search for antibiotics within the genus. Today, 80% of the antibiotics are sourced from the genus *Streptomyces*, actinomycetes being the most important (Wright 2010).

Table 1.2 Some clinically important antibiotics

Antibiotic	Producer organism	Activity	Site or mode of action
Penicillin	*Penicillium chrysogenum*	Gram-positive bacteria	Wall synthesis
Cephalosporin	*Cephalosporium acremonium*	Broad spectrum	Wall synthesis
Griseofulvin	*Penicillium griseofulvum*	Dermatophytic fungi	Microtubules
Bacitracin	*Bacillus subtilis*	Gram-positive bacteria	Wall synthesis
Polymyxin B	*Bacillus polymyxa*	Gram-negative bacteria	Cell membrane
Amphotericin B	*Streptomyces nodosus*	Fungi	Cell membrane
Erythromycin	*Streptomyces erythreus*	Gram-positive bacteria	Protein synthesis
Neomycin	*Streptomyces fradiae*	Broad spectrum	Protein synthesis
Streptomycin	*Streptomyces griseus*	Gram-negative bacteria	Protein synthesis
Tetracycline	*Streptomyces rimosus*	Broad spectrum	Protein synthesis
Vancomycin	*Streptomyces orientalis*	Gram-positive bacteria	Protein synthesis
Gentamicin	*Micromonospora purpurea*	Broad spectrum	Protein synthesis
Rifamycin	*Streptomyces mediterranei*	Tuberculosis	Protein synthesis

Sneader (2005); Walsh and Wright (2005); Finch et al. (2003)

The world's demand for antibacterials (antibiotics) is steadily growing. Since their discovery in the twentieth century, antibiotics have substantially reduced the threat of infectious diseases. The use of these "miracle drugs," combined with improvements in sanitation, housing, food, and the advent of mass immunization programs, led to a dramatic drop in deaths from diseases that were once widespread and often fatal. Over the years, antibiotics have saved lives and eased the suffering of millions. By keeping many serious infectious diseases under control, these drugs also contributed to the increase in life expectancy during the latter part of the twentieth century.

The increasing resistance of pathogenic organisms, leading to severe forms of infection that are difficult to treat, has further complicated the situation, as in the case of carbapenem-resistant *Klebsiella pneumoniae*, and other microorganisms. Infections caused by resistant bacteria do not respond to treatment, resulting in prolonged illness and greater risk of death (Butler and Cooper 2011). Treatment failures also lead to long periods of infectivity with high rates of resistance, which increase the number of infected people circulating in the community and thus expose the population to the risk of contacting a multidrug-resistant strain (Butler and Cooper 2011).

As bacteria become resistant to first-generation antibiotics, treatment has to be changed to second- or third-generation drugs, which are often much more expensive and sometimes toxic. For example, the drug needed to treat multidrug-resistant *Streptococcus pneumoniae*, *Staphylococcus aureus*, *Klebsiella pneumoniae*, and *Mycobacterium tuberculosis* and can cost 100 times more than first-generation drugs used to treat non-resistant forms. Most worrisome is that resistance to virtually all antibiotics has increased (Tuon et al. 2011).

According to Nikaido, 100,000 tons of antibiotics are produced annually, which are used in agriculture, food, and health (Nikaido 2009). Their use has impacted populations of bacteria, inducing antibiotics resistance (Wright 2010). This resistance may be due to genetic changes such as mutation or acquisition of resistance genes through horizontal transfer, which most often occurs in organisms of different taxonomy (Sahoo et al. 2010). Mutations can cause changes at the site of drug action, hindering the action of the antibiotic. Most of the resistance genes are in the same cluster as the antibiotic biosynthesis gene. In nature, the main function of antibiotics is to inhibit competitors, which are induced to inactivate these compounds by chemical modification (hydrolysis), and changes in the site of action and membrane permeability. A recent study carried out with *Streptomyces* from urban soil showed that most strains are resistant to multiple antibiotics, suggesting that these genes are frequent in this environment. Many resistance genes are located on plasmids (plasmid A), which can be passed by conjugation to a susceptible strain; these plasmids are stable and can express the resistance gene (Bosso et al. 2010). The susceptibility to a particular antibiotic can be affected by the physiological state of the bacteria and the concentration of the antibiotic; this may be observed in biofilms through a mechanism known as persister formation—small subpopulations of bacteria survive the lethal concentration of antibiotic without any specific resistance mechanisms although this mechanism does not produce high-level resistance (Bosso et al. 2010).

Microorganisms growing in a biofilm are associated with chronic and recurrent human infections and are resistant to antimicrobial agents (Costerton et al. 1987). The spread of resistant strains is not only linked to antibiotic use, but also to the migration of people, who disperse resistant strains among people in remote communities where the use of antibiotics is very limited. Due to the difficulty of obtaining new antibiotics, the drug industry has made changes to existing antibiotics; these semisynthetics are more efficient and less susceptible to inactivation by enzymes that cause resistance (Yang et al. 2016). This practice has become the strategy for the current antibiotics used today and is known as the second, third, and fourth generation of antibiotics.

1.5 Future of Antibiotics Production from Soil

The first new antibiotic to be discovered in nearly 30 years has been hailed as a "paradigm shift" in the fight against the growing resistance to drugs. Teixobactin has been found to treat many common bacterial infections such as tuberculosis, septicemia, and *Clostridium difficile* and could be available within 5 years. But more importantly, it could pave the way for a new generation of antibiotics because of the way it was discovered. Scientists have always believed that the soil was teeming with new and potent antibiotics because bacteria have developed novel ways to fight off other microbes. But 99% of microbes will not grow in laboratory conditions leaving researchers frustrated that they could not get to the life-saving natural drugs (Hunter 2015).

Now a team from Northeastern University in Boston, Massachusetts, has discovered a way of using an electronic chip to grow the microbes in the soil and then isolate their antibiotic chemical compounds (Piddock 2015). They discovered that one compound, Teixobactin, is highly effective against common bacterial infections *Clostridium difficile*, *Mycobacterium tuberculous*, and *Staphylococcus aureus* (Yang et al. 2016). The rise in antibiotic resistance is a threat to modern healthcare as we know it so this discovery could potentially help to bridge the ever increasing gap between infections and the medicines we have available to treat them.

1.6 Conclusion

There are several sources from where antibiotics can be discovered but soil still remains the most important target for a large group of researchers in their efforts to discover novel antimicrobial metabolite that have pharmaceutical values. This is because many microbes especially bacteria that inhabit in soil have the ability to produce biologically active secondary metabolites such as useful antibiotics.

References

Allen HK, Donato J, Wang HH, Cloud-Hansen KA, Davies J, Handelsman J (2010) Call of the wild: antibiotic resistance genes in natural environments. Nat Rev Microbiol 8:251–259

Alvarez Lerma F, Sierra Camerino R, Alvarez Rocha L, Rodriguez Colomo O (2010) [Antibiotic policy in critical patients]. Med Intensiva 34:600–608

Aminov RI (2009) The role of antibiotics and antibiotic resistance in nature. Environ Microbiol 11:2970–2988

Anderson AS, Wellington EM (2001) The taxonomy of streptomyces and related genera. Int J Syst Evol Microbiol 51:797–814

Baltz RH (2008) Renaissance in antibacterial discovery from actinomycetes. Curr Opin Pharmacol 8:557–563

Baron AL (1950) Handbook of antibiotics. Reinhold, New York

Belgrader P, Benett W, Hadley D, Richards J, Stratton P, Mariella R Jr, Milanovich F (1999) PCR detection of bacteria in seven minutes. Science 284:449–450

Bentley SD, Chater KF, Cerdeno-Tarraga AM, Challis GL, Thomson NR, James KD, Harris DE, Quail MA, Kieser H, Harper D, Bateman A, Brown S, Chandra G, Chen CW, Collins M, Cronin A, Fraser A, Goble A, Hidalgo J, Hornsby T, Howarth S, Huang CH, Kieser T, Larke L, Murphy L, Oliver K, O'Neil S, Rabbinowitsch E, Rajandream MA, Rutherford K, Rutter S, Seeger K, Saunders D, Sharp S, Squares R, Squares S, Taylor K, Warren T, Wietzorrek A, Woodward J, Barrell BG, Parkhill J, Hopwood DA (2002) Complete genome sequence of the model actinomycete Streptomyces coelicolor A3(2). Nature 417:141–147

Berdy J (1974) Recent developments of antibiotic research and classification of antibiotics according to chemical structure. Adv Appl Microbiol 18:309–406

Berdy J (1980) Recent advances and prospects of antibiotic research. Process Biochem 15:28–35

Bosso JA, Mauldin PD, Salgado CD (2010) The association between antibiotic use and resistance: the role of secondary antibiotics. Eur J Clin Microbiol Infect Dis 29:1125–1129

Brotz-Oesterhelt H, Brunner NA (2008) How many modes of action should an antibiotic have? Curr Opin Pharmacol 8:564–573

Burkholder PR, Evans AW, McVeigh I, Thornton HK (1944) Antibiotic activity of lichens. Proc Natl Acad Sci U S A 30:250–255

Butler MS, Buss AD (2006) Natural products—the future scaffolds for novel antibiotics? Biochem Pharmacol 71:919–929

Butler MS, Cooper MA (2011) Antibiotics in the clinical pipeline in 2011. J Antibiot (Tokyo) 64:413–425

Cabello FC (2004) [Antibiotics and aquaculture in Chile: implications for human and animal health]. Rev Med Chil 132:1001–1006

Ceylan O, Okmen G, Ugur A (2008) Isolation of soil Streptomyces as source antibiotics active against antibiotic-resistant bacteria. EurAsian J Biosci 2:73–82

Chopra I, Hesse L, O'Neill AJ (2002) Exploiting current understanding of antibiotic action for discovery of new drugs. J Appl Microbiol 92(Suppl):4S–15S

Costerton JW, Cheng KJ, Geesey GG, Ladd TI, Nickel JC, Dasgupta M, Marrie TJ (1987) Bacterial biofilms in nature and disease. Annu Rev Microbiol 41:435–464

Davies J (1990) What are antibiotics? Archaic functions for modern activities. Mol Microbiol 4:1227–1232

Demain AL, Fang A (2000) The natural functions of secondary metabolites. Adv Biochem Eng Biotechnol 69:1–39

Demain AL, Sanchez S (2009) Microbial drug discovery: 80 years of progress. J Antibiot (Tokyo) 62:5–16

Desai MJ, Armstrong DW (2003) Separation, identification, and characterization of microorganisms by capillary electrophoresis. Microbiol Mol Biol Rev 67:38–51. Table of contents

Falconer SB, Brown ED (2009) New screens and targets in antibacterial drug discovery. Curr Opin Microbiol 12:497–504

Finch R, Greenwood D, Norrby SR, Whitley RJ (2003) Antibiotic and chemotherapy: anti-infective agents and their use in therapy. Churchill Livingstone, New York

Fischbach MA, Walsh CT (2009) Antibiotics for emerging pathogens. Science 325:1089–1093

Fleming A (1980) Classics in infectious diseases: on the antibacterial action of cultures of a penicillium, with special reference to their use in the isolation of B. influenzae by Alexander Fleming, Reprinted from the British Journal of Experimental Pathology 10:226-236, 1929. Rev Infect Dis 2:129–139

Franklin TJ, Snow GA (1981) Biochemistry of antimicrobial action. Chapman and Hall, New York, pp 1–56

Gaskins HR, Collier CT, Anderson DB (2002) Antibiotics as growth promotants: mode of action. Anim Biotechnol 13:29–42

Gonzalez del Val A, Platas G, Basilio A, Cabello A, Gorrochategui J, Suay I, Vicente F, Portillo E, Jimenez del Rio M, Reina GG, Pelaez F (2001) Screening of antimicrobial activities in red, green and brown macroalgae from Gran Canaria (Canary Islands, Spain). Int Microbiol 4:35–40

Gottlieb D (1967) Antibiotics and cell metabolism. Hindustan Antibiot Bull 10:123–134

Hmmond SM, Lambert PA (1978) Antibiotics and antimicrobial action. Edward Arnold, London, pp 5–52

Hunter P (2015) Antibiotic discovery goes underground: the discovery of teixobactin could revitalise the search for new antibiotics based on the novel method the researchers used to identify the compound. EMBO Rep 16:563–565

Jarman KH, Cebula ST, Saenz AJ, Petersen CE, Valentine NB, Kingsley MT, Wahl KL (2000) An algorithm for automated bacterial identification using matrix-assisted laser desorption/ionization mass spectrometry. Anal Chem 72:1217–1223

Koberl M, Schmidt R, Ramadan EM, Bauer R, Berg G (2013) The microbiome of medicinal plants: diversity and importance for plant growth, quality and health. Front Microbiol 4:400

Kummerer K (2004) Resistance in the environment. J Antimicrob Chemother 54:311–320

Kummerer K (2009) Antibiotics in the aquatic environment—a review—part I. Chemosphere 75:417–434

Kurosawa K, Ghiviriga I, Sambandan TG, Lessard PA, Barbara JE, Rha C, Sinskey AJ (2008) Rhodostreptomycins, antibiotics biosynthesized following horizontal gene transfer from Streptomyces padanus to Rhodococcus fascians. J Am Chem Soc 130:1126–1127

Lancini G, Parenti F (1982) Antibiotics an integrated view. Springer, New York, pp l–241

Locci R (1989) Streptomyces and related genera In: Williams ST, Sharpe ME, Holt J (eds) Bergey's manual of systematic bacteriology, vol 4. Williams and Wilkins, Baltimore, pp 2451–2508

Martinez JL, Fajardo A, Garmendia L, Hernandez A, Linares JF, Martinez-Solano L, Sanchez MB (2009) A global view of antibiotic resistance. FEMS Microbiol Rev 33:44–65

Moore KS, Wehrli S, Roder H, Rogers M, Forrest JN Jr, McCrimmon D, Zasloff M (1993) Squalamine: an aminosterol antibiotic from the shark. Proc Natl Acad Sci U S A 90:1354–1358

Nikaido H (2009) Multidrug resistance in bacteria. Annu Rev Biochem 78:119–146

Okafor N (1987) Antibiotics and anti-tumour agents. In: Industrial microbiology, pp 336–369

Omura S, Ikeda H, Ishikawa J, Hanamoto A, Takahashi C, Shinose M, Takahashi Y, Horikawa H, Nakazawa H, Osonoe T, Kikuchi H, Shiba T, Sakaki Y, Hattori M (2001) Genome sequence of an industrial microorganism Streptomyces avermitilis: deducing the ability of producing secondary metabolites. Proc Natl Acad Sci U S A 98:12215–12220

Oskay M (2009) Antifungal and antibacterial compounds from Streptomyces strains. Afr J Biotechnol 8:3007–3017

Oskay M, Tamer A, Azeri C (2004) Antibacterial activity of some actinomycetes isolated from farming soils of Turkey. Afr J Biotechnol 3:441–446

Perez MJ, Falque E, Dominguez H (2016) Antimicrobial action of compounds from marine seaweed. Mar Drugs 14. doi: 10.3390/md14030052

Piddock LJ (2015) Teixobactin, the first of a new class of antibiotics discovered by iChip technology? J Antimicrob Chemother 70:2679–2680
Procopio RE, Silva IR, Martins MK, Azevedo JL, Araujo JM (2012) Antibiotics produced by Streptomyces. Braz J Infect Dis 16:466–471
Qin S, Li J, Chen HH, Zhao GZ, Zhu WY, Jiang CL, Xu LH, Li WJ (2009) Isolation, diversity, and antimicrobial activity of rare actinobacteria from medicinal plants of tropical rain forests in Xishuangbanna, China. Appl Environ Microbiol 75:6176–6186
Queener SW, Sebek OK, Vezina C (1978) Mutants blocked in antibiotic synthesis. Annu Rev Microbiol 32:593–636
Rath CM, Janto B, Earl J, Ahmed A, Hu FZ, Hiller L, Dahlgren M, Kreft R, Yu F, Wolff JJ, Kweon HK, Christiansen MA, Hakansson K, Williams RM, Ehrlich GD, Sherman DH (2011) Meta-omic characterization of the marine invertebrate microbial consortium that produces the chemotherapeutic natural product ET-743. ACS Chem Biol 6:1244–1256
Rinehart KL (1992) Secondary metabolites from marine organisms. Ciba Found Symp 171:236–249. Discussion 249–254
Ruan J (2013) [Bergey's manual of systematic bacteriology (second edition) Volume 5 and the study of Actinomycetes systematic in China]. Wei Sheng Wu Xue Bao 53:521–530
Saenz AJ, Petersen CE, Valentine NB, Gantt SL, Jarman KH, Kingsley MT, Wahl KL (1999) Reproducibility of matrix-assisted laser desorption/ionization time-of-flight mass spectrometry for replicate bacterial culture analysis. Rapid Commun Mass Spectrom 13:1580–1585
Sahoo KC, Tamhankar AJ, Johansson E, Lundborg CS (2010) Antibiotic use, resistance development and environmental factors: a qualitative study among healthcare professionals in Orissa, India. BMC Public Health 10:629
Shlaes DM (2010) Antibiotics: the perfect storm. Springer, Dordrecht
Sköld O (2006) [Antibiotika och antibiotika resistens]. Studentlitteratur
Sneader W (2005) Drug discovery: a history. Wiley, Chichester
Strobel G, Daisy B (2003) Bioprospecting for microbial endophytes and their natural products. Microbiol Mol Biol Rev 67:491–502
Thomashow LS, Bonsall RF, Weller DM (1997) Antibiotic production by soil and rhizosphere microbes *in situ*. In: Hurst CJ, Knudsen GR, MJ MI, Stetzenbach LD, Walter MV (eds) Manual of environmental microbiology. American Society of Microbiology, Washington
Tuon FF, Kruger M, Terreri M, Penteado-Filho SR, Gortz L (2011) Klebsiella ESBL bacteremia-mortality and risk factors. Braz J Infect Dis 15:594–598
Walsh CT, Wright G (2005) Introduction: antibiotic resistance. Chem Rev 105:3
Williams ST, Goodfellow M, Alderson G, Wellington EMH, Sneath PHA, Sackin MJ (1983) Numerical classification of Streptomyces and related genera. J Gen Microbiol 129:1743–1813
Wolff M (2004) Use and misuse of antibiotics. Time to evaluate it beyond humans. Rev Med Chil 132:909–911
Wright GD (2010) Antibiotic resistance in the environment: a link to the clinic? Curr Opin Microbiol 13:589–594
Yang H, Chen KH, Nowick JS (2016) Elucidation of the teixobactin pharmacophore. ACS Chem Biol 11:1823–1826

Chapter 2
Antibiotics Resistance Genes

Ahmad Hasnain, Wajid Nasim, Hussani Mubarak, Nosheen Mirza, Sohaib Khan, Xiaomei Su, Sarfraz Ahmed, and Muhammad Zaffar Hashmi

2.1 Introduction

2.1.1 Antibiotic Resistance

The discovery that antibiotics can treat bacterial infections dramatically changed human health, and many once deadly infections are now curable. Yet often we hear about bacteria that are no longer killed effectively by antibiotics. These bacteria are

A. Hasnain
Department of Geography, Bahauddin Zakariya University, Multan, Pakistan

W. Nasim
Department of Environmental Sciences, COMSATS Institute of Information Technology, Vehari, Pakistan

H. Mubarak
Department of Soil and Environmental Sciences, Ghazi University, Dera Ghazi Khan, Pakistan

N. Mirza
Department of Environmental Sciences, COMSATS Institute of Information Technology, Abbottabad 22060, Pakistan

S. Khan
Department of Anthropology, Bahauddin Zakariya University, Multan, Pakistan

X. Su
College of Geography and Environmental Science, Zhejiang Normal University, Yingbin Road 688#, Jinhua 321004, China

S. Ahmed
Department of Biochemistry, Bahauddin Zakariya University, Multan 60800, Pakistan

M.Z. Hashmi (✉)
Department of Meteorology, COMSATS Institute of information Technology, Islamabad, Pakistan
e-mail: hashmi_qau@yahoo.com

M.Z. Hashmi et al. (eds.), *Antibiotics and Antibiotics Resistance Genes in Soils*, Soil Biology 51, DOI 10.1007/978-3-319-66260-2_2

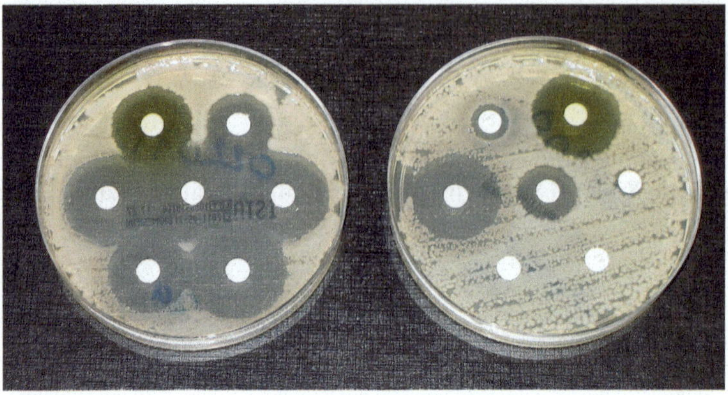

Fig. 2.1 Antibiotic resistance tests: Bacteria are streaked on dishes with white antibiotic impregnated disks. Clear rings, such as those on the *left*, show that bacteria have not grown—indicating that the bacteria are not resistant. Those on the *right* are fully susceptible to only three of the seven antibiotics tested (adapted from wikipedia)

known as antibiotic resistant (Fig. 2.1), and they are a growing problem in medicine (Yang et al. 2010; Yun-jian and Dong-ke 2008).

Accidently, antibiotic (penicillin) was discovered by Alexander Fleming in 1929, and by the 1940s, penicillin was available for medical use and was successfully used to treat infections in soldiers during World War II (Bennett and Chung 2001; Shore and Pruden 2009). Wherein, the dispersal of "foreign genes" into the environment occur through—"horizontal gene transfer" and "vertical gene flow" by seed dispersal, pollen flow considered as major concern. However, there are multiple national and international monitoring programs for drug-resistant threats, including methicillin-resistant *Staphylococcus aureus* (MRSA), vancomycin-resistant *S. aureus* (VRSA), extended spectrum beta-lactamase (ESBL), vancomycin-resistant *Enterococcus* (VRE), multidrug-resistant *A. baumannii* (MRAB).

There is substantial public concern about a potential spread of ARGs from transgenic plants into the soil and intestinal bacteria (Rodriguez-Mozaz et al. 2015; Tang et al. 2015; Zhu et al. 2013). Antibiotics have been detected in different environmental compartments such as groundwater of farms, in aquatic and soil environments (Martinez 2009). Historic evidence for antibiotic-resistant bacteria being a product of human activity is suggested by the study of Datta and Hughes (1983), which found that from a collection of *Enterobacteriaceae*, isolated between 1917 and 1954, 24% carried conjugative plasmids but only 2% were tetracycline resistant and all isolates were from the genera Proteus. None of the Salmonella, Shigella, Escherichia, or Klebsiella isolates were positive for tetracycline resistance (Tcr) (Datta and Hughes 1983). However, by the mid-1950s Tcr and multidrug-resistant Escherichia coli and Shigella were described, which was later determined to be due to the presence of plasmid-mediated antibiotic resistance (Akasaki et al. 1963). A lack of tetracycline resistance genes was also found in early *enterococci* (Atkinson et al. 1997) and *Neisseria gonorrhoeae* (Cousin et al. 2003). These studies suggest that antibiotic resistance genes were acquired as a result of increased antibiotic use by humans in the last 60 years. Forty different tetracycline resistance (*tet*) genes with three specific mechanisms (i.e., target modification with ribosomal protection protein, antibiotic efflux pumps, and antibiotic inactivation) have been characterized to date (Roberts 2005a). Four sulfonamide resistance (*sul*) gene types, including *sul*1, *sul*2, *sul*3, and *sul*A, have also been studied (Pei et al. 2006).

2.2 Mechanism of Antibiotics Resistance

There are a number of different ways that bacteria can become resistant to antibiotics. The first mechanism is due to random chromosomal mutations that lead to changes in the gene product that altered or eliminated the expression of a protein (Box 2.1). A second mechanism is by acquisition of new DNA (deoxyribonucleic acid) that is available to a limited number of bacteria that are naturally transformable. These bacteria have receptors that allow them to take up DNA from related strains or species and integrate this foreign DNA, which may be parts of genes, complete genes, or even defined elements into their genome.

Box 2.1. Mechanism of Antibiotics Resistance. Source: Penesyan et al. (2015)

Bacteria resist the effect of antibiotics by using the following genetic strategies, with thousands of variations:

- Producing destructive enzymes to neutralize antibiotics
- Modifying antimicrobial targets, by mutation, so that drugs cannot recognize them
- Removing antimicrobial agents by pumping them out (efflux)
- Preventing antibiotics from entering by creating a "biofilm" or otherwise reducing permeability
- Creating bypasses that allow target by antibiotics

The integration of new pieces of a gene creates a mosaic gene composed of the host's and foreign DNA, and this mosaic protein is able to reduce the antibiotic susceptibility of the host bacteria. Some species of bacteria are able to acquire foreign DNA by transduction, which uses bacteria phage for transmission of the DNA. However, the most common way bacteria become antibiotic resistant is by acquisition of new genes associated with mobile elements (plasmids, transposons, and integrons). These mobile elements may carry genes for metal resistance, use of alternative carbon sources, and/or classical virulence genes as well as a variety of different antibiotic resistance genes. Mobile elements are the main driving force in horizontal gene transfer between strains, species, and genera. They are normally

Table 2.1 Tetracycline resistance genes unique to environmental bacteria[a]

Efflux 12/27 (44%)	Ribosomal protection 3/12 (25%)	Enzymatic 2/3 (66%)
tet A(P), *tet*(V), *tet*(30), *tet*(35), *tet*(33), *tet*(39), *tet*(41), *tet*(42), *tet*(43), *otr*(B), *otr*(C), *tcr3*	*tetB*(P), *otr*(A), *tet*	*tet*(X)[b], *tet*(34)

Adapted from Roberts (2011)

[a]Species and genera which are primarily found outside of humans and animals rather than in or on animals or humans, even if on occasion they cause infections

[b]The *tet*(X) is only functional in environmental (aerobic) *Spingohucttrium* spp. though it is also found in (anaerobic) *Bacteroides* spp.

responsible for the rapid spread of particular elements throughout bacterial communities around the world. Horizontal gene transfer is associated with three primary mechanisms: (a) Conjugation, plasmid transfer from one bacterium to another; (b) transduction, viral mediated (phage) gene transfer; and (c) transformation, the uptake of naked DNA via the cell wall, and the incorporation of that DNA into the existing genome or plasmids (Kumarasamy et al. 2010; Levy 2002). The tet genes listed in Table 2.1 are associated with conjugative, nonconjugative, and mobilizable plasmids, transposons, and conjugative transposons (Fig. 2.2).

2.2.1 Intrinsic Resistance

In some cases, a type of bacteria will survive antibiotic treatment and multiply because it is intrinsically resistant. For example, although many types of bacteria have cell walls, some don't. An antibiotic like penicillin that prevents cell-wall building can't harm a bacterium that doesn't build a cell wall in the first place (Fig. 2.3).

2.2.2 Acquired Resistance

Bacteria can also acquire resistance. This happens when a type of bacteria changes in a way that protects it from the antibiotic. Bacteria can acquire resistance in two ways: either through a new genetic change that helps the bacterium survive, or by getting DNA from a bacterium that is already resistant.

2.2.3 Genetic Change

So how can a simple DNA change protect bacteria from antibiotics? Remember, DNA provides instructions to make proteins, so a change in DNA can cause a

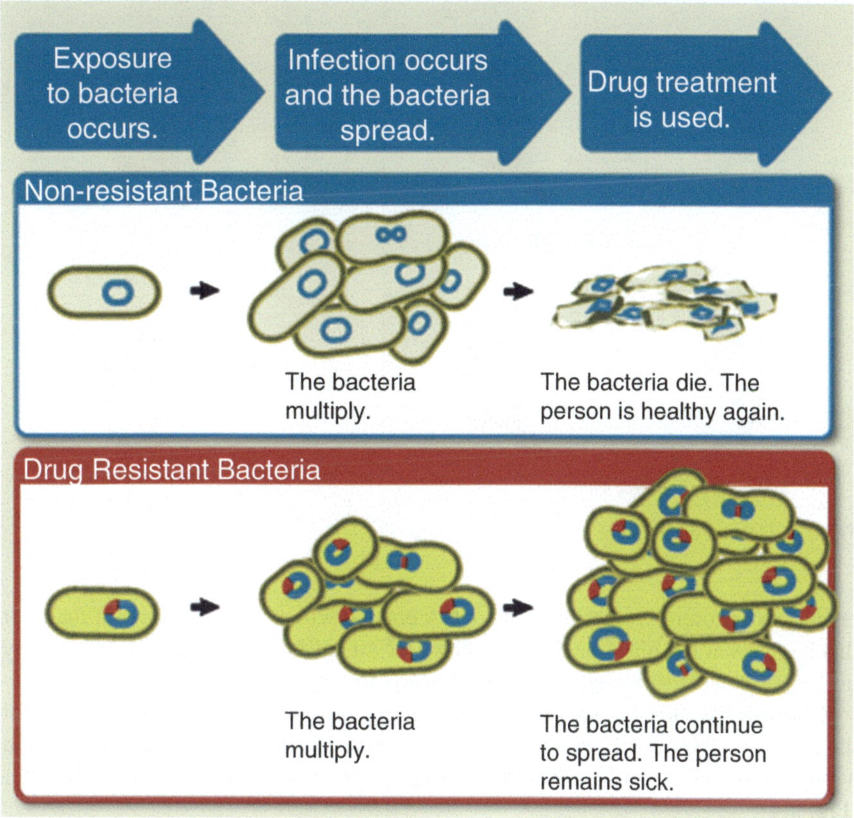

Fig. 2.2 Diagram showing the difference between non-resistant bacteria and drug-resistant bacteria. Non-resistant bacteria multiply, and upon drug treatment, the bacteria die. Drug-resistant bacteria multiply as well, but upon drug treatment, the bacteria continue to spread (adapted from wikipedia)

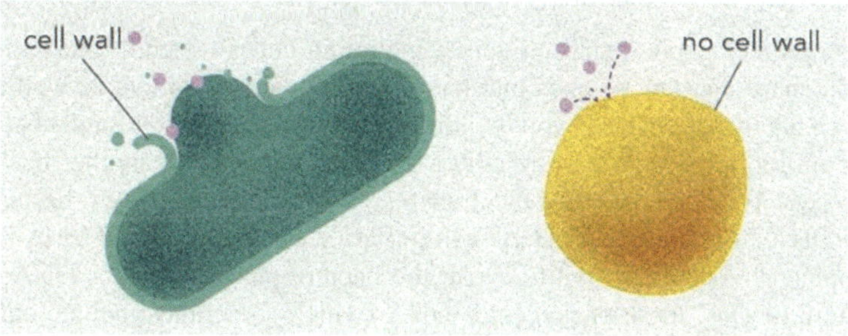

Fig. 2.3 Intrinsic resistance

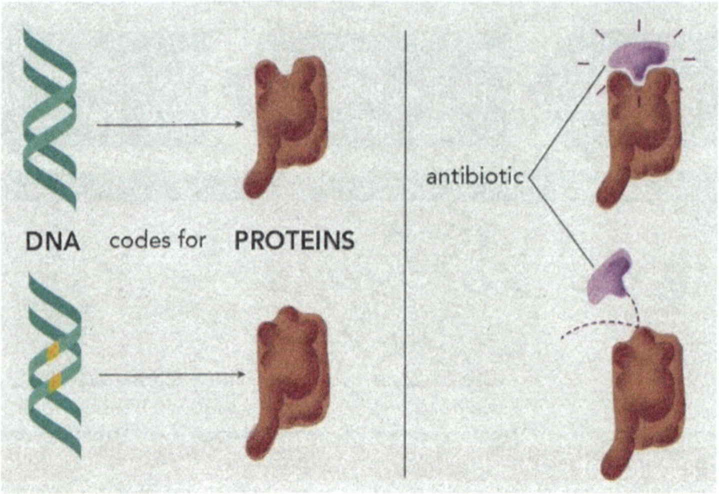

Fig. 2.4 Genetic change

change in a protein. Sometimes, this DNA change will affect the protein's shape. If this happens at the place on the protein where an antibiotic acts, the antibiotic may no longer be able to recognize where it needs to do its job.

Changes like this can prevent an antibiotic from getting into the cell, or prevent the antibiotic from working once it's inside. Once a change occurs, it can spread in a population of bacteria through processes like reproduction or DNA transfer (Fig. 2.4).

2.2.4 DNA Transfer

Bacteria are very good at sharing genes, including genes for antibiotic resistance. They can share resistance genes that have been in the population, as well as new genetic changes that occur. If you explored Agent Antibiotic, you saw a bacterium with an antibiotic resistance gene give a copy of that gene to another bacterium. This process is called lateral gene transfer. There are other ways bacteria can transfer DNA, for example, bacteria can get infected with a type of virus called a bacteriophage? As part of its life cycle, the bacteriophage packages DNA. When the bacterium dies, these packages of DNA (which sometimes include antibiotic resistance genes) are released and can be taken up and used by other bacteria (Fig. 2.5).

Fig. 2.5 DNA transfer

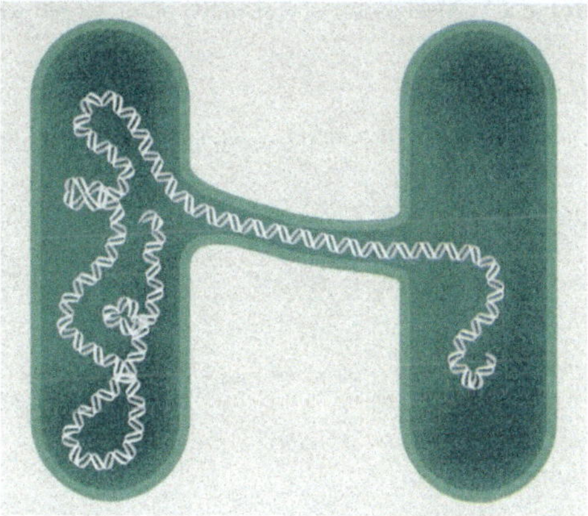

2.3 Tetracycline Resistance Genes

Tetracyclines are one of the oldest classes of antibiotics used and the first broad-spectrum class of antibiotics. Tetracyclines interact with the bacterial ribosomes by reversibly attaching to the ribosome that blocks protein synthesis. Tetracyclines are active against a wide range of Gram-positive, Gram-negative, anaerobic, and aerobic bacteria, cell-wall-free microbes, intercellular bacteria, and protozoan parasites. Tetracyclines are relatively safe, and the older compounds are inexpensive and have been widely used in clinical, veterinary, and agricultural purposes for 60 years (Roberts 2005b). For this chapter, Gram-positive bacteria will also include cell-wall-free Mycoplasma, Ureaplasma, as well as Mycobacterium, Nocardia, and Streptomyces. The first Tcr bacteria were identified in isolates from the 1950s (Watanabe et al. 1972). Bacteria may become resistant to tetracyclines, by mutation, while the majority of bacteria become tetracycline resistant because they acquire new genes that (a) pump tetracycline out the cell (efflux); (b) protect the ribosome from the action of tetracyclines; or (c) enzymatically deactivate tetracyclines (Table 2.1).

2.3.1 *Efflux*

The first tetracycline-resistant efflux proteins were identified in the 1950s in Japan where they were later hypothesized to be located on conjugative plasmids (Watanabe 1963). Today, there are 27 genetically distinct efflux genes characterized coding for drug-Hþ energy-dependent transmembrane sequence (TMS) proteins that span the lipid bilayer of the inner cell membrane 9–14 times. These proteins have been divided into seven different groups based on the number of TMS present (9–14), the G þ C% (guanine–cytosine) of the gene, and similarities to other tet efflux genes (Thaker et al. 2010). These efflux proteins normally export tetracycline and doxycycline but not

Table 2.2 Mechanism of resistance of tet and otr genes

Efflux (27)	Ribosomal protection (12)	Enzymatic (3)	Unknown[a]
tet(A), *tet*(B), *tet*(C), *tet*(D), *tet*(E)	*tet*(*M*), *tet*(O), *tet*(S), *tet*(W), *tet*(32)	*tet*(X)[c]	*tet*(U)
tet(G), *tet*(H), *tet*(J), *tet*(V), *tet*(Y)	*tet*(Q), *tet*(T), *tet*(36)	*tet*(34)	
tet(Z), *tet*(30), *tet*(31), *tet*(33)	*tet*(A), *tet B*(P)[b], *tet*(44), *tet*	*tet*(37)[c]	
tet(35)[d]			
tet(39), *tet*(41)			
tet(*K*), *tet*(L), *tet*(38)			
tet A(P), *tet*(40), *tet*(42), *tet*(43)			
otr(B), *otr*(C), *tcr3*			

Adapted from Roberts (2011)
[a]*tet*(U) has been sequenced but does not appear to be related to either efflux or ribosomal protection proteins
[b]*tet B*(P) is not found alone and *tet A*(P) and *tet B*(P) are counted as one operon
[c]*let*(X) and *tet*(37) arc unrelated but both arc NADP-requiring oxidoreductases
[d]Not related to other *tet* efflux genes

minocycline or tigecycline (a newer glycylcycline) out of the cell. The one exception is the Gram-negative tet(B) gene that exports tetracycline, doxycycline, and minocycline and confers resistance in the host bacterium to all three tetracyclines. The efflux genes are the most commonly found tet genes in aerobic and facultative Gram-negative bacteria (Tables 2.2 and 2.3). Twelve (41%) of the efflux genes [tetA(P), tet(V), tet(Z), tet(30), tet(33), tet(35), tet(39), tet(41), tet(42), otr(B), otr(C), tcr] are unique to environmental bacteria tetracycline-resistant genes coding for efflux proteins are the most commonly found tet genes among Gram-negative aerobic and facultative bacteria. Fifty-five Gram-negative and 25 Gram-positive genera carry these genes (Table 2.3). Of the 76 Gram-negative genera known to carry tetracycline resistance genes, 27 (36%) of these genera carry only efflux genes, of which 13 carry a single efflux gene and 14 carry multiple efflux genes. Of the 47 Gram-positive genera, only 9 (19%) carry efflux genes with 8 carrying a single efflux gene and Nocardia carrying 2 efflux genes (Table 2.3). The tet(B) gene is the most common efflux gene among Gram-negative genera and has been identified in 31 genera, while the tet(A) gene is found in 20, tet(C) gene in 10, tet(D) gene in 16, tet(E) gene in 10, tet(G) gene in 13, the tet(H) gene in 8, and the tet(35) in two Gram-negative genera. The tet(K) gene is found in 12 Gram-positive genera and the otr(B) gene is found in Mycobacterium and Streptomyces. The tet(L) gene is found in 14 Gram-negative and 19 Gram-positive genera, the tet(39) gene is found in 4 Gram-negative and 3 Gram-positive genera, while the tet(42) gene is found in 4 Gram-positive and 2 Gram-negative genera (Table 2.3). Twelve (44%) of the efflux genes including the tet(J), tetA (P) tet(V), tet(Y), tet(Z), tet(30),tet(31), otr(C), tcr, tet(33), tet(40), and tet(41)are found in a single genera (Table 2.3). The tet(43) gene has been isolated from metagenomic DNA library and has yet to be identified in a specific species or genus (Fig. 2.6) .

Table 2.3 Distribution of tet resistance genes among Gram-negative and Gram-positive bacteria

One gene		Two or more genes	
Efflux			
Gram-negative			
***n* = 13**		***n* = 14**	
Aggregatibacter	*tet*(B)	*Alcaligenes*	*tet*(A)(E)(39)
Agrabacterium	*tet*(30)	*Bordetella*	*tet*(A)(C)
Alteromoits	*tet*(D)	*Brevundimonsa*	*tet*/(B)(G)
Brevundimonas	*tet*(39)	*Halomonas*	*tet*(C)(D)
Chlamydia	*tet*(C)	*Mannheimia*	*tet*(B)(G)(H)(L)
Chryseobacterium	*tet*(A)	*Morganella*	*tet*(D)(J)(L)
Erwinia	*tet*(B)	*Moraxella*	*tet*(B)(H)
Francisella	*tet*(C)	*Ochrobactrum*	*tet*(G)(L)
Hisiophilus	*tet*(H)	*Plesiomonas*	*tet*(A)(B)(D)
Laribacter	*tet*(A)	*Roseobacter*	*tet*(B)(C)(E)(G)
Rahnella	*tet*(L)	*Salmonella*	*tet*(A)(B)(C)(D)(G)(L)
Sporosarcina	*tet*(L)	*Stenotrophomonas*	*tet*(35)(39)
Treponema	*tet*(B)	*Variovorax*	*tet*(A)(L)
		Yersinia	*tet*(B)(D)
Gram-positive			
***n* = 8**		***n* – 1**	
Cellulosimicrobium	*tet*(39)	*Nocardia*	*tet*(K)(L)
Geobacillus	*tet*(L)		
Lysinibacillus	*tet*(39)		
Micrococcus	*tet*(42)		
Oceanobacillus	*tet*{L)		
Pediococcus	*tet*(L)		
Vagococcus	*tet*(L)		
Virgibacillus	*tet*(L)		
Ribosomal protection and/or efflux/enzymatic			
Gram-negative			
***n* = 12**		***n* = 37**	
Acidaminococcus	*tet*(W)	*Acinetobacter*	*tet*(A)(B)(G)(H)(L)(M)(39)
Brachybacterium	*tet*(M)	*Actinobacillus*	*tet*(B)(H)(L)(O)
Eikenella	*tet*(M)	*Aeromonas*	*tet*(A)(B)(C)(D)(E)(M)(Y)(31)
Capnocytophaga	*tet*(Q)	*Anaerovibrio*	*tet*(O)(Q)
Chryseobacterium	*tet*(A)	*Bacteroides*	*tet*(M)(Q)(W)(X)(36)
Hafnia	*tet*(M)	*Butyrivibrio*	*tet*(O)(W)
Kingella	*tet*(M)	*Campylobacter*	*tet*(O)(44)
Lawsonia	*tet*(M)	*Citrobacter*	*tet*(A)(B)(C)(D)(L)(M)(O)(S)(W)
Pseudoalteromonas	*tet*(M)	*Edwardsiella*	*tet*(A)(D)(M)
Ralstonia	*tet*(M)	*Enterobacter*	*tet*(A)(B)(C)(D)(G)(L)(M)(39)
Rhanela	*tet*(M)	*Escherichia*	*tet*(A)(B)(C)(D)(EKG)(L)(M)(W)(Y)

(continued)

Table 2.3 (continued)

One gene		Two or more genes	
Spingobacterium	*tet*(X)	*Flavobacterium*	*tet*(A)(E)(L)(M)
		Fusobacterium	*tet*(G)(L)(M)(O)(Q)(W)
		Gallibacterium	*tet*(B)(H)(K)(L)(31)
		Haemophilus	*tet*(B)(K)(M)
		Klebsiella	*tet*(A)(B)(C)(D)(M)(S)(W)
		Kurthia	*tet*(L)(M)
		Megasphaera	*tet*(O)(W)
		Mitsuokella	*tet*(Q)(W)
		Neisseria	*tet*(B)(M)(O)(Q)(W)
		Pantoea	*tet*(B)(M)
		Paenibacillus	*tet*(M)(L)(42)
		Pasteurella	*tet*(B)(D)(H)(G)(L)(M)(O)
		Porphyromonas	*tet*(Q)(W)
		Prevotella	*tet*(M)(Q)(W)
		Providencia	*tet*(B)(E)(G)(M)[j](39)
		Photobacterium	*tet*(B)(D)(M)(Y)
		Pseudomonas	*tet*(A)(B)(C)(E)(G)(M)(34)(L)(X)(42)
		Psychrobacter	*tet*(H)(M)(O)
		Proteus	*tet*(A)(B)(C)(E)(G)(L)(J)(M)
		Selenomonas	*tet*(M)(Q)(W)
		Serratia	*tet*(A)(B)(C)(E)(M)(34)(41)
		Shewanella	*tet*(D)(G)(M)
		Shigella	*tet*(A)(B)(C)(D)(M)
		Subdoligranulum	*tet*(Q)(W)
		Veillonella	*tet*(A)(L)(M)(S)(Q)(W)
		Vibrio	*tet*(A)(B)(C)(D)(E)(G)(M)(34)(35)
Gram-positive/cell-wall-free/others			
***n* = 15**		***n* = 23**	
Abiotrophia	*tet*(M)	*Actinomyces*	*tet*(L)(M)(W)
Afipia	*tet*(M)	*Aerococcus*	*tet*(M)(O)
Anaerococcus	*tet*(M)	*Arthrobacter*	*tet*(M)(33)
Arcanobacterium	*tet*(W)	*Bacillus*	*tet*(K)(L)(M)(W)(39)(42) *otr*(A)
Amycolatopsis	*tet*(M)	*Bifidobacterium*	*tet*(L)(M)(O)(W)
Bacterionema	*tet*(M)	*Clostridium*	*tet*(K)(L)(M)(O)(P)(Q)(W)(36)(40)(44)
Brachybacterium	*tet*(M)	*Corynebacterium*	*tet*(M)(Z)(33)
Catenibacterium	*tet*(M)	*Enterococcus*	*tet*(K)(L)(M)(O)(S)(T)(U)
Erysipelothrix	*tet*(M)	*Eubacterium*	*tet*(K)(M)(O)(Q)(32)
Granulicatella	*tet*(M)	*Gardnerella*	*tet*(M)(Q)
Finegoldia	*tet*(M)	*Gemellu*	*tet*(M)(O)
Mycoplasma	*tet*(M)	*Granulicatella*	*tet*(M)(O)
Roseburia	*tet*(W)	*Lactobacillus*	*tet*(K)(L)(M)(O)(S)(Q)(W)(Z)(36)
Ruminococcus	*tet*(Q)	*Lactococcus*	*tet*(M)(S)

(continued)

Table 2.3 (continued)

One gene		Two or more genes	
Sporosarcina	*tet*(M)	*Listeria*	*tet*(K)(L)(M)(S)
Ureaplasma	*tet*(M)	*Microbacterium*	*tet*(M)(42)
		Mobiluncus	*tet*(O)(Q)
		Mycobacterium	*tet*(K)(L)(M)(V) *otr*(A)(B)
		Paenibacillus	*tet*(L)(M)(42)
		Peptostreptococcus	*tet*(K)(L)(M)(O)(Q)
		Staphylococcus	*tet*(K)(L)(M)(O)(S)(U)(W)(38)(42)
		Streptococcus	*tet*(K)(L)(M)(O)(Q)(T)(U)(W)(32)
		Streptomyces	*tet*(K)(L)(M)(W) *otr*(A)(B)(C) *tet*3 *tet*

Adapted from Roberts (2011)

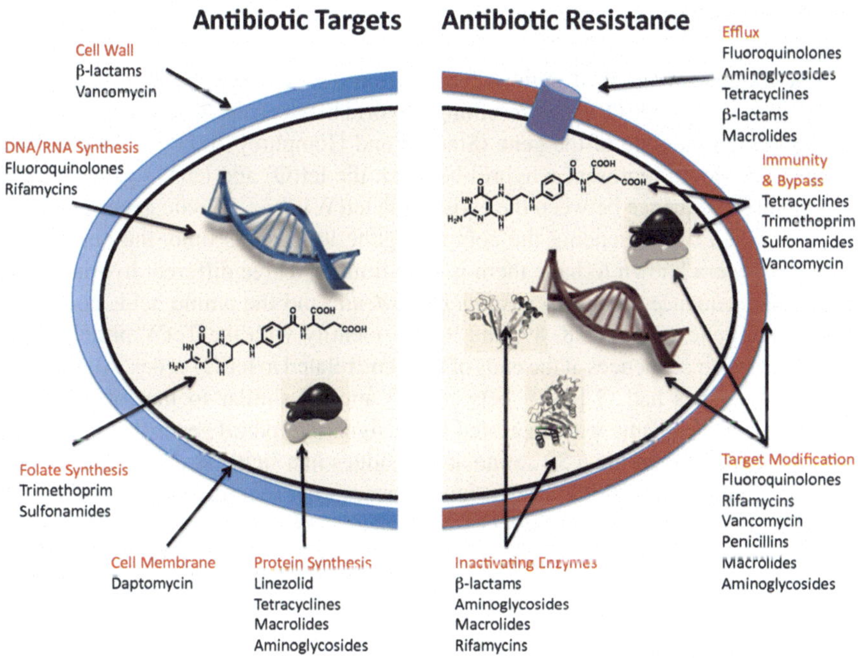

Fig. 2.6 Mechanisms used by common antibiotics to deal with bacteria and ways by which bacteria become resistant to them (adapted from wikipedia)

2.3.2 *Ribosomal Protection*

Twelve ribosomal protection genes have been characterized, of which three (25%) [tetB(P), otr(A), tet] are unique to environmental bacteria (Tables 2.2 and 2.3). The genes have been divided into three base groups related to their amino acid sequences rather than G þ C% content as is done with the efflux genes (Thaker et al. 2010). The ribosomal protection genes code for cytoplasmic proteins of B

72.5 kDa in size that protect the ribosomes from the action of tetracycline in vitro and in vivo. Unlike the efflux genes, the ribosomal protection genes confer resistance to tetracycline, doxycycline, and minocycline but not tigecycline (Roberts 2005a). Forty-nine Gram-negative genera have been characterized that carry at least one ribosomal protection tet gene(s). Of these, 12 (24%) Gram-negative genera carry a single ribosomal protection gene, while the remaining genera carry multiple or ribosomal protection or efflux and ribosomal protection genes. Thirty-eight Gram-positive genera carry ribosomal protection genes, of which 15 carry a single gene and 23 carry one or more ribosomal protection and/or both ribosomal protection and efflux tet genes (Table 2.4).

2.3.3 *Mosaic*

Mosaic tet genes consist of regions from two known tet genes with a descriptive designation such as tet(O/W) representing a hybrid between the tet(O) at one end and tet(W) at the other end of the gene (Stanton and Humphrey 2003). A tet(W/O/W) designation would represent a hybrid between the tet(O) and tet(W) genes with a partial tet(O) sequence between the ends of the tet(W) gene. Mosaic genes can only be determined by sequencing the complete gene and at this time, the number of different genera known to have them is very limited. Three different hybrid genes have been sequenced from *Megasphaera elsdenii*, and the amino acids coded by these three genes share 95.8, 89, and 91.9% identity with the TetW protein with 13–43% of their sequences at the ends of the gene related to tet(O) genes. All three of the mosaic genes had G þ C% between 50 and 55 similar to that of other tet (W) genes. A new name was suggested for designating hybrid genes that coded for proteins made of more than 50 amino acid residues in a single stretch that are from different genes (Levy 2006). The various mosaic genes identified are tet(O/W), tet (O/W/O), tet(O32/O), and tet(O/W/32/W/O).

2.3.4 *Enzymatic*

Three genes that code for inactivating enzymes have been identified, tet (X) (Bacteroides, Pseudomonas, Spingobacterium), tet(34) (Pseudomonas, Serratia, Vibrio), and tet(37) (metagenomic). These three tet genes are found only in Gram-negative species. Six of the seven genera that carry one of these inactivating tet genes may carry efflux and/or ribosomal protection tet genes, thus their contribution to bacterial Tcr compared to the efflux and ribosomal protection tet genes is unclear (Table 2.1). Perhaps as more environmental bacteria are characterized, more genera carrying one of these tet genes may be found and/or other inactivating tet genes will be identified.

Table 2.4 Tetracyline resistance genes linked to other genes

Gene	Linkage	Phenotype/element
Efflux		
tet(A)	bla_{TEM}	β-lactamase
	str A, str B	Streptomycin
	sul2	Sulfamethoxazole
	floR	Florfenicol/chloramphenicol
	SGI1	*Salmonella* genomic island 1
	mer operon	Mercury
	Tn*21*, Tn *1721*	Transposon
tet(*B*)	bla_{TEM}	β-lactamase
	cat A	Chloramphenicol
	tel(M)	Tetracycline
	str A, str B	Streptomycin
	Sul1. sul2	Sulfamethoxazole
	mer operon	Mercury
	int 1	Class I integron
	Tn *10*	Transposon carrying bla_{TEM}
	SGI1	*Salmonella* genomic island 1
tet(G)	*aad A2, aad B*	Aminoglycoside
	dfr A	Trimethoprim
	flo R	Florfenicol chloramphenicol
	sul 1	Sulfamethoxazole
	cml A9	Chloramphenicol
	SGI1	*Salmonella* genomic island 1
	qacE∆1	Detergent resistance
tet(H)	*sul2*	Sulfamethoxazole
	str A, str B	Streptomycin
tet(*K*)	*mec A*	Methicillin
	dfr K	Trimethoprim
	mer operon	*Mercury*
	pT181	*S. aureus* plasmid
	p1258	*V. aureus* plasmid with *mer* operon
	SCC*mec* element III	One of the characterized *mec A* elements
	Tn*554*	Transposon carrying *erm*(*A*) [MLS_B]
tet(L)	*dfrK*	Trimethoprim
tet(33)	*aadA9*	Aminoglycoside
	IS*6100*	Insertion sequence
tet(40)	*tet*(O/32/O)	Tetracycline (mosaic gene)
Ribosomal protection		
tet(*M*)	*erm*(B)	MLS_B
	mef(A), *msr*(D)	Macrolide
	aph A-3	Kanamycin
	tet(B)	Tetracyclinc

(continued)

Table 2.4 (continued)

Gene	Linkage	Phenotype/element
	mer operon	Mercury
	Tn*917*	Transposon carrying *erm*(B)
	Tn*916*-Tn*1545*	Transposon family
tet(O)	*mef*(A), *msr*(D)	Macrolide
tet(Q)	*erm*(B), (F), (G)	MLS_B
	mef(A), *msr*(D)	Macrolide
	rte ABC	excision
	CTnDOT, Tn*4351*, Tn*4400*	*Bacteroides* conjugative transposons
tet(S)	Tn*916S*	Transposon
tet(W)	Tn*B1230*	*Bifidobacterium* transposon
	ATE-1,-2,-3	*Arcanobacterium* transposon
Enzymatic		
tet(X)	*enn*(F)	MLS_B

Adapted from Roberts (2011)

2.3.5 *Unknown*

The tet(U) gene produces a small protein (105 amino acids) that confers low-level tetracycline resistance (Chopra and Roberts 2001). The TetU protein has 21% similarity over its length to the TetM protein, but it does not include the consensus GTP-binding sequences, which are thought to be very important for tetracycline resistance in ribosomal protection proteins. The tet(U) gene has been identified in a vancomycin- and tetracycline-resistant S. aureus strain that did not carry the tet(K), tet(L), tet(M), or tet(O) genes. From the same patient, vancomycin-resistant enterococci were cultured that carried both the tet(U) and tet(L) genes and a few isolates also carried the tet(K) and/or tet(M) genes (Weigel et al. 2004). The tet(U) gene has also been identified in Enterococcus spp. The importance of the tet(U) gene is unclear since both Enterococcus and Staphylococcus isolates are able to carry a variety of efflux and ribosomal protection tet genes.

2.4 Sulfonamide Resistance Genes

The sulfonamides, the first antimicrobials developed for large-scale introduction into clinical practice (in 1935), target dihydropteroate synthase. Their serendipitous discovery (the antibacterial activity was seen initially in vivo when the active compound was released as part of a dye) pales only in comparison with that of Fleming's chance discovery of penicillin (Levy 2002). Two sul genes (sulI and sulII) and one genetic element associated with mobile antibiotic resistance genes [class 1 integron (intI1)] in eight livestock farms in Hangzhou, eastern China was investigated (Cheng et al. 2013).

2.5 Resistance Rates and Trends

Antibiotic resistance patterns of individual pathogens to the drugs used to treat them vary considerably between and within countries. These differences are driven by different patterns of antibiotic use, distinct national disease burdens, disparities in access to first- and second-line treatments, and the burden of co-infections, particularly malaria, the human immunodeficiency virus (HIV), and tuberculosis (O'Neill 2014). Resistance rates have also been correlated with seasonal antibiotic use: in the United States, spikes of resistant *E. coli* correlated significantly with seasonal highs in aminopenicillin and fluoroquinolone prescriptions, lagging by 1 month (Sun et al. 2012). Some antibiotic-resistant infections, such as *H. influenzae* in children under five, have higher mortality rates compared with susceptible infections (27 versus 7% mortality). However, this increased risk of death is not universal: in the case of healthcare-associated infections, antibiotic resistance does not greatly increase mortality or length of hospital stay due to bloodstream infections or pneumonia (Lambert et al. 2011). Antibiotic-resistant infections also contribute to the financial burden on healthcare systems. In Europe, they cost an estimated.

1.5 billion euros annually, including healthcare expenditures and productivity losses (i.e., both direct and indirect costs) (EMA and ECDC 2009). In the United States, the annual cost to the healthcare system is as much as $20 billion, and productivity losses total another $35 billion (CDC 2013). High-income regions and countries. In the United States, CDC (2013) has estimated that more than 2 million infections and 23,000 deaths are due to antibiotic resistance each year. In Europe, an estimated 25,000 deaths are attributable to antibiotic-resistant infections (EMA and ECDC 2009). Resistance of *Streptococcus pneumoniae* invasive isolates to antibiotics has declined in the United States, from 34 to 17% from 1999 to 2013 for penicillins, and from 15 to 8% from 1999 to 2012 for third-generation cephalosporins. From 1999 to 2012, resistance to microclines increased from 23 to 34%, but fluoroquinolone resistance remained stable, at 2%. Among *E. coli* and *K. pneumoniae* isolates, resistance to third-generation cephalosporins and fluoroquinolones increased steadily: for third-generation cephalosporin resistance in *E. coli*, from 2 to 12%, and in *K. pneumoniae*, from 8 to 19%; for fluoroquinolone resistance in *E. coli*, from 5 to 30%, and in *K. pneumoniae*, from 7 to 18%. Among *E. faecium* invasive isolates, vancomycin resistance increased from 65 to 76%. Compared with other high-income countries, the United States has higher rates of resistance to many Gram-positive bacteria, including VRE and MRSA (CDDEP 2015). Low- and middle-income regions and countries *K. pneumoniae* is the most commonly reported Gram-negative pathogen in Asia and Africa, making up nearly half of all Gram-negative infections in neonates. In Asia, median resistance of *K. pneumoniae* to ampicillin was 94%, and to cephalosporins, 84%; in Africa, it was 100 and 50%, respectively. Multidrug resistance appeared in 30% of strains in Asia and 75% of strains in Africa (Le Doare et al. 2014). In sub-Saharan Africa, rates of multidrug resistance exceeding 50% have been reported in invasive typhoidal and nontyphoidal Salmonella infections. Resistance to the drugs used to treat multidrug-resistant Salmonella, such as fluoroquinolones, is also increasing (Kariuki et al. 2015). Invasive nontyphoidal Salmonella infections are responsible for more than 600,000 deaths per year, 55% of them in Africa (Kariuki et al. 2015). Patterns of antibiotic resistance differ slightly in Latin America and the Caribbean, where prevalence of community-associated *Enterobacteriaceae* infections is higher than in the rest of the world, especially in urinary tract infections caused by *E. coli* and intra-abdominal infections caused by *E. coli* and *Klebsiella* spp. These infections show increasing resistance to trimethoprim-sulfamethoxazole, quinolones, and second-generation cephalosporins. In 2009, rates of resistance in urinary tract *E. coli* isolates reached 71% in women and 85% in men, with the highest rates occurring in Argentina and Peru (Salles et al. 2013). In Latin America and the Caribbean in 2013, resistance in community *S. pneumoniae* isolates was generally low to penicillins but ranged from 0% in Bolivia to 97% in Chile. No resistance was detected to vancomycin, and very low resistance was detected in some countries to third-generation cephalosporins. Resistance in *E. faecium* hospital isolates was higher than for *E. faecalis*. Resistance in *E. faecium* was high to ampicillins and vancomycin, reaching 100% resistance to ampicillins in Ecuador, El Salvador, and Paraguay. Paraguay also had the highest resistance to vancomycin, at 75%. *E. faecalis* resistance to ampicillin ranged from 0 to 15%, and resistance to vancomycin ranged from 0 to 22% (PAHO forthcoming). In Nepal, resistance rates exceeded 50% for *S. pneumoniae* and *K. pneumoniae* isolates to commonly used treatments, having increased from 2000 to 2008. Resistance of *Salmonella typhi* and *Salmonella paratyphi* strains have also increased since 1998 to the present, and in *E. coli*, from 2006 to 2010. Resistance rates were above 50% to all drugs tested in *E. coli* urinary tract infections, and high resistance rates were detected in gonorrheal infections.

2.6 Global Patterns and Emerging Threats

The most recent worldwide estimates of global antibiotic resistance, published by the World Health Organization (WHO) in 2014, list *Escherichia coli*, *Klebsiella pneumoniae*, and Staphylococcus aureus as the three agents of greatest concern, associated with both hospital- and community-acquired infections. In five of the six WHO regions, some countries reported *E. coli* resistance of more than 50% to fluoroquinolones and third-generation cephalosporins. *K. pneumoniae* resistance rates to third-generation cephalosporins are above 30% in most WHO member countries and exceed 60% in some regions (WHO 2014). MRSA resistance rates exceed 20% in all WHO regions and are above 80% in some regions (WHO 2014). *Streptococcus pneumoniae*, *nontyphoidal Salmonella*, *Shigella* spp., and *Neisseria gonorrhoeae* were also identified as community-acquired infections of high global concern. High rates of resistance to first- and second-line drugs are already increasing reliance on last-resort drugs, such as carbapenems (WHO 2014). This report provides an overview of the best available data on antibiotic resistance rates worldwide, drawing from Resistance Map (www.resistancemap.org, a global database of antibiotic use and resistance information, developed by the Center for Disease Dynamics, Economics and Policy [CDDEP]), WHO, national sources, and scientific publications.

2.7 Conclusions

Antibiotic pollution has been detected around the world in almost all compartments of the environment. The impact of antibiotics and antibiotic resistance gene pollution has become a major concern lately, and it is essential to understand the interaction of antibiotics with ecosystems. The release of antibiotics into the environment resulted in developing its resistance gene and other resistance genetic material (integrons, transposons, etc.). These resistance genes were also found in human pathogens and pristine environment, and now these genes can persist and spread even in the absence of antibiotics. Resistance among common pathogens causing community- and hospital-associated infections is increasing worldwide though regional patterns of resistance vary. Antibiotic resistance patterns follow patterns in antibiotic use: for newer antibiotics, lower resistance levels are reported, particularly in developing countries, where new drugs may be unaffordable for most. More comprehensive data collection and systematic examination and dissemination of existing data are needed to complete the global picture of antibiotic resistance. Seventeen (39%) of the 43 known tet genes including 12 (44%) of the efflux, 3 (25%) of the ribosomal protection, and 2 (66%) of the enzymatic tet genes are assigned to unique to environmental bacteria. It is possible that over time, these "unique tet genes" will move into bacteria associated with animals and/or humans as the tet(X) gene suggests. To a large extent what is in the environmental bacterial population remains largely unexplored. However, over time and as more of these environments are explored, it is clear that the number of new environmental genera of bacteria carrying tet genes will continue to increase, as will the number of tet genes identified from metagenomic DNA preparations from all types of ecosystems. Unfortunately, it is likely that human activities will continue to pollute the environment, which will make distinction between environmental and nonenvironmental impacts increasingly difficult.

References

Akasaki K, Karasawa K, Watanabe M, Yonehara H, Umezawa H (1963) Monazomycin, a new antibiotic produced by a Streptomyces. J Antibiot 16:127–131

Atkinson BA, Abu-Al-Jaibat A, LeBlanc DJ (1997) Antibiotic resistance among enterococci isolated from clinical specimens between 1953 and 1954. Antimicrob Agents Chemother 41:1598–1600

Bennett JW, Chung K-T (2001) Alexander Fleming and the discovery of penicillin. Adv Appl Microbiol 49:163–184

Centers for Disease Control and Prevention (CDC) (2013) Antibiotic resistance threats in the United States. CDC, Atlanta

Center for Disease Dynamics, Economics & Policy (CDDEP) (2015) Drug resistance index. Retrieved from http://www.cddep.org/projects/resistance_map/drug_resistance_index_urinary_tract_and_skin_infections. Accessed 20 Aug 2015

Cheng W, Chen H, Su C, Yan S (2013) Abundance and persistence of antibiotic resistance genes in livestock farms: a comprehensive investigation in eastern China. Environ Int 61:1–7

Chopra I, Roberts MC (2001) Tetracycline antibiotics: mode of action, applications, molecular biology and epidemiology of bacterial resistance. Microbiol Mol Biol Rev 65:232–260

Cousin SL Jr, Whittington WL, Roberts MC (2003) Acquired macrolide resistance genes and the 1 bp deletion in the mtrR promoter in Neisseria gonorrhoeae. J Antimicrob Chemother 51:131–133

Datta N, Hughes VM (1983) Plasmids of the same Inc groups in Enterobacteria before and after the medical use of antibiotics. Nature 306:616–617

European Medicines Agency (EMA) and European Centre for Disease Prevention and Control (ECDC) (2009) The bacterial challenge: time to react a call to narrow the gap between multidrug-resistant bacteria in the eu and development of new antibacterial agents. ECDC/EMEA, Stockholm

Kariuki S, Gordon MA, Feasey N, Parry CM (2015) Antimicrobial resistance and management of invasive salmonella disease. Vaccine 33(Supp 3):C21–C29

Kumarasamy KK, Toleman MA, Walsh TR, Bagaria J, Butt F, Balakrishnan R, Chaudhary U, Doumith M, Giske CG, Irfan S (2010) Emergence of a new antibiotic resistance mechanism in India, Pakistan, and the UK: a molecular, biological, and epidemiological study. Lancet Infect Dis 10:597–602

Lambert ML, Suetens C, Savey A, Palomar M, Hiesmayr M, Morales I, Agodi A et al (2011) Clinical outcomes of health-care-associated infections and antimicrobial resistance in patients admitted to European intensive-care units: a cohort study. Lancet Infect Dis 11(1):30–38. https://doi.org/10.1016/S1473-3099(10)70258-9

Le Doare K, Bielicki J, Heath PT, Sharland M (2014) Systematic review of antibiotic resistance rates among gram-negative bacteria in children with sepsis in resource-limited countries. J Pediatr Infect Dis Soc 4(1):11–20. https://doi.org/10.1093/jpids/piu014

Levy S (2002) Active efflux, a common mechanism for biocide and antibiotic resistance. J Appl Microbiol 92:65S–71S

Levy SB (2006) Mechanisms for resistance in soil. Science 312:529

Martinez JL (2009) Environmental pollution by antibiotics and by antibiotic resistance determinants. Environ Pollut 157:2893–2902

O'Neill J (2014) Antimicrobial resistance: tackling a crisis for the health and wealth of nations. Review on Antimicrobial resistance (December), London

Pan American Health Organization (PAHO) (Forthcoming) Informe Anual de la Red de Monitoreo/Vigilancia de la Resistencia a los Antibióticos y de Infecciones Asociadas a la Atención de la Salud 2014. Washington

Pei R, Kim S-C, Carlson KH, Pruden A (2006) Effect of river landscape on the sediment concentrations of antibiotics and corresponding antibiotic resistance genes (ARG). Water Res 40:2427–2435

Penesyan A, Gillings M, Paulsen IT (2015) Antibiotic discovery: combatting bacterial resistance in cells and in biofilm communities. Molecules 20:5286–5298

Roberts MC (2005a) Update on acquired tetracycline resistance genes. FEMS Microbiol Lett 245:195–203

Roberts MC (2005b) Tetracycline resistance due to ribosomal protection proteins. In: Frontiers in antimicrobial resistance. American Society of Microbiology, Washington, DC, pp 19–28

Roberts MC (2011) Mechanisms of bacterial antibiotic resistance and lessons learned from environmental tetracycline-resistant bacteria. In Antimicrobial resistance in the environment. pp 93–121. doi:https://doi.org/10.1002/9781118156247.ch7

Rodriguez-Mozaz S, Chamorro S, Marti E, Huerta B, Gros M, Sànchez-Melsió A, Borrego CM, Barceló D, Balcázar JL (2015) Occurrence of antibiotics and antibiotic resistance genes in hospital and urban wastewaters and their impact on the receiving river. Water Res 69:234–242

Salles MJC, Zurita J, Mejia C, Villegas MV (2013) Resistant Gram-Negative Infections in the Outpatient Setting in Latin America. Epidemiol Infect 141(12):2459–2472. https://doi.org/10.1017/S095026881300191X

Shore L, Pruden A (2009) Introduction. In: Hormones and pharmaceuticals generated by concentrated animal feeding operations. Springer, New York, pp 1–5

Stanton TB, Humphrey SB (2003) Isolation of tetracycline-resistant Megasphaera elsdenii strains with novel mosaic gene combinations of tet(O) and tet(W) from swine. Appl Environ Microbiol 69:3874–3882

Sun L, Klein EY, Laxminarayan R (2012) Seasonality and temporal correlation between community antibiotic use and resistance in the United States. Clin Infect Dis 55(5):687–694. doi:10.1093/cid/cis509

Tang X, Lou C, Wang S, Lu Y, Liu M, Hashmi MZ, Liang X, Li Z, Liao Y, Qin W (2015) Effects of long-term manure applications on the occurrence of antibiotics and antibiotic resistance genes (ARGs) in paddy soils: evidence from four field experiments in south of China. Soil Biol Biochem 90:179–187

Thaker M, Spanogiannopoulos P, Wright GD (2010) The tetracycline resistome. Cell Mol Life Sci 67:419–431

Watanabe T (1963) Infective heredity of multiple drug resistance in bacteria. Bacteriol Rev 27:87–115

Watanabe T, Ogata Y, Chan RK, Botstein D (1972) Specialized transduction of tetracycline resistance by phage P22 in Salmonella typhimurium: I. Transduction of R factor 222 by phage P22. Virology 50:874–882

Weigel LM, Donlan M, Shin DH, Jensen B, Clark NC, McDougal LK, Zhu W, Musser KA, Thompson J, Kohlerschmidt D, Dumas N, Linberger RJ, Patel JB (2004) High-level vancomycin-resistant Staphylococcus aureus isolates associated with a polymicrobial biofilm. Antimicrob Agents Chemother 51:231–238

World Health Organization (WHO) (2014) Antimicrobial resistance: global report on surveillance. WHO, Geneva

Yang S-C, Chang W-J, Chang Y-H, Tsai Y-S, Yang T-P, Juan C-W, Shiau M-Y (2010) Prevalence of antibiotics resistance and OXA carbapenemases genes in multidrug-resistant Acinetobacter baumannii isolates in central Taiwan. Eur J Clin Microbiol Infect Dis 29:601–604

Yun-jian H, Dong-ke C (2008) 2007 Mohnarin report: antibiotics resistance of non-fermenting Gram negative bacilli. Chin J Antibiot 33(10):597–601+S6

Zhu Y-G, Johnson TA, Su J-Q, Qiao M, Guo G-X, Stedtfeld RD, Hashsham SA, Tiedje JM (2013) Diverse and abundant antibiotic resistance genes in Chinese swine farms. Proc Natl Acad Sci USA 110:3435–3440

Chapter 3
A Review on Antibiotics Consumption, Physico-Chemical Properties and Their Sources in Asian Soil

Sija Arun, Moitraiyee Mukhopadhyay, and Paromita Chakraborty

3.1 Introduction

Antibiotics are the widely used antimicrobial drugs for prevention and treatment of bacterial infections by killing or inhibiting the growth of bacteria. A large variety of antibiotics are extensively used across the globe to treat human and animals. Antibiotics are used not only to treat individual animal with bacterial infections, but also to promote the growth of livestock. Consumption of antibiotics by livestock was reported to be 63,200 tons in 2010, which is more than the total human consumption worldwide (Van Boeckel et al. 2015). To meet the projected population of 8.5 billion in 2030, the consumption of antibiotics may rise by two-thirds and reach 105,600 tons (United Nations 2015). This increase may be due to the increase in the number of livestock production raised in large scale to meet the increasing demand (Van Boeckel et al. 2015).

The original version of this chapter was revised. An erratum to this chapter can be found at DOI 10.1007/978-3-319-66260-2_22.

S. Arun
Department of Civil Engineering, SRM University, Kattankulathur, Kancheepuram District 603203, Tamil Nadu, India

M. Mukhopadhyay
SRM Research Institute, SRM University, Kattankulathur, Kancheepuram District 603203, Tamil Nadu, India

P. Chakraborty (✉)
Department of Civil Engineering, SRM University, Kattankulathur, Kancheepuram District 603203, Tamil Nadu, India

SRM Research Institute, SRM University, Kattankulathur, Kancheepuram District 603203, Tamil Nadu, India
e-mail: paromita.c@res.srmuniv.ac.in

M.Z. Hashmi et al. (eds.), *Antibiotics and Antibiotics Resistance Genes in Soils*, Soil Biology 51, DOI 10.1007/978-3-319-66260-2_3

Antibiotic consumption increased by more than 30% between 2000 and 2010, from approximately 50 to 70 billion standard units. This is based on the data from 71 countries including most highly populated countries (Van Boeckel et al. 2015). Out of the total consumption, about 20% of antibiotics are used in hospitals and other healthcare clinics in most countries. Remaining 80% of antibiotics that are used in the community are either prescribed by healthcare providers or purchased directly by consumers over the counter without prescription (Kotwani and Holloway 2011). There is a probability that more than half of this usage by the community may end up in the wastewater treatment plants and finally to the surrounding environment thereby leading to antibiotic resistance in the ambient environment.

It is estimated that India alone consumed 12.9 billion antibiotic pills in 2010 followed by China (10 billion) and the USA (6.8 billion). Antibiotic usage in India alone has risen by 62%, from 8 billion pills in 2001 to 12.9 billion in 2010. Globally, antibiotics usage increased by 30% in the first decade of twenty-first century.

Antibiotics used by the human and animals are reaching the environment through excreta. Human excreta in the form of wastewater are treated in wastewater treatment plants (WWTPs), but 100% removal of antibiotics is simply not possible in conventional WWTPs. Some percentage of antibiotics gets partitioned in the sludge. So the sludge and final effluent from WWTPs contain antibiotics. The sludge is used as manure and the effluent is discharged into natural resources. Liquid manure from the livestock farming is also used for soil enrichment. These acts as the routes for the entry of antibiotics in different environmental matrices. Hence, it is important to understand the behavior of these compounds and their fate in the environment. This chapter aims to review the physico-chemical properties of antibiotics, their consumption pattern across the globe, and potential sources in soil.

3.2 Antibiotics: How Does It Work?

Effects of antibiotics include killing the bacteria by blocking the critical physiological processes in bacteria and prevent further prolification, thereby helping the body's natural immune system to fight against the bacterial infection. Antibiotics work differently for different types of bacteria. Antibiotics like amoxicillin and gentamicin targets an extensive range of bacteria and hence are called broad-spectrum antibiotics. Antibiotics like penicillin affect only a few strains of bacteria and are hence categorized as narrow-spectrum antibiotics. There are wide range of antibiotics working in different manner against the bacteria. For example, as shown in Fig. 3.1, penicillin a β-lactam destroys bacterial cell walls, while other antibiotics can affect the way the bacterial cell work.

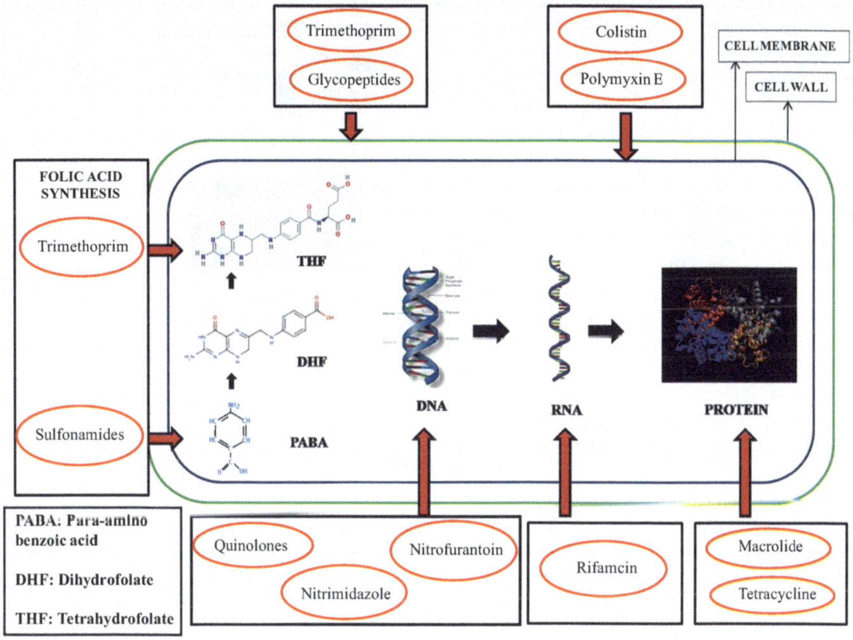

Fig. 3.1 Hypothetical schematic representation showing the mode of action of different antibiotics against bacteria

3.3 Consumption of Antibiotics

Consumption of antibiotics is broadly divided into three categories viz., (a) combating microbial infections, (b) agricultural usage, and (c) livestock infections and production. Consumption pattern of these categories around the world are described below.

3.3.1 *Combating Microbial Infections*

During 2000 and 2010, the rate of consumption of antibiotics has increased by 30%. Nearly three-fourth of this increase was contributed by Russia, India, South Africa and China. With 12.9×10^9 units, India consumed the maximum number of antibiotics in 2010, followed by China and the USA. Antibiotic consumption was found to be seasonally varying in most countries. Two last-resort classes of antibiotic drugs with increasing consumption rate were carbapenems (45%) and polymyxins (13%) (Haley and Morrill 2015).

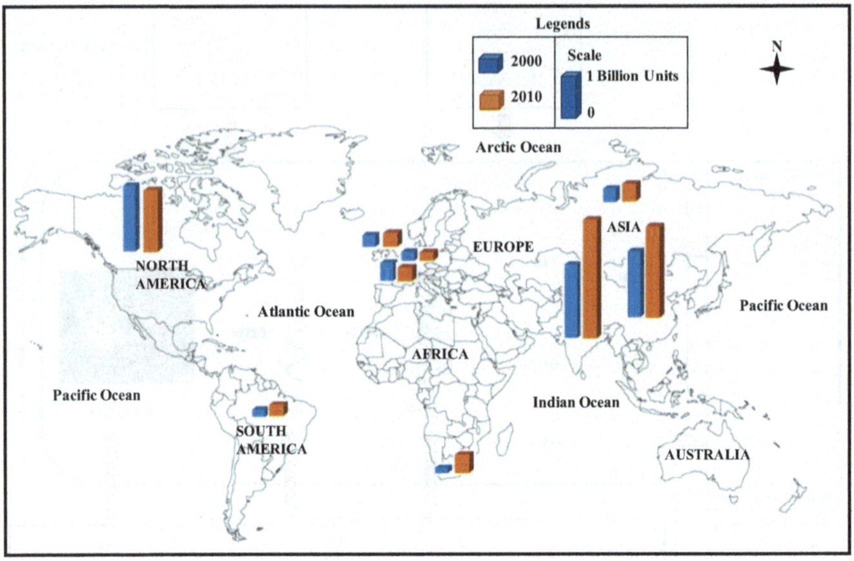

Fig. 3.2 Consumption of antibiotics by different countries across the globe during 2000 and 2010 (data courtesy-Van Boeckel et al. 2015)

Penicillins and cephalosporins accounted for more than 60% of the global consumption in 2010. These old antibiotics are still used for treating infections and their use increased by 40% during the last decade. Fluoroquinolones and macrolides consumption also increased by 30%. But there is not much increase in the consumption of tetracycline, trimethoprim and narrow spectrum of penicillin from 2000 to 2010 (Van Boeckel et al. 2015).

From Fig. 3.2 it is evident that high income countries like France, Germany and the USA realized the threat caused by the antibiotics and they have decreased their consumption since 2000 (Van Boeckel et al. 2015). But in developing nations like China, India and South Africa, antibiotic consumption has drastically increased from 2000 to 2010. In South Africa, the consumption has doubled. Variation in the consumption of antibiotics followed seasonal pattern. There was a sharp increase in antibiotic consumption during winter when compared to summer associated with the increased rate of diseases during winter.

3.3.2 Livestock and Agricultural Usage

As the population increases, the demand for animal food also increases. Antibiotics are therefore extensively used to promote the growth of poultry and aquaculture farms. It was estimated that 63,200 tons of antibiotics were consumed by livestock in the year 2010 (Van Boeckel et al. 2015) which accounts for more than 60% of the estimated 100,000 tons of antibiotics produced annually across the globe (Bbosa and Mwebaza 2013). With this projected increase, the global antibiotic consumption in livestock production may rise up to 105,600 tons by 2030. Sixty-six percent of this increase was due to the raise in the number of animals and the remaining 34% was due to the shift from extensive to intensive farming (Van Boeckel et al. 2015). The analysis of consumption pattern of antibiotics in 2010 showed that China consumed the major part of the total consumption of antibiotics in livestock followed by the USA, Brazil, Germany and India. This is because the developed countries like Germany reduced their consumption after understanding the threat caused by antibiotic resistance. With projected rise in population by 13% in 2030, the rate of consumption of antibiotics is also expected to double in developing nations like India, China and South Africa (Van Boeckel et al. 2015).

3.4 Physico-Chemical Characteristics of Antibiotics

Physico-chemical properties of the listed group of antibiotics have been given in Table 3.1.

3.4.1 Tetracycline and Sulfonamides

Amphoteric antibiotics such as tetracycline and sulfonamides have p*K*a values 3 and 2 respectively. Though tetracycline is sparingly soluble in water, the solubility of the corresponding hydrochloride is much higher. On the other hand, sulfonamides are insoluble in water and undergoes protonation at pH 2–3 and deprotonation at pH 5–11 (Gao et al. 2012a, b). Depending on the dissociation constant (K_d) value of tetracycline (range: 417–1026 mL/g) and sulfonamides (range: 0.9–18.1 mL/g), it has been reported that tetracycline has a higher tendency to migrate in soil than that of sulfonamides (Hu et al. 2010a, b).

3.4.2 Fluoroquinolones

Fluoroquinolones, the representative quinolone derivatives with fluorine at the sixth position, are used as antibacterial agents with a broad range of therapeutic activity (Tokura et al. 1996). The quinolones are amphoteric and with a few exceptions, generally exhibiting poor water solubility at slightly acidic or alkaline condition (pH 6–8). Although the impact on therapeutic efficacy is not clear, they appear to act as weak bases and are much less effective in acidic than in nonacidic condition. The quinolone nucleus contains a carboxylic acid group at position 3 and an exocyclic oxygen at position 4 (hence the term 4-quinolones), which is the active

Table 3.1 Physico-chemical properties and structure of antibiotics

Class	Molar mass (g/mol)	Water solubility (mg/L)	Log Kow	pKa	Henry's constant (Pal/mol)	Structure
Tetracycline	444.5–527.6	230–52,000	−1.3 to 0.05	3.3/7.7/9.3	1.7×10^{-23}–4.8×10^{-22}	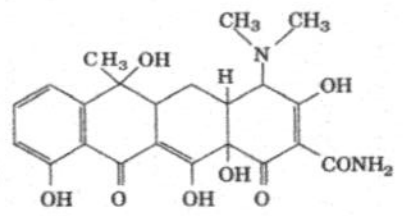
Sulfonamides	172.2–300.3	7.5–1500	−0.1 to 1.7	2–3/4.5–10.6	1.3×10^{-12}–1.8×10^{-8}	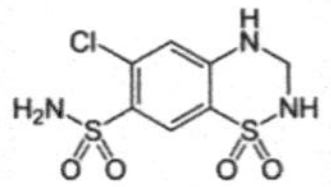
Aminoglycosides	332.4–615.6	10–500	−8.1 to −0.8	6.9–8.5	8.5×10^{-12}–4.1×10^{-18}	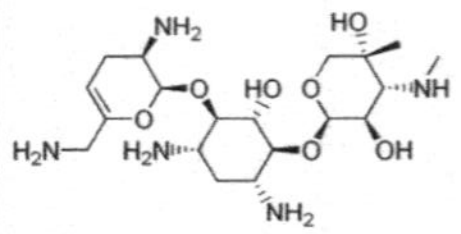
ß-Lactam	334.4–470.3	22–10,100	0.9–2.9	2.7	2.5×10^{-19}–1.2×10^{-12}	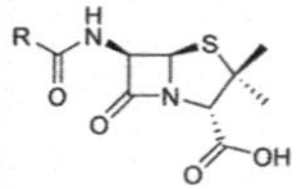
Macrolides	687.9–916.1	0.45–15	1.6–3.1	7.7–8.9	7.8×10^{-36}–2.0×10^{-26}	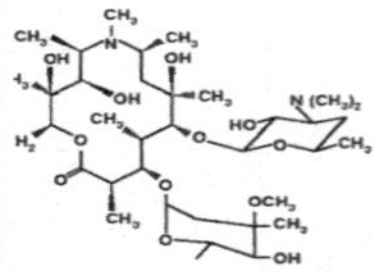
Fluoroquinolones	229.5–417.6	3.2–17,790	−1.0 to 1.6	8.6	5.2×10^{-17}–3.2×10^{-8}	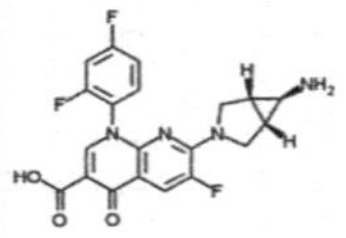

DNA-gyrase binding site. It exhibits its antibacterial activity by preventing the unwinding and duplication of the DNA.

3.4.3 Macrolides

Macrolides are a group of antibiotics that can work on a broad range of bacteria primarily by inhibiting the protein synthesis in bacteria. It has the inability to stay activated in highly alkaline or acidic condition; it can stay activated only in pH ranging from 4 to 10. The presence of one more deoxy sugar attached to the lactone ring enhances the possibility of macrolides to undergo a varied number of chemical reactions. The macrolides typically have a large lactone ring in their structure and are much more effective against Gram-positive than Gram-negative bacteria. They contain a dimethyl amino group which makes them basic. They are sparingly soluble in water but dissolves in polar organic solvents.

3.4.4 Aminoglycoside

Aminoglycoside antibiotics consist of an aminocyclitol group, with amino sugars attached to the aminocyclitol ring in glycosidic linkage. The basic nature and high water solubility characters of the antibiotics come from the amino groups and the hydroxyl groups on the sugar, respectively. Removal of the hydroxyl group from the amino sugars (e.g., tobramycin) increases the lipid solubility thereby increasing the activity as it can get absorbed in the lipid layer of the human body. One major drawback of aminoglycoside drug is its p*Ka* value ranging from 8 to 10 which limits the movement of this drug in human body.

3.4.5 β-Lactam

These antibiotics are characterized by the four-membered β-lactam ring which is the active component in the drug and acts on bacteria by inhibiting the cell wall synthesis.

3.5 Sources of Antibiotics in Asian Soil

Antibiotics that are consumed by human beings and animals are not completely absorbed or metabolized; therefore, they are released to the environment through their excreta. Conventional WWTPs does not completely remove these kind of compounds. In some areas, treated water is used for irrigating land wheras sludge from the treatment plant is used as manure. Manure from the livestock farms are also used as fertilizer. So eventually, 30–50% of the used antibiotics reach different environmental matrices like surface water, groundwater and soil. Some studies also demonstrated the risk associated with the consumption of fresh vegetables grown in soil which is amended with antibiotic laden manure (Kumar et al. 2005). It is therefore very important to know the sources, environmental concentrations and risks associated with the presence of antibiotics in soil as soil can act as a sink for organic compounds. Antibiotics may enter into the soil from different pathways:

(i) Sludge from WWTPs used as manure
(ii) Livestock excreta used as manure
(iii) Treated water used for irrigation
(iv) Surface runoff from solid waste municipal dumpsite
(v) Application of antibiotics in aquaculture farms

(i) Sludge from WWTPs used as manure

For a long period of time, sewage sludge was used as fertilizer because of high nitrogen and phosphorous content. But later it was found that sewage sludge contains heavy metals, pharmaceutical and personal care products, phthalates, dioxins, polycyclic aromatic hydrocarbons, flame retardants and other endocrine disrupters. Wastewater to be treated in the plant includes industrial effluents, storm water runoff, washings from agricultural fields and domestic wastewater. Wastewater from the WWTPs may enter the farmlands and might pose adverse impact on health of both human and animals via exposure through the food chain. Levels of antibiotics in sludge from different studies from China have been given in Table 3.2. In sludge, oxytetracycline was found to be dominant.

(ii) Livestock excreta used as manure

Antibiotics are used in livestock farms to suppress parasites, to treat diseases caused by bacteria, and to promote growth of livestock production (Kim and Carlson 2007). The percent of absorption and metabolism of antibiotics in animal body is

Table 3.2 Antibiotics in the sludge from wastewater treatment plant used as manure

Compounds	Sludge(μg/g)	Country	Reference
Tetracycline			
Tetracycline	2174.46	Sheyang, China(NE)	An et al. (2015)
Oxytetracycline	7369.67	Sheyang, China(NE)	An et al. (2015)
Chlortetracycline	3843.79	Sheyang, China(NE)	An et al. (2015)
Doxycycline	2104.27	Sheyang, China(NE)	An et al. (2015)
Sulfonamides			
Sulfamethoxazole	665	Sheyang, China(NE)	An et al. (2015)
Sulfadiazine	50.32	Sheyang, China(NE)	An et al. (2015)
Sulfamerazine	37.21	Sheyang, China(NE)	An et al. (2015)
Sulfadimidine	27.14	Sheyang, China(NE)	An et al. (2015)

Table 3.3 Antibiotics from livestock excreta used as manure

Compounds	Livestock manure (mg/g)	Country	Reference
Tetracycline			
Tetracycline	56.95	Sheyang, China(NE)	An et al. (2015)
	43.5	North China	Hu et al. (2010a, b)
	3.5	Beijing	Li et al. (2015)
Oxytetracycline	47.25	Sheyang, China(NE)	An et al. (2015)
	183.5	North China	Hu et al. (2010a, b)
	23.271	Beijing	Li et al. (2015)
Chlortetracycline	143.97	Sheyang, China(NE)	An et al. (2015)
	26.8	North China	Hu et al. (2010a, b)
	26.218	Beijing	Li et al. (2015)
Doxycycline	6.5	Sheyang, China(NE)	An et al. (2015)
Fluoroquinolones			
Enrofloxacin	8.684	Beijing	Li et al. (2015)
Ciprofloxacin	4.3	North China	Hu et al. (2010a, b)
	9.342	Beijing	Li et al. (2015)
Norfloxacin	4.187	Beijing	Li et al. (2015)
Ofloxacin	15.7	North China	Hu et al. (2010a, b)
Perfloxacin	24.7	North China	Hu et al. (2010a, b)
Lomefloxacin	0.038	Beijing	Li et al. (2015)
Macrolide			
Roxithromycin	0.067	Beijing	Li et al. (2015)
Lincomycin	3.8	North China	Hu et al. (2010a, b)
Sulfonamides			
Sulfamethoxazole	18.5	Sheyang, China(NE)	An et al. (2015)
	5.7	North China	Hu et al. (2010a, b)
	0.102	Beijing	Li et al. (2015)
Sulfadiazine	4.98	Sheyang, China(NE)	An et al. (2015)
	0.022	Beijing	Li et al. (2015)
Sulfamerazine	4.59	Sheyang, China(NE)	An et al. (2015)
Sulfadimidine	1.95	Sheyang, China(NE)	An et al. (2015)
Sulfamethazine	0.061	Beijing	Li et al. (2015)
Sulfadoxin	32.7	North China	Hu et al. (2010a, b)
Sulfachloropyridazine	2.76	North China	Hu et al. (2010a, b)

less. Approximately 30–90% of the antibiotics used are being excreted through urine and feces and subsequently released into the surrounding environment (Heberer 2002; Bound and Voulvoulis 2004; Kwon et al. 2011). Different studies across the globe reporting the level of antibiotics in livestock have been given in Table 3.3. When comparing the antibiotic concentration in sludge and livestock manure, the latter showed higher values. For example, tetracycline content in sludge and in livestock manure were 2.17 mg/kg and 56.95 mg/kg, respectively (Jing An et al. 2015). So it is evident that antibiotics used for veterinary purposes may affect the human health via food chain.

(iii) Treated water used for irrigation

Recycling of treated water for irrigation is an old practice. Regulation of treated water for irrigation is limited to very few parameters even in the USA. Most of the antibiotic compounds are not metabolized in the body and unused drugs are directly disposed off into the sewage collection system (Kummerer and Henninger 2003). Various processes in the WWTPs do not completely remove the antibiotics and therefore get discharged into the environment (Kummerer and Henninger 2003). Antibiotics are considered as pseudo-persistent contaminants. Antibiotics do not degrade easily hence with the increasing consumption, the concentration of these pseudo-persistent contaminants are increasing in the environment (Shi et al. 2012). The antibiotics concentration in wastewater from different studies have been given in Table 3.4. Azithromycin, clarithromycin, ciprofloxacin, erythromycin, norfloxacin, ofloxacin, sulfamethoxazole, and trimethoprim are the antibiotics normally detected in WWTP effluents (Fatta-Kassinos et al. 2011; Li and Zhang 2011; Sim et al. 2011; Ghosh and Mandal 2010; Watkinson et al. 2007). Some studies have detected antibiotics, such as erythromycin, sulfamethoxazole, trimethoprim and tetracycline in

Table 3.4 Wastewater containing antibiotics used for irrigation

Compounds	Wastewater (ng/l)	Country	Reference
Tetracycline			
Tetracycline	560	New York	Batt et al. (2006)
	48,000	Wisconsin, USA	Karthikeyan et al. (2006)
Oxytetracycline	47,000	Wisconsin, USA	Karthikeyan et al. (2006)
Fluoroquinolones			
Enrofloxacin	250	Wisconsin, USA	Karthikeyan et al. (2006)
Ciprofloxacin	10	USA	He et al. (2015)
	970	New York	Batt et al. (2006)
	310	Wisconsin, USA	Karthikeyan et al. (2006)
Norfloxacin	250	Wisconsin, USA	Karthikeyan et al. (2006)
Ofloxacin	9	USA	He et al. (2015)
Macrolide			
Erithromycin	3900	Wisconsin, USA	Karthikeyan et al. (2006)
Roxithromycin	1500	Wisconsin, USA	Karthikeyan et al. (2006)
Salfonamides			
Sulfamethoxazol	156	South Africa	Rahzia et al. (2012)
	1340	New York	Batt et al. (2006)
	310	Wisconsin, USA	Karthikeyan et al. (2006)
Sulfamethazine	50	China	Sun et al. (2014)
	300	Wisconsin, USA	Karthikeyan et al. (2006)

wastewater irrigated soils (Chen et al. 2011; Kinney et al. 2006). Variability in distribution of antibiotics concentration depending on the land usage pattern shows that the presence of antibiotics depends on varied soil characteristics like microbial diversity, moisture content and soil temperature.

(iv) Surface runoff from solid waste dumpsite
After expiry date, all the unused antibiotics are improperly dumped into solid waste disposal sites in developing nations. In most of the developing countries, the animal waste from slaughtering houses, its excreta, the unused medicines are collected by municipality and finally reaches solid waste dumpsite. During rainy season, all these wastes are washed away and the surface runoff can ultimately end up in the surface water. Leachate containing such leftover antibiotics in the solid waste stream of the municipal dumpsites can penetrate into the groundwater. In some countries, organic solid waste is converted into manure by composting. The amount of removal of the antibiotics by composting is unknown. The final product from composting is generally used as manure in agriculture. So this is a possible source for entry of antibiotics in soil.

(v) Application of antibiotics in aquaculture farms
Antibiotics such as tetracycline and oxolinic acid are extensively used in aquaculture farms (Table 3.5). The amount and type of antibiotics used for aquaculture varies with local and government regulations and farming practices. So the overall consumption pattern of antibiotics in aquaculture differs from one country to another. It has been found that antibiotic usage in Norway for aquaculture is 1 g per metric ton while in Vietnam it is 700 g per metric ton (Defoirdt et al. 2011). Unlike prophylactic use of antibiotics metaphylactic usage is commonly practiced for aquacultural purposes. In metaphylactic usage, the entire population is exposed to the medicine even if it is not required by some part of the population.

3.6 Occurrence of Antibiotics in Asian Soil

From the sources discussed previously in this chapter, it is evident that antibiotics can end up in the soil, and finally via plant uptake, such chemicals may enter the food chain. Despite such existing sources, limited data is available on the level of antibiotics in Asian soil. Antibiotic levels in soil in China and Korea among Asian countries have been given in Table 3.5. All the tetracycline compounds showed highest range in Shenyang region, China. Agriculture, animal husbandry, and agricultural product processing units were dominant in north-eastern part of Shenyang city. Sulfonamide compounds also showed very high concentration in Shenyang city. Sulfonamides and tetracycline are the two antibiotics most commonly used to promote growth in livestock production. So it is evident that the high concentration of these two compounds in Shenyang city is due to the use of these

Table 3.5 Occurrence of antibiotics in Asian soil

Compounds	Soil (μg/kg)	Country	References
Tetracycline			
Tetracycline	976.17	Shenyang, China	An et al. (2015)
	20.9–105	Northern China	Hu et al. (2010a, b)
	20.83–177.64	Korea	Awad et al. (2014)
	22	Beijing, China	Li et al. (2015)
Oxytetracycline	2.5–50 (μg/g)	Denmark	Rabùlle and Spliid (2000)
	124–2683	Northern China	Hu et al. (2010a, b)
	1398.47	Shenyang, China	An et al. (2015)
	0.09–0.71	Korea	Awad et al. (2014)
	423	Beijing, China	Li et al. (2015)
Chlortetracycline	1590.16	Shenyang, China	An et al. (2015)
	33.1–1079	Northern China	Hu et al. (2010a, b)
	0.07–0.85	Korea	Awad et al. (2014)
	120	Beijing, China	Li et al. (2015)
Doxycycline	870.45	Shenyang, China	An et al. (2015)
Fluoroquinolones			
Enrofloxacin	2–200 (μg/g)		Nowara et al. (1997)
	389	Beijing, China	Li et al. (2015)
Ciprofloxacin	250	Shenyang, China	An et al. (2015)
	10.3–30.1	Northern China	Hu et al. (2010a, b)
	253	Beijing, China	Li et al. (2015)
Norfloxacin	69	Beijing, China	Li et al. (2015)
Ofloxacin	0.6–1.6	Northern China	Hu et al. (2010a, b)
Perfloxacin	ND	Northern China	Hu et al. (2010a, b)
Lomefloxacin	10	Beijing, China	Li et al. (2015)
Macrolide			
Erythromycin	ND	Beijing, China	Li et al. (2015)
Roxithromycin	5.7	Beijing, China	Li et al.(2015)
Lincomycin	1.1–11.7	Northern China	Hu et al. (2010a, b)
Salfonamides			
Sulfamethoxazole	671.52	Shenyang, China	An et al. (2015)
	0.1–0.9	Northern China	Hu et al. (2010a, b)
	0.5–1.1	Korea	Awad et al. (2014)
	1.2	Beijing, China	Li et al. (2015)
Sulfadiazine	760.09	Shenyang, China	An et al. (2015)
	0.6	Beijing, China	Li et al. (2015)
Sulfamerazine	311.26	Shenyang, China	An et al. (2015)
Sulfadimidine	11.45	Shenyang, China	An et al. (2015)
Sulfamethazine	0.2–25 (μg/g)		
	ND–1.11	Korea	Awad et al. (2014)
Sulfathiazole	0.04–0.38	Korea	Awad et al. (2014)

two antibiotics in livestock production. As compared to China, the antibiotic concentration in Korean soil is less. This may be because, since 2005, South Korea is gradually decreasing the use of antibiotic growth promoters in livestock production.

3.7 Conclusion

Like several other organic pollutants, anitibiotics also reach soil mostly after partially treated or untreated antibiotics present in wastewater stream. Studies across the globe have reported the uptake of antibiotics by plants. The pathway of these pollutants generally depends on its physio-chemical properties. Studies have shown the development of antibiotic resistance genes in soil, plants, and humans. Further studies are required to find the fate of these pollutants in the environment and the associated risk due to the occurrence of these antibiotics in the environment.

Acknowledgement Authors would like to acknowledge the SRM Excellence Grant given to Dr. Paromita Chakraborty.

References

An J, Chen H, Wei S, Gu J (2015) Antibiotic contamination in animal manure, soil, and sewage sludge in Shenyang, northeast China. Environ Earth Sci 74(6):5077–5086

Awad YM, Kim S-C, Abd El-Azeem SAM, Kim K-H, Kim K-R, Kim K, Jeon C, Lee SS, Ok YS (2014) Veterinary antibiotics contamination in water, sediment, and soil near a swine manure composting facility. Environ Earth Sci 71(3):1433–1440

Batt AL, Bruce IB, Aga DS (2006) Evaluating the vulnerability of surface waters to antibiotic contamination from varying wastewater treatment plant discharges. Environ Pollut 142:295–302

Bbosa GS, Mwebaza N (2013) Global irrational antibiotics/antibacterial drugs use: a current and futurehealth and environmental consequences. Microbial pathogens and strategies for combating them: science, technology and education. Formatex, Badajoz

Bound JP, Voulvoulis N (2004) Pharmaceuticals in the aquatic environment-a comparison of risk assessment strategies. Chemosphere 56(11):1143–1155

Chen F, Ying G-G, Kong L-X, Wang L, Zhao J-L, Zhou L-J, Zhang L-J (2011) Distribution and accumulation of endocrine-disrupting chemicals and pharmaceuticals in wastewater irrigated soils in Hebei, China. Environ Pollut 159(6):1490–1498

Defoirdt T, Sorgeloos P, Bossier P (2011) Alternatives to antibiotics for the control of bacterial disease in aquaculture. Curr Opin Microbiol 14(3):251–258

Fatta-Kassinos D, Meric S, Nikolaou A (2011) Pharmaceutical residues in environmental waters and wastewater: current state of knowledge and future research. Anal Bioanal Chem 1:251–275

Gao P, Mao D, Luo Y, Wang L, Xu B, Xu L (2012a) Occurrence of sulfonamide and tetracycline-resistant bacteria and resistance genes in aquaculture environment. Water Res 46(7):2355–2364

Gao L, Shi Y, Li W, Niu H, Liu J, Cai Y (2012b) Occurrence of antibiotics in eight sewage treatment plants in Beijing, China. Chemosphere 86(6):665–671

Ghosh K, Mandal S (2010) Antibiotic resistant bacteria in consumable fishes from Digha coast, West Bengal, India. Proc Zool Soc 63(1):13–20

Haley J, Morrill KLL (2015) Over consumption of antibiotics. Lancet Infect Dis 15:377–378

He K, Soares AD, Adejumo H, McDiarmid M, Squibb K, Blaney L (2015) Detection of a wide variety of human and veterinary fluoroquinolone antibiotics in municipal wastewater and wastewater-impacted surface water. J Pharm Biomed Anal 106:136–143

Heberer T (2002) Occurrence, fate, and removal of pharmaceutical residues in the aquatic environment: a review of recent research data. Toxicol Lett 131(1–2):5–17

Hu X, Luo Y, Zhou Q (2010a) Simultaneous analysis of selected typical antibiotics in manure by microwave assisted extraction and LC–MS. Chromatographia 71:217–223

Hu X, Zhou Q, Luo Y (2010b) Occurrence and source analysis of typical veterinary antibiotics in manure, soil, vegetables and groundwater from organic vegetable bases, northern China. Environ Pollut 158(9):2992–2998

Karthikeyan K, Meyer MT (2006) Occurrence of antibiotics in wastewater treatment facilities in Wisconsin, USA. Sci Total Environ 361:196–207

Kim S-C, Carlson K (2007) Temporal and spatial trends in the occurrence of human and veterinary antibiotics in aqueous and river sediment matrices. Environ Sci Technol 41:50–57

Kinney CA, Furlong ET, Werner SL, Cahill JD (2006) Presence and distribution of wastewater-derived pharmaceuticals in soil irrigated with reclaimed water. Environ Toxicol Chem 25:317–326

Kotwani A, Holloway K (2011) Trends in antibiotic use among outpatients in New Delhi, India. BMC Infect Dis 11:99

Kummerer K, Henninger A (2003) Promoting resistance by the emission of antibiotics from hospitals and households into effluent. Clin Microbiol Infect 9:1203–1214

Kwon SI, Owens G, Ok YS, Lee DB, Jeon W-T, Kim JG, Kim K-R (2011) Applicability of the Charm II system for monitoring antibiotic residues in manure-based composts. Waste Manag 31(1):39–44

Li B, Zhang T (2011) Mass flow and removal of antibiotics in two municipal wastewater treatment plants. Chemosphere 83:1284–1289

Li C, Chen J, Wang J, Ma Z, Han P, Luan Y, Lu A (2015) Occurrence of antibiotics in soils and manures from greenhouse vegetable production bases of Beijing, China and an associated risk assessment. Sci Total Environ 521–522:101–107

Nowara A, Burhenne J, Spiteller M (1997) Binding of fluoroquinolone carboxylic acid derivatives to clay minerals. J Agric Food Chem 45(4):1459–1463

Rabùlle M, Spliid NH (2000) Sorption and mobility of metronidazole, olaquindox, oxytetracycline and tylosin in soil. Chemosphere 40:715–722

Rahzia H, Pool EJ (2012) The effectiveness of sewage treatment processes to remove faecal pathogens and antibiotic residues. J Environ Sci Health 47:289–297

Shi Y, Gao L, Li W, Liu J, Cai Y (2012) Investigation of fluoroquinolones, sulfonamides and macrolides in long-term wastewater irrigation soil in Tianjin, China. Bull Environ Contam Toxicol 89:857–861

Sim W-J, Lee J-W, Lee E-S, Shin S-K, Hwang S-R, Oh J-E (2011) Occurrence and distribution of pharmaceuticals in wastewater from households, livestock farms, hospitals and pharmaceutical manufactures. Chemosphere 82(2):179–186

Sun Q, Lv M, Hu A, Yang X, Yu C-P (2014) Seasonal variation in the occurrence and removal of pharmaceuticals and personal care products in a wastewater treatment plant in Xiamen, China. J Hazard Mater 277:69–75

Tokura Y, Nishijima T, Yagi H, Furukawa F, Takigawa M (1996) Photohaptenic properties of fluoroquinolones. Photochem Photobiol 64(5):838–844

United Nations, Department of Economic and Social Affairs, Population Division (2015) World population prospects: the 2015 revision, key findings and advance tables. Working Paper No. ESA/P/WP.241

Van Boeckel TP, Gandra S, Ashok A, Caudron Q, Grenfell BT, Levin SA, Laxminarayan R (2014) Global antibiotic consumption 2000 to 2010: an analysis of national pharmaceutical sales data. Lancet Infect Dis 14:742–750

Van Boeckel TP, Gandra S, Ashok A, Caudron Q, Grenfell BT, Levin SA, Laxminarayan R (2015) Global antibiotic consumption 2000 to 2010: an analysis of national pharmaceutical sales data. Lancet Infect Dis 14(8):742–750

Watkinson AJ, Murby EJ, Costanzo SD (2007) Removal of antibiotics in conventional and advanced wastewater treatment: implications for environmental discharge and wastewater recycling. Water Res 41(18):4164–4176

Chapter 4
Entry Routes of Veterinary Antibiotics in the Environment

Reep P. Tasho and Jae Young Cho

Abbreviations

CAFOs	Concentrated animal feeding operation
CDDEP	Center for Disease Dynamics, Economics and Policy
ESVAC	European Surveillance of Veterinary Antimicrobial Consumption
FAO	Food and Agriculture Organization of the United Nations
FAOSTAT	Food and Agriculture Organization of the United Nations Statistical Databases
FEDESA	European Federation of Animal Health
K_d value	Sorption coefficient
K_{ow}	Octanol-water partition coefficient
OTC	Oxytetracycline
PCU	Population correction unit
VA's	Veterinary antibiotics

4.1 Introduction

Veterinary antibiotics (VA's) are ubiquitous molecules that have, in recent years, risen major concern in environmental safety. VA's are antimicrobial agents that have been exploited for other purposes, such as growth enhancement in livestock. Therefore, their use and discharge into the environment, mainly anthropogenic, have also significantly increased which is a major cause for concern. Alexander Fleming discovered antibiotics for the first time in the 1920s. Since then its use for

The original version of this chapter was revised. An erratum to this chapter can be found at DOI 10.1007/978-3-319-66260-2_22.

R.P. Tasho (✉) • J.Y. Cho
Department of Agriculture Chemistry, Chonbuk National University, 3-2, Room 106, Jeonju-si, Jeollabuk-do 561-756, Republic of Korea
e-mail: reeplepcha@gmail.com; soilcosmos@jbnu.ac.kr

M.Z. Hashmi et al. (eds.), *Antibiotics and Antibiotics Resistance Genes in Soils*, Soil Biology 51, DOI 10.1007/978-3-319-66260-2_4

disease treatment and prevention in all fields of medicine, including veterinary sciences, has rapidly increased. However, it was not until 1950s that they started being used as growth promoters and enhancers in livestock. A group of American scientists found that apart from treating diseases VA's also promoted growth in livestock (Ogle 2013). This discovery proved to be almost as essential as the initial one that opened floodgates to the massive production, overuse, and misuse of these biomolecules. In intensive livestock farming, antibiotic usage in feeds has continued for more than 60 years. According to the study by Wang and Tang (2010), the total amount of annual use of antibiotics, including medical and VA's, has reached 100,000–200,000 tons worldwide. The quantity of VA's used in food animal production is thought to be very high, by some estimates even comparable to quantities used in human medicine.

There are several entry ways of VA's into the environment which includes both direct (animal droppings) and indirect methods (field application, waste disposal). VA's are not entirely digested within the animal gut, as a result, get excreted out of the animal system (parent compounds or breakdown metabolites). Thereon, these molecules find its way into the environment either directly or indirectly. Thus, recognizing and acknowledging the current demand, the areas, and frequency of use as well as understanding the properties of VA's is imperative to comprehend the various routes of VA's entry in the environment.

4.2 Veterinary Antibiotics

Antibiotics are naturally occurring compounds, semisynthetic and synthetic, applied parentally, orally, or topically as an antimicrobial agent. Over 150 antibiotics are in use today more than 90% of which are natural products of bacteria and fungi (molds) or semisynthetic modifications of natural compounds. However, a few such as sulfonamides are completely synthetic (von Nussbaum et al. 2006). The term veterinary antibiotics are used to define antimicrobial agents used specifically for the treatment and prophylaxis of animal diseases. Around 2000 veterinary pharmaceuticals have been manufactured from 400 plus active chemical ingredients to treat different animals including pigs, cattle, horses, sheep, goats, birds, fish, deer, cats, and dogs (FDA 2012). Table 4.1 lists some of the commonly used VA's.

Description of herd or flock antibiotics depends on their use for therapy, disease prevention, control, and growth promotion (National Committee for Clinical Laboratory Standards 2002). Therapy is the administration of antibiotics to an animal, or group of animals, exhibiting clear clinical disease. Control is the management of an antimicrobial to herd or flock animals, where the baseline morbidity and mortality have exceeded. For prevention/prophylaxis, antimicrobials are given to healthy animals that are considered to be at the risk contamination but before the onset of expected disease symptoms. Lastly, growth promotion requires the administration of antimicrobials, over time as feed additives, for enhancing the animal's physiological performance. However, the distinction between the purpose of said antibiotic as a growth-promoting agent or for prophylactic applications is

Table 4.1 Commonly used veterinary antibiotics and their characteristics in soil

Drug	Class	Effect	Half-life (days)	K_d value (L kg^{-1})	Water solubility (mg L^{-1})
Oxytetracycline Chlortetracycline Tetracycline	Tetracycline	Growth promoter, veterinary medicine	10–50 10–50 10–50	>200 >200 >200	>200 >200 >200
Sulfamethazine Sulfadiazine	Sulfonamide	Disease treatment	>50 >50	<5 <5	5–200 5–200
Tylosin Erythromycin	Macrolide	Disease treatment	<10 10–50	5–200 0–20	5–200 0–20
Neomycin Streptomycin	Aminoglycoside	Treatment and control of bacterial enteritis	na na	>200 >200	>200 >200
Penicillin	β-lactam	Disease treatment, growth enhancement	10–50	na	5–200
Monensin	Ionophore	Weight gain in cattle	>50	0–20	5–200
Virginiamycin	Peptide	Promotes growth of poultry	NA	>200	na
Enrofloxacin	Fluoroquinolones	Promotes growth of poultry	>50	>200	5–200

na data not recorded; *NA* not available

Sarmah et al. (2006); Thiele-Bruhn (2003); Tolls (2001); Song and Guo (2014)

often not easy. As evidenced in 2001, out of the 23 antibacterial products with US regulatory approval marketed for feed additive applications, only 15 had growth promotion label claims (Phillips et al. 2004). The immoderate appliance of antibiotics results in environmental contamination with original substances or derivatives.

The physical and chemical properties of VA's are an important determinant of their fate in the environment. They determine the interaction of this pharmaceutics with the soil particles. Being organic compounds VA's show a broad range of functional groups (ionic, amphiphilic, or amphoteric) that make their adsorption to the soil particles relatively easy. *These properties influence the characteristics of VA's in the environment, classified mainly by:*

4.2.1 Degradation Rate

It can also be interpreted as the half-life of individual VA's in soil. Half-life is the time required for the inactivation/degradation of half the antibiotics present in excreta, soil, or water, represented in days. The type of VA's, prevailing climatic conditions (especially temperature), type of soil, and other environmental factors influence the half-life. Degradation rate can be high (deterioration in ≤10 days), mediate (≤10–50 days), and low (≥50 days).

4.2.2 *Soil Mobility*

The sorption coefficient or K_d value is a measure of the tendency of a chemical to bind to soil particles. Different antibiotics have different adsorption affinity for the solid phase depending upon their K_d value. The K_d value ranges from >200 L kg^{-1} (high), 5 to 200 L kg^{-1} (mediate), and 0 to 5 L kg^{-1} (low). It is an important determinant of soil mobility which is an indicator of the potential of VA's to move through soil into groundwater and surface runoff.

4.2.3 *Water Solubility*

As a measure for general hydrophobicity, the octanol-water partition coefficient (K_{ow}) is often used. It is the ratio of a compound's concentration in a mixture of two immiscible phases in equilibrium and therefore is a measure of the difference in the solubility of the compound. VA's usually have log K_{ow} values of less than five which indicates that they are relatively non-hydrophobic (Tolls 2001). Water solubility for the majority of antibiotics exceeds 1 g L^{-1} making them relatively hydrophilic. The range of the water solubility ranges from >200 mg L^{-1} for high, 5 to 200 mg L^{-1} for mediate, and 0 to 20 mg L^{-1} for low solubility.

Thus, based on the physicochemical properties, VA's that are bioavailable for a longer period have:

(a) Low degradation rate
(b) High water solubility
(c) Low K_d (sorption coefficient) value

These characteristics effectively increase their chances of entering into different environmental compartments.

4.3 Veterinary Antibiotics and Intensive Farming Operations

Veterinary antibiotics are an integral part of intensive farming operations including both livestock farming and aquaculture. According to FAOSTAT (2016), livestock provides 14% of the total calories (kcal) and 33% of protein in people's diets. Industrial livestock operations make up 74% of the total global poultry production, 40% of pig meat, and 68% of eggs, as reported by FAO. Farm animals also significantly contribute to food security by helping combat micronutrients deficiency, or "hidden hunger," by providing people with essential vitamins and minerals. Also, animal manure for fertilization purposes helps boost crop productivity improving the economy and food security. Many farmers use antibiotics to optimize production in intensive agriculture to satisfy the increasing protein (animal) demands of the growing population. The performance of the animal regarding productivity and labor decreases when they are diseased. Thus, antibiotics are administered, often in an uncontrolled manner, to raise the animal efficiency which in turn raises the profit margin.

There is also an over-reliance on intensive fish farming to meet the global demand making aquaculture a booming industry worldwide. Therefore, antibiotics are widely used in the production of farmed seafood. The major producers are China delivering 80–90% of the world's shrimp and carnivorous fish, and Chile renowned for salmon cultivation (Marshall and Levy 2011). Such farming operations use at least a dozen antibiotics including a significant amount of quinolones (Marshall and Levy 2011). The metaphylactic use, i.e., orally treating an entire fish population even if only a small percentage are affected is common. However, the prophylactic use of antibiotics is rare in aquaculture. The sick fish in such case will usually not eat the medicine, so the medication, in reality, is to protect the healthy ones until the sick fish die and the infection subsides. Therefore, the infection is rarely completely cleared after the treatment.

Hence, an over-reliance pressurizes industrial farming to produce and provide food at the global level leading to the misuse of VA's on a large scale. Also, the high densities required to make industrial livestock operations profitable often exacerbate the use of VA's in a vast majority of farming operations. However, very few data regarding quantities of specific antibiotics employed in specific species of food animals are available publicly worldwide. Usually, concentration limit of the used antibiotics in the environment is not regulated, even though numerous negative implications of such antibiotics in the environment have been discussed.

Only a rough guestimate of the annual use of antibiotics can be made: in 1999 in the EU and Switzerland 13,288 tons of antibiotics were in use of which 29% were veterinary medicine, 6% growth promoters, and 65% were used in human medicine (FEDESA 2001). Also, approximately 70%, out of the total 16,200 tons, of antibiotics used in a year is for livestock farming. Though based on incomplete data, 80% of the antimicrobials used in the United States are for veterinary purposes and not for humans; of which 90% are for non-therapeutic purposes. Before the ban of growth promoters in Europe pigs and poultry were administered the majority of antibiotics administered in agricultural livestock production, whereas other species received only 1% of prescriptions (Ungemach 2000). In 2010, an estimated 63,151 tons of antimicrobials was consumed by livestock across the globe. By 2030, a 67% increase from 63,151 ± 1560 to 105,596 ± 3605 tons is expected (Van Boeckel et al. 2015). The rise in the number of animals raised for food production accounts for two-third of the projected increase while the shift from small-scale to industrial-scale production accounts for the remaining one-third (Van Boeckel et al. 2015). In 2003, China produced 28,000 and 10,000 tons of penicillin and oxytetracycline (OTC), a 60 and 65% of the total global output making them the highest producer and consumer of antibiotics (Yang et al. 2010). Figure 4.1 depicts the current top ten countries with highest antibiotic consumption.

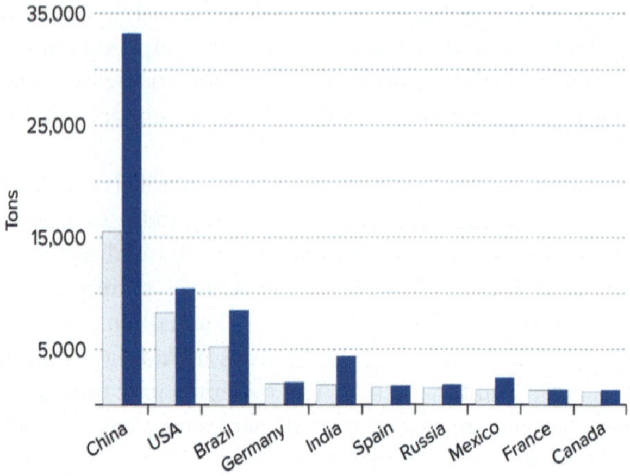

Fig. 4.1 Top ten countries with highest antibiotic consumption in 2010 with an estimated projected increase for 2030. Image source: CDDEP (2015)

The different aqua-farming practices, local and national regulations and government enforcement ability all determine the type and amount of VA's use in aquaculture. In the recent years, some countries have introduced more strict regulations concerning antimicrobial resistance and residues in food. However, the majority of aquaculture production takes place in countries with "permissive regulations" with limited environmental monitoring. Thus, the overall use varies widely among countries. According to the findings of Defoirdt et al. (2011), VA's use ranges from 1 g mt^{-1} ton of production in Norway to 700 g mt^{-1} in Vietnam.

Tetracycline followed by sulfonamides and macrolides are the most commonly used VA's. Their use accounts for approximately 90% of the total antibiotics used in the UK and 50% in Korea (Kim et al. 2011). According to ESVAC (2016), of the overall sales of antimicrobials in the 29 European countries, the largest amounts expressed as a proportion of mg PCU^{-1} were accounted for by tetracycline (33.4 %), penicillin (25.5%), and sulfonamides (11.0 %). However, as a result of the differences in the composition of the animal population and the production systems in different countries, it is likely to have different reported sales and sales patterns for VA's. However, since the prohibition of growth promoters in 2006, an actual decline in antibiotics used in agriculture has been recorded.

4.4 Introduction of Veterinary Antibiotics in the Environment

Industrial farming operations are unfortunately very resource intensive and pollute the soil, ground and surface water, emit greenhouse gasses, and also contribute to antibiotic resistance. The introduction of VA's in the environment occurs mainly through direct or indirect contamination of soil/water by animal manure loaded with undigested antibiotic compounds and metabolites (Fig. 4.2). Direct entry takes

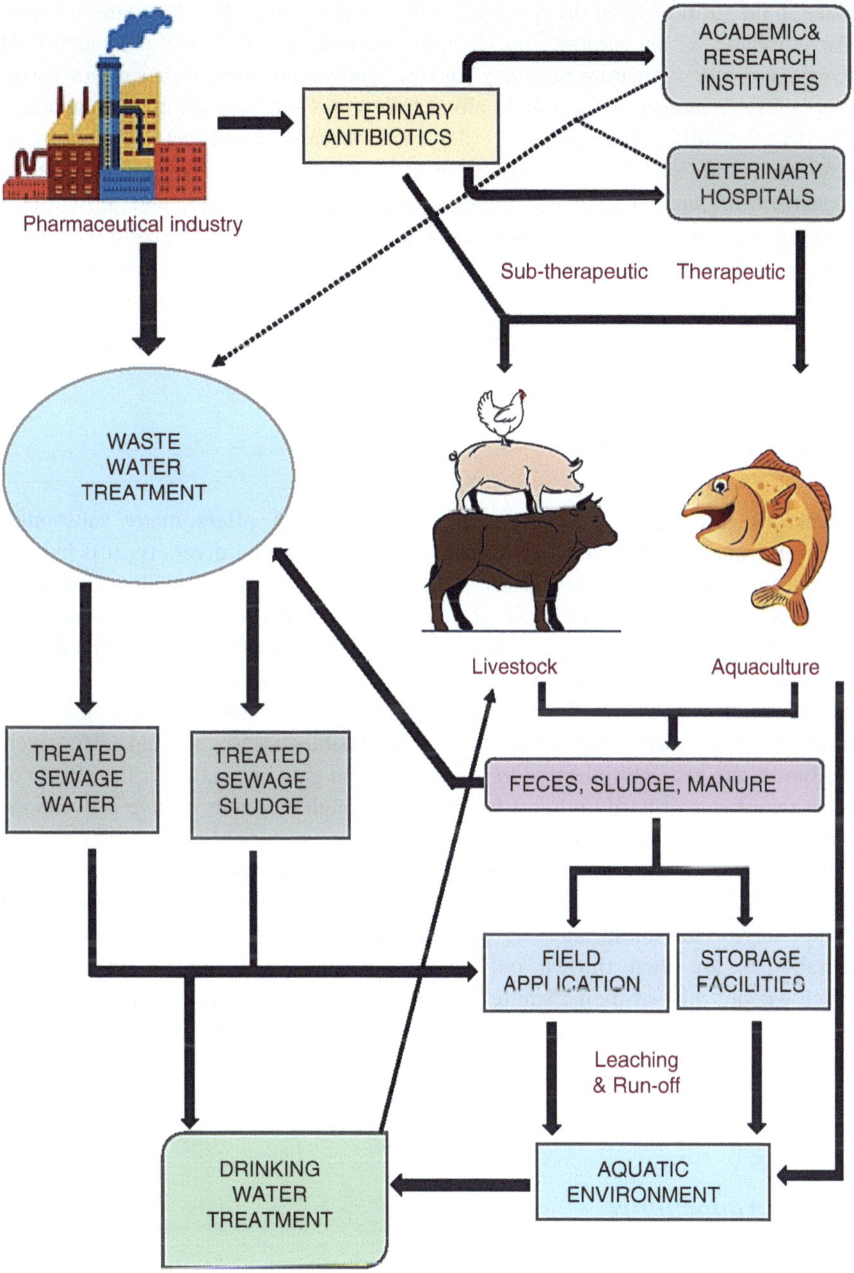

Fig. 4.2 Different entry routes of veterinary antibiotics in the environment

place in the form of fecal droppings of cattle on grassland while the cattle graze or use grassland as an outdoor run. Indirect methods of entry include the use of livestock waste as manure or slurry for fertilization purposes. Other minor entryways of VA's into the environment are as follows—(1) pharmaceutical companies, (2) drug manufacturing process, (3) waste storage structures, (4) hospitals, academic, and research institutes. However, in the scheme of today's global scenario with an overgrowing reliance on intensive livestock and agricultural farming practices to produce food; the untreated and unsafe disposition of farm animal excreta either as waste or as manure is the primary cause of VA's entry into the environment. *Listed below are some of the direct and indirect methods of VA's entry in the environment:*

4.4.1 Grazing Animals

Numerous research suggests that responsible grazing offers many substantial environmental benefits. Grazing increases plant and animal diversity. It is helpful in controlling invasive plant species, erosion, wildfire threats, and habitat fragmentation. Grazing helps improve vegetation and preserve open space meanwhile providing food for consumers. Therefore, many industrial as well as household livestock owners are opting for more natural feeding method of grazing. VA's are administered in the daily feed or water of animals for numerous reasons. Most likely, VA's are used to prevent animals from possible disease outbreaks. However, antibiotics are also administered to promote weight gain or to counter the effects of other treatments or hard to digest feeds. Even though the latter motive is questionable, many would agree that treating a sick animal with medicine is also crucial. Thus, many if not all animals are exposed to VA's at therapeutic or subtherapeutic dosage. As mentioned earlier, the animal gut is unable to digest the VA's completely. The undigested or partially digested VA's and its breakdown metabolites are then thrown out during excretion. Grazing animals excrete much—if not all—of their manure directly onto the land. Depending on the sex, type, and age of the animal, around 29.5 kg of dung and 13.2 l of urine is produced daily. Therefore, a hoard of grazing animals can release a significant amount of VA's that can directly enter the terrestrial environment Animals.

4.4.2 Aquaculture

One of the most unfortunate aspects of aquaculture production system is the use of open net-cages placed directly in the ocean. Farm waste, chemicals, disease, and parasites are potentially released directly into the surrounding waters. Aquaculture waste is largely liquid but may also contain unconsumed fish feed, dead fish, plankton, fecal matter, eggshells, and chemicals, including undigested antibiotics.

When over half a million or more farmed fishes are penned in a small area, fish farm waste can build up rapidly thereby disturbing the ocean bottom and surrounding ecosystems, especially in shallow waters or areas that do not flush well. Regulatory measures such as setting the cages in places with strong currents to wash away the effluent and moving the cages yearly have been effective in deconcentrating the bulk of waste discharge in any one area. However, the risk of antibiotic contamination is still an alarming issue. Also, because of the high densities at which fishes are farmed, disease and parasite outbreaks in fish farms can spread rapidly. Fish farmers, therefore, administer antibiotics and other chemicals in fish feed to combat these outbreaks. However, the effect of the drugs on the ecosystems around the cages, as well as residual antibiotics winding up on consumers' plates is a risky possibility. These antibiotics not only promote resistant bacteria in the farmed fish but also threaten to transmit resistance to wild fish populations and the broader environment.

4.4.3 Manure Application in the Fields

The application of animal excreta, potentially loaded with VA's, as manure for fertilization without proper treatment before field application, is also a major cause for concern. Sufficient nutrients and high organic matter content make animal manure an excellent choice as a fertilizer. Crop yield with the application of animal waste (liquid manure) at agronomics rate is equivalent to that of chemical fertilizers (Schmitt et al. 1995). In the context of today's global scenario where organic agricultural practices predominate, the application of animal manure functions as the principal source of antibiotic entry into the environment. Since 1990 the worldwide organic food market has grown rapidly, reaching $63 billion in 2012 (Helga et al. 2013). In 2011, of the total world farmland, approximately 0.9%, i.e., 37,000,000 ha constituted of organically farmed land (Willer and Kilcher 2011). Due to its ease of availability and other beneficial aspects, livestock manure is an excellent choice for organic farmers. According to the policies, the administration of medicine to organic livestock is only for treatment when they are sick and not for growth enhancement (FAO 2001).

The concentration of antibiotics in manure slurry varies from a trace to as high as 216 mg L^{-1} (Kumar et al. 2005). Detection of oxytetracycline concentrations 6, 7, and <5 μg kg^{-1} at the soil depth of 0, 30, and 60 cm depth after cattle manure application at 96 Mg ha^{-1} is reported by De Liguoro et al. (2003). In the same study, tylosin concentration of <10 μg kg^{-1} was detected as well. In fields amended with swine liquid manure, Hamscher et al. (2002) reported tetracycline concentrations of 86, 199, and 172 mg kg^{-1} of soil at 0–10, 10–20, and 20–30 cm depths. A large number of antibiotics present in manure can potentially transport and persist in the environment after field application (Gavalchin and Katz 1994). It is important to note that in comparison to the farmer's households a higher concentration of antibiotic residues is reported in animal manures from industrial livestock farms.

In the administered animals, the level of VA's residue varies from pig manure > chicken manure > cow manure. Furthermore, between the administered animals the level of VA's residue varies according to the type of animal: pig > chicken > cow (Zhang et al. 2008b).

4.4.4 Land Input of Farm Animal Waste

Each year tons of meat is produced to satisfy customers worldwide, which corresponds to the production of tons of animal waste annually worldwide. Livestock produces an estimated one billion tons of urine and feces annually that potentially may contain serious amounts of antibiotic metabolites which if not treated properly can cause serious environmental health issues. During the period required for swine to attain the market size of 114 kg (5–6 months), 1.5 tons of fresh manure is produced on average (Richert et al. 1995). Among the major livestock, pigs generate the largest amount of waste around 13.4 million tons/year. Beef, dairy, and chicken account for about 7.5, 7.1, and 4.9 million tons/year, respectively. In 2007, 2.2 billion head of livestock and poultry generated approximately 1.1 billion tons of manure. In South Korea, an annual 46.5 million tons of animal manure were produced from 3.3 million cattle, 7 million pigs, and 140 million poultry birds (Ahn et al. 2011). Studies show that as much as 90% of some antibiotics may be excreted as their parent compound (Phillips et al. 2004; Kumar et al. 2005). The nature of the antibiotic used, the animal species it is administered to, the type of application procedure used, and the duration of antibiotic administration determines the excretion rate. Interestingly, the level of VA's residue in animal excreta shows significant statistical difference depending on the sampling sites and the type of animal species (Zhao et al. 2010). Safe and proper disposal of animal excreta, produced daily, from intensive livestock farms that breed thousands of cattle, often in small areas, is a challenge.

Concentrated livestock farming practices that produce tons of animal excreta every day, often dump animal slurry into the surrounding areas from where the antibiotics can easily migrate into the different environmental compartments. For the disposal of large amounts of animal waste generated from the livestock farms, land input is the most common and easiest. It also saves the cost of collection, transport, and proper treatment of wastes for the farmer. *Concentrated Animal Feeding Operation* (CAFOs) workers in the United States and Korea often apply their effluents into the fields by signing contracts with nearby farm owners. However, in most cases, it is not required by law to treat livestock waste. Therefore, studying and evaluating the fate of antibiotics in livestock farms from administration to excretion, waste collection to disposition, taking into account potential soil–water transport is necessary.

4.4.5 Waste/Manure Storage

Waste storage conserves nutrients contained in the manure which is valuable during extended periods of bad weather, winter months when many forage crops are dormant, that make application impractical. It also gives producers the option to use manure as and when crops need or can best utilize the nutrients. However, accumulating manure in a small area can be risky to the environment unless done properly. Improperly managed manure can contaminate both surface and groundwater with nutrients, disease-causing organisms, and even antibiotics. VA's have relatively slow degradation rate during storage (Boehm 1996 and Migliore et al. 1995). After excretion, the antibiotic metabolites can transform back into their parent compound (Langhammer 1989). For instance, some antibiotics after their transformation into conjugates such as acetylated metabolites become inactive and analytically absent. However, the originally active form, after the cleavage of acetyl groups, can release in manure (Christian et al. 2003).Thus, some effective biosolids such as the active form of some antibiotics could be released into the environment through leaching and runoff.

Storage facilities store livestock wastes in three forms: solid, semi-solid, and liquid. Solid waste storage utilizes walls and slabs. For semi-solid waste, pumps or scrapers are used to move manure into containment areas and may separate solids from liquids. Tanks, pits, waste storage ponds, or lagoons are used in liquid waste facilities to hold manure. Over an extended time, the waste storage facilities itself may also incur environmental damage increasing the risk of waste leakage. The side walls of the storage pits may crack and erode due to weathering, wave action, or wetting and drying cycles. Wastes can seep into the underlying soil and contaminate groundwater with potential active VA's or its metabolites. Also, improper maintenance of waste storage facilities may cause leaking and contaminate soil. Infrequent emptying of storage pits causes overflowing endangering surface water. Pipes and pumps for moving wastes must be carefully installed and checked at regular intervals for leakage. Steel and concrete structures should also be routinely checked for cracks or the loss of watertight seals and must be repaired immediately.

4.4.6 Wastewater or Sewage Treatment Plants

Sewage treatment also referred to as wastewater treatment is the process of removing contaminants from household sewage as well as industrial wastewater. The treatment process implies physical, chemical, and biological processes to eliminate contaminants and produce an environmentally safe treated wastewater (or treated effluent). Usually, a semi-solid waste or slurry is formed as a by-product (sewage sludge) which undergoes further treatment before being suitable for disposal or land application. A large part of the administered VA's end up in wastewater. There are several ways in which VA's find their way into wastewater.

Those not broken down by the human body are passed to wastewater, expired antibiotics from homes and hospitals also get dumped into wastewater, and there is the discharge of antibiotic materials from pharmaceutical companies.

However, not all treatment facilities can remove VA's from the waste. The type of treatment technology used as well as the nature of the drug governs the efficiency of removal. Surprisingly, new findings show that wastewater treatment is designed to break down biological substances but not antibiotics. In a study by Hendricks and Pool 2012, no significant difference in fluoroquinolone concentrations between raw wastewater and treated sewage effluents from three treatment plants was found. Thus, high loads of VA's can still be discharged into the environment even after treatment. Also, from lab findings, chlorine is not breaking down antibiotics but is actually creating even stronger antibiotics than the first doxycycline (University of North Carolina Press 2015).

4.4.7 Leaching and Runoff

Leaching and runoff are the main pathways of VA's entry into the aquatic ecosystem. Leaching and runoff are the methods of VA's accession to the soil from manure/slurry storage and land application of manure (Kemper 2008). In the soil, antibiotic degradation, runoff, and sorption to soil particles may take place. Subsequently, both surface water and groundwater may contain antibiotic compounds. The physicochemical properties and the amount of precipitation are the major determinants of VA's runoff/leaching potential (Taylor et al. 2011). Highly mobile antibiotics have low K_d value (sorption coefficient). As a result, their presence in surface and groundwater is higher than antibiotics with a high K_d value (Taylor et al. 2011).

4.4.8 Academic and Research Institutes

The prolific use of VA's opens up research opportunities for the study of both beneficial and negative aspects of these biomolecules. Studies on novel drug design/discovery and efficient mode of action may only be exceeded by those done to study its impact on the ecosystem and the effective elimination/control measures. Such an extensive study requires the use of tons of VA's on a global scale. Therefore, carelessness at even secondary levels can potentially release untreated and active experimental drugs into the environment. Proper handling of antibiotics under strict supervision, especially during discarding, is imperative to prevent the discharge of VA's. Most of the research facilities do have provisions for proper storing and discarding of drugs. However, not undertaking a timely and routine investigation of the quantity and type of antibiotics in storage, as well as in waste, can lead to overstocking and overflow. Also, ignoring the routine checkup of the temporary waste storage and transportation units for damage may delay fixing, causing antibiotic leakage.

4.4.9 Veterinary Hospitals

The use of VA's in veterinary hospitals is given. It is also the preferred choice for disease prevention and cure. It is also imperative for hospitals to be stocked with all types and forms of VA's to be used as and when needed. Therefore, mishandling and carelessness on the part of hospital staff during storage, use, and disposal can release untreated VA's in the environment. Recommending VA's without optimizing the dosage required for treatment can lead to overuse and misuse of VA's. Failure to set up proper waste disposal may cause discarding of expired VA's, empty containers, and used syringes properly directly into the environment. Misrecording and not maintaining logs of the VA's both in storage and in use, their purchase and expiration date, quantities sold and in stock are common and more times than not overlooked. Therefore, routine legalized checks of hospitals and pharmacies to ensure proper and up-to-date records of VA's sales and diagnosis is a must.

4.4.10 Miscellaneous

Other minor entries of VA's from agriculture sources into the environment may occur through exhaust air of animal stable and ventilation containing antibiotics in dust form (Hamscher et al. 2003). Stormwater runoff is another likely entry route of VA's. As rainfall travels over roofs, streets, gardens, and other outdoor areas, it may pick up various contaminants including animal waste loaded with VA's. The runoff water flows into the stormwater system that flows directly into local waterways, at most times, without receiving any treatment. Improper handling and unsafe transportation can also contribute to the antibiotic release. The released antibiotics get adsorbed to colloids or soil particles or get dissolved which is then transported to the surface and groundwater (Krapac et al. 2004).

4.5 Conclusion

There are many pathways for the entry of VA's into the environment. However, fertilization with antibiotics containing animal manures, biosolids, sewage sludge, and sediments seems to be the most dominating pathway for their entry into the tertiary environment. Apart from these, reclaimed water from wastewater treatment plants, surface water, or groundwater that is frequently polluted by antibiotics is also a potential source of VA's entry in the environment. The residual antibiotics find its way into the soil from where it makes its way into the aquatic ecosystem. Farm soil and groundwater serve as two main reservoirs of residual antibiotics. Manure and waste slurries also potentially contain many antibiotics that can transport and persist in the environment after field application. Intensive livestock farming, therefore, may lead to the accumulation of animal wastes within relatively small geographical areas (Ekunwa et al. 2006).

The judicious use of VA's for the right reasons is paramount for limiting their entry into the environment. Antibiotic-mediated growth promotion is the focus of most legal and regulatory efforts to reduce the misuse of VA's. Recent analyses suggest that growth promoters have a smaller effect on animal growth than assumed, particularly in production systems that are otherwise optimized (Laxminarayan et al. 2015). For preventing the entry and mobility of active VA's in the environment focus should be driven towards the treatment and stabilization of VA's at the source of its entry. Therefore, future research needs to concentrate on optimizing pre-treatment methods such as composting, anaerobic fermentation, and irradiation. The development of a standard pre-treatment method that is both efficient and affordable needs to be prioritized. Also, the countries with least efficient farming systems have the highest expected increases in food demand and simultaneous VA's usage. Therefore, emphasis should be given to improving productivity without the use of antibiotic growth promoters. Improved surveillance, legal regulation, and public awareness are also essential in limiting the entry of these biomolecules into the environment (Tasho and Cho 2016).

References

Ahn HK, Kim JH, Kwag JH, Jeong KH, Choi DY, Yoo YH (2011) Sustainable animal waste management and utilization in Korean livestock farms. In: Proceedings international seminar on sustainable resource management of livestock wastes for Asian small-scale farmers, July 25–29, Ho Chi Minh City, Vietnam

Boehm R (1996) Auswirkungen von ruckstanden von antiifektiva in tierischenausscheidungen auf die gullebehandlung und den boden. Dtsch Tieraerztl Wochenschr 103:264–268

CDDEP, Center for Disease Dynamics, Economics & Policy (2015) Drug Resistance Index. Accessed 20 Aug 2015

Christian T, Schneider R, Färber HA, Skutlarek D, Meyer MT, Goldbach H (2003) Determination of antibiotic residues in manure, soil, and surface waters. Acta Hydrochim Hydrobiol 31:36–44

Defoirdt T, Sorgeloos P, Bossier P (2011) Alternatives to antibiotics for the control of bacterial disease in aquaculture. Curr Opin Microbiol 14(3):251–258

De Liguoro MV, Cibin F, Capolongo F, Halling-Sorensen B, Montesissa C (2003) Use of oxytetracycline and tylosin in intensive calf farming: Evaluation of transfer to manure and soil. Chemosphere 52:203–212

Ekunwa PA, Soniregun OO, Oyedeji JO (2006) Economics of small scale deep litter system of egg production in Oredo L.G.A of Edo State, Nigeria. Int J Poult Sci 1:81–83

ESVAC, European Surveillance of Veterinary Antimicrobial Consumption (2016) Sales of veterinary antimicrobial agents in 29 European countries in 2014 European Medicines Agency (EMA/61769/2016)

FAO (2001) Guidelines for the production, processing, labelling and marketing of organically produced foods. Gl: 32–1999, Rev. 1–2001
FAOSTAT (2016) Food and Agriculture Organization of the United Nations, Rome, Italy
FDA (2012) The Green Book—FDA approved products. Food and Drug Administration, Rockville
FEDESA, European Federation of Animal Health (2001) Antibiotic use in farm animals does not threaten human health. FEDESA/FEFANA. FEDESA, Brussels (Press release, 13 July)
Gavalchin J, Katz SE (1994) The persistence of fecal-borne antibiotics in soil. J AOAC Int 77:481–485
Hamscher G, Sczesny S, Hoper H, Nau H (2002) Determination of persistent tetracycline residues in soil fertilized with liquid manure by high-performance liquid chromatography with electrospray ionization tandem mass spectrometry. Anal Chem 74:1509–1518
Hamscher G, Pawelzick HT, Sczesny S, Nau H, Hartung J (2003) Antibiotics in dust originating from a pig-fattening farm: a new source of health hazard for farmers? Environ Health Perspect 111:1590–1594
Helga W, Julia L, Robert H (2013) The world of organic agriculture: statistics & emerging trends. Research Institute of Organic Agriculture (FiBL) and the International Federation of Organic Agriculture Movements (IFOAM)
Hendricks R, Pool EJ (2012) The effectiveness of sewage treatment processes to remove faecal pathogens and antibiotic residues. J Environ Sci Health A Tox Hazard Subst Environ Eng 47 (2):289–297
Kemper N (2008) Veterinary antibiotics in the aquatic and terrestrial environment. Ecol Indic 8:1–13
Kim KR, Owens G, Kwon SI, So KH, Lee DB, Ok YS (2011) Occurrence and environmental fate of veterinary antibiotics in the terrestrial environment. Water Air Soil Pollut 214:163–174
Krapac IG, Koike S, Meyer MT, Snow DD, Chou SFJ, Mackie RI, Roy WR, Chee-Sanford JC (2004) Long-term monitoring of the occurrence of antibiotic residues and antibiotic resistance genes in groundwater near swine confinement facilities 158-172. In: Proceedings of the 4th international conference on pharmaceuticals and endocrine disrupting chemicals in water, Minneapolis, MN, Oct 2004. National Groundwater Association
Kumar K, Gupta SC, Baidoo SK, Chander Y, Rosen CJ (2005) Antibiotic uptake by plants from soil fertilized with animal manure. J Environ Qual 34:2082–2085
Langhammer JP (1989) Untersuchungen zum Verbleib antimikrobiell wirsamer Arzneistoffe als Ruckstande in Gulle und im landwirtschaftlichen Umfeld. PhD-Dissertation, Universitat Bonn
Laxminarayan R, van Boeckel T, Teillant A (2015) The economic costs of withdrawing antimicrobial growth promoters from the livestock sector. OECD Food, Agriculture and Fisheries Papers (78)
Marshall BM, Levy SB (2011) Food animals and antimicrobials: impacts on human health. Clin Microbiol Rev 24(4):718–733
Migliore L, Civitareale C, Brambilla G, Delupis GDD (1995) Toxicity of several important agricultural antibiotics in Artemia. Water Res 31:1801–1806
National Committee for Clinical Laboratory Standards (2002) Performance standards for antimicrobial susceptibility testing: twelfth informational supplement. M100-S12. NCCLS, Wayne, PA
Ogle M (2013) In meat we trust: an unexpected history of carnivore America. ISBN 978-0151013401
Phillips I, Casewell M, Cox T, De Groot B, Friis C, Jones R, Nightingale C, Preston R, Waddell J (2004) Does the use of antibiotics in food animals pose a risk to human health? A critical review of published data. J Antimicrob Chemother 53:28–52
Richert BT, Tokach MD, Goodband RD, Nelssen JL (1995) Assessing producer awareness of the impact of swine production on the environment. J Ext [Online] 33(4)

Sarmah AK, Meyer MT, Boxall ABA (2006) A global perspective on the use, sales, exposure pathways, occurrence, fate and effects of veterinary antibiotics (VAs) in the environment. Chemosphere 65:725–759

Schmitt MA, Evans SD, Randall GW (1995) Effect of liquid manure application methods on soil nitrogen and corn grain yields. J Prod Agric 8:186–189

Song W, Guo M (2014) Residual veterinary pharmaceuticals in animal manures and their environmental behaviors in soils. In: He Z, Zhang H (eds) Applied manure and nutrient chemistry for sustainable agriculture and environment. Springer, Dordrecht, pp 23–52

Tasho RP, Cho JY (2016) Veterinary antibiotics in animal waste, its distribution in soil and uptake by plants: a review. Sci Total Environ 563–564:366–376

Taylor NGH, Verner-Jeffreys DW, Baker-Austin C (2011) Aquatic systems: maintaining, mixing and mobilising antimicrobial resistance? Trends Ecol Evol 26(6):278–284

Thiele-Bruhn S (2003) Pharmaceutical antibiotic compounds in soils—a review. J Plant Nutr 166 (2):145–167

Tolls J (2001) Sorption of veterinary pharmaceuticals in soil: a review. Environ Sci Technol 35:3397–3406

Ungemach FR (2000) Figures on quantities of antibacterials used for different purposes in the EU countries and interpretation. Acta Vet Scand Suppl 93:89–98

University of North Carolina at Charlotte (2015) Wastewater treatment may be creating new antibiotics. Science Daily, May 28

Van Boeckel TP, Brower C, Gilbert M, Grenfell BT, Levin S, Robinson TP (2015) Global trends in antimicrobial use in food animals. Proc Natl Acad Sci USA 112(18):5649–5654

von Nussbaum F, Brands M, Hinzen B (2006) Medicinal chemistry of antibacterial natural products exodus or revival? Angew Chem Int Ed Engl 45:5072–5129

Wang M, Tang JC (2010) Research of antibiotics pollution in soil environments and its ecological toxicity. J Agro Environ Sci 29(suppl):261–266

Willer H, Kilcher L (2011) The world of organic agriculture. statistics and emerging trends. FiBL, Bonn; IFOAM, Frick

Yang YD, Chen DH, Huang MH (2010) The source of antibiotics in the environment and progress of its ecological impact research. Environ Sci Manag 35(1):140–143

Zhang HM, Zhang MK, Gu GP (2008) Residues of tetracyclines in livestock and poultry manures and agricultural soils from North Zhejing provinces. J Ecol Rural Environ 24(3):69–73

Zhao L, Dong YH, Wang H (2010) Residues of veterinary antibiotics in manures from feedlot livestock in eight provinces of China. Sci Total Environ 408:1069–1075

Chapter 5
Monitoring of Antibiotics and Antibiotic Resistance Genes in Agroecosystems

Sarfraz Ahmed, Muhammad Ibrahim, Rabia Khushi Muhammad, Iqra Naseem, Rubbea Idrees, Muhammad Zaffar Hashmi, and Ateeq-ur- Rehman

5.1 Introduction

Antibiotics are drugs used to execute or hinder microorganisms. Antibiotic agents that eliminate microorganisms are called "bactericidal," while those that hinder their growth are called as "bacteriostatic" (Kumar et al. 2012). Different types of phenols, acids, and aromatic compounds can be utilized as antibiotic agents (Tran-Thanh et al. 2010). It is estimated that 210,000 tons of antibiotics is produced every year, of which 48% are utilized as a part of farming and domesticated animal enterprises (Li et al. 2015). A large amount of antibiotics as dynamic pharmaceutical fixings has been utilized as a part of creature cultivation and on fish farms to achieve advance development or controlled diseases. It has been reported that >80% of the antibiotics utilized are discharged as dynamic metabolites in excrements or urine waste (Awad et al. 2015). The extensive utilization of antibiotics is characteristically connected to the occurrence of bacterial resistance against these compounds (Williams-Nguyen et al. 2016).

The original version of this chapter was revised. An erratum to this chapter can be found at DOI 10.1007/978-3-319-66260-2_22.

S. Ahmed (✉) • M. Ibrahim • R.K. Muhammad • I. Naseem • R. Idrees
Department of Biochemistry, Bahauddin Zakariya University, Multan 60800, Pakistan
e-mail: sarfraz.ahmed@bzu.edu.pk; ibrahimjam@bzu.edu.pk; rabiyakhushi04@gmail.com; iqra.chauhdary10@yahoo.com; rubbeaidrees@yahoo.com

M.Z. Hashmi
Department of Meteorology, COMSATS Institute of Information Technology, Islamabad, Pakistan

A. Rehman
Department of Biochemistry, Bahauddin Zakariya University, Multan 60800, Pakistan

Department of Plant Pathology, Bahauddin Zakariya University, Multan 60800, Pakistan
e-mail: ateequrrehman@bzu.edu.pk

M.Z. Hashmi et al. (eds.), *Antibiotics and Antibiotics Resistance Genes in Soils*, Soil Biology 51, DOI 10.1007/978-3-319-66260-2_5

Antibiotic resistance is the ability of microbes to survive potentially in the presence of antibiotics at concentrations that would otherwise result in their death (Franklin et al. 2016). Release of antibiotics into the environment prompts the strains of pathogenic antibiotic-resistant bacteria (Chee-Sanford et al. 2001). The understanding of detection and estimation of antibiotic deposits antagonistically show how these compounds affect the human well-being, animal health, and different ecosystems (Aga et al. 2016). Antibiotic resistance is presently considered a potential ecological threat in the United States and by global specialists. In 2014, a board of trustees of the US official branch issued an alarming report encouraging quick activity to address the approaching antibiotic resistance crisis (Williams-Nguyen et al. 2016). The utilization of extensive amounts of these anti-infection drugs brings up serious concerns about the arrival and pervasiveness of antitoxin safe microorganisms and antibacterial resistance genes (ARGs) in nature. The presence of antibiotic residues, antibiotic resistance genes (ARGs), and antibiotic-resistant bacteria (ARB) in agroecosystems has turned into a noteworthy range of research (Franklin et al. 2016). Agriculture ecosystems are of special interest for checking the potential for antibiotic resistance that spreads through the environment resulting in contribution to human exposure. The agroecosystems incorporate domesticated animals, related administration frameworks, biosolids, manure sources, soil, neighboring water bodies, and products of the soil developed with anti-infection utilization. Furthermore, shifts in indigenous aquatic and terrestrial microbial populations are possible, which are affected by the release and/or enrichment of antibiotics, ARB, and ARGs into the environment. The use of antibiotic drugs is believed to selectively enrich ARB and ARGs. But, still gaps remain at large in investigating the relationship between antibiotic drugs, ARB, and ARGs in diverse agroecosystems. In these ecosystems, subtle direct links between the presence or absence of antibiotics and the occurrence of antibiotic resistance have become challenging in recent era. The prime challenge in this regard is the natural phenomenon of ARGs due to feature intrinsic resistance within the microorganisms. Bacteria can acquire resistance through horizontal gene transfer (HGT) of mobile genetic elements (MGEs) that contain not only ARGs but also several other functional genes. The wide spread of resistance is largely due to HGT and span growth of microorganisms; both of these factors facilitate the advantageous or disadvantageous spontaneous mutations and genetic elements (Normark and Normark 2002). These elements create a number of challenges in determining the bioavailability, environmental fate, and effects of antibiotics, ARB, and ARGs in agroecosystems. Molecular strategies which target DNA, RNA, and other subatomic segments of bacterial cells show certain points of interest for describing and evaluating markers of antibiotic resistance and their lateral gene transfer (Luby et al. 2016).

Recently, a special report by World Health Organization (WHO 2014) suggests that antibiotic flop is already a global concern and that investigation of antibiotic drugs, ARB, and ARGs in the environment is a critical area for further research. Agroecosystems with nontherapeutic use of antibiotics are of special interest, and mitigation efforts to prevent antibiotics, antibiotic residues or metabolites, ARB, and ARGs from reaching the environment are limited or absent. Unlike human wastewaters and biosolids that encounter treatment before land application, it is uncommon for animal wastes to be applied to land. As a result, systemized full-scale investigations are necessary to evaluate the impact of agroecosystems on the spread of antibiotic resistance in the wide-ranging environments to elucidate the potential influence of these systems on the development, movement, and survival or continuance of ARB and ARGs. This chapter evaluates the presence of antibiotic residues released into the agroecosystems and identifies ARGs in environmental components such as sediment and soil, possibly affected by therapeutic or nontherapeutic-based compost facility.

5.2 Monitoring of Antibiotics

From the time when penicillin was discovered in 1928 by Alexander Fleming, the antibiotics became broadly accessible for utilization by human and in veterinary medicine in the 1940s (Knapp et al. 2009). Antibiotics have led to the elimination of various diseases in developed countries, but this efficiency of antibiotics led to their overuse. In past, antibiotics were only consumed for therapeutic purposes, but afterward they were used for both the therapeutic and nontherapeutic purposes. The therapeutic purposes take account of its use in dealing with diseases in human as well as in animals, while nontherapeutic purposes include its employment as a food additive for reducing mortality rate and promoting growth of animals to augment food production for humans since the 1950s (Chee-Sanford et al. 2009; Halling-Sørensen et al. 1998; Kumar et al. 2005, 2012; Kümmerer 2009). For instance, antibiotics such as *bacitracin*, *lincomycin*, and *neomycin* are used to promote weight gain of chicken and swine, which were ranged from 4 to 10 mg/L. But, recently their 10–20-fold enhanced level has been reported. The use of antibiotics as growth promoters has been prohibited in Europe since 2006 (Chee-Sanford et al. 2009; Kumar et al. 2012).

The excreted antibiotic residues from therapeutic or nontherapeutic use are released to the surrounding environment or ecosystems, resulting in elevation of antibiotic concentrations (Pei et al. 2006). The antibiotic residues of four antibiotic groups, sulfonamides (SAs), tetracyclines (TCs), macrolides (MLs), and ionophores, are detectable in water and sediment near the mixed landscape of different watersheds (Awad et al. 2015). The existence of antibiotic residues along water ecosystem is more crucial as they are thought to be highly mobile (Awad et al. 2015). The excessive use of antibiotics may lead to the appearance of resistant genes. If antibiotic resistance continues to rise, valuable treatments for a huge number of infectious diseases in human and animal may be jeopardized.

To understand the relationship between antibiotics and corresponding ARGs, seasonal monitoring of therapeutic or nontherapeutic antibiotics is needed due to climate change features and overused annual consumption of antibiotics around the world. The high-intensity temperature, rainfall, and difference between summer and winter seasons due to the geographical monsoon impact should also be considered. The climate change can lead to the mobilization of antibiotics owing to the contamination of surrounding ecosystems or environment. A continuous monitoring of antibiotics should be performed near agroecosystems with concentrated animal or other farming operations (therapeutic or nontherapeutic). The antibiotics could be detected in agroecosystem or its components such as sediment and soil.

5.3 Antibiotic Resistance

Antibiotic resistance (AR) is the capability of a bacterium to survive and grow in the presence of an antibiotic at a concentration at which its growth is usually inhibited (Fig. 5.1). Antibiotic resistance occurs when an antibiotic becomes ineffective to efficiently suppress the growth of bacteria or due to exposure to megadoses of antibiotics (Franklin et al. 2016). When antibiotics are used, there is a greater possibility of destruction of most susceptible bacteria, and survival of resistant bacteria can lead to an increased antibiotic resistance and bacterial population (Kumar et al. 2012). AR may be intrinsic or acquired. Intrinsic AR occurs when there is no antibiotic target present in bacterial genome (Alekshun and Levy 2007; Pawlowski et al. 2016), whereas acquired AR occurs due to mutation in the bacterial chromosomes (Alekshun and Levy 2007). AR may also occur by exchange of genetic material between different bacteria through plasmids (Baker-Austin et al. 2006; Witte 2004).

Antibiotic resistance gene (ARG) is a gene that confers resistance to a number of antibiotics or antibiotic classes (Franklin et al. 2016). Some bacterial species have natural ARGs for particular antibiotics, and in some species genes are transferred from non-disease-causing species to pathogenic species (Kumar et al. 2012). Antibiotics such as penicillin and erythromycin, which were considered as the most efficient antibiotics against various bacterial species have now become less effective due to enhanced bacterial resistance. A bacterium carrying several resistant genes is known as multidrug-resistant (MDR) or superbug (Kumar et al. 2012). It has been reported that about half million cases of multidrug-resistant tuberculosis (MDR-TB) are reported from all over the world each year. And this MDR-TB could not be treated by previous antibiotics that were used for the treatment of normal TB. In the United States, about two million people are infected by ARB and about

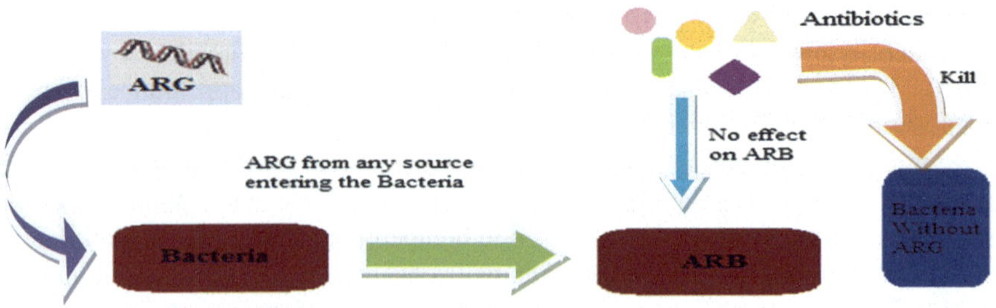

Fig. 5.1 Transfer of antibiotic resistance genes (ARGs) in bacteria and its effect

23,000 people expire every year. There are ample chances that this threat of antibiotic resistance can spread to multi-environments or ecosystems through gene transfer. There are various national and international programs regarding the threats of ARB. *Methicillin*-resistant *Staphylococcus aureus*, vancomycin-resistant *S. aureus*, and vancomycin-resistant *Enterococcus* are drug-resistant bacteria reported.

It has established that release of antibiotics into the environment leads to the pathogenic antibiotic-resistant bacterial strains (Chee-Sanford et al. 2001). For example, TC resistance genes have been reported in water samples collected from wastewater treatment plants near swine production facilities in the United States (Chee-Sanford et al. 2001). Antibiotic resistance genes (ARGs) have been isolated against MLs, TCs, and streptomycin in bacteria isolated from farmlands treated with swine manure slurry (Sengeløv et al. 2003). It has been reported that dissemination of ARGs severely degrades environments biochemically and should be recognized as a pollutant (Rysz and Alvarez 2004). Once antibiotics or their residues enter bacterial cells via passive diffusion, they inhibit bacterial growth (Schnappinger and Hillen 1996). Tetracyclines including oxytetracycline (OTC), chlortetracycline (CTC), minocycline (MNC), and doxycycline (DXC) inhibit protein synthesis in Gram-positive and Gram-negative bacteria by preventing the binding of aminoacyl-tRNA molecules to the 30S ribosomal subunit (Schnappinger and Hillen 1996). Bacterial resistance to all these antibiotics occurs by two mechanisms: (1) conferring of bacterial resistance and/or (2) the multi-antibiotic resistance pump (Schnappinger and Hillen 1996).

Antibiotic research on resistance genes has been confined to culturable bacteria isolated from wastewaters of pharmaceutical origin. The cultural isolation method is the most commonly employed method; however, only a fraction of actual microbiota in systems containing ARGs can be determined through this method (Amann et al. 1995). According to ARG occurrence in environments affected by animal waste, the polymerase chain reaction (PCR) method is highlighted to quantify genes conferring resistance to selective antibiotics. Several studies have attempted to quantify ARGs by isolating DNA (Auerbach et al. 2007). It has been shown that isolated DNA from five long-term soil series (over 60 years) was very informative regarding ARG abundance and their resistance to antibiotics, revealing that ARGs have increased sharply in the environment from 1940 to 2008. Primers with unique designs are needed to detect antibiotic bacterial resistance (Knapp et al. 2009). Bacterial resistance to different types of antibiotics was primarily mediated by synthetic primers, such as tet(G), tet(A)-(E), tet(M), tet(Q), tet(O), and tet(S), for TCs (Auerbach et al. 2007; Burdett 1991) and sul(I) and sul(II) for SAs.

5.4 Transfer of Antibiotics and Antibiotic Resistance Genes in Agroecosystem

There are numerous ways of antibiotic access to ecosystem. They may enter the ecosystem via seepage during transport, by dumping of drugs in water during their synthesis, by disposal of drug containers, and through waste material containing drugs (Boles and Wells 2010; Chee-Sanford et al. 2009; Pillai 2011; Subedi and Kannan 2014; Youngquist 2014). Some of the antibiotics that are employed for therapeutic or nontherapeutic purposes by animals are not digested and absorbed in the gastrointestinal tract, and these antibiotics are entirely removed from the body via feces referred to as manure. About 75% of the antibiotics utilized by agricultural animals are removed into the ecosystem by several ways (Chee-Sanford et al. 2009; Kumar et al. 2005; Sarmah et al. 2006; Tasho and Cho 2016). When the antibiotic-contaminated water is applied for irrigation or animal manure is exploited as fertilizer, antibiotics penetrate into the soil. The topmost layer of the soil is rich in bacteria, thus incessant exposure of antibiotics to soil bacteria results in generation of antibiotic resistance in bacteria. On the other hand, the consumption of high doses of antibiotics to promote weight gain in animals directs to the generation of antibiotic resistance genes in bacterial species. These ARGs are eliminated along with fecal material and get absorbed into the soil from manure during agriculture practice. These ARGs enter into the soil bacteria to bring about ARB (Chee-Sanford et al. 2009; Franklin et al. 2016). These ARB enter the crops or vegetables and are finally ingested by human in the course of food supply (Fig. 5.2). They can also be transferred from animals to human via dairy products or taken as a meat source (Vieira et al. 2011). It has been studied that the direct contact with cattle can also lead to the spread of ARB from animals to humans (Casey et al. 2013; Chang et al. 2015).

Increased use of antibiotics coupled with advancements in technology has resulted in more frequent detection of antibiotic compounds and, to a lesser extent, their metabolites in a variety of agroecosystem compartments including soil, water, sediment, and biota (Pruden et al. 2006; Zhang et al. 2013). In spite of advancements in detection methods, limited data are available on the fate and occurrence of antibiotics in agroecosystems, as well as their temporal and spatial distribution. A causal model has been proposed for agroecosystems to explore the effects of antibiotic drugs, ARB, and ARGs (Williams-Nguyen et al. 2016). The causal model takes a One Health approach by describing the key interactions between antibiotics, ARB, and ARGs as well as their resulting interactions within the agroecosystem. This model is based on three specific conclusions: (1) ecosystem function, (2) human health, and (3) agricultural system productivity. To estimate expected concentrations of antibiotics at the landscape scale, a predictive modeling has been presented as an alternative to large-scale and high-cost monitoring programs (Boxall et al. 2003). These models are counted on accurate antibiotic usage data as well as mechanistic knowledge of the metabolism, transport, fate, landscape, and hydrologic processes. But, usage data is not universally available

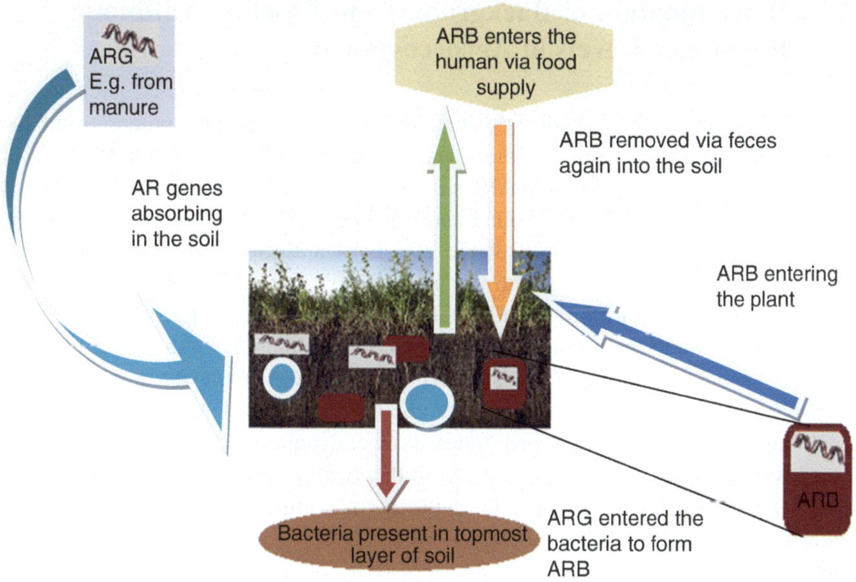

Fig. 5.2 Transfer of ARGs in agroecosystem and from agriculture to human via food supply

with the lack of data for a number of medically important antibiotics. A number of pathways have been suggested for the release of antibiotics into agroecosystems, and their effect on ARB and ARGs depends on these specific pathways. The application of wastewater and land manure solids is a common route for antibiotics to enter the environment. The administration of manure with or without antibiotics is thought to increase both ARB and ARGs (Franklin et al. 2016; Pruden et al. 2006; Udikovic-Kolic et al. 2014); however, data availability is limited with inconsistent results (Franklin et al. 2016; McLain and Williams 2014), elucidating the need for further research. Biosolids, manure, and wastewater effluents sustain ARB and ARGs, but preexisting or intrinsic resistance in populations of native bacteria further enhances the complexity to modeling efforts. The resistance in soils that is induced by the release of ARB, ARGs, and trace levels of antibiotics would establish relationship between antibiotic use and antibiotic resistance present in agroecosystems. Antibiotic-resistant bacteria with intrinsic ARGs pose a direct threat to agroecosystem health based on the extensive studies of pathogenic bacteria, while ARGs are thought to pose an indirect risk.

5.5 Determination of Background and Baseline Antibiotic Resistance Levels in Agroecosystem

Research indicates that antibiotic resistance in an ancient phenomenon and the utilization of anthropogenic antibiotic agents additionally impact the presence of antibiotic compounds, ARB, and ARGs in a domain. Subsequently, it is emphasized that the determination of background and baseline levels of antibiotic resistance is crucial for the precise appraisal of the effects of anthropogenic contributions to agroecosystems. All-around acknowledged meanings of background and baseline levels for antibiotics are not found in literature. However, background is characterized as the focus in a domain not affected by human activities, and baseline includes range of antibiotic drugs, ARB, as well as ARG levels present at the beginning of survey in an environment. Standardization of antibiotic resistance found in agroecosystems against foundation and pattern levels will (1) permit assessment of noteworthy modifications in the event of ARB or potentially ARGs within the review, (2) enhance the capacity to compare the results within the studies, and (3) distinguish interfaces between agricultural or ecological treatments and activities.

5.6 Culture-Based Methods for Detection of Antibiotic Resistance in Agroecosystem

Various culture-based strategies exist for investigating antibiotic resistance in samples collected in agroecosystem. These methods target microorganism isolation on general or selective media assessing their growth in response to specific concentrations of antibiotics. Culture-based techniques give chances to interface phenotypic and genotypic attributes and evaluate ARG exchange potential, considering more prominent comprehension of general resistance designs and recognizable proof of different antibiotic resistance within the single organism. Culture-based methods commonly involve isolating target bacteria on general or selective media and assessing growth in response to specific concentrations of antibiotics. The microorganisms that are normally focused in culture-based studies to assess antibiotic resistance in agroecosystems are microbial groups that are clinically relevant and simple to culture. These objective microorganisms are additionally indicators of water quality. Generally, the most widely recognized organisms focused for ecological investigation are *Escherichia coli*, *Enterococcus* spp., *Salmonella* spp., and *Staphylococcus* spp. Recent research has recommended the expansion of *Aeromonas* spp., *Klebsiella pneumonia*, and *Pseudomonas aeruginosa*. Before testing of antibiotic resistance, identification of bacterial isolates is basic. When target living beings have been effectively separated and distinguished, antibiotic resistance testing can be performed by means of three common techniques; (a) broth and agar dilution, (b) agar disk diffusion, and (c) E-tests. The choice of strategy relies on the extent of research; however, other considerations include lab limitation and whether subjective and quantitative outcomes are desired or not. Agar disk diffusion method reports number of isolates that are resistant, while broth microdilution techniques are more quantitative and deliver MIC50 values that represent to the fixation at which ≥50% of the isolates in the population are restrained. Given the quantitative nature of this technique, analysts are urged to not overemphasize MIC50 values in small test populations (10–30 isolates), when a couple of strains with high MIC qualities may skew the MIC50 (Franklin et al. 2016). Epsilometer test (E-test) methods for antibiotic susceptibility testing of selected microorganisms use the agar dilution method as a standard to determine the MIC. The E-test provides a direct quantification of antimicrobial or antibiotic susceptibility of microorganisms (Nachnani et al. 1992). The culture-based methods have multifold advantages. In particular, isolation of bacteria is key in understanding the phenotypic characteristics of isolates and their resistance patterns. Majority of the projects around the world for antibiotic resistance monitoring are isolate based.

5.7 Molecular Aspects of Antibiotic Resistance

Antibiotic resistance is characterized as phenotypic property, the capacity of a bacterial cell to survive and develop within the sight of an antimicrobial fixation that is inhibitory to susceptible cells. Antibiotic resistance genes encode the capacity for bacterial cells to develop in the presence of antibiotics. There are many means by which cells can accomplish this, including biofilms formation to offer physical assurance from the assault. A schematic presentation of molecular aspects of cell's machinery that give resistance is depicted in Fig. 5.3. Antibiotic resistance genes are heritable and are likewise generally equipped for being shared among microscopic organisms, so specific monitoring can give data about the expansion of antibiotic resistance in a framework. Consequent to extraction of DNA from interest sample, ARGs are regularly recognized by PCR-based strategies (Luby et al. 2016). Other regular focuses for antibiotic resistance investigation incorporate RNA and proteins, which can track the expression of antibiotic resistance components. Nonetheless, RNA- and protein-based strategies are testing procedures, and, therefore, DNA-based techniques are most preferred for tracking ARGs (Franklin et al. 2016). Monitoring RNA rather than DNA is one means by which it can be anticipated that ARG is available inside a practical and dynamic bacterium. This is commonly proficient by specifically extracting RNA from a specimen, rather than, or couple with, DNA extraction. RNA can then be subjected to reverse transcription followed by the same downstream investigations ordinarily connected for ARGs and horizontal gene transfer markers. Other method for evaluating RNA expression exploits the use of reporter gene, for example, the green fluorescence protein, which affirms that a quality resistance is being initiated in an in vivo framework (Binh et al. 2008; Musovic et al. 2010). Proteins are the molecular markers that most nearly reflect cell work. It has been reported that techniques such as Western blots

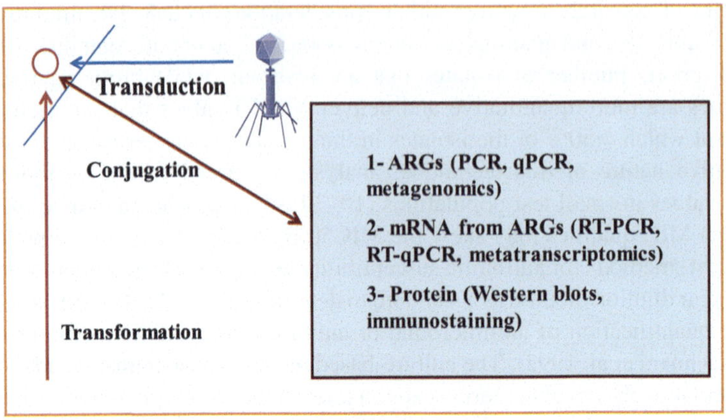

Fig. 5.3 Overview of molecular aspects of antibiotics

and immunostaining can be applied to identify particular proteins related with antibiotic resistance. For instance, carbapenemase activity presence can be affirmed in *Acinetobacter baumannii* to comprehend conditions that control real expression of resistance.

5.8 Molecular Methods for Assessing the Antibiotic Resistance and ARGs

The strategies for recognizing, describing, and measuring molecular targets relating to resistance are as follows.

5.8.1 Detection of Antibiotic Resistance Genes in Agroecosystems

5.8.1.1 Sampling

Sampling can be done by collecting water from a source exposed to ARB or by collecting sediment or soil rich in animal manure. For example, methods for assessing the antibiotic resistance and ARGs from sampling to an end of study can be explained by a study conducted in Korea. In March 2009, samples were drawn from the sites located in Hongcheon, Gangwon Province, Korea. These sites of sampling were thought to be influenced by antibiotic release from a swine manure composting facility located at 37° 34 ft and 28 in. toward north and 127° 52 f. and 26 in. toward east. The average temperature recorded was 17.5°C and total precipitation was 95.3 mm. Description of sampling sites is provided in Table 5.1.

Table 5.1 Description of sampling

Samples	Sample site description
Sediment #1	Site located 0.2 km away from a swine manure composting facility
Sediment #2	Site located 0.5 km away from a swine manure composting facility
Sediment #3	Site located 1.0 km away from a swine manure composting facility
Sediment #4	Site located 1.5 km away from a swine manure composting facility
Soil #1	Rice paddy soil treated with swine manure and have a distance of 2.0 km from a swine manure composting facility
Soil #2	Rice paddy soil influencing antibiotics via irrigation and have a distance of 2.0 km from a swine manure composting facility

Awad et al. (2014); Ok et al. (2011)

Sample collection was based on the distance of sediments from the composting facility, beside the Naerincheon River. Paddy soil collection was done from two sites:

1. The soil which was directly applied with swine manure for agricultural practice as soil #1.
2. The soil which was only irrigated by Naerincheon River as a water source as soil #2.

Before analysis, samples must be air-dried. The study was a part of an inclusive monitoring of antibiotics in water, sediment, and soil near swine composting facility (since April 2008) (Awad et al. 2014; Ok et al. 2011).

5.8.1.2 Antibiotic Extraction and Quantification

After sampling, the collected samples were firstly processed for the withdrawal of antibiotic residues. The methods described by Kim and Carlson (2007) and Ok et al. (2011) and others can be used for this extraction. Then these antibiotic residues were quantified by high-performance liquid chromatography-tandem mass spectrometry (HPLC-MS/MS) (API 3000, Applied Biosystems, Foster City, CA, USA).

Extraction

For the extraction of antibiotics such as TCs and SAs, 1 g of sample along with 200 μL of 5% Na_2EDTA and 20 mL McIlvaine buffer, i.e., a citrate-phosphate buffer (mixture of 0.2 M disodium hydrogen phosphate and 0.1 M citric acid) at pH 4, were added to a 50 mL polypropylene centrifuge tube and centrifuged at 4000 rpm for 15 min (Centrifuge FLETA 5, Hanil Science Industry, Seoul, Korea). A 0.2 μm glass fiber filter was used to filter the supernatant. The extraction process was repeated again and again. Consequently the extracts obtain were then combined in a vial for solid-phase extraction (SPE). SPE was carried out to keep the antibiotics on the cartridge for their useful extraction with methanol (Kim and Carlson 2005). Due to broad range of pH, hydrophilic-lipophilic balanced cartridges are used for the extraction of antibiotics (Kim and Carlson 2005).

Quantification

Electrospray ionization can be applied for the quantification of antibiotics by means of HPLC-MS/MS in positive mode. The conditions previously maintained for quantification of antibiotics for HPLC-MS/MS spectroscopy are given in Table 5.2.

5.8.1.3 Heterotrophic Plate Counts on Antibiotic-Selective Media

Firstly, the sample was diluted by sterilized water and briskly stirred for 30 min. Then it was further diluted to 100-folds. Secondly, it was directly spread on an agar media, say R2A agar media (Difco, Sparks, MD, USA), that enclosed a variety of antibiotics or no antibiotic as a control to count and separate resistant bacteria. The concentration fivefold to the reported average LD_{50} value was considered for the water-soluble antibiotics such as TC, CTC, and STZ, while the maximum concentration which dissolved in water after adding to melted agar was considered for the insoluble antibiotics such as SMX, SMT, and OTC (Pei et al. 2006). Finally treated plates were incubated for a week or less. At the end of culture period, colony-forming units (CFUs) were counted (Pepper and Gerba 2009).

Table 5.2 Conditions for HPLC-MS/MS spectroscopy

HPLC conditions		MS conditions	
Column temp.	15°C	Ion source	ESI, positive
Column flow rate	300 μL/min	Spray voltage	4500 V
Injection volume	20 μL	Vaporizer temp.	320°C
Mobile phase	A (99.9% water + 0.1% formic acid) B (99.9% CAN + 0.1 % formic acid)	Drying gas flow	10.0 L/min
		Drying gas and nebulizer gas	Nitrogen gas
Gradient	A→ 96% + B→ 4% (0 min) A→ 70% + B→ 30% (29 min) A→ 96% + B→ 4% (30 min)	Sheath gas pressure	40 psig
		Aux gas pressure	5 psig

Awad et al. (2014); Ok et al. (2011)

5.8.1.4 DNA Extraction and Purification

FastDNA spin kit can be used to extort DNA from 0.5 g of sample. Then purification of the extracted DNA was carried out using a Geneclean spin kit to reduce PCR hang-up. The amount of DNA before or after purification and recovery was determined (Awad et al. 2015).

5.8.1.5 Primer Design

On the basis of gene bank database, particular primers for nucleotide sequences encoding specific ARGs were designed (Awad et al. 2015). Seven sets of primers generated by Awad et al.'s work in 2015 are given in Table 5.3 (Awad et al. 2015).

5.8.1.6 Detection of ARGs Using Qualitative PCR

ARGs that code for ribosomal protection were detected using PCR. A Bio-Rad kit can be used in a reaction mixture containing 2.5 mM dNTP having final volume of 20 μL. Amplification can be done using PTC-100 thermal cycler. Then annealing was carried out at different temperatures for different ARGs. A Gel Doc 1000 apparatus (Bio-Rad) can be used to visualize PCR products on agarose gel (Awad et al. 2015). Results were represented as peaks indicating the concentration of particular ARGs and ARB in the collected sample.

Table 5.3 Primers designed for TC- and SA-resistant genes

Gene	Primer	Sequences	Annealing temp. (°C)	Amplicon size (bp)
tet(S)	tetS-FW[a] tetS-RV[b]	GAAAGCTTACTATACAGTAGC AGGAGTATCTACAATATTTAC	50	169
tet(T)	tetT-FW tetT-RV	AAGGTTTATTATATAAAAGTG AGGTGTATCTATGATATTTAC	46	169
otr(A)	otrA-FW otrA-RV	GGCATTCTGGCCCACGT CCCGGGGTGTCGTAAGG	66	212
sul(I)	sul(I)-FW sul(I)-RV	CGCACCGGAAACATCGCTGCAC TGAAGTTCCGCCGCAAGGCTCG	55.9	163
sul(II)	sul(II)-FW sul(II)-RV	TCCGGTGGAGGCCGGTATCTGG CGGGAATGCCATCTGCCTTGAG	60.8	191
sul(III)	sul(III)-FW sul(III)-RV	TCCGTTCAGCGAATTGGTGCAG TTCGTTCACGCCTTACACCAGC	60	128
sul(A)	sulA-FW sulA-RV	TCTTGAGCAAGCACTCCAGCAG TCCAGCCTTAGCAACCACATGG	60	229

Awad et al. (2014); Ok et al. (2011)
[a]Represents forward
[b]Represents reverse

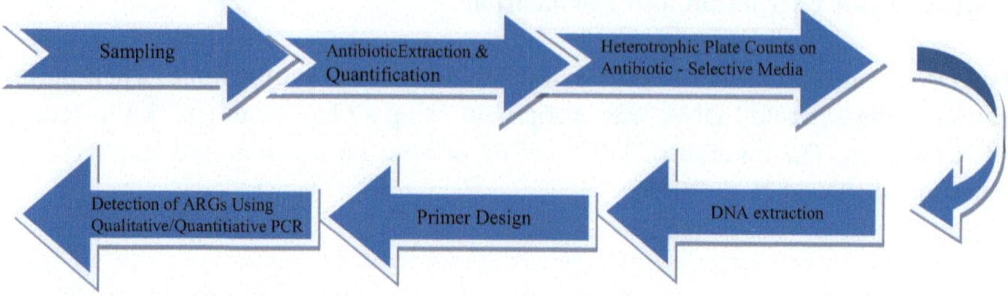

Fig. 5.4 Flow chart for molecular detection of ARGs in agroecosystem

A flow chart is presented in Fig. 5.4 to show the molecular detection of antibiotics and ARGs from the sample to an end of study in agroecosystems.

5.8.2 Polymerase Chain Reaction-Based Assays

- **Conventional polymerase chain reaction**

  Polymerase chain reaction has turned into a famous technique for identifying ARGs of interest for ecological examples since it is exceptionally delicate, gives generally quick outcomes (2–3 h), and yields coordinate data about the DNA sequence of intrigue. It is a protein-based assay utilizing oligonucleotide preliminaries that are integral to the flanking region of the objective to intensify target genes or fragments of genes. Polymerase chain reaction has been effectively and broadly connected to identify ARGs in samples from agroecosystems since 2001 (Chee-Sanford et al. 2001). At the point when reporting PCR information for sample from agriculture ecosystem, in addition to checking expected item measurement by gel electrophoresis, it is firmly prudent to grouping the PCR items from a delegate subset of sample to guarantee that the PCR is amplifying the planned target. In addition to confirm the specificity of the objective, arrangement inconstancy among ARG variations may likewise be of premium and can be investigated by contrasting and GenBank or different databases (Garder et al. 2014; Koike et al. 2007).
- **Real-time quantitative polymerase chain reaction**

  Quantitative PCR (qPCR) gives the benefits of PCR, furthermore yielding quantitative data about the abundance of the objective ARG. For qPCR, the response itself is modified to incorporate either probes which give fluorescence when bound to the objective DNA, or it gives color, for example, SYBR Green, that binds to twofold stranded DNA. Quantitative PCR has been adjusted to manage longer formats (up to 6 kb), which is fundamental to capture harm occasions. It must be applied if the expectation is to survey the effects of different types of medicine on the DNA itself (Luby et al. 2016).

5.8.3 Horizontal Gene Transfer Assays

Horizontal gene transfer can be concentrated through location and quantification of specific marker genes connected with versatile hereditary components which include direct assays of transfer by means of conjugation, transduction, retrospective genome, and/or metagenome examinations (Luby et al. 2016).

5.8.3.1 Conjugation, Transduction, and Transformation Assays

Bacterial conjugation refers to the exchange of conjugative transposable hereditary components or plasmids through close physical contact between a contributor and beneficiary bacterium. Transduction is a virtual trade of DNA that is for the most part constrained between firmly related strains and has just been shown for a couple of organism varieties. Transduction happens when the virus exchanges bacterial DNA from its past host into another bacterial host, which then fuses the new genes into its genome to continue the repeats (Luby et al. 2016). Transformation is the capacity of microscopic organisms to take up and consolidate extracellular or foreign DNA, which is central to the idea of ARGs as toxins (Dodd 2012). Likewise with conjugation and transduction, common change affects antibiotic resistance since it can bring about exchange of ARGs among different sorts of microorganisms and in this manner may upgrade their perseverance and dispersal. The capacity to be changed (i.e., competence) has by and large been thought to be restricted, particularly as a characteristic procedure; it has now been exhibited for no less than 80 strains of microscopic organisms (Johnston et al. 2014; Lorenz and Wackernagel 1994).

5.8.4 Metagenomic Methods

Metagenomics is the study of the metagenome, and metagenome is the collective genome of microorganisms from an environmental sample to give an information on ecology and the microbial diversity of a specific environment. Metagenomics has proven to be a powerful tool to investigate the metabolic profiling, ecology, and comparison of complex microbial communities. The key feature of genomic analysis is that it is performed using metagenomic libraries designed from total DNA isolated from a particular niche rather than a laboratory culture. The metagenomic analyses potentially investigate all the genetic resources present in an environment. Samples used in metagenomics are taken directly from the environment such as soil, water, and hot spring. Once the bacteria have been isolated from the ecosystem, the DNA can be extracted using extraction protocols following purification. The DNA sample is then analyzed using fragment analyses or DNA sequencing techniques and/or ultrahigh-throughput sequencing technologies. Next-generation sequencing can be even more useful to determine the sequence of a communal genome. The analysis of such datasets is aimed at determining and comparing the biological diversity and the functional activity of different microbial communities. The metagenomic methods may include taxonomical approach, discovery rate datasets, functional assessment, metadata analysis, taxonomy-guided capture of reads, and comparative visualizations. Sequence-based metagenomic analyses depend on comparisons with databases of known genomic sequences, while functional analyses depend on screening libraries on the basis of the phenotype-cloned DNA that can confer to host bacteria. This is the reason that functional analyses allow the identification of novel genes with functions that otherwise could not have been predicted from their DNA sequences. The computational analysis is also important for metagenomic analysis not because of the large amount of metagenomic data but also for the new questions introduced by metagenomic projects such as community-realized functions, simultaneous assembly of multiple genomes, and host-microbe interactions.

Recent metagenomic studies have uncovered an extensive variety of ARGs and mobile hereditary components from different ecosystem (Bengtsson-Palme et al. 2014; Kristiansson et al. 2011). Excrement and its effects on soil are of special interest for agricultural environments. Soil metagenomes have been revealed to contain a very assorted pool of ARGs which include the most widely recognized sorts of resistances found in other ecological metagenomes (Donato et al. 2010; Nesme and Simonet 2015). Since metagenomics gives data on the aggregate genes present, thus examination of information sets can give expansive logical data for the identification of ARGs. For instance, identification of horizontal gene transfer markers through plasmids and transposons can give intimations about how ARGs may have spread starting with one environment then onto the next (Bengtsson-Palme et al. 2014; Nesme and Simonet 2015; Ochman et al. 2000). Metagenomics has likewise been connected to give a thorough correlation of ARGs and markers of horizontal gene transfer in excrement and agrarian soils (Durso et al. 2012).

5.8.5 Culture-Based Methods: Advantages and Challenges

Most culture-based assays are time-consuming and recover a small subset of the total bacterial community; however, they have distinct advantages such as direct identification and analysis of antibiotic resistance in individual bacterial isolates (McLain et al. 2016). Culture-based techniques also provide opportunities to link phenotypic and genotypic characteristics by assessing ARG transfer potential. They give greater understanding of overall resistance patterns as well as identification of multiple antibiotic resistance in single organisms. Standard clinical classification protocols exist to categorize bacterial isolates as susceptible, resistant, and intermediate to antibiotics based on the bacterial growth at defined antibiotic concentrations which are called as breakpoints. These breakpoints measured as MICs are used to determine specific dosage formulations for antibiotic treatments in clinical settings. The microorganisms that are commonly used in culture-based studies for the evaluation of antibiotic resistance in agroecosystems are microbial groups which are clinically relevant and easy to culture. The results of culture-based methods have been found to be reliable and comparable in clinical settings.

The choice of culture-based methods predominantly depends on the scope of research or studies; however, laboratory limitations exist that either qualitative or quantitative results are desired (Franklin et al. 2016). Similarly, questions remain that how many isolates are necessary per sample and how many samples within an agroecosystem must be analyzed to produce a representative dataset for a precise analysis of antibiotic resistance (Persoons et al. 2011). Culture-based methods have other limitations that include inherent culture bias. Most of the bacterial species in water or soil are not capable of being cultivated; thus, culture-based approaches apply only to a small subset of the microbial species and do not provide the full spectrum of diversity present in environmental samples. When culture- and molecular-based techniques are used to identify ARB simultaneously, the results have been found to be different (Garcia-Armisen et al. 2013). Another notable limitation shows that culture methods cannot identify bacteria isolates that are in the viable but nonculturable state. This state has important implications with regard to antibiotic resistance as bacteria become resistant to antibiotics, yet they have a potential to eventually return to be pathogenic and metabolically active (Franklin et al. 2016). Another potential culture bias regarding antibiotic resistance is the presence of persister cells which are dormant variants of regular cells and are highly tolerant to antibiotics (Franklin et al. 2016). Even with their limitations, culture-based methods are the basis of international surveillance efforts to monitor antibiotic resistance, and standardized molecular methods cannot replace them at full-fledge scale. Future assessment of antibiotic resistance in the environment will depend on standardized methods and techniques that will incorporate both culture- and molecular-based procedures.

5.8.6 Molecular Methods: Advantages and Challenges

Molecular methods offer the distinct advantages, for instance, they provide direct information about the extractable pool of DNA, RNA, and proteins within a sample. The isolated DNA, RNA, and protein can be sequenced and directly compared against publicly available databases. Utilization of molecular methods also helps to avoid biases associated with culture-based methods. Molecular methods offer a means of tracking the fate of various antibiotic resistance indicators in and between agrosystems for the analysis of antibiotic resistance. The usage of molecular methods as a measure of antibiotic resistance analysis requires certain knowledge such as properly framed research questions, familiarity with common molecular techniques based on specific molecular targets, and awareness of advantages and disadvantages of various methods for the accurate interpretation of data (Luby et al. 2016).

Traditional PCR that is one of the most popular methods for the detection of known ARGs in environmental samples is highly sensitive and provides relatively rapid results and direct information about the DNA sequence of interest. However, challenges and limitations exist for applying PCR to samples from agroecosystems. One of the most significant challenges is that PCR is dependent on the extraction of DNA which should be optimized for the matrix of interest to capture clean DNA from as many different kinds of bacteria as possible and applied consistently across samples intended for comparison. Furthermore, sequencing a subset of the PCR products obtained during analysis is advisable to verify that PCR is amplifying the intended product. Direct polymerase chain reaction (PCR)-based and metagenomic techniques exhibit great promise to characterize ARG diversity and abundance in complex environment. However, these methods do not enable functional validation of identified resistance mechanisms and generally cannot correlate between specific ARGs and bacterial phyla (Franklin et al. 2016; Nordmann et al. 2012).

Real-time quantitative PCR (qPCR) provides the same benefits as PCR, yielding additional information about the copy number of a particular ARG. Determination and reporting of limits of quantification are critical in qPCR as a standard quantitative method. Further, normalization to 16S rRNA genes is believed to aid for minor variations in extraction efficiency as well as providing information about the proportion of total bacteria carrying ARGs in the samples (Franklin et al. 2016; Heuer et al. 2011). The development of qPCR arrays is a promising way to analyze multiple targets. However, it can be best used as a screening tool as the limit of detection is higher than traditional qPCR. One major drawback of PCR-based methods is that sequences for the genes of interest must be known and selected ahead of time, which may bias the results by overlooking key genetic elements associated with antibiotic resistance.

Direct assays, for example, conjugation, transduction, and transformation, are useful to determine mechanisms of action, transfer rates of ARGs on mobile elements, and host ranges including the identification of functionality of ARGs. Horizontal gene transfer allows bacteria to share ARGs through MGEs (mobile genetic elements) such as plasmids, integrons, and transposons (Franklin et al. 2016; Klumper et al. 2015; Nandi et al. 2004). But, these analyses need recipient cells to be culturable, which limits their application in agroecosystem research. The use of a reporter gene such as green fluorescent protein can reduce the need for culturing and selection steps by confirming that the genes of interest are actually being expressed under the specific conditions of the study (Klumper et al. 2015). It has also been observed that RNA- and protein-based methods are challenging techniques in handling; therefore, DNA-based methods are generally preferred for tracking ARGs. The development of next-generation DNA sequencing methods has led to a new era of molecular characterization of environmental and agroecosystems. These technologies overcome the need for PCR and give a broad snapshot of the MGEs, ARGs, various other functional genes, and virulence genes in the samples of interest (Bengtsson-Palme et al. 2014; Franklin et al. 2016). These also provide broad contextual information beyond identification of ARGs and other targets of interest.

Identification of HGT elements provides information on how ARGs may pass from one environment to another (Nesme and Simonet 2015). Identification of genes of interest from metagenomic datasets is facilitated by publicly available databases and tools. But, several challenges are associated with data analysis, for example, further development of approaches and consensus in the scientific community that standardized analysis would be beneficial or not. Combining molecular- and culture-based methods can present some advantages to assist in linking genotype with phenotype. Regardless of the method or combination of methods selected, experimental design is predominant and must be carefully planned to address the research questions and problems.

5.8.7 *Impact of Antibiotics and Antibiotic Resistance Genes and Challenges*

Antibiotics are being considered as pervasively occurring persistent contaminants in agroecosystems, and their ecological risks are a growing problem with respect to agro-environmental quality. Many antibiotics are chronic with longer half-lives in soil. Long-term accumulation of these antibiotics and their metabolites in agroecosystems are bioactive and ecotoxic to soil microorganisms and crops, particularly bacteria (Baguer et al. 2000).

Some negative impacts of antibiotics on agroecosystems are:

(a) Inhibition of seed germination and crop growth
(b) Hampered microbial activity and soil enzymatic activity
(c) Antibiotic accumulation in crop biomass
(d) Leaching and runoff diffusion into groundwater

Besides deleterious effects on soil microflora and crop growth, antibiotics can provoke resistant pathogens through long-time exposure due to genetic variation and transfer. As novel persistent pollutants, antibiotics were included in limitations issued by the Food and Agriculture Organization (FAO) in the United States (Du and Liu 2012) There is a wide consensus that antibiotic use enriches microscopic organisms conveying the antibiotic resistance genes (ARGs). These antibiotic-resistant microorganism and antibiotic resistance genes from agricultural settings can be physically exchanged to humans. Complicating the estimation of agricultural antibiotic resistance and its potential for affecting human well-being are physical, synthetic, spatial, transient, and biological complexities of common frameworks and the "many ecologies" of resistance. Continued development of sensitive and robust analytical methods will permit improved measurement of bioavailable fractions of antibiotics and improve risk analysis. Large-scale efforts involving multiple agencies and university research groups would be valuable in attempting to unify information and approaches to improve fate and risk assessment of antibiotics in agroecosystems (Franklin et al. 2016).

The health effects in humans who have exposure to low levels of antibiotics and ARB from environmental sources are unknown. The pathway for human exposure to antibiotics may include ingestion of contaminated water, food, and inhalation of contaminated dust particles. Antibiotic sediments measurement has been taken into account in food crops, water sources, and animal-based food products but often at lower levels (Franklin et al. 2016). The effects of long-term chronic exposures to low levels of antibiotics in humans need to be investigated; however, available data suggest that low levels of antibiotics may promote ARB and ARGs (Lin et al. 2014). Studies have also shown a toxic response to antibiotics in fish invertebrates and cyanobacteria. The impact of antibiotics on ecosystem function primarily focuses on soil microorganisms. Research has shown that antibiotics in the environment have the potential to alter the microbial biomass, community structure, and functional endpoints such as substrate-induced respiration, ammonification, iron reduction, nitrification, and mineralization (Franklin et al. 2016; Guo et al. 2015; Solis et al. 2011; Toth et al. 2011). These alterations in soil microbial function may ultimately affect higher-level organisms and ecosystem processes. The effects of ARB or ARGs on ecosystem function have not been well studied. Various hypotheses have been proposed for alterations in microbial function, diversity and composition (Martinez 2009), as well as effects on wildlife health (Franklin et al. 2016). However, more evidence is essential for their thorough evaluation. The effects of single antibiotic compounds on different agroecosystems have been selectively characterized. Toxicological effects of mixtures have not well investigated with limited research on antibiotic mixtures. It has been reported that combinations of these compounds and their metabolites can often result in synergistic, additive, and antagonistic effects enriching ARB (Franklin et al. 2016; Liu et al. 2014). Predicting the biological effects of these mixtures is challenging as alterations in the composition of these compounds can change mixture toxicity from synergistic to antagonistic level (Liu et al. 2014). The mixtures of contaminants found in agroecosystems urge for future evaluation of environmental risk assessment of antibiotics and their adverse effects. Lastly, the possible links between the occurrence of ARB in the environment and agricultural systems have yet to be determined.

The ARB from environment spread to agricultural systems quoting the documented links between wildlife (e.g., birds) and common food-borne pathogens in agricultural products (Greig et al. 2015), indicating transfer to animals and crops within an agroecosystem. Detection and measurement of antibiotics and their residues are essential for understanding their potential to affect ecosystem function, agricultural systems, and animal and human health. The potential for antibiotics to have adverse impact in agroecosystems is directly related to its original use, in vivo, and its environmental persistence and inherent biological activity. Not only antibiotic compounds and their wide range are of concern, but their metabolites transformation products may also affect biological activity and therefore need to be elucidated while conducting environmental studies.

The development of sensitive analytical methods is essential to measure the concentrations of antibiotics in complex environmental samples. In recent years, the technology has made possible to detect antibiotic limits in the picogram per gram or parts per trillion ranges. But, difficulties in separating antibiotics and their degradation products or metabolites from complex matrices (e.g., soils, manures, and wastewaters) still limit the ability to measure accuracy and reproducibility (Wilga et al. 2008). A greater challenge in this context is the determination of the ecological implications and significance of the bioavailable fractions of antibiotics at their predicted environmental concentrations. Unfortunately, absolute recovery of multiple antibiotic residues from environmental matrices is typically not possible, even using improved analytical techniques; the fraction recovered from soil or other samples does not necessarily correspond with the fraction that plants or microbes are exposed to in the environment (Naidu et al. 2008).

Quantitative analysis requires elaborate extraction and cleanup procedures to lessen interferences. The choice of extraction and cleanup techniques for aqueous samples is solid-phase extraction (SPE) because of improved specificity, selectivity, and reproducibility, minimal organic solvent consumption, shorter sample preparation time, and ease of operation and automation (Franklin et al. 2016). Preparation of semisolid and solid samples such as manure or soil is extremely challenging due to high concentrations of natural organic matter. Instrumental analyses using high-performance liquid chromatography and tandem mass spectrometry (LC/MS/MS) have been established as the primary analytical tools for quantification of antibiotics. High-resolution instruments such as quadrupole time-of-flight and Orbitrap MS (Thermo Fisher Scientific) are best suited for identification of unknown samples, whereas triple quadrupole gives high selectivity for antibiotic detection (Johnson et al. 1990). Ion trap mass spectrometry helps to identify transformation products, which is critical, as many transformation products preserve antimicrobial properties (Díaz-Cruz and Barceló 2007). Currently, standard methods do not exist for the detection of antibiotics in environmental samples (Franklin et al. 2016) because methods are not yet standardized and well-described procedures including details of validation are needed to make comparisons between studies. Furthermore, procedures and methods to determine the limits of detection variations should have been the subject of environmental literature (Franklin et al. 2016). Enzyme-linked immunosorbent assay (ELISA) is going to be used as a screening tool and a semiquantitative method for the determination of total analyte concentrations within a class of antibiotics (Aga et al. 2005). The value of this technique is that the ELISA has the potential to estimate bioavailability regardless of a compound's structure, while targeted analysis using LC/MS/MS would not be able to detect an unknown transformation product. Bioreporters, which are genetically engineered cells, can be capable of producing detectable signals in the presence of a target compound and thus can be useful alternatives to chemical analyses (Franklin et al. 2016). Continued development of sensitive and robust analytical methods will permit improved measurements of bioavailable fractions of antibiotics and their residues with improved risk assessments.

At present, the pathways that permit antibiotic compounds, ARB, and ARGs to move through the environment are not completely understood. However, this data is not complete and additional research is important to completely clarify the current reservoir of antibiotic-related contaminants in the environment, while additionally recognizing those that are not known. All-around well-developed standard techniques for exact investigation of antibiotics, ARB, and ARGs from ecological specimen are uncommon. While techniques have been developed for examination of antibiotic and antibiotic resistance in clinical settings, these strategies cannot promptly be connected in environmental settings. Standard techniques have not been developed for antibiotic research in environment; most labs must build up their own particular strategies. This restrains the capacity to make correlations between samples analyzed in different research centers and impedes risk assessment analysis. Surveillance programs for monitoring antibiotics and antibiotic resistance in the environment are lacking to date. The development and execution of these sorts of projects at nearby, national, and worldwide level would give long-term, comprehensive data about how and where antibiotics and antibiotic resistance are affecting agroecosystem. These projects would give information about the general effects within the agroecosystem to help in determining areas that require additional research focus. Surveillance data would also help in recognizing the reservoirs of antibiotics, ARB, and ARGs, pathways that permit these contaminants into and out of agroecosystem (Franklin et al. 2016).

5.9 Conclusion

The interpretation of antibiotics and antibiotic resistance in agroecosystems is an important field of research in which a one health strategy is required to entirely understand the health implications of antibiotic drugs, ARB, and ARGs in the ecosystem. Since the use of antibiotics is not declining and the prevalence of antibiotic resistance is on the augment in human and animal populations, a greater conception of the transport and fate of antibiotics, ARB, and ARGs in the environment is critical to conclude the probable risks and impacts on human, animals, and ecological health. Food production systems and biosolid applications are recognized as remarkable input sources of antibiotic-related contaminants, while the direct and indirect impacts on agroecosystems are not known. Development of standard approaches and their implementation among the scientific community is needed for an accurate identification and quantification of antibiotics, ARB, and ARGs in soil, water, manure, and other environmental matrices. Additional focus on standard research approaches and execution step in obtaining the reliable data are essential to provide an inclusive assessment of antibiotics and antibiotic resistance in agroecosystems.

References

Aga DS, O'Connor S, Ensley S, Payero JO, Snow D, Tarkalson D (2005) Determination of the persistence of tetracycline antibiotics and their degradates in manure-amended soil using enzyme-linked immunosorbent assay and liquid chromatography-mass spectrometry. J Agric Food Chem 53:7165–7171

Aga DS, Lenczewski M, Snow D, Muurinen J, Sallach JB, Wallace JS (2016) Challenges in the measurement of antibiotics and in evaluating their impacts in agroecosystems: a critical review. J Environ Qual 45:407–419

Alekshun MN, Levy SB (2007) Molecular mechanisms of antibacterial multidrug resistance. Cell 128:1037–1050

Amann RI, Ludwig W, Schleifer K-H (1995) Phylogenetic identification and in situ detection of individual microbial cells without cultivation. Microbiol Rev 59:143–169

Auerbach EA, Seyfried EE, McMahon KD (2007) Tetracycline resistance genes in activated sludge wastewater treatment plants. Water Res 41:1143–1151

Awad YM, Kim S-C, El-Azeem SAA, Kim K-H, Kim K-R, Kim K, Jeon C, Lee SS, Ok YS (2014) Veterinary antibiotics contamination in water, sediment, and soil near a swine manure composting facility. Environ Earth Sci 71:1433–1440

Awad YM, Kim KR, Kim S C, Kim K, Lee SR, Lee SS, Ok YS (2015) Monitoring antibiotic residues and corresponding antibiotic resistance genes in an agroecosystem. J Chem 2015:1–7

Baguer AJ, Jensen J, Krogh PH (2000) Effects of the antibiotics oxytetracycline and tylosin on soil fauna. Chemosphere 40:751–757

Baker-Austin C, Wright MS, Stepanauskas R, McArthur JV (2006) Co-selection of antibiotic and metal resistance. Trends Microbiol 14:176–182

Bengtsson-Palme J, Boulund F, Fick J, Kristiansson E, Larsson DGJ (2014) Shotgun metagenomics reveals a wide array of antibiotic resistance genes and mobile elements in a polluted lake in India. Front Microbiol 5:648

Binh CT, Heuer H, Kaupenjohann M, Smalla K (2008) Piggery manure used for soil fertilization is a reservoir for transferable antibiotic resistance plasmids. FEMS Microbiol Ecol 66:25–37

Boles TH, Wells MJ (2010) Analysis of amphetamine and methamphetamine as emerging pollutants in wastewater and wastewater-impacted streams. J Chromatogr A 1217:2561–2568

Boxall ABA, Fogg LA, Kay P, Blackwel PA, Pemberton EJ, Croxford A (2003) Prioritisation of veterinary medicines in the UK environment. Toxicol Lett 142:207–218

Burdett V (1991) Purification and characterization of Tet (M), a protein that renders ribosomes resistant to tetracycline. J Biol Chem 266:2872–2877

Casey JA, Curriero FC, Cosgrove SE, Nachman KE, Schwartz BS (2013) High-density livestock operations, crop field application of manure, and risk of community associated methicillin-resistant Staphylococcus aureus infection in Pennsylvania. JAMA Intern Med 173:1980–1990

Chang Q, Wang W, Regev-Yochay G, Lipsitch M, Hanage WP (2015) Antibiotics in agriculture and the risk to human health: how worried should we be? Evol Appl 8:240–247

Chee-Sanford JC, Aminov RI, Krapac IJ, Garrigues-Jeanjean N, Mackie RI (2001) Occurrence and diversity of tetracycline resistance genes in lagoons and groundwater underlying two swine production facilities. Appl Environ Microbiol 67:1494–1502

Chee-Sanford JC, Mackie RI, Koike S, Krapac IG, Lin Y-F, Yannarell AC, Maxwell S, Aminov RI (2009) Fate and transport of antibiotic residues and antibiotic resistance genes following land application of manure waste. J Environ Qual 38:1086–1108

Díaz-Cruz MS, Barceló D (2007) Recent advances in LC-MS residue analysis of veterinary med icines in the terrestrial environment. TrAC Trends Anal Chem 26:637–646

Dodd MC (2012) Potential impacts of disinfection processes on elimination and deactivation of antibiotic resistance genes during water and wastewater treatment. J Environ Monit 14: 1754–1771

Donato JJ, Moe LA, Converse BJ, Smart KD, Berklein FC, McManus PS, Handelsman J (2010) Metagenomics reveals antibiotic resistance genes encoding predicted bifunctional proteins in apple orchard soil. Appl Environ Microbiol 76(13):4396–4401

Du L, Liu W (2012) Occurrence, fate, and ecotoxicity of antibiotics in agro-ecosystems. A review. Agron Sustain Dev 32:309–327

Durso LM, Miller DN, Wienhold BJ (2012) Distribution and quantification of antibiotic resistant genes and bacteria across agricultural and non-agricultural metagenomes. PLoS One 7:e48325

Franklin AM, Aga DS, Cytryn E, Durso LM, McLain JE, Pruden A, Roberts MC, Rothrock MJ, Snow DD, Watson JE, Dungan RS (2016) Antibiotics in agroecosystems: introduction to the special section. J Environ Qual 45:377–393

Garcia-Armisen T, Anzil A, Cornelis P, Chevreuil M, Servais P (2013) Identification of antimicrobial resistant bacteria in rivers: insights into the cultivation bias. Water Res 47:4938–4947

Garder JL, Moorman TB, Soupir ML (2014) Transport and persistence of tylosin-resistant enterococci, genes, and tylosin in soil and drainage water from fields receiving Swine manure. J Environ Qual 43:1484–1493

Greig J, Rajic A, Young I, Mascarenhas M, Waddell L, LeJeune J (2015) A scoping review of the role of wildlife in the transmission of bacterial pathogens and antimicrobial resistance to the food Chain. Zoonoses Public Health 62:269–284

Guo R, Xie W, Chen J (2015) Assessing the combined effects from two kinds of cephalosporins on green alga (Chlorella pyrenoidosa) based on response surface methodology. Food Chem Toxicol 78:116–121

Halling-Sørensen B, Nielsen SN, Lanzky P, Ingerslev F, Lützhøft HH, Jørgensen S (1998) Occurrence, fate and effects of pharmaceutical substances in the environment-A review. Chemosphere 36:357–393

Heuer H, Schmitt H, Smalla K (2011) Antibiotic resistance gene spread due to manure application on agricultural fields. Curr Opin Microbiol 14:236–243

Johnson JV, Yost RA, Kelley PE, Bradford DC (1990) Tandem-in-space and tandem-in-time mass spectrometry: triple quadrupoles and quadrupole ion traps. Anal Chem 62:2162–2172

Johnston C, Martin B, Fichant G, Polard P, Claverys J-P (2014) Bacterial transformation: distribution, shared mechanisms and divergent control. Nat Rev Microbiol 12:181–196

Kim S-C, Carlson K (2005) LC–MS2 for quantifying trace amounts of pharmaceutical compounds in soil and sediment matrices. TrAC Trends Anal Chem 24:635–644

Kim S-C, Carlson K (2007) Temporal and spatial trends in the occurrence of human and veterinary antibiotics in aqueous and river sediment matrices. Environ Sci Technol 41:50–57

Klumper U, Riber L, Dechesne A, Sannazzarro A, Hansen LH, Sorensen SJ, Smets BF (2015) Broad host range plasmids can invade an unexpectedly diverse fraction of a soil bacterial community. ISME J 9:934–945

Knapp CW, Dolfing J, Ehlert PA, Graham DW (2009) Evidence of increasing antibiotic resistance gene abundances in archived soils since 1940. Environ Sci Technol 44:580–587

Koike S, Krapac IG, Oliver HD, Yannarell AC, Chee-Sanford JC, Aminov RI, Mackie RI (2007) Monitoring and source tracking of tetracycline resistance genes in lagoons and groundwater adjacent to swine production facilities over a 3-year period. Appl Environ Microbiol 73: 4813–4823

Kristiansson E, Fick J, Janzon A, Grabic R, Rutgersson C, Weijdegård B, Söderström H, Larsson DJ (2011) Pyrosequencing of antibiotic-contaminated river sediments reveals high levels of resistance and gene transfer elements. PLoS one 6:e17038

Kumar K, Gupta SC, Chander Y, Singh AK (2005) Antibiotic use in agriculture and its impact on the terrestrial environment. Adv Agron 87:1–54

Kumar RR, Lee JT, Cho JY (2012) Fate, occurrence, and toxicity of veterinary antibiotics in environment. J Korean Soc Appl Biol Chem 55:701–709

Kümmerer K (2009) Antibiotics in the aquatic environment—a review—part I. Chemosphere 75: 417–434

Li C, Chen J, Wang J, Ma Z, Han P, Luan Y, Lu A (2015) Occurrence of antibiotics in soils and manures from greenhouse vegetable production bases of Beijing, China and an associated risk assessment. Sci Total Environ 521–522:101–107

Lin D, Chen K, Li R, Liu L, Guo J, Yao W, Chen S (2014) Selection of target mutation in rat gastrointestinal tract E. coli by minute dosage of enrofloxacin. Front Microbiol 5:468

Liu Y, Zhang J, Gao B, Feng S (2014) Combined effects of two antibiotic contaminants on Microcystis aeruginosa. J Hazard Mater 279:148–155

Lorenz MG, Wackernagel W (1994) Bacterial gene transfer by natural genetic transformation in the environment. Microbiol Rev 58:563–602

Luby E, Ibekwe AM, Zilles J, Pruden A (2016) Molecular methods for assessment of antibiotic resistance in agricultural ecosystems: prospects and challenges. J Environ Qual 45:441–453

Martinez JL (2009) Environmental pollution by antibiotics and by antibiotic resistance determinants. Environ Pollut 157:2893–2902

McLain J, Williams C (2014) Sustainability of water reclamation: long-term recharge with reclaimed wastewater does not enhance antibiotic resistance in sediment bacteria. Sustainability 6:1313

McLain JE, Cytryn E, Durso LM, Young S (2016) Culture-based methods for detection of antibiotic resistance in agroecosystems: advantages, challenges, and gaps in knowledge. J Environ Qual 45:432–440

Musovic S, Dechesne A, Sorensen J, Smets BF (2010) Novel assay to assess permissiveness of a soil microbial community toward receipt of mobile genetic elements. Appl Environ Microbiol 76:4813–4818

Nachnani S, Scuteri A, Newman MG, Avanessian AB, Lomeli SL (1992) E-test: a new technique for antimicrobial susceptibility testing for periodontal microorganisms. J Periodontol 63: 576–583

Naidu R, Semple KT, Megharaj M, Juhasz AL, Bolan NS, Gupta SK, Clothier BE, Schulin R (2008) Bioavailability: definition, assessment and implications for risk assessment. Dev Soil Sci 32:39–51

Nandi S, Maurer JJ, Hofacre C, Summers AO (2004) Gram-positive bacteria are a major reservoir of Class 1 antibiotic resistance integrons in poultry litter. Proc Natl Acad Sci USA 101: 7118–7122

Nesme J, Simonet P (2015) The soil resistome: a critical review on antibiotic resistance origins, ecology and dissemination potential in telluric bacteria. Environ Microbiol 17:913–930

Nordmann P, Poirel L, Dortet L (2012) Rapid detection of carbapenemase-producing Enterobacteriaceae. Emerg Infect Dis 18:1503–1507

Normark BH, Normark S (2002) Evolution and spread of antibiotic resistance. J Intern Med 252: 91–106

Ochman H, Lawrence JG, Groisman EA (2000) Lateral gene transfer and the nature of bacterial innovation. Nature 405:299–304

Ok YS, Kim S-C, Kim K-R, Lee SS, Moon DH, Lim KJ, Sung J-K, Hur S-O, Yang JE (2011) Monitoring of selected veterinary antibiotics in environmental compartments near a composting facility in Gangwon Province, Korea. Environ Monit Assess 174:693–701

Pawlowski AC, Wang W, Koteva K, Barton HA, McArthur AG, Wright GD (2016) A diverse intrinsic antibiotic resistome from a cave bacterium. Nat Commun 7:13803

Pei R, Kim S-C, Carlson KH, Pruden A (2006) Effect of river landscape on the sediment concentrations of antibiotics and corresponding antibiotic resistance genes (ARG). Water Res 40: 2427–2435

Pepper IL, Gerba CP (2009) Cultural methods. In: Environmental microbiology, doi: 10.1016/B978-0-12-370519-8.00010-9

Persoons D, Bollaerts K, Smet A, Herman L, Heyndrickx M, Martel A, Butaye P, Catry B, Haesebrouck F, Dewulf J (2011) The importance of sample size in the determination of a flock-level antimicrobial resistance profile for Escherichia coli in broilers. Microb Drug Resist 17:513–519

Pillai DK (2011) Assessment of pet dogs as a reservoir of antibiotic resistant bacteria. Kansas State University

Pruden A, Pei R, Storteboom H, Carlson KH (2006) Antibiotic resistance genes as emerging contaminants: studies in Northern Colorado. Environ Sci Technol 40:7445–7450

Rysz M, Alvarez PJ (2004) Amplification and attenuation of tetracycline resistance in soil bacteria: aquifer column experiments. Water Res 38:3705–3712

Sarmah AK, Meyer MT, Boxall AB (2006) A global perspective on the use, sales, exposure pathways, occurrence, fate and effects of veterinary antibiotics (VAs) in the environment. Chemosphere 65:725–759

Schnappinger D, Hillen W (1996) Tetracyclines: antibiotic action, uptake, and resistance mechanisms. Arch Microbiol 165:359–369

Sengeløv G, Agersø Y, Halling-Sørensen B, Baloda SB, Andersen JS, Jensen LB (2003) Bacterial antibiotic resistance levels in Danish farmland as a result of treatment with pig manure slurry. Environ Int 28:587–595

Solis Y, Chavarria G, Garcia F, Rodriguez C (2011) Exposure of a tropical soil to MG/KG of oxytetracycline elicits hormetic responses in the catabolic activities of its microbial community. Dose Response 9:434–441

Subedi B, Kannan K (2014) Mass loading and removal of select illicit drugs in two wastewater treatment plants in New York State and estimation of illicit drug usage in communities through wastewater analysis. Environ Sci Technol 48:6661–6670

Tasho RP, Cho JY (2016) Veterinary antibiotics in animal waste, its distribution in soil and uptake by plants: a review. Sci Total Environ 563:366–376

Toth JD, Feng Y, Dou Z (2011) Veterinary antibiotics at environmentally relevant concentrations inhibit soil iron reduction and nitrification. Soil Biol Biochem 43:2470–2472

Tran-Thanh D, Buttars S, Wen Y, Wilson C, Done SJ (2010) Cyclooxygenase-2 inhibition for the prophylaxis and treatment of preinvasive breast cancer in a her-2/neu mouse model. Cancer Prev Res (Phila) 3:202–211

Udikovic-Kolic N, Wichmann F, Broderick NA, Handelsman J (2014) Bloom of resident antibiotic-resistant bacteria in soil following manure fertilization. Proc Natl Acad Sci USA 111: 15202–15207

Vieira AR, Collignon P, Aarestrup FM, McEwen SA, Hendriksen RS, Hald T, Wegener HC (2011) Association between antimicrobial resistance in Escherichia coli isolates from food animals and blood stream isolates from humans in Europe: an ecological study. Foodborne Pathog Dis 8:1295–1301

WHO (2014) Antibiotic resistance: global report on surveillance 2014. World Health Organization, Geneva

Wilga J, Kot-Wasik A, Namiesnik J (2008) Studies of human and veterinary drugs' fate in environmental solid samples—analytical problems. J Chromatogr Sci 46:601–608

Williams-Nguyen J, Sallach JB, Bartelt-Hunt S, Boxall AB, Durso LM, McLain JE, Singer RS, Snow DD, Zilles JL (2016) Antibiotics and antibiotic resistance in agroecosystems: state of the science. J Environ Qual 45:394–406

Witte W (2004) International dissemination of antibiotic resistant strains of bacterial pathogens. Infect Genet Evol 4:187–191

Youngquist CP (2014) Local biosolids compost: agricultural use, fate of antibiotics, and community engagement. Washington State University

Zhang Y, Zhang C, Parker DB, Snow DD, Zhou Z, Li X (2013) Occurrence of antimicrobials and antimicrobial resistance genes in beef cattle storage ponds and swine treatment lagoons. Sci Total Environ 463–464:631–638

Chapter 6
Role of Antibiotics in Climate Change

Rida Akram, Muhammad Zaffar Hashmi, and Wajid Nasim

6.1 Introduction

Climate change is considered as a long-term change in the statistics of the weather. These changes include variations in average values of temperature and precipitation/rainfall. It is estimated that the last decade of the twentieth and beginning of the twenty-first century is the warmest period in the whole global instrumental temperature record (NOAA 2007). It is found that the trend of average temperature and volatile rainfall patterns are increasing faster as predicted National Academy of Science (2001). Most of these changes in climate (90–95%) are caused by human being, and the average global temperature is increased by 1.4–5.8°C between the 1900 and 2100 (IPCC 2001, 2007). It is also estimated that the average global temperature will be increased by 0.3–1.3°C during the next upcoming 30 years (Zwiers 2002).

The original version of this chapter was revised. An erratum to this chapter can be found at DOI 10.1007/978-3-319-66260-2_22.

R. Akram
Department of Environmental Sciences, COMSATS Institute of Information Technology (CIIT), Vehari, Pakistan

M.Z. Hashmi
Department of Meteorology, COMSATS Institute of Information Technology (CIIT), Islamabad, Pakistan

W. Nasim (✉)
Department of Environmental Sciences, COMSATS Institute of Information Technology (CIIT), Vehari, Pakistan

CIHEAM-Institut Agronomique Mediterraneen de Montpellier (IAMM), Montpellier, France

National Research Flagship, CSIRO Sustainable Ecosystems, Toowoomba, QLD, 4350, Australia
e-mail: wajidnaseem2001@gmail.com

M.Z. Hashmi et al. (eds.), *Antibiotics and Antibiotics Resistance Genes in Soils*,
Soil Biology 51, DOI 10.1007/978-3-319-66260-2_6

6.2 Causes of Climate Change

These are the following causes of climate change:

6.2.1 *Greenhouse Gasses and Climate Change*

When sunlight radiations reach earth's surface, there are two possibilities that it can either be reflected to the space or absorbed by the earth's surface. Greenhouse gasses (GHGs) cause the atmosphere to retain the heat and are considered as main contributors to climate change via direct and indirect sources.

- **Direct sources**
  It occurs when gasses (CO_2, CH_4, N_2O, O_3, and water vapors) itself absorbs radiation.
- **Indirect sources**
  It occurs when transformation of chemical substances produces greenhouse gasses as a product (IPCC 2013).

These climatic changes mostly due to the anthropogenic GHG emissions are going to produce a series of serious environmental issues. But it is also fact that agriculture sector is responsible for 30–35% of the global GHG production (Foley et al. 2011).

6.2.2 *Ozone Depletion and Climate Change*

Ozone (O_3) forms a layer in the stratosphere and considered vital to maintain life on earth. It has the ability to absorb the ultraviolet (UV) rays that come from the sun and have damaging effect on the living organisms. Only 3% or fewer amounts of UV rays that come through the O_3 layer seriously cause many health problems like sunburns, skin allergy, and skin cancer. So, if there is no O_3 layer at all, then what will happen? Bitter consequences! And we are not ready to face it.

Chlorofluorocarbons (CFCs) are considered as the main reason for the ozone layer depletion. In the twentieth century, CFCs become most demanding in market for its use in refrigerators and aerosol products. In start, it is considered that CFCs are harmless, and mostly don't have the ability to easily react with other substances. But chemical stability permits CFCs to stay for a long duration that is enough for its drift into the stratosphere after the emission. When reached to stratosphere, CFCs react with O_3 and cause its depletion. So, with ozone depletion excessive amount of sun radiation enters and enhances the overall temperature of the earth, ultimately bringing severe changes in climate (Ozone Depletion and Climate Change 2011).

6.2.3 *Chemistry in Climate Change*

In stratosphere, CFCs break down as it become exposed to UV rays. Free chlorine radicals ($Cl^•$) are produced, and taken as chemically reactive substance. During the process of decomposition of O_3, $Cl^•$ behave as a strong catalyst, change two O_3 molecules into three O_2 molecules. Also lose UV absorbing ability during this decomposition reaction. The produced catalysts mostly not undergo in another chemical reaction, the same $Cl^•$ continues the further destruction of O_3 (Eq. 6.1).

$$\begin{aligned} CFCl_3 + \text{UV light} &\rightarrow CFCl_2 + Cl^• \\ Cl^• + O_3 &\rightarrow ClO + O_2 \\ ClO + O &\rightarrow Cl^• + O_2 \end{aligned} \tag{6.1}$$

On polar sides, the stratosphere is cool enough to make polar stratospheric clouds (PSCs) and provides optimal conditions for the $Cl^•$ to produce ClO (chlorine monoxide). There was a catalytic cycle of $Cl^•$ and ClO that continues the O_3 depletion. Antarctica is a more suitable region for O_3 depletion than the Arctic, just because of lowest temperatures and its wind system that prevent the drifting of O_3-depleting substances out of the area (Ozone Depletion and Climate Change 2011).

6.3 Methane Production in Livestock and Climate Change

Methane (CH_4) has 25 times more ability to trap heat than CO_2 in the atmosphere. In last 250 years, overall concentration of CH_4 increases by 160% (IPCC 2013; CDIAC 2015). Methane gas production from the ruminants is considered as natural and unavoidable process during the digestion of feed. Mostly, CH_4 leaves the ruminant body via expired air and small friction approximately 2% forms in the large intestine. These animals stomach have millions of microorganisms like bacteria, fungi, and protozoa that break down the food, mainly consisted of carbohydrates (e.g., starch). Enzymatic activity breaks the starch, but the presence of microorganisms is essential for the breakdown of cellulose. These microorganisms break the carbohydrates present in the feed to fatty acids, acetic, propionic, and butyric acid. With the formation of acetic acid and butyric acid, hydrogen ions release. Hydrogen ions have damaging effect to cattle, but CH_4 producing bacteria start the metabolism of hydrogen ions to CH_4 and H_2O (Berglund et al. 2008). According to the estimation, 6.5% of gross energy present in the diet is lost as CH_4. Cow yielding 9000 kg milk annually produces approximately 120–130 kg CH_4/year (IPCC 2006).

Global emission of these GHG mostly from the livestock production system was 18% of total anthropogenic GHG emissions and considered nearly equal to emissions from the global transportation systems (Asner and Archer 2010). In 2050, the direct GHG emission rate from production of milk, meat, and egg has the ability to increase approximately 39% above the 2000 years levels (Pelletier and Tyedmers 2010).

Cows belong to the class of ruminants, having four-chambered stomach. They regurgitate their feed as cud before the chewing and eating it again. The digestion of plants and grass is difficult; that's why these ruminants have four chambers, so food is initially digested in the first two chambers and then digested food (cud) is brought back and rechewed, and finally it goes to stomach and intestine to digest again. Microbes that are present break the food and produce CH_4 as a by-product (Pradhan 2015).

Cows produce a massive amount of CH_4 through belching and lesser amount via flatulence. The agricultural CH_4 production could increase by the 60% by 2013. Worldwide 1.5 billion cows and billions of grazing animals produce dozens of GHGs especially CH_4, 2/3 of all NH_3 comes from cow. Some expert's reports indicate that a cow emits 100–200 liters a day (about 26–52 gallons), while others say it is up to 500 l (132 gallons) a day. That is a lot of CH_4 and a comparable to the pollution produced by a car (Lean 2006).

Beef cattle are relatively considered as a large source of GHG emissions (Crosson et al. 2011). Direct emissions of GHG from the beef cattle production was estimated by the EPA (environmental protection agency), mostly based on three most obvious localized sources, that are, (1) CH_4 from enteric fermentation, (2) CH_4 from manure decomposition, and (3) NO from manure (both direct and indirect) (EPA 2010). Beef cattle produce a significant amount of CO_2 through respiration and enteric fermentation (Table 6.1).

6.4 Role of Antibiotics in Climate Change

Antibiotics are used to increase the milk and meat production rate as well as to maintain health of livestock. Previous studies show that there is no relationship between use of antibiotics and climate change. But recent some reports show that dosing livestock animal with antibiotics increases GHG emissions, especially from the cow dung. The antibiotics boost the production rate of CH_4 in cows. Clear damaging impacts of antibiotics were measured and producing 1.8 times more CH_4. Methane generated by cattle is released as burps (release of gas from the digestive tract), and antibiotics are considered to increase burped CH_4 as well (Harrabin 2016).

Farmers feeding their livestock antibiotics and may be doing more than the level of creating drug resistance microbes. The excessive use of antibiotics boosts up the GHG emissions. Tetracycline is taken as a commonly used antibiotic. A 3-day treatment was used to measure the amount of CH_4 produced within the cattle manure. It is observed that the emission of planet warming CH_4 from the manure of antibiotic-dosed cattle was 80% higher than the manure of untreated cattle. This increase in CH_4 may be due to the increase in CH_4 producing microbes that is

Table 6.1 Livestock emissions of CH_4 and N_2O in the USA (1990 and 2008)

Gas/animal type	1990 Tg[a] CO_2 Eq.% of total		2008 Tg CO_2 Eq.% of total	
CH_4 from manure				
Total US livestock	29.3	100.0%	45.0	100.0%
Dairy cattle	10.2	34.8	19.4	43.1
Beef cattle	2.6	8.6%	2.5	5.6%
Sheep	0.1	0.3%	0.8	1.8%
Poultry	2.8	9.6%	2.6	5.8%
NO from manure				
Total US livestock	14.4	100.0%	17.1	100.0%
Beef cattle	6.3	43.8%	7.4	43.3%
Dairy cattle	5	34.7%	5.5	32.2%
Sheep	0.1	0.7%	0.3	1.8%
Poultry	1.5	10.4%	1.8	10.5%
CH_4, enteric fermentation				
Total US livestock	132.0	100.0%	140.6	100.0%
Beef cattle	94.5	71.6%	100.8	71.7%
Dairy cattle	32.0	24.2%	33.1	23.5%
Sheep	1.9	0.0%	1	0.7%

[a]One teragram is equal to 1012 g, or 1 million metric tons

present in the digestive system of cattle treated with tetracycline, due to the suppression of antibiotic susceptible bacteria. The CH_4 emission from the cud chewing livestock worldwide account for approximately 4% of the GHG emission related to anthropogenic activity (Perkins 2016).

If this hypothesis is correct, then antibiotics (tetracycline) have the same effect on the livestock as well as on direct gaseous emissions. Because the livestock will generate CH_4 (a potent GHG) that, in turn, contributes to climate change. Tetracycline changes the microbial competition inside the intestine of cow and hampers the balance. In addition, the same effect occurs in case of belching that would be a cause of great concern. Cattle is a known source of CH_4 which is taken as more potent GHG than CO_2. Cow dung fed with tetracycline was compared with the cow dung not fed with antibiotic to measure the amount of CO_2, N_2O, and CH_4. Activity of gut microbes that is known as archaea produces CH_4 in cow intestine. These gut microbes flourish in air free (anaerobic) condition. This study shows abrupt alteration in the microbiota of the cow's dung and enhances the rate CH_4 emission. Tetracycline may also increase CH_4 amount of cow farts and burps (Roy 2016) (Fig. 6.1).

Fig. 6.1 Antibiotics increase CH_4 amount of cow farts and burps

6.5 Effect of Antibiotics on Soil Microbes Combat Against Climate Change

Antibiotics are introduced to agro-ecosystem through land application of manure and associated with potential health. Antibiotics have negative impacts on soil microbes and also bring changes in working of these microbes (Unger et al. 2012).

In addition to the development of antibiotic resistance, the use of antibiotics increasingly disrupts the ecology of the microbes, and microbes may not perform vital functions such as nutrient recycling (Guarner and Malagelada 2003). Dung beetle plays a key role in recycling of nutrients and reduction of CH_4 emission by break down of cow pats. This is done by reducing the anaerobic archaea and oxygenation of cow dung. The dung beetle also alters the microbiota. So, increase in antibiotics feeding cause increase in CH_4 emission (Roy 2016).

These antibiotics change the microbes present in digestive system of dung beetles, which are considered vital in carbon cycling and also improving soil. A recent study shows that methanotrophs have the ability to use large amounts of copper for the purpose of CH_4 oxidation. Copper is taken as a vital element and used for the biological CH_4 oxidation for over 30 years. This information is helpful to make new approaches for exploiting the bacteria in laboratory as well as in environment. New copper storage proteins (Csp) was identified and present in wide range of bacteria, and these proteins have the ability to store metal in a way that was not seen previously. So, methanotrophs are the biological mechanism for avoiding too much amount of CH_4 from the environment by consuming it for carbon and energy. For the oxidation of CH_4, methanotrophs use an enzyme (methane monooxygenase) that requires copper/iron to work (Mathewson 2015; Reay 2003; Singh et al. 2010) (Fig. 6.2).

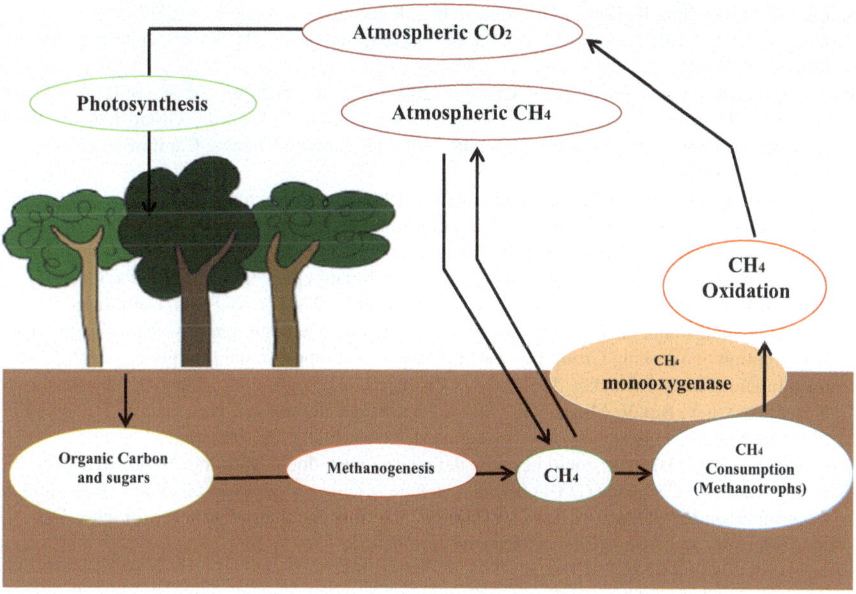

Fig. 6.2 Microbial control on GHG emission (modified after Singh et al. 2010)

6.6 Conclusion

It is concluded that by minimizing the negative impact of anthropogenic activities in environment less use of antibiotics can help combat against climate change. Further investigation is also required to explain the whole processes that how use of antibiotics in livestock increases the rate of CH_4 gas production.

References

Asner G, Archer S (2010) Livestock and the global carbon cycle. Island Press, Washington, DC

Berglund M, Cederberg C, Clason C, Henriksson M, Torner L (2008) Jordbrukets klimatpåverkan—underlag för att beräkna växthusgasutsläpp och nulägesanalyser av exempelgårdar. Delrapport 1 i Jokerprojektet, Hushållningssällskapet Halland. Manuskript

Carbon Dioxide Information Analysis Center (2015) Recent greenhouse gas concentrations

Crosson P, Shalloo L, O'Brien D, Lanigan GJ, Foley PA, Boland TM, Kenny DA (2011) A review of whole farm systems models of greenhouse gas emissions from beef and dairy cattle production systems. Anim Feed Sci Technol 166–167:29–45

Environmental Protection Agency (2010) Inventory of U.S. Greenhouse Gas Emissions and Sinks: 1990–2008. Washington, DC

Foley J, Ramankutty N, Brauman K, Cassidy E, Gerber J, Johnston M, Mueller N, O'Connell C, Ray D, West P, Balzer C, Bennett E, Carpenter S, Hill J, Monfreda C, Polasky S, Rockstrom J, Sheehan J, Siebert S, Tilman D, Zaks D (2011) Solutions for a cultivated planet. Nature 478:337–342

Guarner F, Malagelada JR (2003) Gut flora in health and disease. Lancet 361:512–519
Harrabin R (2016) Cattle drugs could fuel climate change, study suggests. Science and Environment. BBC news report
Intergovernmental Panel on Climate Change (2001) Climate change: the scientific basis. In: Houghton JT, Ding Y, Griggs DJ et al (eds) Contribution of Working Group I to the third assessment report of the Intergovernmental Panel on Climate Change. Cambridge University Press, Cambridge
Intergovernmental Panel on Climate Change (2006) IPCC guidelines for National Greenhouse Gas Inventories, prepared by the National Greenhouse Gas Inventories Programme. Eggleston HS, Buendia L, Miwa K, Ngara T, Tanabe K (eds). IGES, Japan. Published
Intergovernmental Panel on Climate Change (2007) Climate change: the physical science basis. Intergovernmental Panel on Climate Change. Cambridge University Press, Cambridge
Intergovernmental Panel on Climate Change (2013) Climate change: the physical science basis. Contribution of Working Group I to the fifth assessment report of the Intergovernmental Panel on Climate Change. Stocker TF, Qin D, Plattner G-K, Tignor M, Allen SK, Boschung J, Nauels A, Xia Y, Bex V, Midgley PM (eds). Cambridge University Press, Cambridge, pp 1535
Lean G (2006) Cow 'emissions' more damaging to planet than CO_2 from cars
Mathewson S (2015) Bacteria could be a key defense against global warming. Nature World News
National Academy of Science (2001) Climate change science: an analysis of some key questions. Committee on the Science of Climate Change. National Academy Press, Washington DC
National Oceanic and Atmospheric Administration (2007)
Ozone Depletion and Climate Change (2011) Climatesight
Pelletier N, Tyedmers P (2010) Forecasting potential global environmental costs of livestock production 2000-2050. Proc Natl Acad Sci U S A 107:18371–18374
Perkins S (2016) Antibiotics may give cows gas, contribute to climate change. Climate. People and events. doi:10.1126/science.aaf5745
Pradhan R (2015) Do cows contribute more to global warming than cars? Science
Reay DS (2003) Sinking methane. Biologist 50:15–19
Roy R (2016) Antibiotic-fed cows, burping and farting may lead to climate change; Tetracycline releases potent greenhouse gas methane
Singh BK, Bardgett RD, Smith P, Reay DS (2010) Microorganisms and climate change: terrestrial feedbacks and mitigation options. Nat Rev Microbiol 8:779–790
Unger IM, Goyne KW, Kennedy AC, Kremer RJ, Williams CF (2012) Antibiotic effects on microbial community characteristics in soils under conservation management practices. Soil Sci Soc Am J 77:100–112
Zwiers FW (2002) Climate change: the 20-year forecast. Nature 416:690–691

Chapter 7
Potential Dissemination of ARB and ARGs into Soil Through the Use of Treated Wastewater for Agricultural Irrigation: Is It a True Cause for Concern?

Nada Al-Jassim and Pei-Ying Hong

7.1 Introduction

Antibiotic resistance is increasingly being recognized as an emerging contaminant, threatening effective treatment of infections and carrying a great risk to public health. Anthropogenic activities such as the rise in antibiotic use for medical and agricultural purposes are considered a major cause for escalating the threat.

In all cases of usage, antibiotic end up in sewage waters at subtherapeutic levels, that is, in concentrations not high enough to kill bacteria but instead impose a selective pressure to favor the occurrence of antibiotic resistant bacteria (ARB) with their associated antibiotic resistance genes (ARGs) (Pruden et al. 2013). In recent years, WWTPs have been shown to be potential hotspots for ARB and ARGs propagation (Rizzo et al. 2013). Despite undergoing treatment, the treated municipal wastewater can still contain a significant amount of ARB and ARGs. This problem is of particular concern in water-scarce countries with pressing needs to reuse the treated wastewater. Reuse of treated wastewater effluents might impose a potential risk to the public health if ARB and ARGs accumulate in the agricultural soils.

Soils, however, also inherently contain a baseline abundance of ARB and ARGs. It is therefore required to account for how much of the ARB and ARGs in

The original version of this chapter was revised. An erratum to this chapter can be found at DOI 10.1007/978-3-319-66260-2_22.

N. Al-Jassim • P.-Y. Hong (✉)
Biological and Environmental Science & Engineering Division (BESE), King Abdullah University of Science and Technology (KAUST), Thuwal 23955-6900, Saudi Arabia

Water Desalination and Reuse Center (WDRC), King Abdullah University of Science and Technology (KAUST), Thuwal 23955-6900, Saudi Arabia
e-mail: nada.aljassim@kaust.edu.sa; peiying.hong@kaust.edu.sa

M.Z. Hashmi et al. (eds.), *Antibiotics and Antibiotics Resistance Genes in Soils*, Soil Biology 51, DOI 10.1007/978-3-319-66260-2_7

agricultural soils are truly contributed by wastewater during irrigation events, and also which of these ARB and ARGs are potential new threats of concern.

This book chapter aims to address the underlying question of whether the use of treated wastewater for agricultural irrigation can lead to dissemination of ARGs and ARB. To achieve this aim, the chapter starts by first stating that pristine soils or even soils that predate the antibiotic era naturally contain ARB and ARGs. Findings from earlier studies are collated to provide both sides of the argument on whether wastewater reuse can lead to accumulation of ARB and ARGs in agricultural soils. Emphasis is made on the emerging ARGs, particularly the bla_{NDM} gene that confers resistance against carbapenem. Carbapenem is an antibiotic typically used as a last line of defense against Gram-negative bacterial infections (Walther-Rasmussen and Høiby 2007). Bacterial pathogens possessing the bla_{NDM} gene are hence associated with patient morbidity and mortality rates. The fate and persistence of emerging ARGs (e.g., bla_{NDM}) are not studied in depth but the chapter reviews insights that have been gained from studies involving other types of ARGs to discern if horizontal gene transfers are likely in a wastewater-irrigated soil matrix. Finally, the chapter discusses several intervention strategies, namely solar irradiation and phage treatment that can potentially be applied in the agricultural setting to combat against emerging ARB and ARGs threats.

7.2 Pristine Environments Harbor ARBs and ARGs

Natural environments are thought to be the origin of most antibiotic resistance genes and serve as reservoirs for antibiotic resistance (D'costa et al. 2006; Wright 2007). Soil environments are a particularly significant reservoir as they are one of the richest habitats for microbial diversity and abundance (Cytryn 2013). In one study, a majority of the 93 bacterial colonies isolated from a cave that had been secluded for over 4 million years were revealed to be multi-drug resistant. These bacterial isolates demonstrated resistance to a wide range of structurally different antibiotics, including the last-resort antibiotic daptomycin (Bhullar et al. 2012). However, resistance patterns showed relatively little resistance to new classes of synthetic antibiotics compared to natural antibiotics. In another study on deep terrestrial subsurface soil samples, 153 bacterial isolates were tested against 13 antibiotics and results found 70% of these isolates to be resistant to more than one antibiotic, with over 35 isolates resistant to five or more antibiotics (Brown and Balkwill 2009). Most frequently noted resistance was against nalidixic acid, mupirocin, or ampicillin, and to lesser extents, against ciprofloxacin, tetracycline, neomycin, and chloramphenicol. Resistance against rifampin, streptomycin, kanamycin, vancomycin, erythromycin, and gentamicin was also detected.

7.2.1 *Overview of Range of Antibiotic Resistance Classes in Pristine Environments/Soil*

Given the presence of ARB in pristine environments, detection of ARGs is expected too. An analysis of the ARGs distribution in glacier environments showed a widespread distribution of ARGs in samples from various glaciers in Central Asia, North and South America, Greenland, and Africa (Segawa et al. 2013). Reported ARGs included $bla_{\mathrm{TEM\text{-}1}}$, *tetW*, *aac(3)*, *aacC*, and *strA* and even metallo beta-lactamase gene (bla_{IMP}), encompassing ARGs of both clinical and agricultural origins. In another study, soil DNA was cloned into vectors and expressed for the insert genes. It was determined that at least nine clones were resistant to aminoglycosides and one to tetracycline (Riesenfeld et al. 2004). Aminoglycoside resistance genes sequences were further analyzed and six of them resembled genes that express 6′-*N*-aminoglycoside acetyltransferase [AAC(6′)] enzymes. All but one of the aminoglycoside resistance genes encode amino acid sequences that are considerably different (<60% identity) from any previously reported sequences. This indicates that natural soil environments are not only reservoirs for common ARGs, but are also reservoirs for genetically diverse and novel ARGs. Another study used functional metagenomics to study remote Alaskan soils and revealed the presence of diverse beta-lactamases, namely Ambler classes A, C, D (active site serine beta-lactamases), and B (metallo beta-lactamase) (Allen et al. 2009). Class A beta-lactamase were recovered from *Burkholderiapseudomallei*, *Pseudomonas luteola*, and *Yersinia entercolitica*, and these recovered beta-lactamases were distantly related to the clinically relevant CTX-M family. The lone representative of the class D beta-lactamases was linked with a class C beta-lactamase as part of a single open reading frame harboring two full-length genes, making the study one of the first to report a bifunctional beta-lactamase. Class D causes resistance to amoxicillin, ampicillin, and carbenicillin, while class C causes resistance to cephalexin. Class B beta-lactamase in the Alaskan soils fell into one of the three subgroups of known metallo beta-lactamases, but were more closely related to the ancestral beta-lactamases than beta-lactamases isolated in clinical settings. They, however, remained capable of conferring resistance to *E. coli* despite this evolutionary distance, demonstrating that resistance genes residing in the environmental reservoir do pose a threat to human health, especially if they are horizontally transferred to pathogens.

7.2.2 *ARGs Predate the Use of Antibiotics*

Pristine environments are those subjected to minimal perturbation by human activities, but they might still be indirectly subjected to unknown anthropogenic contamination due to weather elements and animal migration. For an assessment of samples free from modern anthropogenic activities and antibiotic influences, insight can be gained from examining pre-antibiotic era bacterial isolates. Retrospective studies have found ARGs in bacterial isolates sampled prior to 1950, with some of the detected resistance elements being able to be transferred via conjugation (Hughes and Datta 1983; Smith 1967). Metagenomic analysis of ancient 30,000-year-old DNA from permafrost sediments also identified a highly diverse collection of genes encoding resistance to beta-lactam, tetracycline, and glycopeptide antibiotics, and confirmed the similarity of a complete *van*A gene to modern variants (D'Costa et al. 2011).These results showed that ARGs exist naturally in the environment even prior to extreme selection pressure imposed by rampant antibiotic use. ARGs appear to facilitate bacterial survival in the natural environment and may be co-selected for by environmental factors like solar radiation or the presence of heavy metals and other toxic compounds in soil (Piepersberg et al. 1988; Nies 2003; Poole 2005). It may also be possible that ARGs occur consequentially from symbiotic relationships shared among different microorganisms. For instance, to defend against antibiotic-producing *Streptomyces*, other bacterial species may have co-evolved resistance against the corresponding antibiotics (D'Costa et al. 2011; Allen et al. 2010).

7.2.3 Baseline Abundance of ARGs in Soil

Despite the vast information related to the diversity of ARB and ARGs that could be recovered from pristine soils or from soils predating the antibiotic era, little information is available on the baseline abundance of ARGs in such samples. Wang and coworkers utilized qPCR arrays to report the relative abundance of a wide diversity of ARGs. Their results found seven ARGs (bla_{TEM}, bla_{SFO}, bla_{FOX}, *cph*A, *mex*F, *opr*D, and *opr*J) that were frequently and evenly represented across all samples, suggesting that the examined Antarctic region is a reservoir for these genes (Wang et al. 2016). However, the study only provides relative abundance of genes and not absolute copy numbers. Instead, such information can be inferred from another study that used qPCR to quantify values of different classes of ARGs in agricultural soil samples collected over the 1940s, 1960s, 1970s, 1980s, and 2000s (Knapp et al. 2010). The study revealed that some genes increased in abundance with time. Specifically, genes *tet*Q, *tet*O, *tet*M, $bla_{\mathrm{TEM\text{-}1}}$ were among ARGs with the highest rates of increment, coinciding with the increase in industrial antibiotic production in the 1950s and the increased use of related antibiotics (i.e., tetracycline and beta-lactam) in the recent years (Van Boeckel et al. 2014). To illustrate, at one of the study's sampling sites (Site C), *tet*Q, *tet*O, and $bla_{\mathrm{TEM\text{-}1}}$ had abundance of $10^{-3.49}$, $10^{-5.47}$, and $10^{-1.85}$ copies, respectively, per gram of dry soil collected in 1942. The abundance of *tet*Q, *tet*O, and $bla_{\mathrm{TEM\text{-}1}}$ increased to $10^{-2.62}$, $10^{-2.65}$, and $10^{-0.50}$ copies, respectively, per gram of dry soil collected in 1975. Although this study only examined archived soils collected from the Netherlands and may not be representative of baseline ARG abundance on a global scale, knowledge of the baseline ARG abundance would allow one to infer the required ARG fluxes from anthropogenic sources to significantly perturb the baseline ARG abundance. This would suggest that irrigation with municipal wastewater and/or manure application, both of which inherently contain high abundances of ARB and ARGs (Da Silva et al. 2006; Munir et al. 2011; Zhu et al. 2013; Munir and Xagoraraki 2011), may be potential contributors of ARGs and ARB to the soils. The following subsection therefore aims to further review potential impact on soils arising from manure application and wastewater irrigation.

7.3 Antibiotic Resistance Genes in Anthropologically Perturbed Soils

Numerous studies showed that application/irrigation of manure and wastewater can lead to potential detrimental impacts on soils. Soils applied with dairy or swine manure were compared to inorganically fertilized soils (Marti et al. 2013) and results showed enrichment of ARB and increment of ARGs abundances in manure-applied soils. However, there was no coherent corresponding increase in abundances enumerated from vegetables grown in the soils. Heuer and coworkers applied manure-containing sulfadiazine, an antibiotic typically used on livestock, to soils and compared the abundance of sulfonamide resistance genes against the non-treated soils. Their findings showed an increase in sulfonamide resistance gene numbers compared to non-treated soils and that sulfonamide resistance genes continued to be detected more than 2 months after manure application (Heuer and Smalla 2007). When sulfadiazine-supplemented manure was applied repeatedly to soils, sulfonamide resistance gene abundances accumulated within the soil samples (Heuer et al. 2011). Although such studies demonstrate an increase in ARGs as a result of manure application, the increase could be due to a higher nutrient input that subsequently changed the microbial community and enriched for bacterial populations that inherently possess the associated ARGs. To address this, a separate study examined soils treated with a single application of manure derived from cows which had not received any antibiotics treatment (Udikovic-Kolic et al. 2014). For over 130 days, the ARG abundances in manure-applied soils were compared to that detected in soils adjusted to the same nutrient input load (i.e., nitrogen, phosphorus, potassium) levels with inorganic fertilizer (i.e., controls). It was reported that manure-applied soils contained a higher abundance of beta-lactam-resistant bacteria with bla_{CEP} (i.e., cephalothin) resistance genes. The increase in abundance for this gene was linked to the enrichment of beta-lactamase-harboring resident soil bacteria. A further identification showed an increase in abundance for *Pseudomonas* spp. and *Janthinobacterium* sp.; both known to harbor beta-lactamases. This suggests that increases in the abundance of ARGs and ARB after manure application can possibly be accounted for by the influx of contaminants associated with the manure. On the same note, a qPCR-based assessment on tetracycline resistance genes and integrase genes found that manure application caused gene abundances to increase by sixfold (Hong et al. 2013). These genes remained above background levels for up to 16 months. Through 16S rRNA gene analysis, the study found that soil microbial communities collected before and after manure application did not change significantly, suggesting that the increase in the ARGs due to manure application possibly arose from the influx of these ARGs from the animal feces and not due to changes in microbial communities.

The impact on soils arising from wastewater irrigation seems to differ from manure application depending on the extent of wastewater treatment received. A recent study analyzed soils that were continuously irrigated with untreated wastewater for 100 years (Dalkmann et al. 2012). Compared to control soils, sulfonamide resistance gene copy numbers increased when normalized to either 16S rRNA genes or per gram of dry soil. The wastewater-irrigated soils were also noted to have increased total number of 16S rRNA gene copies, and this long-term increase in biomass correlated to the increase in absolute concentration of resistance genes in soils.

In contrast, negligible or insignificant detrimental impact was observed when wastewater was first treated prior to reuse. To exemplify, a comparison between treated-wastewater-irrigated and freshwater-irrigated soils found that ARB and ARGs levels were on the whole identical or sometimes even lower in treated-wastewater-irrigated soils (Negreanu et al. 2012). The findings indicate that the high numbers of ARB that entered the soil did not compete successfully against the resident soil bacteria, and hence were unable to survive in the soil environment. Another study irrigated soil microcosms with secondary-treated (i.e., wastewater that received treatment in a biological activated sludge process), chlorinated, or dechlorinated effluents in a single irrigation event, and did not observe any significant changes in the ARG levels compared to microcosms irrigated with deionized water (Fahrenfeld et al. 2013). However, there were elevated levels of sulfonamide resistance genes in soils upon repeated irrigation with secondary-treated wastewater but not with the chlorinated and dechlorinated effluents. A follow-up study monitored abundances of ARB and ARGs in vegetables grown in fields fertilized with digested biosolids or untreated municipal sewage sludge. When compared to inorganic fertilizer (Rahube et al. 2014), results did not show that either treatment had a significant impact on viable coliform ARB, except in one instance where sewage sludge application increased the occurrence frequency of ARB from 46.4 to 79.2%. However, the PCR approach detected gene targets in both treated soils and vegetables grown in them that were not present in inorganically fertilized soils.

7.4 Performance of Wastewater Treatment Plants

These combined reports emphasize the importance of treatments achieving sufficient microbial contaminant removal before wastewater is used to irrigate agricultural soils. A conventional municipal wastewater treatment plant (WWTP) comprises a primary clarifier that serves to provide sedimentation of settleable solid particulates from the raw wastewater (i.e., influent), followed by biological activated sludge process. Within the activated sludge process, microorganisms serve to biodegrade the organic matter, hence reducing the organic and nutrient load. The wastewater generated from the biological activated sludge tank is then channeled to a secondary clarifier to separate the supernatant from the settleable solid particulates. Effluent generated at this point is typically referred to as the secondary-treated wastewater. In most WWTPs, chlorination is performed on the secondary-treated wastewater to achieve an additional inactivation of microbial agents present. In most instances, secondary biological treatment processes can achieve satisfactory treatment with regard to fecal coliforms in wastewater, and generally are able to meet a discharge requirement that includes a permissible level of fecal coliforms in wastewater <1000 CFU/100 mL for restricted irrigation or <2.2 CFU/100 mL for unrestricted irrigation (Al-Jassim et al. 2015; Al-Jasser 2011).

However, secondary treatment processes do not necessarily address specific classes of pathogens and/or emerging microbial contaminants like ARB and ARGs, which are more difficult to remove than fecal coliforms. To illustrate, an investigation of the performance of a full-scale wastewater treatment plant assessed influent, secondary-treated, and chlorinated effluents using culture-based and molecular methods (Al-Jassim et al. 2015). Results found that abundance of regulated contaminants like coliforms and fecal coliforms was effectively reduced and met quality standards for restricted irrigation. However, removal rates of emerging contaminants were lower and that proportions of pathogenic genera and multi-drug-resistant bacteria increased over the treatment schematic. An assessment of the performance of a full-scale and a bench-scale membrane bioreactor for wastewater treatment found variable but never total removal of pathogens from the influent to the effluent, despite the use of microfiltration membranes (Harb and Hong 2016).

Given that a total removal of microbial agents is not likely to be achieved by most WWTPs, more care should be placed to ensure removal of emerging ARB and ARGs that do not constitute a part of the baseline ARGs and ARB in soils, as discussed earlier. The bla_{NDM} is an example of a gene that has thus far not been found to constitute part of the baseline ARGs in soils. It results in the production of New Delhi metallo-beta lactamase (NDM), an enzyme that confers resistance to a wide spectrum of beta-lactams, including carbapenems. A study addressing the occurrence of bla_{NDM} genes at different phases in two wastewater treatment facilities in northern China found that bla_{NDM} genes were detected in the influent, effluent, and chlorinated effluent, which in turn resulted in the discharge of significant levels of these genes to the environment (Luo et al. 2013). The findings from that study therefore indicate that bla_{NDM} is an emerging contaminant of special concern when the treated wastewater is to be reused for agricultural irrigation.

7.4.1 New Delhi Metallo-Beta-Lactamase as an Emerging Contaminant of Special Concern

Carbapenems are beta-lactam antibiotics that have been used to combat severe Gram-negative bacterial infections and represent a last line of defense treatment (Walther-Rasmussen and Høiby 2007). Hence, emergence and global spread of carbapenem resistance in bacteria that would render this last-resort treatment ineffective can be a cause of great concern to public health. Resistance is conferred through carbapenemases, a type of beta-lactamase enzymes categorized into Ambler classes B, A, C, and D (Bush 2010). Class B carbapenemases are metallo-beta-lactamases, MBLs, that use bound zinc atoms in the active site to help ionize and coordinate a nucleophilic hydroxide ion to mediate hydrolysis, while class A, C, and D carbapenemases are serine carbapenemases that use active site serine as a nucleophile (Bush 2010; Wang et al. 1999). The New Delhi Metallo-beta-lactamase is a broad-spectrum beta-lactamase that falls into Ambler class B and is a novel MBL that was identified in 2009 in a Swedish patient of Indian origin who traveled to New Delhi, India, and acquired a carbapenem-resistant *K. pneumoniae* infection (Yong et al. 2009). MBL enzymes exhibit tendency to have a broad-spectrum substrate profile. Biochemical characterization of protein structure of variant NDM-1 has shown that it has an expansive active site with a unique electrostatic profile that leads to accommodation of a wide variety of substrate molecules (King and Strynadka 2011). Furthermore, the protein also exhibits a molecular profile that allows for broad-spectrum antibiotic substrate binding and product release, hence conferring a bacterium with NDM its unique trait of exhibiting broad-spectrum antibiotic resistance (King and Strynadka 2011).

Since discovery, bla_{NDM}-positive infections have been reported in nosocomial environments in numerous countries (including the United States, Canada, the United Kingdom, Germany, Kenya, South Africa, Oman, Pakistan, Honk Kong, Japan, Australia, and more) in all continents except Antarctica (Table 7.1). Besides the NDM-positive Enterobacteriaceae as shown in Table 7.1, bla_{NDM} has also been detected in numerous, including virulent, bacterial species such as *Acinetobacter* spp., *Aeromonascaviae*, *Enterobactercloacae*, *Pseudomonas* spp., and *Vibriocholera* (Kumarasamy et al. 2010; Walsh et al. 2011), some of which are listed in Table 7.1. The encoding gene for bla_{NDM} was initially detected in a 180-kb plasmid for *K. pneumoniae* and a 140-kb plasmid for *E. coli*, of which both were easily transferable and at a high frequency to susceptible *E. coli* J53 (Yong et al. 2009). Since then, the gene has been found in plasmids of various sizes (~50–300 kb) that belonged to different incompatibility (Inc) groups (A/C, FI/FII) (Table 7.1). In 2011, a variant of NDM-1 (designated NDM-2) that differed by a single amino acid was reported (Kaase et al. 2011). In 2013, a review paper reported that a series of further variants (designated NDM-3 to NDM-7) have been reported on the Lahey Clinic beta-lactamase website (http://www.lahey.org/Studies/) (Johnson and Woodford 2013). A subsequent search on the same database showed reports of

Table 7.1 List of bla_{NDM}-carrying plasmids of various sizes, incompatibility groups, and co-resistance isolated in Enterobacteriaceae from different sources

Country of isolation	Source	Plasmid size (kb)	Inc group	NDM variant	Co-resistance[a]	Reference
Escherichia coli						
Australia	Clinical	50	Untypable	NDM-1	NR	Poirel et al. (2010b)
Canada	Clinical	75	Untypable	NDM-1	NR	Peirano et al. (2011a)
Canada	Clinical	129	A/C	NDM-1	bla_{CMY-6}	Mulvey et al. (2011)
Canada	Clinical	130	A/C	NDM-1	bla_{CMY-6}; *rmt*C	Borgia et al. (2012)
China	Clinical	50	Untypable	NDM-1	NR	Ho et al. (2012)
Denmark	Clinical	–	A/C	NDM-1	bla_{CMY-4}; *armA*	Nielsen et al. (2012)
France	Clinical	120	FIA	NDM-4	$bla_{CTX-M-15}$; bla_{OXA-1}; *aac*A4	Dortet (2012)
France	Clinical	150	A/C	NDM-1	bla_{OXA-10}; bla_{CMY-16}	Denis et al. (2012), Poirel et al. (2011c)
France	Clinical	110	F	NDM-1	bla_{OXA-1}; and markers for kanamycin; gentamicin; tobramycin; trimethoprim; and sulfonamide resistance (genes not specified)	Poirel et al. (2010a), Poirel et al. (2011c)
Hong Kong	Clinical	90	L/M	NDM-1	bla_{TEM-1}; bla_{DHA-1}; *aac*C2; *arm*A; *sul*1; *mel*; *mph*2	Ho et al. (2011)
India	Clinical	120	F	NDM-4	*arm*A; and resistance to all aminoglycosides	Nordmann et al. (2012), Poirel et al. (2011c)
India	Community-acquired	87	FII	NDM-1	bla_{OXA-1}; *aac*C2; *aac*C4; *aad*A2; *dfr*A12	Bonnin et al. (2012)

(continued)

Table 7.1 (continued)

Country of isolation	Source	Plasmid size (kb)	Inc group	NDM variant	Co-resistance[a]	Reference
India	Waste seepage	140	A/C	NDM-1	NR	Walsh et al. (2011)
India	Waste seepage	140	A/C	NDM-1	NR	Walsh et al. (2011)
India	Waste seepage	140	Untypable	NDM-1	NR	Walsh et al. (2011)
Japan	Clinical	196	A/C	NDM-1	*bla*$_{TEM-1}$; *bla*$_{CMY-4}$; *aad*A2; *arm*A; *sul*1; *mel*; *mph2*; *dfr*A12	Sekizuka et al. (2011)
New Zealand	Clinical	>100	Untypable	NDM-6	*rmt*C	Williamson et al. (2012)
New Zealand	Clinical	>100	Untypable	NDM-1	NR	Williamson et al. (2012)
New Zealand	Clinical	>100	Untypable	NDM-1	NR	Williamson et al. (2012)
Poland	Clinical	90	FII	NDM-1	*aacA4*; *aacC2*	Fiett et al. (2014)
Saudi Arabia	Wastewater	110	F	NDM-1	*rmt*C; *dhps*	Mantilla-Calderon et al. (2016)
Spain	Clinical	300	HII	NDM-1	*bla*$_{TEM-1}$; *bla*$_{CTX-M-15}$; *bla*$_{DHA-1}$; *arm*A	Sole et al. (2011)
Switzerland	Clinical	130	F	NDM-1	*bla*$_{TEM-1}$; *arm*A	Poirel et al. (2011g)
UK	Clinical	>100	F	NDM-5	*aad*A5; *dfr*A17; *rmt*B	Hornsey et al. (2011)
Klebsiella pneumoniae						
Australia	Clinical	70	Untypable	NDM-1	*bla*$_{CMY-6}$; *aac*-6′-1b; *rmt*C	Sidjabat et al. (2011)

Canada	Clinical	102	A/C	NDM-1	*bla*$_{CMY-6}$	Mulvey et al. (2011)
Canada	Clinical	120	FII	NDM-1	NR	Peirano et al. (2011b)
Canada	Clinical	150	A/C	NDM-1	*bla*$_{SHV-12}$; *arm*A	Tijet (2011), Peirano et al. (2011b)
Canada	Clinical	130	A/C	NDM-1	*bla*$_{CMY-6}$; *rmtC*	Borgia et al. (2012)
China	Clinical	50	Untypable	NDM-1	NR	Ho et al. (2012)
Croatia	Clinical	–	A/C	NDM-1	*bla*$_{CTX-M-15}$; *bla*$_{CMY-16}$; *qnr*A6	Mazzariol et al. (2012)
France	NR	150	Untypable	NDM-1	*rmt*C	Poirel et al. (2011c)
France	Clinical	270; 300	Untypable	NDM-1	*bla*$_{CTX-M-15}$; *bla*$_{OXA-1}$; *aac*(6′)-Ib-like; *arm*A; *qnr*B1	Arpin et al. (2012)
France	Clinical	100	Untypable	NDM-1	NR	Poirel et al. (2011c, d)
Guatemala	Clinical	–	Untypable	NDM-1	*bla*$_{SHV-12}$	Pasteran et al. (2012)
India	Clinical	160	A/C	NDM-1	NR	Kumarasamy and Kalyanasundaram (2012)
India	Clinical	180	Untypable	NDM-1	*arr*-2; *ere*C; *aad*A1; *cml*A7	Yong et al. (2009)
India	Waste seepage	140	Untypable	NDM-1	NR	Walsh et al. (2011)
Kenya	Clinical	120	A/C$_2$	NDM-1	*rmt*C	Poirel et al. (2011e)
Mauritius	Clinical	120	A/C	NDM-1	*bla*$_{CMY-6}$; *rmt*C	Poirel et al. (2012a)

(continued)

Table 7.1 (continued)

Country of isolation	Source	Plasmid size (kb)	Inc group	NDM variant	Co-resistance[a]	Reference
Morocco	Clinical	250	Untypable	NDM-1	$bla_{CTX\text{-}M\text{-}15}$; $bla_{OXA\text{-}1}$	Poirel et al. (2011b)
The Netherlands	Clinical	70	II	NDM-1	NR	Halaby et al. (2012)
New Zealand	Clinical	>100	Untypable	NDM-1	NR	Williamson et al. (2012)
Oman	Clinical	170	L/M	NDM-1	*arm*A	Poirel et al. (2011a)
Oman	Clinical	170	Untypable	NDM-1	*arm*A	Poirel et al. (2011a)
South Korea	Clinical	50; 60; 70; 100	N	NDM-1	NR	Kim et al. (2012)
Spain	Clinical	120	FIB	NDM-1	NR	Oteo et al. (2012)
Switzerland	Clinical	150	A/C	NDM-1	*rmt*A	Poirel et al. (2011g)
Switzerland	Clinical	150	A/C	NDM-1	$bla_{OXA\text{-}10}$; $bla_{CMY\text{-}16}$; *qnr*A6	Poirel et al. (2011g)
Turkey	Clinical	80	FIB	NDM-1	*rmt*B	Poirel et al. (2012b)
Klebsiella oxytoca						
Taiwan	Clinical	–	Untypable	NDM-1	*arm*A; *aac*C2	Lai et al. (2011)
Citrobacter freundii						
France	Clinical	65	Untypable	NDM-1	NR	Poirel et al. (2011f)

India	Waste seepage	140	A/C	NDM-1	NR	Walsh et al. (2011)
Proteus mirabilis						
Switzerland	Clinical	150	A/C	NDM-1	*bla*$_{OXA-10}$; *bla*$_{CMY-16}$; *arm*A	Poirel et al. (2011g)
Providencia stuartii						
Afghanistan	Clinical	178	A/C	NDM-1	*bla*$_{OXA-10}$; *arm*A; *sul*1; *qnr*A1; *aac*(6′); *cml*A7	McGann et al. (2012)
Shigella boydii						
India	Waste seepage	250	Untypable	NDM-1	NR	Walsh et al. (2011)

NR not reported
[a]Co-resistance indicates antibiotic resistance determinants reported in the same plasmid as the *bla*$_{NDM}$ gene

NDM variants that include all the way to NDM-16, suggesting a rapid variation of NDM.

In addition to the wide substrate range and the rapid variation of NDM, the problem arising from NDM is further aggravated by a number of complications. These complications include a lack of standard routine phenotypic tests for MBL detection (Miriagou et al. 2010). A commonly used approach as of now is the use of EDTA as a chelator of zinc to detect loss of MBL activity. Consequently, high prevalence of unrecognized asymptomatic carriers is probable, which would lead to an underestimation of the global dissemination of NDM-harboring bacteria. Given that the bla_{NDM} gene is often encoded in plasmids that are of various types of incompatibility classes, this indicates to the possibility of horizontal gene transfer among many different types of Gram-negative bacteria. Moreover, the scarcity of available effective antibiotics poses challenges to treatment, hence indicating a higher risk of morbidity or mortality for patients who are infected by NDM-positive pathogens.

7.4.2 New Delhi Metallo-Beta-Lactamase in Wastewater

The presence of bla_{NDM}-positive isolates is not restricted to only nosocomial environment. Instead, bla_{NDM}-positive isolates have also been isolated from non-nosocomial environments. To illustrate, a *K. pneumoniae* carrying the $bla_{\mathrm{NDM\text{-}1}}$ gene was isolated from river water in Hanoi, Vietnam (Isozumi et al. 2012). Various reports also implicate wastewaters as reservoirs for bacterial isolates carrying bla_{NDM} genes. Bacterial species carrying NDM have been isolated from waste seepage and tap water sampled from New Delhi, India, city center and surrounding areas (Walsh et al. 2011), and from untreated wastewater in Jeddah, Saudi Arabia (Mantilla-Calderon et al. 2016).

In one of these studies, in-depth genomic characterization of the bla_{NDM}-positive *E. coli* that was isolated from wastewater influent showed that this bacterium possessed a mosaic of traits representative of different *E. coli* pathotypes (Mantilla-Calderon et al. 2016). Furthermore, the isolate was demonstrated to internalize into mammalian cells, and has a genome encoding for various virulence traits. The non-chromosomal genome of this bacterium also includes at least one plasmid that encodes for the bla_{NDM} gene, suggesting possible exchange of carbapenemase genes between this isolate with other competent recipients. Besides the presence of viable NDM-positive bacteria in wastewater, $bla_{\mathrm{NDM\text{-}1}}$ genes were also shown to be present at significant numbers in municipal wastewaters, which include wastewater discharged from hospitals. Untreated hospital wastewater from two hospitals in Singapore contained 2.29×10^6 gene copies/mL of bla_{NDM} and 4.08×10^7, 1.25×10^6, and 6.19×10^5 gene copies/mL of genes bla_{KPC}, $bla_{\mathrm{CTX\text{-}M}}$, and bla_{SHV}, respectively (Le et al. 2016).Another study that monitored $bla_{\mathrm{NDM\text{-}1}}$ numbers in raw wastewater entering a WWTP in Saudi Arabia reported $3.4 \times 10^4 \pm 2.3 \times 10^4$ copies/m^3 (Mantilla-Calderon et al. 2016). Similarly, significant copy numbers of *bla*$_{NDM-1}$ persisted through several treatment units (including disinfection by chlorination) in two WWTPs in northern China (Luo et al. 2013). Levels present in the effluent discharged from both WWTPs were from $1.3 \times 10^3 \pm 2.3 \times 10^2$ to $1.4 \times 10^3 \pm 2.5 \times 10^2$ copies/mL, representing a range of 4.4–93.2%, respectively, of influent levels.

Collectively, the presence of viable NDM-positive bacteria and the ubiquitous detection of *bla*$_{NDM}$ genes reiterate causes for concern. This is especially in cases of wastewater release into the environment or application of these waters onto soils in agricultural settings.

7.5 Fate and Persistence of ARB and ARGs

Although introduction of ARB and ARGs, particularly those that encode *bla*$_{NDM}$ genes, into soils via wastewater application might carry various risks, the full extent of the potential risks would need to be further elucidated by understanding the fate and persistence of these bacteria in the environment. Upon dissemination into the soil environment, ARGs can be adsorbed or degraded or taken up by competent cells. Similarly, ARB can be adsorbed onto particulates or inactivated or internalized into other hosts. Hence, not all of the ARB and ARGs contributed by the wastewater into the soil matrix would remain available to impose potential public health risks. Conversely, if ARGs or ARB continue to persist or multiply in their copy numbers within the soil environment, the risks would be potentially exponentially amplified. The following subsection aims to elaborate on these various scenarios.

7.5.1 Horizontal Gene Transfer

Horizontal gene transfer, HGT, is a mechanism for exchange of genetic material that can occur via transformation (i.e., uptake of naked DNA by bacteria) or conjugation (i.e., transfer mediated by cell-to-cell junctions and a pore through which DNA can pass) (Thomas and Nielsen 2005). It is now widely recognized that HGT is a major mechanism of bacterial adaptation to clinical antibiotic concentrations. This is even more evident when considering that the most potent ARGs in pathogens are often encoded on mobile genetic elements (Nesme and Simonet 2015; Stokes and Gillings 2011; Schlüter et al. 2007; Djordjevic et al. 2013).In the case of *bla*$_{NDM}$ genes, they are often found on plasmids belonging to different incompatibility groups that have a broad host range and can be replicated in different bacterial lineages (Table 7.1). The *bla*$_{NDM}$ genes are also often found on conjugative plasmids that possess all the genes required for their autonomous transfer (Nesme and Simonet 2015; Carattoli et al. 2012). Further highlighting the risk associated with HGT, many human pathogenic bacteria including representatives of the genera *Campylobacter*, *Haemophilus*, *Helicobacter*, *Neisseria*, *Pseudomonas*, *Staphylococcus*, and *Streptococcus* are naturally transformable (Lorenz and Wackernagel 1994). Soil environments present a large genetic diversity at small spatial scale, and ample opportunities for cell-to-cell contacts, cellular movement, or activity. Soil matrices are therefore considered to be hotspots conducive for the exchange of genetic materials through HGT. A review by Elsas and Bailey names the plant rhizosphere and plant tissue, phyllosphere, manured soil, guts of soil animals, aquatic sediments, sewage, and sludge environments as some of the most prominent hotspots (Van Elsas and Bailey 2002). Collectively, these environmental compartments may contribute to ARG dissemination between bacteria and eventually acquisition by pathogens (Nesme and Simonet 2015).

Natural transformation of naked DNA is dependent upon exposure of bacteria to extracellular DNA molecules in the environment. DNA can enter the environment through release from decomposing cells, disrupted cells, and virus particles, or excreted from living cells (Thomas and Nielsen 2005). This extracellular DNA may (1) persist by binding to soil minerals and humic substances, (2) be degraded by microbial DNases and used as a nutrient for plant and microbial growth, or (3) be incorporated into a bacterial genome as a possible source of genetic instructions (Levy-Booth et al. 2007). Extracellular DNA ranges at approximately 0.03–1 μg/g of material in soil and sediments (Ogram et al. 1987; Selenska and Klingmüller 1992), and in approximately 0.03–88 μg of dissolved DNA per liter of fresh and marine water (DeFlaun and Paul 1989; Karl and Bailiff 1989). Work estimating extracellular DNA in activated sludge found 4–52 mg/g of volatile suspended solids (VSS) in sludge collected from different wastewater treatment plants (Dominiak et al. 2011).

In the environment, various factors can affect transformation and success rate of recombination for this available extracellular DNA. One is that DNA adsorption to soil matrix is influenced by soil characteristics such as concentration of humic substances, soil mineralogy, cation concentration, and soil pH (Levy-Booth et al. 2007). Work by Nielsen et al. found that cell lysates persisted for up to 4 days after incubation in sterile soil and remained accessible for uptake by competent *Acinetobacter* sp. during this period. However, transformation activity was limited to 4–8 h in nonsterile soil because of DNA degradation, loss of DNA stability with temperature, and because DNA no longer maintained by cellular repair mechanisms decays faster (Nielsen et al. 1997a, b, 2000). Nielsen et al. provides a more detailed review of factors affecting stability of extracellular DNA (Nielsen et al. 2007), and readers of this chapter are encouraged to refer to that review paper for more details.

DNA degradation can also take place, resulting in fragmentation of long DNA to shorter sizes of approximately 400 bp. GC content affect DNA degradation kinetics. For example, DNA from high-GC-content Gram-positive *Actinobacteria* was found to persist longer in frozen soil than DNA from low-GC-content Gram-positive *Clostridiaceae* (Hofreiter et al. 2001). Although it has been acknowledged that long fragments may recombine more effectively compared to short linear DNA fragments of a few bp to less than 200 bp, such estimates may not be entirely accurate since recombination events resulting in nucleotide changes of only a few bp can be difficult to distinguish from genetic changes arising from sequential mutations. In this manner, HGT can hence be easily overlooked (Feil and Spratt 2001; Ikeda et al. 2004). Additionally, integration of foreign DNA into genome is influenced by a number of factors including competent cells availability and sequence homology between genomic DNA and foreign strand. This is particularly so if one were to consider that recombination typically occurs between chromosomal DNA and sequence that is less than 25% divergent (Matic et al. 1997; Vulić et al. 1997; de Vries et al. 2001; Majewski and Cohan 1998; Majewski et al. 2000).

Upon natural transformation, DNA may be integrated into the host's chromosome. Foreign DNA in the cytoplasm that is not integrated is degraded quickly by nucleases and enters the internal DNA metabolism cycle since the salvaged nucleoside can be used for the synthesis of nucleotide at a lower ATP cost. Different bacteria have different rates of DNA internalization and success of integration. Under in vitro conditions, DNA uptake occurred at rapid speeds of 100 bp/s and 60 bp/s in *S. pneumoniae* and *A. baylyi* competent bacteria, respectively (Palmen and Hellingwerf 1997; Méjean and Claverys 1993). Successful recombination of internalized DNA, also under optimal in vitro conditions, has been reported at 0.1% of internalized DNA in *A. baylyi* and up to 25–50% of internalized DNA in *B. subtilis* and *S. pneumoniae* (Palmen and Hellingwerf 1997).

Besides natural transformation, conjugative transfer is a process more specifically linked to plasmid acquisition. Plasmids are autonomously replicating genetic elements that can remove the need for a foreign gene to integrate into the recipient chromosome to become established (Thomas and Nielsen 2005). Plasmid conjugation depends on the hosts, and thus the fate of conjugative plasmids depends on host fitness, efficiency of transfer to new hosts, and selective advantages and disadvantages conferred by the plasmids (Van Elsas et al. 2000; Fernandez-Astorga et al. 1992). Different plasmids also have different host ranges, with some exhibiting broader host range (e.g., IncA/C_2, IncL/M, IncN, IncP, IncQ, and IncW incompatibility group plasmids (Novais et al. 2007; Götz et al. 1996)) while others exhibit narrower host range(e.g., IncF, IncH, IncI and IncX (Novais et al. 2007; Suzuki et al. 2010)).

Abiotic factors also affect conjugative plasmid transfer and have been extensively reviewed by Van Elsas et al. (Van Elsas et al. 2000; Van Elsas and Bailey 2002). As examples, extreme pH and temperature values are detrimental to cells, while the presence of nutrients in wastewater and soil might enhance bacterial donor's activity. However, a study assessed conjugation in *E. coli* strains and found that conjugative plasmid transfer can take place within a wide range of conditions (Fernandez-Astorga et al. 1992). Conjugation was not affected in a wide range of pH (6–8.5), low nutrient levels (down to 1 mg of carbon per liter), and low temperatures (8–15°C).

7.5.2 HGT on Plant Surfaces

Plant-associated bacteria have been observed frequently to form assemblages or biofilms. Biofilm formation can be due to passive processes like accumulation of bacterial cells as water moves along plant surfaces, or due to active bacterial attachment and production of exopolymeric substances (Morris and Monier 2003). An example is the genus *Pseudomonas*, which are ubiquitous in the terrestrial ecosystems, and are frequently found in association with plants (Espinosa-Urgel 2004). They aggregate at high cell densities, forming biofilms that are conducive for horizontal gene transfer and plasmid conjugation. Plant components like rhizosphere and phylloplane are hotspots for bacterial metabolic activity and HGT, as are other biofilm-supporting environments, with transconjugant to donor ratios (T/D) as high as 10^{-3} or even 10^{-1} for indigenous or foreign plasmids (Van Elsas and Bailey 2002; Lilley et al. 1994). This is in contrast to bulk environments such as bulk water and bulk soil where plasmid transfer efficacy is lower (T/D $< 10^{-5}$) and usually requires nutrient enrichment (Sørensen and Jensen 1998). In other plant components, like the phytosphere, elevated transfer frequencies have generally been attributed to plant exudates stimulating bacterial metabolic activity (Sørensen and Jensen 1998; Lilley et al. 1994). These observations suggest a likelihood that ARGs, if present in the wastewater that is to be used for agricultural irrigation, can be horizontally transferred to other bacterium attached on plant surfaces, as well as in the soil matrix.

To illustrate more specifically the potential risk from ARG presence in wastewater, the following analysis is presented. Past study has shown that the transformation frequency of antibiotic resistance in native populations of nonsterile sediments were approximately 3×10^{-9} when 10 μg of DNA were added to 1 cm^3 of sediment (ref). Given the abundance of $bla_{\mathrm{NDM-1}}$ gene in WWTP discharge reported in China (approximately 1374 copies/mL) and that $bla_{\mathrm{NDM-1}}$ genes have been located on conjugative plasmids of size 126 kb (assumed average size based on *E. coli* and *K. pneumoniae* plasmids reported in Table 7.1), this would equate to an approximate amount of 0.19 pg/mL of extracellular plasmid DNA (based on Avogadro's constant of 6.022×10^{23} g/molecule). Thus, this would mean that up to 52 $m^3$of treated wastewater would need to be irrigated in order to account for the required amount of DNA to cause a transformation frequency of 3×10^{-9} per cm^3 of sediment.

Hence, it is unlikely that a single event of reusing the treated wastewater would lead to any substantial concerns in terms of horizontal gene transfer. Even if the same plot of land is to be continuously exposed to treated wastewater, and if the ARG continues to persist indefinitely and accumulate in the soils, this would equate to approximately less than one horizontal gene transfer event per cm^3 of soil when conditions are favorable for transformation. This estimate is made on the basis of an estimated cell number of 10^8 cells per gram of soil (Raynaud and Nunan 2014). It is however to be pointed out that these calculated transformation events may not provide accurate estimate of the actual events as the assumed transformation rates did not take into account the variation in natural competence of different bacterial cells. Furthermore, it is likely that the transformation frequency may vary with different physicochemical factors like concentrations of ions, temperature, pH, and natural organic matter content (Lorenz and Wackernagel 1994). Regardless, this estimation suggests that the contribution of ARGs by wastewater to agricultural soil may not be imposing that much of a concern although it cannot be concluded whether the wide multitude of different types of ARGs in the same wastewater would collectively result in a concern or not.

7.5.3 Internalization of Pathogens

Besides HGT of ARGs, pathogenic ARB that may be present in the treated wastewater should also be assessed for their likelihood of internalizing into plants. There are various routes by which bacteria can enter plant tissues. Entry can occur through natural openings in the plant surface (stomata, lenticels, sites of lateral root emergence, etc.) and/or through sites of biological or physical damage (Brandl 2008; Itoh et al. 1998; Kroupitski et al. 2009). Following closure of the guard cells, internalized bacteria can be protected from various sanitizers (Gomes et al. 2009). Bacteria may also be passively carried into the plant tissue with water (e.g., water used to soak seeds, irrigate plants, or to wash produce crops following harvest) (Deering et al. 2012). Bacteria can also be recovered from above-ground portions of the plant following exposure of the roots to water containing the pathogen, indicating that the bacteria can be taken up through the roots and move within the plant (Deering et al. 2012).

Bacteria of concern may actively infect and colonize plant tissues. It has been shown that certain plant pathogenic bacteria like *Pantoea agglomerans* and endophyte *K. pneumoniae* have been associated with opportunistic infections in animals, including humans (Holden et al. 2009). To illustrate further, recent studies on the plant pathogen *Pectobacterium atrosepticum* and the plant-associated *Klebsiella pneumoniae* were shown to share a remarkably high proportion of their genome with the human-pathogenic *K. pneumoniae* (Bell et al. 2004; Fouts et al. 2008; Toth et al. 2006; Holden et al. 2009). Many plant and animal pathogens share a common molecular mechanism, namely the Type III secretion system (TTSS), for attacking their host (Rahme et al. 1995; Staskawicz et al. 2001; Deering et al. 2012). Alternatively, wounding or destruction of living tissue in plants can be mediated by plant pathogen first, which in turn creates a microenvironment that is favorable for the survival and/or replication of human pathogens in the plant tissues.

Many studies have shown that both *Salmonella* spp. and *E. coli* O157:H7 can internalize within a variety of plant tissue types and that there are numerous factors that can influence the extent of internalization (Deering et al. 2012). Lettuce plants that were grown in manure amended with fluorescently marked *E. coli* O157:H7 were shown to harbor bacteria that had internalized into the plant tissue, including the edible parts of it (Holden et al. 2009; Solomon et al. 2002). Long-term persistence of *E. coli* O157:H7 in fresh produce has been demonstrated with carrots and onions grown in artificially contaminated manure compost (Islam et al. 2004). This study showed that the bacteria could be detected from carrots for up to 12 weeks after initial application and in onions for up to 9 weeks. A similar study showed that *S. enterica* could be detected in tomato plants harvested 7 weeks after the seeds were sown in soil artificially contaminated with the bacteria (Barak and Liang 2008).

7.6 Intervention Strategies Needed

Given that the presence of ARGs and ARB in wastewater can be a potential cause of concern during long-term reuse events, this section aims to discuss several natural or low-cost intervention strategies to reduce ARB and ARGs presence in wastewaters. The known effect and limitations of sunlight radiation are discussed, and the idea of water augmentation by bacteriophage therapy to improve ARB reduction is visited.

7.6.1 Solar Inactivation

The biocidal effect of sunlight is attributed to the UV portion of its irradiance (wavelength ranges of UV are: 400–315 nm for UV-A; 315–280 nm for UV-B; and 280–100 nm for UV-C (McGuigan et al. 2012)) that can result in photo-degradation by direct or indirect mechanisms. In direct photoinactivation, components like microbial genome and proteins absorb shorter wavelengths of sunlight radiation, and subsequently degrade (Boehm et al. 2009). Studies have shown that UV-irradiation on growing *Escherichia coli* cultures results in DNA lesions where some of that light is absorbed by the pyrimidine rings of thymine and cytosine bases in the DNA. This leads to the formation of new bonds between adjacent pyrimidine bases, forming pyrimidine dimers (pairs connected by covalent bonds) (Goodsell 2001; McGuigan et al. 2012). These dimers prevent base-pairing with the complementary purines on the other strand of DNA, which changes the shape of the DNA molecule, in turn making it difficult for polymerases to move through the region of the dimer. The end result is a transient block on the essential processes of replication and transcription (Courcelle et al. 2001). In indirect photoinactivation, endogenous (e.g., porphyrins, flavins, quinones, NADH/NADPH, and others (Eisenstark 1987; Jagger 1981; Lloyd et al. 1990; McGuigan et al. 2012; Webb and Brown 1979)) or exogenous molecules (e.g., humic substances and photosynthetic pigments like chlorophyll (Blough and Zepp 1995; McGuigan et al. 2012; Schwartz et al. 2003; Curtis et al. 1992) may absorb UV light and subsequently damage other cellular material through generation of reactive oxygen species(ROS, examples include singlet oxygen, hydroxyl radicals, or alkyl peroxyl radicals) (Pattison and Davies 2006; Santos et al. 2012).

The efficacy of solar photoinactivation on pathogenic waterborne bacteria and pathogen indicators has been variable in study reports. For instance, many studies found rapid inactivation of fecal indicator organisms within a few hours of exposure to natural sunlight, and it was reported that all of the classically defined waterborne pathogenic bacteria were readily amenable to 6 h of solar disinfection under suitable field conditions (Boyle et al. 2008; McGuigan et al. 1998; Ubomba-Jaswa et al. 2009; Wegelin et al. 1994; McGuigan et al. 2012). But on the other hand, numerous studies also reported fecal coliforms showing much slower inactivation rates, and that some indicator bacteria remained detectable after a full day of sunlight exposure (Oates et al. 2003; Rijal and Fujioka 2003; Fisher et al. 2008; Fisher et al. 2012; Sinton et al. 2002; Sommer et al. 1997). A study examined the use of effluent from a municipal WWTP, without and after solar disinfection, prior to its use as irrigation water for cultivated lettuce crops (Bichai et al. 2012). The effluent was from secondary treatment, i.e., after receiving a standard biological treatment (activated sludge) followed by sedimentation in settling ponds. Results of inactivation assays showed that solar disinfection processes can reduce bacterial concentrations from $>10^3$ to 10^4 *E. coli* CFU/mL in real WWTP effluent to <2 CFU/mL. Out of the 16 lettuce samples irrigated with untreated WWTP effluent (i.e., not treated with solar irradiation), 14 samples were contaminated and positive for the presence of *E. coli* 24 h after irrigation. On the other hand, out of 28 lettuce samples irrigated with solar-disinfected WWTP effluent, only two samples were positive, confirming improved safety of irrigation practices due to solar treatment. Positive presence of *E. coli* for one of the two lettuce samples was tracked back to a highly contaminated WWTP effluent with an initial *E. coli* concentration of 1.3×10^4 CFU/mL vs. $2.4–3.8 \times 10^3$ CFU/mL in all other wastewater samples. The other positive sample is speculated to have had incomplete inactivation and/or microbial regrowth, as the wastewater effluent still contained organic carbon and nutrients that can be assimilated to allow for bacterial survival and replication during dark storage. Another study examined the effect of solar disinfection on two antibiotic-resistant *E. coli* isolates from a WWTP effluent (Rizzo et al. 2012). The inactivation rate observed during solar radiation test for both *E. coli* strains investigated, namely 60 and 40% removal after 180 min of irradiation, was quite low compared to previous works on similar inactivation of *E. coli* in confined systems (Malato et al. 2009; Dunlop et al. 2011). The differences might be explained by variation in experimental design, but more importantly, the antibiotic-resistant *E. coli* strains may have characteristics that affected their resistance to photoinactivation, resulting in a lower inactivation rate.

In regard to the effect of solar disinfection on ARGs, information on their inactivation kinetics upon exposure to solar irradiance is lacking, and this knowledge gap requires more in-depth and systematic future studies. Most available literature explores the use of UV-disinfection to reduce ARG loads within WWTPs. For example, one of the early studies was performed by McKinney and Pruden (2012) to explore the use of UV to dimerize ARGs, with the intention of first inactivating these genes prior to their discharge. Their findings revealed that this would require UV doses that are at least 1 order of magnitude higher than those required for inactivation of the associated host bacterial cells. Generally, about 200–400 mJ/cm^2 of UV dosage is required to result in 3–4 log removal of an ARG. This UV dosage is slightly higher than the highest recommended UV dose of 186 mJ/cm^2 to achieve 4-log removal and/or inactivation of viruses (USEPA 2006). The study also found that certain ARGs like *tetA* and *ampC* were significantly harder to inactivate than *mecA* and *vanA*. To illustrate, a UV dose of 186 mJ/cm^2 would only achieve an inactivation of 1–2 log for *tetA* and *ampC*, while the same dose would have inactivated *mecA* and *vanA* by 3–4 log.

The inefficacy in reducing ARGs by UV is repeatedly shown in other studies. To illustrate, an independent study assessed ARG removal by UV (with UV transmittance of 45%, total power of 900 kW, and light intensity >100 mJ/cm^2) in a WWTP using advanced treatment systems, and found no apparent decrease in *tetM*, *tetO*, *tetQ*, *tetW*, *sul1*, *sul2*, and *intI1* genes in total extracted DNA from treated waste samples (Chen and Zhang 2013). In another study that used UV fluence of up to 249.5 mJ/cm^2 on secondary-treated municipal wastewater effluent samples, only 0.58-log removal of *tetX* gene was observed, with a less effective removal (at 0.36–0.4-log) of *sul1*, *tetG*, and *intI1* genes (Zhang et al. 2015). Yet another study that assessed UV/H_2O_2 advanced oxidation processes for disinfection of sterile water spiked with bla_{TEM} gene-carrying *E. coli* found that the treatment could inactivate the tested antibiotic-resistant *E. coli* strain, but did not significantly change the copy number per mL of bla_{TEM} gene (Ferro et al. 2017). All these studies were in contrast with a study that assessed the effect of UV on secondary-treated municipal wastewater effluent samples and surprisingly found 3- and 1.9-log reduction of erythromycin and tetracycline resistance genes, respectively, by the time 5 mJ/cm^{2}fluence is reached (Guo et al. 2013). Another study found that UV/H_2O_2 disinfection achieved a reduction of 2.8–3.5 logs in copy numbers of *sul1*, *tetX*, and *tetG* from secondary-treated municipal wastewater effluent (Zhang et al. 2016). Little explanation was offered for these discrepancies in literature, and comparisons are made harder due to the differences in experimental designs, UV sources, tested water matrices, and ARB/ARGs.

In application, the efficacy of solar photoinactivation can be subjected to various factors. These factors include atmospheric conditions like water vapor, CO_2, ozone, and oxygen, in addition to pollutants in the atmosphere, which can scatter and absorb various portions of the light (McGuigan et al. 2012). Water quality parameters can have a big influence on efficacy of solar disinfection in water bodies with turbidity being one of the important factors, and dissolved solids such as iron can absorb UV light and decrease the UV transmittance (Jones et al. 2014). Exogenous photosensitizers naturally present in surface waters include humic acids and chlorophyll, both of which can absorb sunlight and then react with oxygen to produce ROS (Blough and Zepp 1995; McGuigan et al. 2012; Schwartz et al. 2003). Organic and inorganic matter present in water bodies can cause bacterial growth instead of inactivation (McGuigan et al. 2012) or may generate ROS upon sunlight irradiation (Corin et al. 1996; Rizzo et al. 2012). Water salinity and alkalinity also play a role in the end efficacy of solar irradiation. The presence of ions may help to retain bacterial integrity, and if ions are present in high concentrations, they could have a limiting effect on photoinactivation. To illustrate, UV-A mediates its biological effects on bacteria by ROS like hydrogen peroxide and hydroxyl radicals. If bicarbonates HCO_3^- are present in water, they react with hydroxyl radicals producing $CO_3^{\bullet -}$, which has a slower reaction with organic molecules when compared to $^{\bullet}O$ (Canonica et al. 2005; McGuigan et al. 2012). Also, HCO_3^- induces photo-absorption, which limits the amount of light reaching bacteria in water. Other anions such as phosphates, chloride, and sulfates are shown to be absorbed by bacteria but do not illicit a direct effect on solar inactivation unless in the presence of a photo-catalyst such as titanium dioxide (McGuigan et al. 2012).

The total irradiance dose received by the bacteria influence the extent of photoinactivation damage, and there is evidence that the rate at which that dose is delivered is an equally important factor. Additionally, different portions of UV light also have different effects. UV-A radiation wavelengths bordering on visible light are not sufficiently energetic to directly modify DNA bases, but are able to induce cellular membrane damage through the production of reactive oxygen species (Khaengraeng and Reed 2005; Rizzo et al. 2012; McGuigan et al. 2012). UV-B and UV-C are the more germicidal portions of UV light and represent the most genotoxic wavebands of solar radiation reaching the Earth's surface, causing direct DNA damage by inducing the formation of DNA photoproducts (Pfeifer 1997; Rizzo et al. 2012) as well as indirectly through photosensitization processes (Bolton et al. 2010; Muela et al. 2002; Santos et al. 2012).

Lastly, bacteria may differ in their response to solar irradiance and their capacity to combat its effects. For example, bacteria with larger genome sizes were observed to be more susceptible to UV damage, presumably because larger genomes offered more sites for UV damage (McKinney and Pruden 2012).

7.6.2 Bacteriophages

A potential strategy that might be augmented into existing systems to alleviate ARB and ARG load is the use of bacteriophages as control agents. Bacteriophages are viruses that infect and lyse bacteria and are categorized into virulent (or lytic) and temperate (lysogenic) bacteriophages (Withey et al. 2005). The two categories of viruses differ in their life cycles. During lytic infection, virulent phages inject their nucleic acid into the host cell after attachment. Expression of the phage genome directs the cellular machinery of the host to synthesize new phage capsule material. The resulting phage progeny are released by fatal cell lysis, enabling the lytic cycle to continue as new cells are infected. In contrast, during lysogenic infection, temperate phages' nucleic acid recombines with the host cell genome forming a dormant endogenous phage (known as a prophage). The prophage is reproduced in the host cell line and confers immunity from infection and remains dormant until host conditions deteriorate, perhaps due to depletion of nutrients. Subsequently, the prophages become active. At this point, they initiate the reproductive cycle, resulting in lysis of the host cell (Mason et al. 2011). Bacteriophages, or phages for short, have several characteristics that make them attractive options as therapeutic agents or agents of biocontrol (Jassim and Limoges 2014). Such characteristics include their effectiveness in killing their target bacteria (i.e., host specificity), adaptability, natural residence in the environment and the fact that they are self-replicating and self-limiting (Jassim and Limoges 2014; Sulakvelidze et al. 2001; Jassim et al. 2016).

Bacteriophages can be isolated from the environment. However, bacteriophage isolation is a time-consuming process, and it may be difficult to isolate the desired bacteriophages that demonstrate the right host specificity. Furthermore, bacteria may become desensitized to the isolated phages after long-term exposure and would require repetition of the entire isolation process.

Alternatively, new synthetic phages can be programmed and used. Synthetic phages offer a powerful advantage in their potential to specifically target certain ARB of concern, functioning to sensitize bacteria to antibiotics and selectively killing ARB. A proof-of-concept study utilized temperate phages to deliver a functional CRISPR-associated (Cas) system, otherwise known as interspaced short palindromic repeats that are clustered regularly, into the genome of ARB (Yosef et al. 2015). The delivered CRISPR-Cas system destroyed antibiotic resistance-conferring plasmids via sequence-targeting DNA cleavage. In addition, the CRISPR-Cas system genetically modified lytic phages to kill only antibiotic-resistant bacteria while protecting antibiotic-sensitized bacteria. This linkage between antibiotic sensitization and protection from lytic phages was a key feature of the tested strategy.

Currently, phage treatment is utilized in a number of ways. Most notably, phage therapy is applied in medical settings for the treatment of ARB infections, in veterinary settings for the treatment and prevention of infections in animals, as well as for the treatment of plants (Balogh et al. 2010; Jones et al. 2012). Other examples demonstrating the use of bacteriophages on larger scale include phage application for treatment and preservation of foods, phage treatment of aquaculture and fish, and in wastewater treatment (Araki 1986; Withey et al. 2005; Brockhurst et al. 2006; Goldman et al. 2009). When applied to wastewater treatment processes, phages have been proposed as an eco-friendly tool to control the abundance of filamentous bacteria, which pose bulking and foaming problems in wastewater activated sludge process (ASP) systems (Withey et al. 2005; Khan et al. 2002a, b; Thomas et al. 2002; Weinbauer 2004; Petrovski et al. 2011a, b; Khairnar et al. 2014; Pal et al. 2014). During ASP for sewage treatment, sludge settles in tanks and the supernatant is drained off for further purification. This process is detrimentally affected by filamentous microbes, which because of their filamentous morphologies, have high surface area and low density, hence impeding settleability of biomass (Withey et al. 2005). *Sphaerotilus natans*, a filamentous bacteria, was targeted by specific phages isolated from sewage (Choi et al. 2011). Phage application was observed to reduce sludge volume and produced clearer supernatant after 12 h. In addition, the phages remained stable and active for over 9 months and tolerated temperature and pH fluctuations common to activated sludge processes (Jassim et al. 2016).

Regardless of the potentially promising results, the application of phages in wastewater treatment systems for control of ARB and ARGs is still in need of systematic and in-depth experimentation. Some of the challenges and obstacles in utilizing phage in wastewater treatment and/or application in a field scale are: (1) high concentrations of phages must be used for a successful application; (2) use of polyvalent phages with broader host range could lead to the degradation of useful bacterial populations (e.g., nitrifying populations, phosphate accumulating bacteria, etc.); (3) specific phages must be identified by WWTP operators to target specific undesired bacterial populations; (4) microbial analysis of the system is a prerequisite to phage application as the bacterial population may vary between wastewater treatment plants (Jassim et al. 2016). As of now, there are no studies demonstrating successful phage application directly into the agricultural soil and/or in combination with other intervention strategies. Therefore, it is likely that some other unique factors such as pH, temperature, multiplicity of infection (MOI, ratio of phage to bacterial particles), decay rates under solar irradiation, and so on could affect efficacies of bacteriophages in agricultural settings and would require further examination.

7.7 Concluding Statement

This book chapter discussed the native and introduced resistomes of natural environments. Soil environments are rich natural reservoirs of ARGs and are the source of many clinically relevant ARGs today, including undiscovered ARGs that may impose new health threats. Soils often receive high inputs of clinically relevant ARGs through manure and reused wastewater application. These ARGs, including novel ARGs of pressing concern such as the bla_{NDM} genes/plasmids, confer a wide range of antibiotic resistance, persist through WWTP schematics, and accumulate in soil environments. As soil and wastewater environments are conducive matrices for microbial interaction and horizontal gene transfer, the potential of ARG transfer to new and pathogenic bacteria poses great risks, and intervention strategies are necessary. Solar inactivation is a naturally available resource that has shown to reduce the numbers of ARB in water bodies and can be further exploited to disinfect treated wastewater before irrigation. However, sunlight alone is much less effective in ARG removal, and bacteriophages offer a novel potential strategy to specifically target certain ARB and ARGs in water bodies. This tool still faces many obstacles before it can be applied effectively, and further investigation is required.

References

Al-Jasser AO (2011) Saudi wastewater reuse standards for agricultural irrigation: Riyadh treatment plants effluent compliance. J King Saud Univ Eng Sci 23(1):1–8

Al-Jassim N, Ansari MI, Harb M, Hong P-Y (2015) Removal of bacterial contaminants and antibiotic resistance genes by conventional wastewater treatment processes in Saudi Arabia: is the treated wastewater safe to reuse for agricultural irrigation? Water Res 73:277–290

Allen HK, Moe LA, Rodbumrer J, Gaarder A, Handelsman J (2009) Functional metagenomics reveals diverse β-lactamases in a remote Alaskan soil. ISME J 3(2):243–251

Allen HK, Donato J, Wang HH, Cloud-Hansen KA, Davies J, Handelsman J (2010) Call of the wild: antibiotic resistance genes in natural environments. Nat Rev Microbiol 8(4):251–259

Araki M (1986) Advanced slime control process with bacteriophage. Kogyo Yosui 332:25–30

Arpin C, Noury P, Boraud D, Coulange L, Manetti A, Andre C, M'Zali F, Quentin C (2012) NDM-1-producing Klebsiella pneumoniae resistant to colistin in a French community patient without history of foreign travel. Antimicrob Agents Chemother 56(6):3432–3434

Balogh B, Jones JB, Iriarte F, Momol M (2010) Phage therapy for plant disease control. Curr Pharm Biotechnol 11(1):48–57

Barak JD, Liang AS (2008) Role of soil, crop debris, and a plant pathogen in Salmonella enterica contamination of tomato plants. PLoS One 3(2):e1657

Bell K, Sebaihia M, Pritchard L, Holden M, Hyman L, Holeva M, Thomson N, Bentley S, Churcher L, Mungall K (2004) Genome sequence of the enterobacterial phytopathogen Erwinia carotovora subsp. atroseptica and characterization of virulence factors. Proc Natl Acad Sci U S A 101(30):11105–11110

Bhullar K, Waglechner N, Pawlowski A, Koteva K, Banks ED, Johnston MD, Barton HA, Wright GD (2012) Antibiotic resistance is prevalent in an isolated cave microbiome. PLoS One 7(4): e34953

Bichai F, Polo-López MI, Ibañez PF (2012) Solar disinfection of wastewater to reduce contamination of lettuce crops by Escherichia coli in reclaimed water irrigation. Water Res 46 (18):6040–6050

Blough NV, Zepp RG (1995) Reactive oxygen species in natural waters. In: Foote CS, Valentine JS, Greenberg A, Liebman JF (eds) Active oxygen in chemistry. Springer, Dordrecht, pp 280–333

Boehm AB, Yamahara KM, Love DC, Peterson BM, McNeill K, Nelson KL (2009) Covariation and photoinactivation of traditional and novel indicator organisms and human viruses at a sewage-impacted marine beach. Environ Sci Technol 43(21):8046–8052

Bolton NF, Cromar NJ, Hallsworth P, Fallowfield HJ (2010) A review of the factors affecting sunlight inactivation of micro-organisms in waste stabilisation ponds: preliminary results for enterococci. Water Sci Technol 61(4):885–890

Bonnin RA, Poirel L, Carattoli A, Nordmann P (2012) Characterization of an IncFII plasmid encoding NDM-1 from Escherichia coli ST131. PLoS One 7(4):e34752

Borgia S, Lastovetska O, Richardson D, Eshaghi A, Xiong J, Chung C, Baqi M, McGeer A, Ricci G, Sawicki R, Pantelidis R, Low DE, Patel SN, Melano RG (2012) Outbreak of carbapenem-resistant enterobacteriaceae containing blaNDM-1, Ontario, Canada. Clin Infect Dis 55(11):e109–e117

Boyle M, Sichel C, Fernandez-Ibanez P, Arias-Quiroz GB, Iriarte-Puna M, Mercado A, Ubomba-Jaswa E, McGuigan KG (2008) Bactericidal effect of solar water disinfection under real sunlight conditions. Appl Environ Microbiol 74(10):2997–3001

Brandl M (2008) Plant lesions promote the rapid multiplication of Escherichia coli O157: H7 on postharvest lettuce. Appl Environ Microbiol 74(17):5285–5289

Brockhurst M, Buckling A, Rainey P (2006) Spatial heterogeneity and the stability of host-parasite coexistence. J Evol Biol 19(2):374–379

Brown MG, Balkwill DL (2009) Antibiotic resistance in bacteria isolated from the deep terrestrial subsurface. Microb Ecol 57(3):484

Bush K (2010) Alarming β-lactamase-mediated resistance in multidrug-resistant Enterobacteriaceae. Curr Opin Microbiol 13(5):558–564

Canonica S, Kohn T, Mac M, Real FJ, Wirz J, von Gunten U (2005) Photosensitizer method to determine rate constants for the reaction of carbonate radical with organic compounds. Environ Sci Technol 39(23):9182–9188

Carattoli A, Villa L, Poirel L, Bonnin RA, Nordmann P (2012) Evolution of IncA/C blaCMY-2-carrying plasmids by acquisition of the blaNDM-1 carbapenemase gene. Antimicrob Agents Chemother 56(2):783–786

Chen H, Zhang M (2013) Effects of advanced treatment systems on the removal of antibiotic resistance genes in wastewater treatment plants from Hangzhou, China. Environ Sci Technol 47(15):8157–8163

Choi J, Kotay SM, Goel R (2011) Bacteriophage-based biocontrol of biological sludge bulking in wastewater. Bioengineered Bugs 2(4):214–217

Corin N, Backlund P, Kulovaara M (1996) Degradation products formed during UV-irradiation of humic waters. Chemosphere 33(2):245–255

Courcelle J, Khodursky A, Peter B, Brown PO, Hanawalt PC (2001) Comparative gene expression profiles following UV exposure in wild-type and SOS-deficient Escherichia coli. Genetics 158 (1):41–64

Curtis TP, Mara DD, Silva SA (1992) Influence of pH, oxygen, and humic substances on ability of sunlight to damage fecal coliforms in waste stabilization pond water. Appl Environ Microbiol 58(4):1335–1343

Cytryn E (2013) The soil resistome: the anthropogenic, the native, and the unknown. Soil Biol Biochem 63:18–23

Dalkmann P, Broszat M, Siebe C, Willaschek E, Sakinc T, Huebner J, Amelung W, Grohmann E, Siemens J (2012) Accumulation of pharmaceuticals, Enterococcus, and resistance genes in soils irrigated with wastewater for zero to 100 years in central Mexico. PLoS One 7(9):e45397

Da Silva MF, Tiago I, Veríssimo A, Boaventura RA, Nunes OC, Manaia CM (2006) Antibiotic resistance of enterococci and related bacteria in an urban wastewater treatment plant. FEMS Microbiol Ecol 55(2):322–329

D'costa VM, McGrann KM, Hughes DW, Wright GD (2006) Sampling the antibiotic resistome. Science 311(5759):374–377

D'Costa VM, King CE, Kalan L, Morar M, Sung WW, Schwarz C, Froese D, Zazula G, Calmels F, Debruyne R (2011) Antibiotic resistance is ancient. Nature 477(7365):457–461

Deering AJ, Mauer LJ, Pruitt RE (2012) Internalization of E. coli O157: H7 and Salmonella spp. in plants: a review. Food Res Int 45(2):567–575

DeFlaun MF, Paul JH (1989) Detection of exogenous gene sequences in dissolved DNA from aquatic environments. Microb Ecol 18(1):21–28

Denis C, Poirel L, Carricajo A, Grattard F, Fascia P, Verhoeven P, Gay P, Nuti C, Nordmann P, Pozzetto B (2012) Nosocomial transmission of NDM-1-producing Escherichia coli within a non-endemic area in France. Clin Microbiol Infect 18(5):E128–E130

de Vries J, Meier P, Wackernagel W (2001) The natural transformation of the soil bacteria Pseudomonas stutzeri and Acinetobacter sp. by transgenic plant DNA strictly depends on homologous sequences in the recipient cells. FEMS Microbiol Lett 195(2):211–215

Djordjevic SP, Stokes HW, Roy Chowdhury P (2013) Mobile elements, zoonotic pathogens and commensal bacteria: conduits for the delivery of resistance genes into humans, production animals and soil microbiota. Front Microbiol 4:86

Dominiak DM, Nielsen JL, Nielsen PH (2011) Extracellular DNA is abundant and important for microcolony strength in mixed microbial biofilms. Environ Microbiol 13(3):710–721

Dortet L (2012) New Delhi Metallo-β-Lactamase 4-producing Escherichia coli in Cameroon. Emerg Infect Dis 18(9):1540

Dunlop P, Ciavola M, Rizzo L, Byrne J (2011) Inactivation and injury assessment of Escherichia coli during solar and photocatalytic disinfection in LDPE bags. Chemosphere 85 (7):1160–1166

Eisenstark A (1987) Mutagenic and lethal effects of near-ultraviolet radiation (290-400 nm) on bacteria and phage. Environ Mol Mutagen 10(3):317–337

Espinosa-Urgel M (2004) Plant-associated Pseudomonas populations: molecular biology, DNA dynamics, and gene transfer. Plasmid 52(3):139–150

Fahrenfeld NL, Ma Y, O'Brien M, Pruden A (2013) Reclaimed water as a reservoir of antibiotic resistance genes: distribution system and irrigation implications. Front Microbiol 4:130

Feil EJ, Spratt BG (2001) Recombination and the population structures of bacterial pathogens. Annu Rev Microbiol 55(1):561–590

Fernandez-Astorga A, Muela A, Cisterna R, Iriberri J, Barcina I (1992) Biotic and abiotic factors affecting plasmid transfer in Escherichia coli strains. Appl Environ Microbiol 58(1):392–398

Ferro G, Guarino F, Cicatelli A, Rizzo L (2017) β-lactams resistance gene quantification in an antibiotic resistant Escherichia coli water suspension treated by advanced oxidation with UV/H 2 O 2. J Hazard Mater 323:426–433

Fiett J, Baraniak A, Izdebski R, Sitkiewicz I, Zabicka D, Meler A, Filczak K, Hryniewicz W, Gniadkowski M (2014) The first NDM metallo-beta-lactamase-producing Enterobacteriaceae isolate in Poland: evolution of IncFII-type plasmids carrying the bla(NDM-1) gene. Antimicrob Agents Chemother 58(2):1203–1207

Fisher MB, Keenan CR, Nelson KL, Voelker BM (2008) Speeding up solar disinfection (SODIS): effects of hydrogen peroxide, temperature, pH, and copper plus ascorbate on the photoinactivation of E. coli. J Water Health 6(1):35–51

Fisher MB, Iriarte M, Nelson KL (2012) Solar water disinfection (SODIS) of Escherichia coli, Enterococcus spp., and MS2 coliphage: effects of additives and alternative container materials. Water Res 46(6):1745–1754

Fouts DE, Tyler HL, DeBoy RT, Daugherty S, Ren Q, Badger JH, Durkin AS, Huot H, Shrivastava S, Kothari S (2008) Complete genome sequence of the N 2-fixing broad host range endophyte Klebsiella pneumoniae 342 and virulence predictions verified in mice. PLoS Genet 4(7):e1000141

Goldman G, Starosvetsky J, Armon R (2009) Inhibition of biofilm formation on UF membrane by use of specific bacteriophages. J Membr Sci 342(1):145–152

Gomes C, Da Silva P, Moreira RG, Castell-Perez E, Ellis EA, Pendleton M (2009) Understanding E. coli internalization in lettuce leaves for optimization of irradiation treatment. Int J Food Microbiol 135(3):238–247

Goodsell DS (2001) The molecular perspective: ultraviolet light and pyrimidine dimers. Oncologist 6(3):298–299

Götz A, Pukall R, Smit E, Tietze E, Prager R, Tschäpe H, Van Elsas J, Smalla K (1996) Detection and characterization of broad-host-range plasmids in environmental bacteria by PCR. Appl Environ Microbiol 62(7):2621–2628

Guo M-T, Yuan Q-B, Yang J (2013) Ultraviolet reduction of erythromycin and tetracycline resistant heterotrophic bacteria and their resistance genes in municipal wastewater. Chemosphere 93(11):2864–2868

Halaby T, Reuland AE, Al Naiemi N, Potron A, Savelkoul PH, Vandenbroucke-Grauls CM, Nordmann P (2012) A case of New Delhi metallo-beta-lactamase 1 (NDM-1)-producing Klebsiella pneumoniae with putative secondary transmission from the Balkan region in the Netherlands. Antimicrob Agents Chemother 56(5):2790–2791

Harb M, Hong P-Y (2016) Molecular-based detection of potentially pathogenic bacteria in membrane bioreactor (MBR) systems treating municipal wastewater: a case study. Environ Sci Pollut Res 24(6):5370–5380

Heuer H, Smalla K (2007) Manure and sulfadiazine synergistically increased bacterial antibiotic resistance in soil over at least two months. Environ Microbiol 9(3):657–666

Heuer H, Solehati Q, Zimmerling U, Kleineidam K, Schloter M, Müller T, Focks A, Thiele-Bruhn S, Smalla K (2011) Accumulation of sulfonamide resistance genes in arable soils due to repeated application of manure containing sulfadiazine. Appl Environ Microbiol 77(7):2527–2530

Ho PL, Lo WU, Yeung MK, Lin CH, Chow KH, Ang I, Tong AH, Bao JY, Lok S, Lo JY (2011) Complete sequencing of pNDM-HK encoding NDM-1 carbapenemase from a multidrug-resistant Escherichia coli strain isolated in Hong Kong. PLoS One 6(3):e17989

Ho PL, Li Z, Lai EL, Chiu SS, Cheng VC (2012) Emergence of NDM-1-producing Enterobacteriaceae in China. J Antimicrob Chemother 67(6):1553–1555

Hofreiter M, Serre D, Poinar HN, Kuch M, Pääbo S (2001) Ancient DNA. Nat Rev Genet 2 (5):353–359

Holden N, Pritchard L, Toth I (2009) Colonization outwith the colon: plants as an alternative environmental reservoir for human pathogenic enterobacteria. FEMS Microbiol Rev 33 (4):689–703

Hong P-Y, Yannarell AC, Dai Q, Ekizoglu M, Mackie RI (2013) Monitoring the perturbation of soil and groundwater microbial communities due to pig production activities. Appl Environ Microbiol 79(8):2620–2629

Hornsey M, Phee L, Wareham DW (2011) A novel variant, NDM-5, of the New Delhi metallo-beta-lactamase in a multidrug-resistant Escherichia coli ST648 isolate recovered from a patient in the United Kingdom. Antimicrob Agents Chemother 55(12):5952–5954

Hughes VM, Datta N (1983) Conjugative plasmids in bacteria of the 'pre-antibiotic' era. Nature 302(5910):725–726

Ikeda H, Shiraishi K, Ogata Y (2004) Illegitimate recombination mediated by double-strand break and end-joining in Escherichia coli. Adv Biophys 38:3–20

Islam M, Morgan J, Doyle MP, Jiang X (2004) Fate of Escherichia coli O157: H7 in manure compost-amended soil and on carrots and onions grown in an environmentally controlled growth chamber. J Food Prot 67(3):574–578

Isozumi R, Yoshimatsu K, Yamashiro T, Hasebe F, Nguyen BM, Ngo TC, Yasuda SP, Koma T, Shimizu K, Arikawa J (2012) bla(NDM-1)-positive Klebsiella pneumoniae from environment, Vietnam. Emerg Infect Dis 18(8):1383–1385

Itoh Y, Sugita-Konishi Y, Kasuga F, Iwaki M, Hara-Kudo Y, Saito N, Noguchi Y, Konuma H, Kumagai S (1998) Enterohemorrhagic Escherichia coli O157: H7 present in radish sprouts. Appl Environ Microbiol 64(4):1532–1535

Jagger J (1981) Near-UV radiation effects on microorganisms. Photochem Photobiol 34 (6):761–768

Jassim SA, Limoges RG (2014) Natural solution to antibiotic resistance: bacteriophages 'The Living Drugs'. World J Microbiol Biotechnol 30(8):2153–2170

Jassim SA, Limoges RG, El-Cheikh H (2016) Bacteriophage biocontrol in wastewater treatment. World J Microbiol Biotechnol 32(4):1–10

Johnson AP, Woodford N (2013) Global spread of antibiotic resistance: the example of New Delhi metallo-beta-lactamase (NDM)-mediated carbapenem resistance. J Med Microbiol 62 (Pt 4):499–513

Jones JB, Vallad GE, Iriarte FB, Obradović A, Wernsing MH, Jackson LE, Balogh B, Hong JC, Momol MT (2012) Considerations for using bacteriophages for plant disease control. Bacteriophage 2(4):e23857

Jones LA, Worobo RW, Smart CD (2014) UV light inactivation of human and plant pathogens in unfiltered surface irrigation water. Appl Environ Microbiol 80(3):849–854

Kaase M, Nordmann P, Wichelhaus TA, Gatermann SG, Bonnin RA, Poirel L (2011) NDM-2 carbapenemase in Acinetobacter baumannii from Egypt. J Antimicrob Chemother 66 (6):1260–1262

Karl DM, Bailiff MD (1989) The measurement and distribution of dissolved nucleic acids in aquatic environments. Limnol Oceanogr 34(3):543–558

Khaengraeng R, Reed RH (2005) Oxygen and photoinactivation of Escherichia coli in UVA and sunlight. J Appl Microbiol 99(1):39–50

Khairnar K, Pal P, Chandekar RH, Paunikar WN (2014) Isolation and characterization of bacteriophages infecting nocardioforms in wastewater treatment plant. Biotechnol Res Int 2014:151952

Khan M, Satoh H, Mino T, Katayama H, Kurisu F, Matsuo T (2002a) Bacteriophage-host interaction in the enhanced biological phosphate removing activated sludge system. Water Sci Technol 46(1–2):39–43
Khan MA, Satoh H, Katayama H, Kurisu F, Mino T (2002b) Bacteriophages isolated from activated sludge processes and their polyvalency. Water Res 36(13):3364–3370
Kim MN, Yong D, An D, Chung HS, Woo JH, Lee K, Chong Y (2012) Nosocomial clustering of NDM-1-producing Klebsiella pneumoniae sequence type 340 strains in four patients at a South Korean tertiary care hospital. J Clin Microbiol 50(4):1433–1436
King D, Strynadka N (2011) Crystal structure of New Delhi metallo-beta-lactamase reveals molecular basis for antibiotic resistance. Protein Sci 20(9):1484–1491
Knapp CW, Dolfing J, Ehlert PA, Graham DW (2010) Evidence of increasing antibiotic resistance gene abundances in archived soils since 1940. Environ Sci Technol 44(2):580–587
Kroupitski Y, Golberg D, Belausov E, Pinto R, Swartzberg D, Granot D, Sela S (2009) Internalization of Salmonella enterica in leaves is induced by light and involves chemotaxis and penetration through open stomata. Appl Environ Microbiol 75(19):6076–6086
Kumarasamy K, Kalyanasundaram A (2012) Emergence of Klebsiella pneumoniae isolate co-producing NDM-1 with KPC-2 from India. J Antimicrob Chemother 67(1):243–244
Kumarasamy KK, Toleman MA, Walsh TR, Bagaria J, Butt F, Balakrishnan R, Chaudhary U, Doumith M, Giske CG, Irfan S (2010) Emergence of a new antibiotic resistance mechanism in India, Pakistan, and the UK: a molecular, biological, and epidemiological study. Lancet Infect Dis 10(9):597–602
Lai CC, Lin TL, Tseng SP, Huang YT, Wang JT, Chang SC, Teng LJ, Wang JT, Hsueh PR (2011) Pelvic abscess caused by New Delhi metallo-beta-lactamase-1-producing Klebsiella oxytoca in Taiwan in a patient who underwent renal transplantation in China. Diagn Microbiol Infect Dis 71(4):474–475
Le T-H, Ng C, Chen H, Yi XZ, Koh TH, Barkham TMS, Zhou Z, Gin KY-H (2016) Occurrences and characterization of antibiotic-resistant bacteria and genetic determinants of hospital wastewater in a tropical country. Antimicrob Agents Chemother 60(12):7449–7456
Levy-Booth DJ, Campbell RG, Gulden RH, Hart MM, Powell JR, Klironomos JN, Pauls KP, Swanton CJ, Trevors JT, Dunfield KE (2007) Cycling of extracellular DNA in the soil environment. Soil Biol Biochem 39(12):2977–2991
Lilley AK, Fry JC, Day MJ, Bailey MJ (1994) In situ transfer of an exogenously isolated plasmid between Pseudomonas spp. in sugar beet rhizosphere. Microbiology 140(1):27–33
Lloyd RE, Rinkenberger JL, Hug BA, Tuveson RW (1990) Growing Escherichia coli mutants deficient in riboflavin biosynthesis with non-limiting riboflavin results in sensitization to inactivation by broad-spectrum near-ultraviolet light (320-400 nm). Photochem Photobiol 52 (4):897–901
Lorenz MG, Wackernagel W (1994) Bacterial gene transfer by natural genetic transformation in the environment. Microbiol Rev 58(3):563–602
Luo Y, Yang F, Mathieu J, Mao D, Wang Q, Alvarez P (2013) Proliferation of multidrug-resistant New Delhi metallo-β-lactamase genes in municipal wastewater treatment plants in northern China. Environ Sci Technol Lett 1(1):26–30
Majewski J, Cohan FM (1998) The effect of mismatch repair and heteroduplex formation on sexual isolation in Bacillus. Genetics 148(1):13–18
Majewski J, Zawadzki P, Pickerill P, Cohan FM, Dowson CG (2000) Barriers to genetic exchange between bacterial species: Streptococcus pneumoniae transformation. J Bacteriol 182 (4):1016–1023
Malato S, Fernández-Ibáñez P, Maldonado M, Blanco J, Gernjak W (2009) Decontamination and disinfection of water by solar photocatalysis: recent overview and trends. Catal Today 147 (1):1–59
Mantilla-Calderon D, Jumat MR, Wang T, Ganesan P, Al-Jassim N, Hong P-Y (2016) Isolation and characterization of NDM-positive Escherichia coli from municipal wastewater in Jeddah, Saudi Arabia. Antimicrob Agents Chemother 60(9):5223–5231

Marti R, Scott A, Tien Y-C, Murray R, Sabourin L, Zhang Y, Topp E (2013) Impact of manure fertilization on the abundance of antibiotic-resistant bacteria and frequency of detection of antibiotic resistance genes in soil and on vegetables at harvest. Appl Environ Microbiol 79 (18):5701–5709

Mason KA, Losos JB, Singer SR, Peter H (2011) Biology. McGraw-Hill, New York, p 533

Matic I, Radman M, Taddei F, Picard B, Doit C, Bingen E, Denamur E, Elion J (1997) Highly variable mutation rates in commensal and pathogenic Escherichia coli. Science 277 (5333):1833–1834

Mazzariol A, Bosnjak Z, Ballarini P, Budimir A, Bedenic B, Kalenic S, Cornaglia G (2012) NDM-1-producing Klebsiella pneumoniae, Croatia. Emerg Infect Dis 18(3):532–534

McGann P, Hang J, Clifford RJ, Yang Y, Kwak YI, Kuschner RA, Lesho EP, Waterman PE (2012) Complete sequence of a novel 178-kilobase plasmid carrying bla(NDM-1) in a Providencia stuartii strain isolated in Afghanistan. Antimicrob Agents Chemother 56(4):1673–1679

McGuigan KG, Joyce TM, Conroy RM, Gillespie JB, Elmore-Meegan M (1998) Solar disinfection of drinking water contained in transparent plastic bottles: characterizing the bacterial inactivation process. J Appl Microbiol 84(6):1138–1148

McGuigan KG, Conroy RM, Mosler HJ, du Preez M, Ubomba-Jaswa E, Fernandez-Ibanez P (2012) Solar water disinfection (SODIS): a review from bench-top to roof-top. J Hazard Mater 235–236:29–46

McKinney CW, Pruden A (2012) Ultraviolet disinfection of antibiotic resistant bacteria and their antibiotic resistance genes in water and wastewater. Environ Sci Technol 46(24):13393–13400

Méjean V, Claverys J-P (1993) DNA processing during entry in transformation of Streptococcus pneumoniae. J Biol Chem 268(8):5594–5599

Miriagou V, Cornaglia G, Edelstein M, Galani I, Giske C, Gniadkowski M, Malamou-Lada E, Martinez-Martinez L, Navarro F, Nordmann P (2010) Acquired carbapenemases in Gram-negative bacterial pathogens: detection and surveillance issues. Clin Microbiol Infect 16 (2):112–122

Morris CE, Monier J-M (2003) The ecological significance of biofilm formation by plant-associated bacteria. Annu Rev Phytopathol 41(1):429–453

Muela A, Garcia-Bringas JM, Seco C, Arana I, Barcina I (2002) Participation of oxygen and role of exogenous and endogenous sensitizers in the photoinactivation of Escherichia coli by photosynthetically active radiation, UV-A and UV-B. Microb Ecol 44(4):354–364

Mulvey MR, Grant JM, Plewes K, Roscoe D, Boyd DA (2011) New Delhi metallo-beta-lactamase in Klebsiella pneumoniae and Escherichia coli, Canada. Emerg Infect Dis 17(1):103–106

Munir M, Xagoraraki I (2011) Levels of antibiotic resistance genes in manure, biosolids, and fertilized soil. J Environ Qual 40(1):248–255

Munir M, Wong K, Xagoraraki I (2011) Release of antibiotic resistant bacteria and genes in the effluent and biosolids of five wastewater utilities in Michigan. Water Res 45(2):681–693

Negreanu Y, Pasternak Z, Jurkevitch E, Cytryn E (2012) Impact of treated wastewater irrigation on antibiotic resistance in agricultural soils. Environ Sci Technol 46(9):4800–4808

Nesme J, Simonet P (2015) The soil resistome: a critical review on antibiotic resistance origins, ecology and dissemination potential in telluric bacteria. Environ Microbiol 17(4):913–930

Nielsen KM, Bones AM, Van Elsas J (1997a) Induced natural transformation of Acinetobacter calcoaceticus in soil microcosms. Appl Environ Microbiol 63(10):3972–3977

Nielsen KM, Van Weerelt M, Berg TN, Bones AM, Hagler AN, Van Elsas J (1997b) Natural transformation and availability of transforming DNA to Acinetobacter calcoaceticus in soil microcosms. Appl Environ Microbiol 63(5):1945–1952

Nielsen KM, van Elsas JD, Smalla K (2000) Transformation of Acinetobacter sp. strain BD413 (pFG4ΔnptII) with transgenic plant DNA in soil microcosms and effects of kanamycin on selection of transformants. Appl Environ Microbiol 66(3):1237–1242

Nielsen KM, Johnsen PJ, Bensasson D, Daffonchio D (2007) Release and persistence of extracellular DNA in the environment. Environ Biosaf Res 6(1–2):37–53

Nielsen JB, Hansen F, Littauer P, Schonning K, Hammerum AM (2012) An NDM-1-producing Escherichia coli obtained in Denmark has a genetic profile similar to an NDM-1-producing E. coli isolate from the UK. J Antimicrob Chemother 67(8):2049–2051

Nies DH (2003) Efflux-mediated heavy metal resistance in prokaryotes. FEMS Microbiol Rev 27 (2-3):313–339

Nordmann P, Boulanger AE, Poirel L (2012) NDM-4 metallo-beta-lactamase with increased carbapenemase activity from Escherichia coli. Antimicrob Agents Chemother 56 (4):2184–2186

Novais Â, Cantón R, Moreira R, Peixe L, Baquero F, Coque TM (2007) Emergence and dissemination of Enterobacteriaceae isolates producing CTX-M-1-like enzymes in Spain are associated with IncFII (CTX-M-15) and broad-host-range (CTX-M-1,-3, and-32) plasmids. Antimicrob Agents Chemother 51(2):796–799

Oates PM, Shanahan P, Polz MF (2003) Solar disinfection (SODIS): simulation of solar radiation for global assessment and application for point-of-use water treatment in Haiti. Water Res 37 (1):47–54

Ogram A, Sayler GS, Barkay T (1987) The extraction and purification of microbial DNA from sediments. J Microbiol Methods 7(2–3):57–66

Oteo J, Domingo-Garcia D, Fernandez-Romero S, Saez D, Guiu A, Cuevas O, Lopez-Brea M, Campos J (2012) Abdominal abscess due to NDM-1-producing Klebsiella pneumoniae in Spain. J Med Microbiol 61(Pt 6):864–867

Pal P, Khairnar K, Paunikar W (2014) Causes and remedies for filamentous foaming in activated sludge treatment plant. Global NEST J 16(4):762–772

Palmen R, Hellingwerf KJ (1997) Uptake and processing of DNA by Acinetobacter calcoaceticus—a review. Gene 192(1):179–190

Pasteran F, Albornoz E, Faccone D, Gomez S, Valenzuela C, Morales M, Estrada P, Valenzuela L, Matheu J, Guerriero L, Arbizu E, Calderon Y, Ramon-Pardo P, Corso A (2012) Emergence of NDM-1-producing Klebsiella pneumoniae in Guatemala. J Antimicrob Chemother 67 (7):1795–1797

Pattison DI, Davies MJ (2006) Actions of ultraviolet light on cellular structures. In: Bignold LP (ed) Cancer: cell structures, carcinogens and genomic instability, vol 96. Birkhäuser, Basel, pp 131–157

Peirano G, Ahmed-Bentley J, Woodford N, Pitout JD (2011a) New Delhi metallo-beta-lactamase from traveler returning to Canada. Emerg Infect Dis 17(2):242–244

Peirano G, Pillai DR, Pitondo-Silva A, Richardson D, Pitout JD (2011b) The characteristics of NDM-producing Klebsiella pneumoniae from Canada. Diagn Microbiol Infect Dis 71 (2):106–109

Petrovski S, Seviour RJ, Tillett D (2011a) Characterization of the genome of the polyvalent lytic bacteriophage GTE2, which has potential for biocontrol of Gordonia-, Rhodococcus-, and Nocardia-stabilized foams in activated sludge plants. Appl Environ Microbiol 77 (12):3923–3929

Petrovski S, Seviour RJ, Tillett D (2011b) Prevention of Gordonia and Nocardia stabilized foam formation by using bacteriophage GTE7. Appl Environ Microbiol 77(21):7864–7867

Pfeifer GP (1997) Formation and processing of UV photoproducts: effects of DNA sequence and chromatin environment. Photochem Photobiol 65(2):270–283

Piepersberg W, Distler J, Heinzel P, Perez-Gonzalez J-A (1988) Antibiotic resistance by modification: many resistance genes could be derived from cellular control genes in actinomycetes-A hypothesis. Actinomycetologica 2(2):83–98

Poirel L, Hombrouck-Alet C, Freneaux C, Bernabeu S, Nordmann P (2010a) Global spread of New Delhi metallo-β-lactamase 1. Lancet Infect Dis 10(12):832

Poirel L, Lagrutta E, Taylor P, Pham J, Nordmann P (2010b) Emergence of metallo-beta-lactamase NDM-1-producing multidrug-resistant Escherichia coli in Australia. Antimicrob Agents Chemother 54(11):4914–4916

Poirel L, Al Maskari Z, Al Rashdi F, Bernabeu S, Nordmann P (2011a) NDM-1-producing Klebsiella pneumoniae isolated in the Sultanate of Oman. J Antimicrob Chemother 66 (2):304–306

Poirel L, Benouda A, Hays C, Nordmann P (2011b) Emergence of NDM-1-producing Klebsiella pneumoniae in Morocco. J Antimicrob Chemother 66(12):2781–2783

Poirel L, Dortet L, Bernabeu S, Nordmann P (2011c) Genetic features of blaNDM-1-positive Enterobacteriaceae. Antimicrob Agents Chemother 55(11):5403–5407

Poirel L, Fortineau N, Nordmann P (2011d) International transfer of NDM-1-producing Klebsiella pneumoniae from Iraq to France. Antimicrob Agents Chemother 55(4):1821–1822

Poirel L, Revathi G, Bernabeu S, Nordmann P (2011e) Detection of NDM-1-producing Klebsiella pneumoniae in Kenya. Antimicrob Agents Chemother 55(2):934–936

Poirel L, Ros A, Carricajo A, Berthelot P, Pozzetto B, Bernabeu S, Nordmann P (2011f) Extremely drug-resistant Citrobacter freundii isolate producing NDM-1 and other carbapenemases identified in a patient returning from India. Antimicrob Agents Chemother 55(1):447–448

Poirel L, Schrenzel J, Cherkaoui A, Bernabeu S, Renzi G, Nordmann P (2011g) Molecular analysis of NDM-1-producing enterobacterial isolates from Geneva, Switzerland. J Antimicrob Chemother 66(8):1730–1733

Poirel L, Lascols C, Bernabeu S, Nordmann P (2012a) NDM-1-producing Klebsiella pneumoniae in Mauritius. Antimicrob Agents Chemother 56(1):598–599

Poirel L, Ozdamar M, Ocampo-Sosa AA, Turkoglu S, Ozer UG, Nordmann P (2012b) NDM-1-producing Klebsiella pneumoniae now in Turkey. Antimicrob Agents Chemother 56 (5):2784–2785

Poole K (2005) Efflux-mediated antimicrobial resistance. J Antimicrob Chemother 56(1):20–51

Pruden A, Larsson DJ, Amézquita A, Collignon P, Brandt KK, Graham DW, Lazorchak JM, Suzuki S, Silley P, Snape JR (2013) Management options for reducing the release of antibiotics and antibiotic resistance genes to the environment. Environ Health Perspect 121(8):878

Rahme LG, Stevens EJ, Wolfort SF, Shao J (1995) Common virulence factors for bacterial pathogenicity in plants and animals. Science 268(5219):1899

Rahube TO, Marti R, Scott A, Tien Y-C, Murray R, Sabourin L, Zhang Y, Duenk P, Lapen DR, Topp E (2014) Impact of fertilizing with raw or anaerobically digested sewage sludge on the abundance of antibiotic-resistant coliforms, antibiotic resistance genes, and pathogenic bacteria in soil and on vegetables at harvest. Appl Environ Microbiol 80(22):6898–6907

Raynaud X, Nunan N (2014) Spatial ecology of bacteria at the microscale in soil. PLoS One 9(1): e87217

Riesenfeld CS, Goodman RM, Handelsman J (2004) Uncultured soil bacteria are a reservoir of new antibiotic resistance genes. Environ Microbiol 6(9):981–989

Rijal GK, Fujioka RS (2003) Use of reflectors to enhance the synergistic effects of solar heating and solar wavelengths to disinfect drinking water sources. Water Sci Technol 48 (11-12):481–488

Rizzo L, Fiorentino A, Anselmo A (2012) Effect of solar radiation on multidrug resistant E. coli strains and antibiotic mixture photodegradation in wastewater polluted stream. Sci Total Environ 427:263–268

Rizzo L, Manaia C, Merlin C, Schwartz T, Dagot C, Ploy MC, Michael I, Fatta-Kassinos D (2013) Urban wastewater treatment plants as hotspots for antibiotic resistant bacteria and genes spread into the environment: a review. Sci Total Environ 447:345–360

Santos AL, Gomes NC, Henriques I, Almeida A, Correia A, Cunha A (2012) Contribution of reactive oxygen species to UV-B-induced damage in bacteria. J Photochem Photobiol B Biol 117:40–46

Schlüter A, Szczepanowski R, Pühler A, Top EM (2007) Genomics of IncP-1 antibiotic resistance plasmids isolated from wastewater treatment plants provides evidence for a widely accessible drug resistance gene pool. FEMS Microbiol Rev 31(4):449–477

Schwartz T, Kohnen W, Jansen B, Obst U (2003) Detection of antibiotic-resistant bacteria and their resistance genes in wastewater, surface water, and drinking water biofilms. FEMS Microbiol Ecol 43(3):325–335

Segawa T, Takeuchi N, Rivera A, Yamada A, Yoshimura Y, Barcaza G, Shinbori K, Motoyama H, Kohshima S, Ushida K (2013) Distribution of antibiotic resistance genes in glacier environments. Environ Microbiol Rep 5(1):127–134

Sekizuka T, Matsui M, Yamane K, Takeuchi F, Ohnishi M, Hishinuma A, Arakawa Y, Kuroda M (2011) Complete sequencing of the bla(NDM-1)-positive IncA/C plasmid from Escherichia coli ST38 isolate suggests a possible origin from plant pathogens. PLoS One 6(9):e25334

Selenska S, Klingmüller W (1992) Direct recovery and molecular analysis of DNA and RNA from soil. Microb Releases 1(1):41–46

Sidjabat H, Nimmo GR, Walsh TR, Binotto E, Htin A, Hayashi Y, Li J, Nation RL, George N, Paterson DL (2011) Carbapenem resistance in Klebsiella pneumoniae due to the New Delhi Metallo-beta-lactamase. Clin Infect Dis 52(4):481–484

Sinton LW, Hall CH, Lynch PA, Davies-Colley RJ (2002) Sunlight inactivation of fecal indicator bacteria and bacteriophages from waste stabilization pond effluent in fresh and saline waters. Appl Environ Microbiol 68(3):1122–1131

Smith DH (1967) R factor infection of Escherichia coli lyophilized in 1946. J Bacteriol 94(6):2071

Sole M, Pitart C, Roca I, Fabrega A, Salvador P, Munoz L, Oliveira I, Gascon J, Marco F, Vila J (2011) First description of an Escherichia coli strain producing NDM-1 carbapenemase in Spain. Antimicrob Agents Chemother 55(9):4402–4404

Solomon EB, Yaron S, Matthews KR (2002) Transmission of Escherichia coli O157: H7 from contaminated manure and irrigation water to lettuce plant tissue and its subsequent internalization. Appl Environ Microbiol 68(1):397–400

Sommer B, Marino A, Solarte Y, Salas M, Dierolf C, Valiente C, Mora D, Rechsteiner R, Setter P, Wirojanagud W (1997) SODIS—an emerging water treatment process. Aqua (Oxford) 46 (3):127–137

Sørensen SJ, Jensen LE (1998) Transfer of plasmid RP4 in the spermosphere and rhizosphere of barley seedling. Antonie Van Leeuwenhoek 73(1):69–77

Staskawicz BJ, Mudgett MB, Dangl JL, Galan JE (2001) Common and contrasting themes of plant and animal diseases. Science 292(5525):2285–2289

Stokes HW, Gillings MR (2011) Gene flow, mobile genetic elements and the recruitment of antibiotic resistance genes into Gram-negative pathogens. FEMS Microbiol Rev 35 (5):790–819

Sulakvelidze A, Alavidze Z, Morris JG (2001) Bacteriophage therapy. Antimicrob Agents Chemother 45(3):649–659

Suzuki H, Yano H, Brown CJ, Top EM (2010) Predicting plasmid promiscuity based on genomic signature. J Bacteriol 192(22):6045–6055

Thomas CM, Nielsen KM (2005) Mechanisms of, and barriers to, horizontal gene transfer between bacteria. Nat Rev Microbiol 3(9):711–721

Thomas J, Soddell JA, Kurtböke D (2002) Fighting foam with phages? Water Sci Technol 46 (1–2):511–518

Tijet N (2011) New Delhi metallo-β-lactamase, Ontario, Canada. Emerg Infect Dis 17(2):306

Toth IK, Pritchard L, Birch PR (2006) Comparative genomics reveals what makes an enterobacterial plant pathogen. Annu Rev Phytopathol 44:305–336

Ubomba-Jaswa E, Navntoft C, Polo-Lopez MI, Fernandez-Ibanez P, McGuigan KG (2009) Solar disinfection of drinking water (SODIS): an investigation of the effect of UV-A dose on inactivation efficiency. Photochem Photobiol Sci 8(5):587–595

Udikovic-Kolic N, Wichmann F, Broderick NA, Handelsman J (2014) Bloom of resident antibiotic-resistant bacteria in soil following manure fertilization. Proc Natl Acad Sci USA 111(42):15202–15207

USEPA (2006) Ultraviolet disinfection guidance manual for the final long term 2 enhanced surface water treatment rule. https://nepis.epa.gov/Exe/ZyPDF.cgi?Dockey=600006T3.txt. Date last accessed 4 May 2017

Van Boeckel TP, Gandra S, Ashok A, Caudron Q, Grenfell BT, Levin SA, Laxminarayan R (2014) Global antibiotic consumption 2000 to 2010: an analysis of national pharmaceutical sales data. Lancet Infect Dis 14(8):742–750

Van Elsas JD, Bailey MJ (2002) The ecology of transfer of mobile genetic elements. FEMS Microbiol Ecol 42(2):187–197

Van Elsas J, Fry J, Hirsch P, Molin S (2000) Ecology of plasmid transfer and spread. In: Thomas CM (ed) The horizontal gene pool; Bacterial plasmids and gene spread. Harwood, Amsterdam, pp 175–206

Vulić M, Dionisio F, Taddei F, Radman M (1997) Molecular keys to speciation: DNA polymorphism and the control of genetic exchange in enterobacteria. Proc Natl Acad Sci USA 94 (18):9763–9767

Walsh TR, Weeks J, Livermore DM, Toleman MA (2011) Dissemination of NDM-1 positive bacteria in the New Delhi environment and its implications for human health: an environmental point prevalence study. Lancet Infect Dis 11(5):355–362

Walther-Rasmussen J, Høiby N (2007) Class A carbapenemases. J Antimicrob Chemother 60 (3):470–482

Wang Z, Fast W, Valentine AM, Benkovic SJ (1999) Metallo-β-lactamase: structure and mechanism. Curr Opin Chem Biol 3(5):614–622

Wang F, Stedtfeld RD, Kim O-S, Chai B, Yang L, Stedtfeld TM, Hong SG, Kim D, Lim HS, Hashsham SA (2016) Influence of soil characteristics and proximity to antarctic research stations on abundance of antibiotic resistance genes in soils. Environ Sci Technol 50 (23):12621–12629

Webb RB, Brown MS (1979) Action spectra for oxygen-dependent and independent inactivation of Escherichia coli WP2s from 254 to 460 nm. Photochem Photobiol 29(2):407–409

Wegelin M, Canonica S, Mechsner K, Fleischmann T, Pesaro F, Metzler A (1994) Solar water disinfection: scope of the process and analysis of radiation experiments. Aqua 43(4):154–169

Weinbauer MG (2004) Ecology of prokaryotic viruses. FEMS Microbiol Rev 28(2):127–181

Williamson DA, Sidjabat HE, Freeman JT, Roberts SA, Silvey A, Woodhouse R, Mowat E, Dyet K, Paterson DL, Blackmore T, Burns A, Heffernan H (2012) Identification and molecular characterisation of New Delhi metallo-beta lactamase-1 (NDM-1)- and NDM-6-producing Enterobacteriaceae from New Zealand hospitals. Int J Antimicrob Agents 39(6):529–533

Withey S, Cartmell E, Avery L, Stephenson T (2005) Bacteriophages—potential for application in wastewater treatment processes. Sci Total Environ 339(1):1–18

Wright GD (2007) The antibiotic resistome: the nexus of chemical and genetic diversity. Nat Rev Microbiol 5(3):175–186

Yong D, Toleman MA, Giske CG, Cho HS, Sundman K, Lee K, Walsh TR (2009) Characterization of a new metallo-β-lactamase gene, blaNDM-1, and a novel erythromycin esterase gene carried on a unique genetic structure in Klebsiella pneumoniae sequence type 14 from India. Antimicrob Agents Chemother 53(12):5046–5054

Yosef I, Manor M, Kiro R, Qimron U (2015) Temperate and lytic bacteriophages programmed to sensitize and kill antibiotic-resistant bacteria. Proc Natl Acad Sci USA 112(23):7267–7272

Zhang Y, Zhuang Y, Geng J, Ren H, Zhang Y, Ding L, Xu K (2015) Inactivation of antibiotic resistance genes in municipal wastewater effluent by chlorination and sequential UV/chlorination disinfection. Sci Total Environ 512:125–132

Zhang Y, Zhuang Y, Geng J, Ren H, Xu K, Ding L (2016) Reduction of antibiotic resistance genes in municipal wastewater effluent by advanced oxidation processes. Sci Total Environ 550:184–191

Zhu Y-G, Johnson TA, Su J-Q, Qiao M, Guo G-X, Stedtfeld RD, Hashsham SA, Tiedje JM (2013) Diverse and abundant antibiotic resistance genes in Chinese swine farms. Proc Natl Acad Sci USA 110(9):3435–3440

Chapter 8
Antibiotic Resistance Gene Due to Manure Application

Srujana Kathi

8.1 Introduction

Veterinary manure is a significant pool of antibiotics and antibiotic-resistant bacteria (Yang et al. 2014). Antibiotic resistance has become a most intractable challenge in twenty-first century global public health issue resulting in the release of, antibiotic resistance genes (ARGs) into the receiving environment through waste disposal processes such as manure application on agricultural soil. Growing scientific evidence show that ARGs are emerging environmental contaminants (He et al. 2014). The practice of spreading manure on agricultural soils not only introduces nutrients required for maintaining the soil fertility but also antibiotics, their transformation products (TPs), and antibiotic-resistant bacteria (Ding et al. 2014). They possess a potential for reaching the soil environment where they develop resistance and can impact the soil ecosystem services (Hashmi et al. 2017). An urgent need exists to improve our understanding of the mechanisms associated with the spread and development of ARGs in both clinical and veterinary settings, the human body, as well as in engineered and natural environments (Sanderson et al. 2016). Manure management can impact the persistence, survival, and distribution of bacteria and AR genes in agroecosystems (Durso and Cook 2014).

Vegetables grown in soil receiving raw or digested manure are at risk of contamination with manure-borne antibiotic-resistant bacteria compared to those grown in ground receiving composted manure (Tien et al. 2017). Soil microbiota are a natural pool of antibiotic resistance determinants to antibiotics. Group of genes conferring resistance to antibiotics are referred to as the antibiotic resistome

The original version of this chapter was revised. An erratum to this chapter can be found at DOI 10.1007/978-3-319-66260-2_22.

S. Kathi (✉)
Pondicherry University, Kalapet, Puducherry, India
e-mail: srujanaaramesh@gmail.com

M.Z. Hashmi et al. (eds.), *Antibiotics and Antibiotics Resistance Genes in Soils*,
Soil Biology 51, DOI 10.1007/978-3-319-66260-2_8

(Jechalke et al. 2015). A collection of all the ARGs is referred to as the antibiotic resistome, which includes intrinsic resistance genes, acquired resistance genes, proto-resistance genes, and silent resistance genes (Perry et al. 2014; Cytryn 2013).

8.2 Fate of Veterinary Antibiotics in Manure and Soil

Abundance of ARGs was reported in chicken, pig, and duck manure (Chen and Zhang 2013; He et al. 2014, 2016; Cui et al. 2017). Manure is the primary source by which AR bacteria and AR genes gain entry into the surrounding environment. So, manure is considered as starting point for monitoring AR levels and a critical control point for isolating and remediating AR before it is transported more broadly in agroecosystems (Pruden et al. 2013). Animal manures that gain entry into agricultural fields carry antibiotic-resistant bacteria that can influence the antibiotic resistome of agricultural soils. Application of some antibiotics can help bacteriophage-mediated transfer of ARGs in agricultural soil microbiomes (Ross and Topp 2015). Typical antibiotics that are widely used in livestock production are tetracyclines including chlortetracycline, oxytetracycline, tetracycline, and sulfonamides including sulfamethazine, sulfamethoxazole, and sulfadiazine (Tang et al. 2015). Fifteen tetracycline resistance genes (*tetA*, *tetG*, *tetM*, *tetO*, *tetQ*, *tetW*, and others) were detected in soil samples in the vicinity of nine swine farms located in Beijing, Tianjin, and Jiaxing of China (Wu et al. 2010). The fate of manure-originated antimicrobials applied to the soil will be affected by their sorption properties to soil particles and susceptibility to biotic and abiotic degradation (Joy et al. 2013).

8.3 Implications for Human Health

ARGs are readily caught by human pathogenic bacteria (HPB) to form *superbugs* such as *Salmonella, Bacteroidales, Campylobacter, Shigella, and E. coli* O157:H7 (Fischbach and Walsh 2009; Forsberg et al. 2012; Fang et al. 2014). Approximately 75% of antibiotics are not absorbed by animals and are excreted. Gastrointestinal bacteria are involved in antibiotic resistance selection, which are also excreted in manure (Chee-Sanford et al. 2009). It is a matter of concern to study the direct transfer of food-borne pathogens with clinically relevant resistances and the indirect transfer mediated through bacteria of clinically relevant resistance encoding mobile genetic elements such as plasmids, transposons, and bacteriophages from manure to human beings (Durso and Cook 2014).

These HPB species confer antibiotic resistance and pathogenicity and easily infect humans by contact or by consumption of raw vegetables (Fang et al. 2014). Horizontal transfer of these elements to bacteria adapted to soil or other habitats supports their environmental transmission independent of the original host (Heuer et al. 2011). The World Health Organization (WHO) is developing a 5-year Global

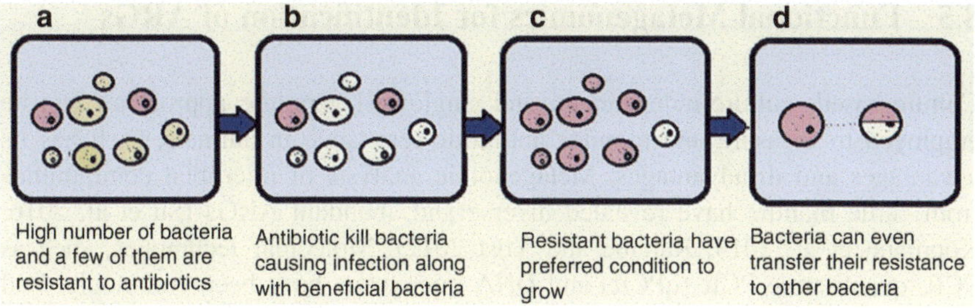

Fig. 8.1 Process of occurrence of antibiotic resistance

Antimicrobial Resistance Surveillance System to support the Global Action Plan on Antibiotic Resistance with a major focus on antibiotic resistance in human pathogens (Su et al. 2016). The ARGs were found in clusters originated from a taxonomically diverse set of species, suggesting that some microorganisms in manure harbor multiple resistance genes. Microbiomes of farm animals are reservoirs of antibiotic resistance genes, which might affect distribution of antibiotic resistance genes in human pathogens (Wichmann et al. 2014). Efforts to understand the missing links in the transfer of antibiotic resistance to human beings are mediated through molecular epidemiology between antibiotic resistance genes in the environment and those in human or animal pathogens and these studies might elucidate key transmission pathways (Smalla et al. 2016) (Fig. 8.1).

8.4 Analytical Methods for Determination of Soil Bound Antibiotics

Measurement of these compounds are limited by the complexity of the sample matrices and the difficultly in eliminating interferences that affect antibiotic detection. Efficient extraction methods combined with high sensitivity analysis by liquid chromatography/mass spectrometry can provide accurate quantification of antibiotics and their transformation products. In order to reflect their bioavailable fractions and effects in the environment, chemical analysis should be accompanied with biological assays (Aga et al. 2016). As a result, changes in soil microbial population such as the ability to degrade contaminants and their role in chemical cycles, such as nitrification might be significantly affected. Persistent antibiotics can accumulate in the top layers of soil, may leach to the ground water, or can be transported to surface waters. Sulfonamide antibiotics are persistent in the environment and do not bind strongly to soil that have been impacted by agricultural and human activities (Wegst-Uhrich et al. 2014).

8.5 Functional Metagenomics for Identification of ARGs

Culture-based, culture-independent, and single-cell genomic approaches can be employed to measure and monitor antibiotic resistance in animals; each has its advantages and disadvantages. Metagenomic analysis of microbial communities from cattle manure have revealed diverse and abundant ARGs (Su et al. 2016; Kopmann et al. 2013; Rahube and Yost 2012). Molecular techniques, such as PCR, quantitative PCR (qPCR) and DNA microarray have been commonly used to determine the fate of environmental ARGs (Li et al. 2015). Amplification-based methods such as PCR and qPCR have several limitations, including low-throughput, limited availability of primers, amplification bias, false-negative results due to inhibition in PCR and false-positive results due to nonspecific amplification. Zhu et al. (2013) reported that high-capacity quantitative PCR arrays have been applied to detect ARGs in manure, compost, and soil to overcome the capacity limitations.

Bacterial groups such as *Escherichia coli*, *Klebsiella pneumonia*, *Aeromonas* spp., *Pseudomonas aeruginosa*, *Enterococcus faecalis*, *Enterococcus faecium*, and genetic determinants like intl1; sul1 and sul2; $bla_{CTX\text{-}M}$ and bla_{TEM}, $bla_{NDM\text{-}1}$, bla_{VIM}, bla_{KPC}, qnrS, aac-(6′)-Ib-cr, vanA, mecA, ermB and ermF, aph have been suggested as possible indicators to assess the antibiotic resistance status in environmental settings (Berendonk et al. 2015). The genome location of resistance genes is likely to shift towards mobile genetic elements such as broad-host-range plasmids, integrons, and transposable elements (Heuer et al. 2011). ARGs can be identified by homology search in the databases, such as the Antibiotic Resistance Gene Database, the National Centre for Biotechnology Information, and the Comprehensive Antibiotic Resistance Database. In large metagenomes, sequence-based homology search has the advantage over automatic annotation of ARGs (McArthur et al. 2013).

Knapp et al. (2011) demonstrated an exponential increase of antibiotic resistance genes in agricultural soils, applying real-time PCR of antibiotic resistance genes to total DNA from archived soils, over six decades of increasing use of antibiotics. The abundances of 18 genes coding for resistance to beta-lactams, erythromycins, or tetracyclines were quantified in archived soil samples taken from five Dutch arable field sites. All three groups of resistance genes exponentially increased in copy numbers relative to 16S rRNA gene copies, which coincided with the dramatic increase in the production of the veterinary antibiotics since 1950. Durso et al. (2012) conducted a study that uses metagenomic sequencing to compare antibiotic resistance gene profiles in fecal, soil, and aquatic samples, including identification of taxa likely to be carrying specific types of resistance genes in each sample. This study raises the question of whether the underlying biology of individual bacterial taxa can contribute to their likelihood of transfer via the food chain. Microorganisms associated with agricultural manure have relative abundance of the clinical class 1 integron-integrase gene, *intI1*. *intI1* is linked to genes conferring resistance to antibiotics found in a wide variety of bacteria. A single DNA sequence variant of *intI1* is now found on a wide diversity of xenogenetic elements which are complex mosaic DNA elements fixed by means of human selection (Gillings et al. 2015).

Manure has been shown to promote the horizontal transfer of antibiotic resistance genes in soil by using vectors such as broad-host-range plasmids. Mainly the plasmid groups *IncP-1* and *IncQ* were highly abundant in 15 field-scale manures as revealed by southern blot hybridization of amplified backbone genes, and *IncN* and *IncW* plasmids were also frequently detected (Heuer et al. 2011). Among plasmids that transfer resistance genes, plasmids the *Inc18*-type has been identified as an important transmitter of antibiotic resistance. The host range of the *Inc18*-type plasmids spans from Gram-positive to Gram-negative bacteria. Enterococcal plasmids were shown to mediate the horizontal transfer of chromosomal regions up to 857 kbp in size (Manson et al. 2010).

The usage of DNA microarrays where detection of a large number of ARGs in a single assay is possible is limited by the possibility for cross-talk of different probes coupled with low-sensitivity restricts its applications for comprehensive surveys of ARGs in complicated environmental samples (Yang et al. 2013).

8.6 Influence of Treatment Strategies on ARGs

Treatment of animal manure before field application, usage of alternative bio-agent for disease treatment, and a well-targeted legalized use of antibiotics are the methods that can be suggested to limit the entry of veterinary antibiotics into the environment. It appears to be a challenging attempt when we try to limit the movement of biosolids in the environment because of their varying physiological interactions. Supervised inoculation of beneficial microorganisms and electron irradiation can be effective remediation strategies (Tasho and Cho 2016). Ye et al. (2016) evaluated the feasibility of using aqueous DNA solution as an agent for soil washing to remove PBDEs, heavy metals, antibiotics, ARG-contaminated soil, and examined the combined effects of ultrasonication and successive washings on the removal efficiency. This strategy was found to be environmentally friendly technology which is prominent for the risk assessment and management of mixed contaminated sites.

Composting has been shown to be effective in significantly reducing the levels of antibiotics in soil (Arikan et al. 2009; Selvam et al. 2012a). For example, Selvam et al. (2012b) manifested the abundance of ARGs and the bacterial diversity during composting of swine manure spiked with chlortetracycline, sulfadiazine,and ciprofloxacin at two different levels. After 28–42 days of composting, among resistance genes of tetracycline (*tet*Q, *tet*W, *tet*C, *tet*G, *tet*Z, and *tet*Y), sulfonamide (*sul*1, *sul*2, *dfr*A1, and *dfr*A7), and fluoroquinolone (*gyr*A and *par*C) reported, only *par*C, were detectable in the composting mass indicating that composting is a potential method of manure management. Polymerase chain reaction-denaturing gradient gel electrophoresis analysis of bacterial 16S rDNA of the composting mass indicated that the addition of antibiotics chlortetracycline, sulfadiazine, and ciprofloxacin, respectively, elicited only a transient perturbation and the bacterial diversity was restored in due course of composting. Two different biochars, namely rice straw biochar and mushroom biochar, had an opposite influence on the fate of ARGs during chicken manure composting (Cui et al. 2016). Cui et al. (2016) reported that in pig manure composting, mushroom biochar addition increased the average removal value of ARGs. The proper biochar should be considered when adding it into different manure compostings. The quality of poultry manure as organic fertilizer after anaerobic thermophilic treatment may raise significantly due to the elimination of ARG and self-transmissible plasmids (Anjum et al. 2017).

8.7 Antibiotic Resistance: Global Challenges

Treatment of antibiotic-resistant infections annually costs between US$21,000 and US$34,000 million in the United States alone and around 1500 million € in Europe (ECDC 2009). A lot of new generation antimicrobials have become ineffective against previously susceptible organisms which is a huge challenge for global health care management, especially for those involved in the development of new antibiotics (Jindal et al. 2015). Roca et al. (2015) suggested that a global and coordinated initiative will be needed to tackle antibiotic resistance to persuade the general population, policy makers, regulatory agencies, pharmaceutical companies, and the scientific community of the advantages of combating the threat of antimicrobial resistance.

There are several public databases and global surveillance projects, such as the Antimicrobial Resistance Global Report on Surveillance from the World Health Organization (WHO); the European Centre for Disease Prevention and Control (ECDC)-based European Antimicrobial Resistance Interactive Database (EARS-Net), EUCAST, and the European Antimicrobial Susceptibility Surveillance in Animals (EASSA) in Europe; the Surveillance Network Database (TSN) in the United States and Australia; and the Study for Monitoring Antimicrobial Resistance Trends (SMART) task force in the Asia–Pacific region. In addition, there are two centralized databases on ARGs: the Comprehensive Antibiotic Resistance Database (CARD) and the Antibiotic Resistance Genes Database (ARDB) (WHO 2014; Liu and Pop 2009; McArthur et al. 2013). The association of these data to both generic like European Molecular Biology Laboratory (EMBL) or GenBank and metagenomics databases such as Metagenomic Rapid Annotations using Subsystems Technology (MG-RAST) would further provide access to related sequence data and metadata from gene fragments to metagenomes, transcriptomes, and proteomes (Berendonk et al. 2015).

The US National Organic Program states that manure should be applied to soil at least 120 days prior to harvest for plants whose edible parts come into contact with soil (Code of Federal Regulations 2013). The application of manure from swine fed subtherapeutic concentrations of antibiotics requires more than 6 months to return to background levels of ARGs; in contrast, ARG levels in dairy manure-amended soils return to backgroundconcentrations within 6 months (Sandberg and LaPara 2016).

Increasing resistance to third-generation cephalosporins has been observed for *E. coli* and *Klebsiella pneumoniae*, with high proportions of these resistant isolates ascertained as extended-spectrum β-lactamase-positive (65–100% for *E. coli* and

75–100% for *K. pneumoniae*) (ECDC 2011). Combined resistance of *E. coli* to aminopenicillins, augmented suggestively in 10 of 28 European countries from 2007 to 2010. Combined resistance of *K. pneumoniae* to third-generation cephalosporins, fluoroquinolones, and aminoglycosides was found in 19% of isolates (Paphitou 2013).

8.8 Conclusions

More comprehensive understanding of the mechanisms associated with the acquisition and spread of antibiotic resistance needs to be achieved in a global perspective. Recent advances in functional screening and the growth of metagenomic databases contribute to characterize the diversity and prevalence of resistance genes present in the antibiotic resistome. It is essential to make concerted effort on the part of academic researchers, industry, and policy makers to battle against transfer of antibiotic-resistant genes from field to human beings.

Acknowledgements The author would like to thank Pondicherry University for providing Internet access tocollect the information essential for writing this book chapter.

References

Aga DS, Lenczewski M, Snow D, Muurinen J, Sallach JB, Wallace JS (2016) Challenges in the measurement of antibiotics and in evaluating their impacts in agroecosystems: a critical review. J Environ Qual 45(2):407–419

Anjum R, Grohmann E, Krakat N (2017) Anaerobic digestion of nitrogen rich poultry manure: impact of thermophilic biogas process on metal release and microbial resistances. Chemosphere 168:1637–1647

Arikan O, Mulbry W, Ingram D, Millner P (2009) Minimally managed composting of beef manure at the pilot scale: effect of manure pile construction on pile temperature profiles and on the fate of oxytetracycline and chlortetracycline. Bioresource Technol 100(19):4447–4453

Berendonk TU, Manaia CM, Merlin C, Fatta-Kassinos D, Cytryn E, Walsh F, Bürgmann H, Sørum H, Norström M, Pons MN, Kreuzinger N (2015) Tackling antibiotic resistance: the environmental framework. Nat Rev Microbiol 13(5):310–317

Chee-Sanford JC, Mackie RI, Koike S, Krapac IG, Lin YF, Yannarell AC, Maxwell S, Aminov RI (2009) Fate and transport of antibiotic residues and antibiotic resistance genes following land application of manure waste. J Environ Qual 38(3):1086–1108

Chen H, Zhang M (2013) Occurrence and removal of antibiotic resistance genes in municipal wastewater and rural domestic sewage treatment systems in eastern China. Environ Int 55:9–14

Code of Federal Regulations (2013) Title 7, Part 205

Cui E, Wu Y, Zuo Y, Chen H (2016) Effect of different biochars on antibiotic resistance genes and bacterial community during chicken manure composting. Bioresource Technol 203:11–17

Cui E, Wu Y, Jiao Y, Zuo Y, Rensing C, Chen H (2017) The behavior of antibiotic resistance genes and arsenic influenced by biochar during different manure composting. Environ Sci Pollut Res Int 24(16):14484–14490

Cytryn E (2013) The soil resistome: the anthropogenic, the native, and the unknown. Soil Biol Biochem 63:18–23

Ding GC, Radl V, Schloter-Hai B, Jechalke S, Heuer H, Smalla K, Schloter M (2014) Dynamics of soil bacterial communities in response to repeated application of manure containing sulfadiazine. PLoS One 9(3):e92958
Durso LM, Cook KL (2014) Impacts of antibiotic use in agriculture: what are the benefits and risks? Curr Opin Microbiol 19:37–44
Durso LM, Miller DN, Wienhold BJ (2012) Distribution and quantification of antibiotic resistant genes and bacteria across agricultural and non-agricultural metagenomes. PLoS One 7:e48325
European Centre for Disease Prevention and Control (ECDC) (2009) European Medicines Agency (EMEA) technical report, The bacterial challenge: time to react. Available at: http://www.ema.europa.eu/docs/en_GB/document_library/Report/2009/11/WC500008770.pdf. Accessed 29 Apr 17
European Centre for Disease Prevention and Control (ECDC) (2011) Antimicrobial resistance surveillance in Europe 2010. Annual report of the European Antimicrobial Resistance Surveillance Network (EARS-Net). ECDC, Stockholm
Fang H, Wang H, Cai L, Yu Y (2014) Prevalence of antibiotic resistance genes and bacterial pathogens in long-term manured greenhouse soils as revealed by metagenomic survey. Environ Sci Technol 49(2):1095–1104
Fischbach MA, Walsh CT (2009) Antibiotics for emerging pathogens. Science 325(5944):1089–1093
Forsberg KJ, Reyes A, Wang B, Selleck EM, Sommer MO, Dantas G (2012) The shared antibiotic resistome of soil bacteria and human pathogens. Science 337(6098):1107–1111
Gillings MR, Gaze WH, Pruden A, Smalla K, Tiedje JM, Zhu YG (2015) Using the class 1 integron-integrase gene as a proxy for anthropogenic pollution. ISME J 9(6):1269–1279
Hashmi MZ, Mahmood A, Kattel DB, Khan S, Hasnain A, Ahmed Z (2017) Antibiotics and antibiotic resistance genes (ARGs) in soil: occurrence, fate, and effects. In: Xenobiotics in the soil environment. Springer, Cham, pp 41–54
He LY, Liu YS, Su HC, Zhao JL, Liu SS, Chen J, Liu WR, Ying GG (2014) Dissemination of antibiotic resistance genes in representative broiler feedlots environments: identification of indicator ARGs and correlations with environmental variables. Environ Sci Technol 48 (22):13120–13129
He LY, Ying GG, Liu YS, Su HC, Chen J, Liu SS, Zhao JL (2016) Discharge of swine wastes risks water quality and food safety: antibiotics and antibiotic resistance genes from swine sources to the receiving environments. Environ Int 92:210–219
Heuer H, Schmitt H, Smalla K (2011) Antibiotic resistance gene spread due to manure application on agricultural fields. Curr Opin Microbiol 14(3):236–243
Jechalke S, Cook KL, Smalla K (2015) Approaches to assess the effects and risks of veterinary antibiotics applied with manure to soil. In: Antimicrobial resistance—an open challenge. InTech, Rijeka
Jindal AK, Pandya K, Khan ID (2015) Antimicrobial resistance: a public health challenge. Med J Armed Forces India 71(2):178–181
Joy SR, Bartelt-Hunt SL, Snow DD, Gilley JE, Woodbury BL, Parker DB, Marx DB, Li X (2013) Fate and transport of antimicrobials and antimicrobial resistance genes in soil and runoff following land application of swine manure slurry. Environ Sci Technol 47(21):12081–12088
Knapp CW, SM MC, Singh BK, Campbell CD, Hudson G, Graham DW (2011) Antibiotic resistance gene abundances correlate with metal and geochemical conditions in archived Scottish soils. PLoS One 6(11):e27300
Kopmann C, Jechalke S, Rosendahl I, Groeneweg J, Krögerrecklenfort E, Zimmerling U, Weichelt V, Siemens J, Amelung W, Heuer H, Smalla K (2013) Abundance and transferability of antibiotic resistance as related to the fate of sulfadiazine in maize rhizosphere and bulk soil. FEMS Microbiol Ecol 83(1):125–134
Li B, Yang Y, Ma L, Ju F, Guo F, Tiedje JM, Zhang T (2015) Metagenomic and network analysis reveal wide distribution and co-occurrence of environmental antibiotic resistance genes. ISME J 9(11):2490–2502
Liu B, Pop M (2009) ARDB—antibiotic resistance genes database. Nucleic Acids Res 37(suppl 1): D443–D447

Manson JM, Hancock LE, Gilmore MS (2010) Mechanism of chromosomal transfer of *Enterococcus faecalis* pathogenicity island, capsule, antimicrobial resistance, and other traits. Proc Natl Acad Sci USA 107(27):12269–12274

McArthur AG, Waglechner N, Nizam F, Yan A, Azad MA, Baylay AJ, Bhullar K, Canova MJ, De Pascale G, Ejim L, Kalan L (2013) The comprehensive antibiotic resistance database. Antimicrob Agents Chemother 57(7):3348–3357

Paphitou NI (2013) Antimicrobial resistance: action to combat the rising microbial challenges. Int J Antimicrob Agents 42:S25–S28

Perry JA, Westman EL, Wright GD (2014) The antibiotic resistome: what's new? Curr Opin Microbiol 21:45–50

Pruden A, Larsson DJ, Amézquita A, Collignon P, Brandt KK, Graham DW, Lazorchak JM, Suzuki S, Silley P, Snape JR, Topp E (2013) Management options for reducing the release of antibiotics and antibiotic resistance genes to the environment. Environ Health Perspect 121 (8):878

Rahube TO, Yost CK (2012) Characterization of a mobile and multiple resistance plasmid isolated from swine manure and its detection in soil after manure application. J Appl Microbiol 112 (6):1123–1133

Roca I, Akova M, Baquero F, Carlet J, Cavaleri M, Coenen S, Cohen J, Findlay D, Gyssens I, Heure OE, Kahlmeter G (2015) The global threat of antimicrobial resistance: science for intervention. New Microbes New Infect 6:22–29

Ross J, Topp E (2015) Abundance of antibiotic resistance genes in bacteriophage following soil fertilization with dairy manure or municipal biosolids, and evidence for potential transduction. Appl Environ Microbiol 81(22):7905–7913

Sandberg KD, LaPara TM (2016) The fate of antibiotic resistance genes and class 1 integrons following the application of swine and dairy manure to soils. FEMS Microbiol Ecol 92(2):fiw001

Sanderson H, Fricker C, Brown RS, Majury A, Liss SN (2016) Antibiotic resistance genes as an emerging environmental contaminant. Environ Rev 24(2):205–218

Selvam A, Xu D, Zhao Z, Wong JW (2012a) Fate of tetracycline, sulfonamide and fluoroquinolone resistance genes and the changes in bacterial diversity during composting of swine manure. Bioresource Technol 126:383–390

Selvam A, Zhao Z, Wong JW (2012b) Composting of swine manure spiked with sulfadiazine, chlortetracycline and ciprofloxacin. Bioresource Technol 126:412–417

Smalla K, Simonet P, Tiedje J, Topp E (2016) Editorial: Special section of FEMS microbiology ecology on the environmental dimension of antibiotic resistance. FEMS Microbiol Ecol 92 (11):fiw172

Su JQ, Cui L, Chen QL, An XL, Zhu YG (2016) Application of genomic technologies to measure and monitor antibiotic resistance in animals. Ann N Y Acad Sci 1388(1):121–135

Tang X, Lou C, Wang S, Lu Y, Liu M, Hashmi MZ, Liang X, Li Z, Liao Y, Qin W, Fan F (2015) Effects of long-term manure applications on the occurrence of antibiotics and antibiotic resistance genes (ARGs) in paddy soils: evidence from four field experiments in south of China. Soil Biol Biochem 90:179–187

Tasho RP, Cho JY (2016) Veterinary antibiotics in animal waste, its distribution in soil and uptake by plants: a review. Sci Total Environ 563:366–376

Tien YC, Li B, Zhang T, Scott A, Murray R, Sabourin L, Marti R, Topp E (2017) Impact of dairy manure pre-application treatment on manure composition, soil dynamics of antibiotic resistance genes, and abundance of antibiotic-resistance genes on vegetables at harvest. Sci Total Environ 581:32–39

Wegst-Uhrich SR, Navarro DA, Zimmerman L, Aga DS (2014) Assessing antibiotic sorption in soil: a literature review and new case studies on sulfonamides and macrolides. Chem Cent J 8(1):5

Wichmann F, Udikovic-Kolic N, Andrew S, Handelsman J (2014) Diverse antibiotic resistance genes in dairy cow manure. MBio 5(2):e01017–e01013

World Health Organization (2014) Antimicrobial resistance: 2014 global report on surveillance. World Health Organization

Wu N, Qiao M, Zhang B, Cheng WD, Zhu YG (2010) Abundance and diversity of tetracycline resistance genes in soils adjacent to representative swine feedlots in China. Environ Sci Technol 44(18):6933–6939

Yang Y, Li B, Ju F, Zhang T (2013) Exploring variation of antibiotic resistance genes in activated sludge over a four-year period through a metagenomic approach. Environ Sci Technol 47 (18):10197–10205

Yang Q, Ren S, Niu T, Guo Y, Qi S, Han X, Liu D, Pan F (2014) Distribution of antibiotic-resistant bacteria in chicken manure and manure-fertilized vegetables. Environ Sci Pollut Res 21 (2):1231–1241

Ye M, Sun M, Wan J, Zhao Y, Xie S, Tian D, Hu F, Li H, Zong L, Kengara FO, Xin J (2016) Feasibility of an enhanced washing process to extract PBDEs/heavy metals/antibiotics from antibiotic resistance gene-affected soil with aqueous DNA followed by microbial augmentation. J Soil Sediments 16(3):954–965

Zhu YG, Johnson TA, Su JQ, Qiao M, Guo GX, Stedtfeld RD, Hashsham SA, Tiedje JM (2013) Diverse and abundant antibiotic resistance genes in Chinese swine farms. Proc Natl Acad Sci USA 110(9):3435–3440

Chapter 9
Antibiotics and Antibiotics Resistance Genes Dissemination in Soils

Eddie Cytryn, Zdzisław Markiewicz, and Magdalena Popowska

9.1 Introduction

Soil plays an important role in the ecosystem. It functions as a nutrition source, provides habitat for plants and other organisms, and also serves as an enormous bioreactor, where processes of pollutant decomposition and nutrient modification are carried out. It is also a place of interactions between autochthonic microorganisms and antibiotics, which flow into the soil with natural fertilizers and wastewaters. In alkaline or neutral soils, arable land with humus, at a depth of 5–30 cm, there are 3–15 tons of bacteria per 1 ha. In 1 g of soil, there can be billions of bacteria. There are an estimated 60,000 different bacterial species. Only 1% of all bacteria can be cultured in laboratory conditions; most of them are unknown. Only through metagenomic studies can we learn about their diversity and species richness. Most live in the top 10–15 cm of soil, where organic matter is present. Soil bacteria carry out many important processes in the environment: decomposition of organic materials (e.g., *Bacillus subtilis* and *Pseudomonas fluorescens*; a number of decomposers can break down pesticides and toxic pollutants in soil), nitrogen fixation (*Rhizobium*—this form of nitrogen fixation can add the equivalent of more than 100 kg of nitrogen per hectare per year, or *Azotobacter*, *Azospirillum*, *Agrobacterium*,

The original version of this chapter was revised. An erratum to this chapter can be found at DOI 10.1007/978-3-319-66260-2_22.

E. Cytryn
Institute of Soil, Water and Environmental Sciences, Volcani Center, Agricultural Research Organization, Bet Dagan, Israel
e-mail: eddie@volcani.agri.gov.il

Z. Markiewicz • M. Popowska (✉)
Department of Applied Microbiology, Institute of Microbiology, Faculty of Biology, University of Warsaw, Warsaw, Poland
e-mail: z.markiewicz@gmail.com; magdapop@biol.uw.edu.pl

M.Z. Hashmi et al. (eds.), *Antibiotics and Antibiotics Resistance Genes in Soils*, Soil Biology 51, DOI 10.1007/978-3-319-66260-2_9

Gluconobacter, *Flavobacterium*, and *Herbaspirillum* are all examples of free-living, nitrogen-fixing bacteria, often associated with nonlegumes), nitrification (*Nitrosolobus multiformis*-, *Nitrosomonas europaea*-, *Nitrococcus mobilis*-, and *Nitrobacter winogradskyi*-nitrifying bacteria oxidize ammonium (NH_{4+}) to nitrite (NO_2–) and then to nitrate (NO_{3-})), denitrification (*Pseudomonas*-, *Bacillus*, *Alcaligenes*-, and *Flavobacterium*-denitrifying bacteria convert nitrate to nitrogen (N_2) or nitrous oxide (N_2O) gas), breakdown of humates and humic acids in soil (Actinobacteria), sulfur oxidation (*Thiobacillus* bacteria can convert sulfides into sulfates—a form of sulfur accessible for plants), degradation of starch (*Bacillus polymyxa*, *B. subtilis*, *B. macerans*, *B. stearothermophilus*, *Pseudomonas stutzeri*, *Clostridium* spp.), urea decomposition (*Urobacterium*, *Urobacillus*, *Sporosarcina*, *Proteus*, *Bacillus* sp., *Corynebacterium* sp., *Clostridium* sp.), ammonification (*Bacillus*, *Proteus*, *Pseudomonas*), or proteolysis (*Pseudomonas fluorescens*, *P. aeruginosa*, *Bacillus mycoides*, *Clostridium sporogenes*).

In the soil there are also plant pathogens, including *Zymomonas* spp., *Erwinia* spp., and *Agrobacterium* spp., and human and animal pathogens, e.g., *Listeria monocytogenes*, *Salmonella* spp., *Shigella* sp., *Yersinia pestis*, *Bordetella pertussis*, *Bacillus anthracis*, *Clostridium botulinum* (Buscot and Varma 2005; Nannipieri et al. 2003).

The above data indicate that many soil microbial functions may be essential for efficient plant growth. For example, the rate of soil organic material decomposition by soil microorganisms governs the supply of organic and inorganic forms of nitrate and nitrogen to plants (Herridge et al. 2008; Dungait et al. 2012). Inhibition of these processes by antibiotics may therefore affect the access of crop plants to nitrogen, an element essential to efficient plant growth. The impact of antibiotics on nitrogen-fixing plant symbionts, such as *Rhizobia*, which supply up to 80% of total nitrogen in pasture legume plants (Xie et al. 2012), is also possible (Kleineidam et al. 2010; Cevheri 2012). Thus, antibiotic contamination of agricultural soils can have serious implications for agriculture.

Antibiotics have revolutionized medicine, but the increasing intake and inappropriate use of these drugs, especially in nonhospital treatment, agriculture, veterinary medicine, and, until recently, animal husbandry (growth stimulators), resulted in rapid escalation of bacterial resistance (Gootz 2010). Shortly after antibiotics went into common usage, antibiotic-resistant bacteria (ARB) and multidrug resistance (MDR) emerged and rapidly spread (Aminov 2009). The only way to fight this process effectively is to determine the role of antibiotic resistance in biology and evolution of bacteria and the means of its dissemination and also to identify genes and mechanisms of resistance. Currently used antibiotics act extensively on bacterial cells, targeting crucial life processes, such as DNA replication, RNA synthesis, cell wall synthesis, or protein synthesis. Drug performance is a very complex process. It starts from physical interaction of the molecule with its molecular target, which leads to growth inhibition (bacteriostatic action) and/or cell death (bactericidal action) (Chopra and Brennan 1998; Brötz-Oesterhelt and Brunner 2008). It would seem that we have "an ideal weapon" against pathogenic bacteria. However, there are multiple antibiotic resistance genes (sometimes several genes determining resistance to one antibiotic), while only a few mechanisms of antibiotic action in the bacterial cell.

Over 20,000 genes are recognized as potentially coding resistance to antibiotics of nearly 400 different types (Liu and Pop 2009). Infections caused by ARB are extremely difficult or at times even impossible to cure. As a result of the omnipresence of antibiotics and other pharmaceuticals, questions rise regarding the direction of resistance transfer, dependence on industrial pollution, and prevention possibilities.

Resistance dissemination is driven mainly by antibiotic resistance genes' (ARGs) presence on mobile genetic elements (MGEs), which comprise, i.e., plasmids, genomic islands, and transposable elements (Frost et al. 2005). These small DNA molecules are transferred between cells by mechanisms of horizontal gene transfer (HGT), using one of three processes: conjugation (bacterium-bacterium), transformation (bacterium-exogenous DNA), or transduction (bacterium-bacteriophage). However, to maintain the acquired genetic information within the cell and express encoded proteins, selective pressure in the surrounding environment is necessary. Production of certain proteins related to specific resistance mechanisms brings about a considerable energy input for the cell. Only when the pressure occurs, that is, when antibiotic is present in the environment, the cell would try to survive at any cost and therefore undertake costly protein synthesis and maintenance of the acquired genetic material (Levy and Marshall 2004). What should be emphasized, a mechanism for co-selection has been observed, which is connected with a prevalent pollution of the environment by heavy metals and detergents (Martínez 2008, 2009).

Numerous research results indicate that wastewater treatment plants and soil are hotspots for antibiotic resistance dissemination, both considered significant reservoirs of antibiotic resistance genes (Piotrowska and Popowska 2014, 2015; Piotrowska et al. 2017; Miller et al. 2016; Popowska et al. 2012). This is caused by the presence of antibiotics, as well as antibiotic-resistant bacteria, including human and animal pathogens, within these areas. Both habitats share the characteristic of suitable reproduction conditions, resulting in high bacterial number and encouraging genetic transfer by HGT (Zhang et al. 2011; Ding and He 2010; Thanner et al. 2016).

Most antibiotics are not easily degradable in soil, and they remain active for a long time (Thiele-Bruhn 2003). Resistant and multidrug-resistant bacteria, resistance genes, and antibiotics themselves may spread through surface water, posing a real threat to human and animal health.

Bacterial antibiotic resistance is widespread, yet particularly soil bacteria carry multiple distinct resistance determinants against natural and synthetic antibiotics, commonly used in treatment (D'Costa et al. 2006; D'Costa et al. 2007; Wright 2010). What is interesting, antibiotic resistance genes identified in the soil turned out to be identical or very similar to those of clinical significance, found in resistant human pathogens (Forsberg et al. 2012). Many studies demonstrate that environmental resistance dissemination poses a constant threat to clinical antibiotic use, even in the case of novel synthetic compounds, while the emergence of pathogens resistant to a new drug is still likely.

Metagenomic and functional analyses proved the existence of resistance genes in uncharted lands (D'Costa et al. 2011; Bhullar et al. 2012), which clearly states that antibiotic resistance surpasses the clinical use. Nevertheless, unlike natural/inner resistance, developed by bacterial populations before the antibiotic era, acquired resistance is a consequence of human actions. The pool of resistance genes related to plasmids and other mobile genetic elements hosted by bacterial pathogens and environmental strains as well indicates a potential existing throughout the world, ready to transfer in case selective pressure occurs, such as the presence of antibiotics or other compounds, including heavy metals.

Antimicrobials used in human and animal therapy reach the natural environment through manure, wastewater, agriculture, etc. To estimate the occurrence and range of ARB and ARGs in environmental strains, numerous studies have been carried out. The results demonstrate unambiguously that the natural environment is the largest and oldest repository of potential ARGs and that soil bacteria form a reservoir of resistance-determining factors, which may be mobilized to transfer to pathogenic bacteria (Aminov and Mackie 2007; Allen et al. 2010; D'Costa et al. 2011; Knapp et al. 2010). One of excellent examples is the resistance to tetracycline, an antibiotic from the tetracyclines group, an important class of broad-spectrum antimicrobials against numerous pathogens. Rising prevalence of resistance among opportunistic pathogens widespread in soils, e.g., *Pseudomonas aeruginosa*, *Acinetobacter* spp., *Burkholderia* spp., and *Stenotrophomonas* spp., as a result of selective pressure, has had an enormous impact on the clinical use of these pharmaceuticals (Thaker et al. 2010; Popowska et al. 2010). Furthermore, resistance to many other antibiotics, such as aminoglycosides, macrolides, fluoroquinolones, sulfonamides and β-lactams, and other drugs containing telithromycin, has already been identified in soil (Allen et al. 2010; Popowska et al. 2012; Riesenfeld et al. 2004).

Those determinants are located mainly on MGEs, which ensure their dissemination by HGT (Martínez 2009; Stokes and Gillings 2011). Multiple studies showed that the frequency of resistance plasmids occurrence in soils is very high. Among the replicons contributing to antimicrobials resistance, representatives of P, Q, N, and W incompatibility groups were identified. IncP-1 plasmids may serve as an example of such genetic elements (Popowska and Krawczyk-Balska 2013). Plasmids carrying resistance genes have been isolated from pathogenic bacteria from genera *Escherichia*, *Salmonella*, *Shigella*, *Klebsiella*, *Aeromonas*, and *Pseudomonas*, which reside in soil and water (Stokes and Gillings 2011). What is important, those replicons encode factors determining resistance to at least one heavy metal (Ni, Cd, Co, Cu, Hg, Pb, Zn) and antimicrobials from different groups, i.e., tetracyclines, quinolones, aminoglycosides, sulfonamides, β-lactams, and chemotherapeutic agents (Sen et al. 2011; Seiler and Berendonk 2012). The alternative mechanism resulting in antibiotic resistance—mutations—should also be borne in mind, especially those in genes encoding transcriptional and translational apparatus, which lead to changes in global metabolism (Derewacz et al. 2013). There is abundant evidence concerning HGT between environmental bacteria and clinical pathogens, obtained by high-throughput functional metagenomic approach

(Forsberg et al. 2012). It was shown that, i.e., multidrug-resistant soil bacteria carry gene cassettes, encoding resistance to five classes of antimicrobials (β-lactams, aminoglycosides, amphenicols, sulfonamides, and tetracyclines), with high nucleotide identity to genes from diverse human pathogens. Because of this, it is essential to examine the level of antibiotic resistance in soil environment, with an emphasis on determining the latent potential of the resistance mechanisms, including novel, but clinically important mechanisms, or novel MDR pathogens.

9.2 Antibiotics in Soils

It seems that in nature, the role of antibiotics in soil is rather different from what could be expected. Some of them, for instance, rifampicin, gentamicin, streptolydigin, or some beta-lactams, have been found to modulate the transcription of virulence—or motility-associated genes—which may play a role in the maintenance of microbial communities in the soil (Yim et al. 2006, 2007). Beta-lactams were shown to strongly enhance the expression of the alpha-toxin coding *hla* gene of *Staphylococcus aureus*, whereas the macrolide erythromycin and certain aminoglycoside antibiotics reduced the expression of this gene. Subinhibitory concentrations of the beta-lactam ampicillin were found to cause genomic rearrangements in some pathogenic enteric bacteria, with subsequent dissemination of strains having increased virulence. It would seem that sub-MIC concentrations of antibiotic molecules resulting in behavioral changes of microbes affected would be environmentally relevant and of greater importance than cidal amounts. In this case antibiotics serve as signal molecules in inter-bacterial communication, bringing about changes in the characteristics of a microbial community (Sengupta et al. 2013). There are many examples of signaling activity. Subinhibitory concentrations of various antibiotics have been shown to induce different states in bacteria like the SOS response, biofilm formation, or changes in primary metabolism, which may even include tolerance to antibiotics (Anderson and Hughes 2012; Bernier and Surette 2013; Tsui et al. 2004; Yim et al. 2011). Antibiotics have been found to have sex pheromone activity, stimulating bacterial conjugation, and in the case of *Streptococcus pneumoniae*, a competence-stimulating peptide involved in genetic transformation can stimulate the production of various bacteriocins. Subsequent DNA exchange is enhanced by the secretion of the bacteriocins, suggesting that co-stimulation of bacteriocins with competence provides an adaptive advantage in the exchange of genetic material (Wholey et al. 2016). Another example of the role of low doses of antibiotics in microbial communities is the inhibition of the outgrowth of *Clostridium difficile* spores by fidaxomicin and vancomycin (Babakhani et al. 2011; Allen et al. 2013).

Antibiotic molecules have also been shown to play a significant role in quorum sensing. The dialkylresorcinols (DARs), including 2-hexyl-5-propyl-alkylresorcinol (HPR), produced by many bacteria, such as *Pseudomonas* aurantiaca, have moderate antibacterial and antifungal properties, and some of their analogs, like resorstatin, can act as free radical scavengers (Kato et al. 1993). The DARs have also been recently shown to be implicated as novel bacterial signaling molecules that are sensed by a LuxR homologous receptor. They are endogenously produced in a specific subset by the insect and human pathogenic bacterium *Photorhabdus asymbiotica* as well as by many other bacterial species (Brameyer et al. 2015). Molecules with antibiotic activity that also play a role as intracellular signaling molecules are synthesized by a wide number of bacterial species. A good example are the quinolones belonging to the family of 2-alkyl-4-quinolones (AQs) produced by *Pseudomonas aeruginosa*. Besides having antimicrobial properties, compounds of this family act as quorum-sensing signal molecules, controlling the expression of many virulence genes as a function of cell population density (Heeb et al. 2011; Rampioni et al. 2016).

Antibiotics can enter the environment through very many different routes. They can be released from a hospital setting as hospital wastewater effluent. Sludge from wastewater treatment plants can be dispersed on fields as a fertilizer or released directly into surface waters. Antibiotics are also used therapeutically or as growth promoters in livestock husbandry and poultry breeding. Manure from antibiotic-fed animals exacerbates the resistance spread, as demonstrated by the high levels in manure-amended vegetable garden soils. The spread of resistance and multiresistant strains of pathogens and opportunistic bacteria that can infect humans and animals is aided and enhanced by the fact that they are frequently carried on mobile genetic elements, notably plasmids and transposons, that can be transferred not only among bacteria of the same species, but among different species creating an interesting but at the same time dangerous soil resistome (Popowska et al. 2012). Many other studies have been carried out with similar results. It has been shown, for instance, that ARGs are more abundant in rivers downstream of wastewater purification plants, pharmaceutical industry or other anthropogenically impacted areas such as farms or orchards where antibiotics are used. Some examples of such activities are horrifying, like the discarding by pharmaceutical producers of ciprofloxacin in excess of 50 kg a day into rivers in central India (Fick et al. 2009). It is highly probable that ARGs and resistant bacteria mix with the indigenous flora. Such environments are very likely "hotspots" where new resistant strains can be readily created by HGT (Walsh 2013; Berglund 2015; Blazquez et al. 2012).

Antibiotics, and resistance genes therefore, occur naturally in soil due to the "arms race" between microbial species competing for nutrients. Almost 50% of *Actinomycetes* isolated from soil are capable of synthesizing antibiotics, which provide a natural antibiotic residue in soils, but the use of antibiotics to promote livestock growth boosts the resistance pool to a whole new level, as demonstrated by the differences in antibiotics and ARG levels between agricultural and forested soils. Manure from antibiotic-fed animals exacerbates the resistance spread, as demonstrated by the high levels in manure-amended vegetable garden soils. The spread of resistance and multiresistant strains of pathogens and opportunistic bacteria that can infect humans and animals is aided and abetted by the fact that they are frequently carried on mobile genetic elements, notably plasmids and transposons, that can be transferred not only among bacteria of the same species, but among different species creating an interesting but at the same time dangerous soil resistome (Popowska et al. 2012). Many other studies have been carried out with similar results. It has been shown, for instance, that ARGs are more abundant in rivers downstream of wastewater purification plants, pharmaceutical industry, or other anthropogenically impacted areas such as farms or orchards where antibiotics are used. Some examples of such activities are horrifying, like the discarding by pharmaceutical producers of ciprofloxacin in excess of 50 kg a day into rivers in central India (Fick et al. 2009).

The examples presented above clearly show and stress several aspects of antibiotic production and resistance to these compounds. It is indisputable that antibiotics have been produced by microorganisms for a very long time, well before the appearance of humans. Currently, over 80% of the antibiotics used in clinical practice are of soil origin. The soil is very rich in antibiotic-producing organisms producing a rich variety of antibiotic molecules, many of which have not yet been identified. The same is true for antibiotic degrading molecules involved in resistance mechanisms. Structure-based phylogeny of serine and metallo-β-lactamases, for example, established that these ancient enzymes originated more than 2 billion years ago, with some serine β-lactamases being present on plasmids for millions of years (Hall and Barlow 2004; Garau et al. 2005). Phylogeny of the β-lactamase and housekeeping genes is highly compatible in *Klebsiella oxytoca* implying that these genes have been evolving in this host for over 100 million years (Fevre et al. 2005). A similar phylogenetic analysis of β-lactamases in metagenomic clones derived from the 10,000 years old cold-seep sediments in the Bismarck Archipelago area indicated that most of the diversity of these enzymes is the result of ancient evolution (Song et al. 2005; Davies and Davies 2010).

In the following sections, we present current knowledge on antibiotics in soil, antibiotic-producing bacteria in soil, dissemination of antibiotics in soil, and contribution of agriculture and animal husbandry in this phenomenon; and finally, we present a scheme of general sources of antibiotics ARB and ARG in wastewater treatment plants and soil, in an attempt to better understand the complexity of the problem of antibiotic resistance in soil.

9.2.1 Antibiotic-Producing Bacteria in Soil

It is to be kept in mind that soil is inhabited by microorganisms able to synthesize antibiotics (Thiele-Bruhn 2003). Considering quantity and variety of bacteria species, part of which is capable of production of several dozen of biologically active compounds, a chemical diversity of soil microbial compounds may be well estimated (Wright 2010). For example, bacteria of *Actinobacteria* class produce millions of biologically active molecules (Allen et al. 2010; Weber et al. 2015). As proven, over half of these bacteria possess the capacity to synthesize antimicrobials. Most of these are found in the rhizosphere in concentrations up to 5 μg/g of soil

Table 9.1 Bacteria- and fungi-producing antibiotics

Bacteria	Antibiotic	Fungi	Antibiotic
Streptomyces spp.		*Penicillium* spp.	
S. griseus	Streptomycin	*P. notatum*	Penicillin
S. spectabilis	Spectinomycin		
S. erythreus	Erythromycin		
S. aureofaciens	Tetracycline		
S. venezuelae	Chloramphenicol		
S. orientalis	Vancomycin		
S. teichomyceticus	Teicoplanin		
Micromonospora spp.		*Cephalosporium* spp.	Cephalosporin
M. purpurea	Gentamicin		
Bacillus spp.			
B. licheniformis	Bacitracin		
B. brevis	Gramicidin		
B. polymyxa	Polymyxin		

Data selected based on Weber et al. (2015)

(Thiele-Bruhn 2003). Among them, clinically important antibiotics have been identified.

Generally, microorganisms are the main antibiotic-producing organisms (Table 9.1). More than 60% of whole production is attributed to microbes, including over 50% of the most popular antimicrobial pharmaceuticals, such as tetracyclines, gentamycin, or erythromycin, synthesized by *Actinobacteria* from *Actinomycetales* order (*Streptomyces*, *Micromonospora*, *Saccharopolyspora* genus) (Procópio et al. 2012), the rest being synthesized mainly by bacilli from the genus *Bacillus* (Mannanov and Sattarova 2001), e.g., gramicidin-producing *Bacillus brevis* (Baltz 2007). Fungi generate about 16–18% of all antibiotics, with imperfect fungi accounting for 10–12%, while *Basidiomycota* and Ascomycota for about 6%. The remnant is produced by lichens, plants, and animals.

The question therefore arises, what is the role of antibiotic production and resistance to these compounds in a natural environment, such as the soil? Fortunately, the problems of drug production by microorganisms and microbial drug and multidrug resistance, though difficult, are amenable to experimental studies which are beginning to shed some light on the issue. However, most of the information we have on antibiotic production and resistance to antibiotics is from detailed studies embracing pure cultures of a single species or even strain, and this knowledge can by no means be extrapolated to what goes on in the natural environment, such as soil. This research, though some progress has been made, can regrettably be said to be still in its infancy.

Antibiotics in the soil are produced by microorganisms as secondary metabolites at concentrations far lower than those used, e.g., in medicine. Even though their production does not seem to be a prerequisite for survival, in most cases, it involves many steps and products of antibiotic resistance genes (ARGs) as well as the natural expenditure of energy, which shows that these antibiotics must serve some important purpose. In nature microbial communities are composed of very many diverse species which exist as a community or network of cells that frequently have to compete for nutrients. When the latter become scarce, the microbes begin to secrete an abundance of secondary metabolites, including those with antibiotic activity. Davies (2011) coined the phrase parvome for this enormous array of mostly low molecular weight compounds. The molecules forming the parvome have only recently begun to be studied, with particular focus on those with antibiotic activity. Specific examples of the use of antibiotic molecules produced by members of one microbial species to eliminate competing bacteria are few and far between (e.g., Neeno-Eckwall et al. 2001; Cafaro and Currie 2005), this most likely because these exist in the environment in amounts far below their minimum inhibitory concentrations (MICs) against other species. Moreover, most of the very many microorganisms that produce antibiotics also carry genes coding resistance determinants for protection. In fact, antibiotic producers may have been the original sources for many of the antibiotic resistance genes circulating in the clinic environment (Ilumeniuk et al. 2002; Perry and Wright 2013).

9.2.2 *Dissemination of Antibiotics in Soil*

As a consequence of human actions in natural ecosystems such as soil, an increase in antibiotic concentration has been observed (Allen et al. 2010). Most of the antibiotics, delivered to prevent or cure human and animal infections, after being metabolized, are released into the environment in often unaltered form (Martínez 2009). The sources of antibiotics, ARB, and ARGs in soil are illustrated in Fig. 9.1. The average antibiotic level in soil ranges between 0.8 and 2700 μg/kg; the highest levels are detected in soils fertilized with manure (De La Torre et al. 2012). Antibiotics are washed off with groundwater and move forward.

Some antibiotics, e.g., amoxicillin and erythromycin, are used to eliminate human pathogens and in veterinary medicine as well, but they also were applied as growth promoters (Table 9.2). On January 1, 2006, a ban on merchandising the feed supplemented with antibiotics as growth stimulators became effective (http://europa.eu). Until 2006, 90% of antibiotics used in agriculture had been destined for growth stimulation and only 10% for fighting bacterial diseases. Statistical data indicate that in the last 50 years, over one million of tons of antibiotics were introduced into the environment, 50% derived from veterinary medicine and agriculture. Unfortunately, numerous news bulletins and journalistic provocations revealed the existence of a black market in this area. Greed for money drives the reckless use of antibiotics in agriculture (Allen et al. 2010). Antibacterial compounds are still applied in fish farming (carp, salmon, trout), dispersed over farmlands and orchards (streptomycin, tetracycline), and used for improving the freshness of vegetables, fruits, and flowers (McManus et al. 2002; Berger et al. 2010; Gillings 2013).

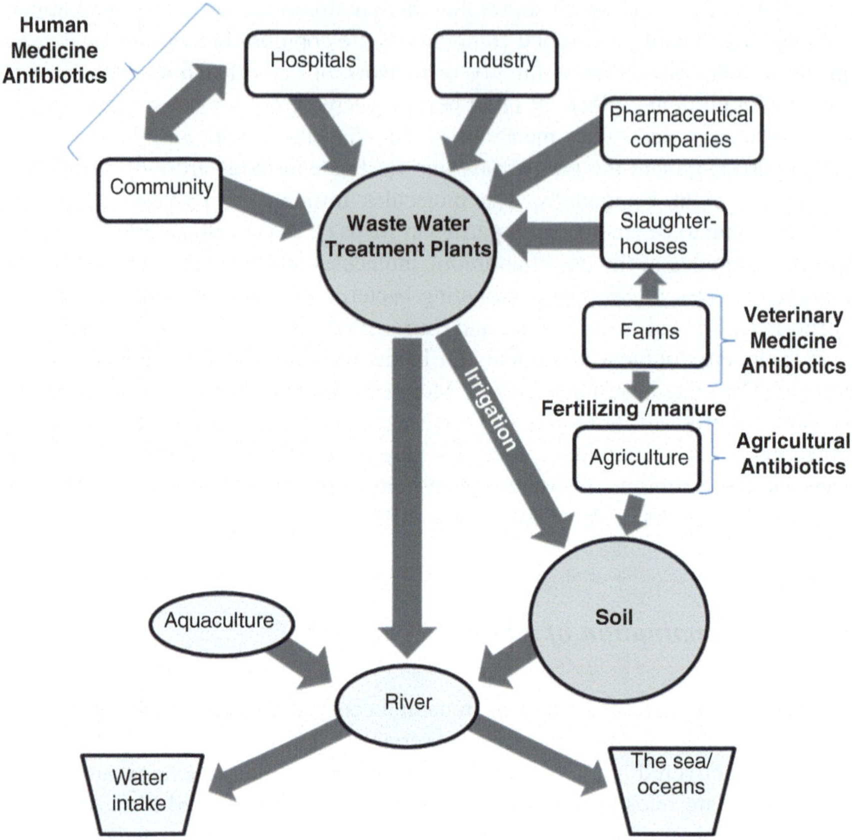

Fig. 9.1 Sources of antibiotics, ARB, and ARG in waste water treatment plants and soil

Application of antibiotics in agriculture promotes resistance, which is important for therapeutical reasons not only in animals, but also in humans, as in veterinary medicine one uses antibiotics of the same classes as in human treatment (Table 9.3). This concerns especially the pathogens transmitted via food, e.g., *Campylobacter jejuni*, *Escherichia coli*, *Salmonella*, and *Enterococcus faecium*. The same strains may hence colonize animals and humans, yet resistance genes disseminate easily between closely related species. Additionally, antibiotics may remain within animal cells and, as food pollution, stimulate the development of allergic reactions and resistance within human microbiota (Martínez 2009).

As a result of exposure to antibiotics, the structure, as well as activity of the microbial population (i.e., denitrification, nitrification, enzyme activity), may reshape. Most commonly used indicators for changes within the population are the balance between bacteria and fungi and the balance between Gram-positive and Gram-negative bacteria. The first one is closely connected with ecosystem performance. Many researchers have proven the increase in fungi number in relation to

Table 9.2 List of the most important antibiotics added to feed in recent decades

Antibiotic	Active against	Antibiotic	Active against
Aminoglycosides[a] (hygromycin)	Gram-negative bacteria, fungi, and higher eukaryotic cells	Beta-lactams[a] (penicillin)	Gram-positive and Gram-negative bacteria
Aminocoumarins (novobiocin)	Staphylococci—sensitive to novobiocin	Polyene antibiotic[a] (nystatin)	Fungi
Diterpene antibiotic (tiamulin)	Gram-positive bacteria, mycoplasmas	Polyether antibiotic (monensin, salinomycin)	Gram-positive bacteria, parasitic protozoa of the genus *Coccidia* and *Eimeria* causing coccidiosis
Glycopeptides[a] (bambermycin)	Gram-positive bacteria	Polypeptides (bacitracin)	Gram-positive bacteria
Ionophores (laidlomycin, lasalocid, monensin, narasin, salinomycin)	Gram-positive bacteria	Streptogramins (virginiamycin)	Gram-positive cocci
Lincosamides[a] (lincomycin)	Gram-positive and Gram-negative bacteria, mycoplasmas, mycobacteria, some rickettsia, and chlamydia	Tetracyclines[a] (chlorotetracycline, oxytetracycline	Gram-positive and Gram-negative bacteria, *Chlamydia*, *Mycoplasma*, and *Rickettsia*
Macrolides[a] (oleandomycin, tilmicosin, tylosin, spiramycin)	Gram-positive and Gram-negative bacteria, *Mycoplasmas*, *Mycobacteria*, some *Rickettsia*, and *Chlamydia*	Quinoxalines (carbadox)	Gram-positive and Gram-negative bacteria
Erythromycin antibiotic complex (avilamycin)	Gram-positive bacteria		

Based on National Research Council (1980) and FDA (2015)
[a]Antibiotics used to treat people and animals

Table 9.3 Antibiotic classes used in veterinary medicine and in medicine

Antibiotic class	Antibiotic class
Aminocoumarins	Nitrofurans
Aminoglycosides	Polypeptides
Aminopenicillins	Streptogramins
Amphenicols	Sulfonamides
Beta-lactams	Tetracyclines
Cephalosporins	Quinolones
Cyclopolypeptides	Nitroimidazole[a]
Lincosamides	Trimetoprim[a]
Macrolides	

Based on Marshall and Levy (2011)
[a]Chemotherapeutic

bacteria after enriching the soil with antibiotic-containing fertilizer. Then again, different classes of antibiotics have a diverse influence on Gram-positive and Gram-negative bacteria. It is related, among other factors, to differences in cell wall composition. The outcome depends also on soil properties, microorganism type, and antibiotic concentration (Ding and He 2010). Significantly, the effect of antibiotic action in natural water ecosystems is less visible than in soil (Zhang et al. 2009). It presumably results from lower concentration of these compounds due to considerable dilution (0.003–1.9 μg/l) (Kümmerer 2009). It is also harder to detect the microbes in such environment to determine or compare the structure of the bacterial population. On the other hand, lower concentration and short-time exposure makes the effects less predictable, but it does not necessarily mean that the impact on water ecosystems is limited in comparison to soils. What is more, it was shown that even sublethal dose of antibiotic affect microorganisms. In contrast, in fish ponds, as a consequence of antibiotic abuse, these compounds reach enormous concentrations up to several hundreds of mg per kg in the sediment.

As a result, the number of bacteria decreases, coupled with resistance dissemination, and altogether this alters the functioning of the ecosystem. It should be noted that human existence is inseparable from water. The presence of antibiotics in soil and water ponds promotes the growth of resistant strains, as well as disturbs the structure and physiology of microbial populations (Martínez 2009). Moreover, it poses a serious threat of extensive antibiotic and/or resistance gene dissemination (Ding and He 2010; Popowska et al. 2010).

Bacteria in natural ecosystems form mostly complex communities, in which the nutrient and chemical resources are shared, with metabolic products and secondary metabolites, such as antibiotics, among them. In such intricate habitat-like soil, comprising antibiotic-producing microorganisms, bacteria are naturally exposed to lethal and nonlethal concentrations of these compounds. Nonlethal antibiotic doses may alter the expression of genes associated with various life functions, such as metabolism, regulation, virulence, DNA repair, or stress response (Goh et al. 2002; Tsui et al. 2004; Davies et al. 2006; Yim et al. 2006, 2007, 2011; Blazquez et al. 2012), as well as modify the behavior of bacteria within the biofilms or play role in its formation (Baquero et al. 2013). These observations suggest that antibiotics in natural ecosystems, such as soil, act as signaling molecules in regulatory pathways, not as antimicrobial particles (Yim et al. 2007; Aminov 2009; Allen et al. 2010).

In vitro studies enable to determine the action of antibiotics on bacterial cells (pure cultures), which may be of bactericidal or bacteriostatic character. Minimal inhibitory concentration (MIC) or minimal bactericidal concentration (MBC) may be calculated. Antimicrobial properties of antibiotics may be verified in vitro and in vivo as well, but biological role of antibiotic resistance and of the compounds themselves in natural ecosystems is still unclear. Further understanding of interactions between bacteria, mediated by antibiotics and other molecules, would be of vital importance for developing novel strategies to fight antibiotic resistance.

9.2.3 Contribution of Agriculture

In apple and pear orchards, streptomycin and oxytetracycline are routinely used for the prevention of fire blight disease (causative agent *Erwinia amylovora*). The use of streptomycin is strictly controlled within the EU, and it is authorized only for use on a yearly basis. Streptomycin, initially used in plant breeding in the USA, was replaced by oxytetracycline because of the development of resistance among *E. amylovora* strains, isolated from apple orchards. In other countries, such as Israel or Japan, oxolinic acid is applied (Shtienberg et al. 2001); in Mexico and Middle America, gentamicin stands as the main treatment for this and other vegetable diseases (Stockwell and Duffy 2012). It was not until the problem of resistance emerged that the discussion about the use of antimicrobials in agriculture was initiated, although the amount of antibiotics used in plant breeding is low compared with human and veterinary medicine and animal production (McManus et al. 2002).

Irrigation with wastewater effluent also poses a potential route of entry of antibiotics into soil ecosystems. Multiple studies point out the presence of ARGs and ARB in wastewater effluent (Auerbach et al. 2007; Manaia et al. 2010; LaPara et al. 2011; Munir et al. 2011; Piotrowska and Popowska 2015; Piotrowska et al. 2017), but no attention has been devoted to the impact of residing antibiotics on ARGs or ARB level (Negreanu et al. 2012; McLain and Williams 2014). Previous studies indicate that irrigation does not seem to impact AR levels in the soil microbiome (Gatica and Cytryn 2013). At the same time, researchers conclude that further studies aimed at assessing the scope of horizontal gene transfer between effluent-associated ARB and soil bacteria need to be further conducted.

Microcosm studies of soil systems indicate that antibiotic exposure may induce changes in microbial biomass, community structure, and outcomes of functional endpoints, such as substrate-induced respiration, iron reduction, N-mineralization, nitrification, or the potential to degrade other anthropogenic substances (Williams-Nguyen et al. 2016). Antibiotics transferred into the soil may also affect the microorganisms and other biota inhabiting the niche. Research suggests that these interactions also affect mixture toxicity (Majewsky et al. 2014; Aga et al. 2016). Likewise, data indicate that wildlife such as birds and bats can accumulate nonantibiotic pharmaceutical compounds through the food chain and can also be affected (Bean et al. 2014); however, the importance of accumulation of antibiotics in tissues of wildlife species is not known.

9.2.4 Contribution of Animal Husbandry

The antimicrobials, used in the treatment of human diseases from the mid-1940s, were also introduced in veterinary medicine shortly afterwards. Even though some drugs are exclusively designed for veterinary use, most belong to the same antimicrobial classes as those used in human medicine with identical or very similar structures (Heuer et al. 2011). Currently the application of livestock manure in agriculture is occurring at large scale, causing serious threats of antibiotic dissemination in the environment as excreta of farm animals may contain high doses of antibiotics (Marshall and Levy 2011; Lathers 2001; Sarmah et al. 2006). Also, aquatic environments are contaminated by land application of antibiotics in agriculture (Anjum and Krakat 2015).

In the last dozen or so years, a large number of antibiotics from a wide range of classes, including fluoroquinolones (ciprofloxacin, enrofloxacin, norfloxacin), lincosamides (lincomycin), macrolides (erythromycin, tilmicosin, tylosin,), sulfonamides (sulfadiazine, sulfamethazine, sulfamethoxazole), tetracyclines (chlortetracycline, doxycycline, oxytetracycline, tetracycline), thiamphenicol analogs (chloramphenicol, florfenicol), and other classes (monensin, trimethoprim), have been detected in agroecosystems across the world. The research was conducted in North and South America, Europe, and Asia and in a variety of environmental compartments, including soils, surface waters, sediments, and biota such as plants and earthworms (Williams-Nguyen et al. 2016).

All growth promoters in the feed of food-producing animals were banned in the European Union (EU) countries from January 1, 2006 (http://europa.eu). Avoparcin, spiramycin, tylosin, and virginiamycin were banned from use for growth promotion in 1997 and 1998, respectively. In the USA, politicians are discussing the introduction of a similar ban on the use of antimicrobials in animal husbandry for growth promotion (http://www.govtrack.us/congress/bills/109/s742). Despite these bans, in some parts of the world, medically important antibiotics are still routinely fed to livestock as a precaution to increase profits and to ward-off potential bacterial infections in stressed and crowded livestock and aquaculture environments (Cantas et al. 2013).

Upon a request from the European Commission, the European Medicines Agency (EMA) has created the European Surveillance of Antimicrobial Consumption (ESVAC) program (http://www.ema.europa.eu/ema/index.jsp?curl=pages/regulation/document_listing/document_listing_000302.jsp). ESVAC project collects information on how antimicrobial medicines are used in animals across the EU. This type of information is essential to identify possible risk factors that could lead to the development and spread of antimicrobial resistance in animals. EMA started this project in April 2010 and is involved in monitoring and evaluating the risks of veterinary use of antibiotics in animals and looking in particular at the risk of the development of antimicrobial resistance in animals and the transmission of resistance to humans. During 2014, in 29 European countries, the sale of antimicrobial drugs for therapeutic use as veterinary medicine varied from 3.1 to 418.8 mg/kg biomass (Sales of veterinary antimicrobial agents in 29 European countries in 2014, Sixth ESVAC report). The largest amounts, expressed as a proportion of mg/PCU, were accounted for by tetracyclines (33.4%), penicillins (25.5%) and sulfonamides (11.0%). From the antimicrobial classes listed in the third World Health Organization (WHO) list of critically important antimicrobials (CIAs) with the highest priority in human medicine, the sales for food-producing animals of third- and fourth-generation cephalosporins, fluoroquinolones, and macrolides accounted for 0.2%, 1.9% and 7.5%, respectively, of the total sales in the 29 countries participating in ESVAC in 2014. Overall, the sales of polymyxins (mg/PCU) accounted for 6.6% of the total sales, with colistin representing more than 99% of the sales of polymyxins.

One of the concluding remarks was: "The substantial decline in the sales of antimicrobials for food-producing species observed for some countries indicates that there is also a potential for a decrease in other countries."

9.3 Antibiotic Degradation Rates in the Environment

Most of the antibiotics routinely used in medicine is only partially metabolized (Kümmerer 2004). Metabolites often remain in the environment, and their inactivation is impossible (Thiele-Bruhn 2003). Antibiotics are introduced into the fields in the form of natural fertilizers (utilized in animal husbandry) or by watering the plants with wastewaters. Eventually they persist in the soil, sediment, or groundwater. Through sewage systems, antimicrobials are released into the water resources. From the soil surrounding the fish ponds with large-scale antibiotic usage, they may be rinsed with rainwater and end up in water systems (Fig. 9.1).

The rate of degradation of antimicrobials in the environment varies and is dependent on a range of environmental conditions, for example, antibiotic concentration, chemical structure of the compound, composition and structure of soil/sediment, humic acid content, humidity, pH, temperature, sorption capacity, chemical composition of the environment, presence of other sources of carbon, presence of inorganic matter, and availability of oxygen and microorganisms that support biodegradation (Kümmerer 2004, 2009). Table 9.4 shows the time needed for decomposition of selected antibiotics. These are degraded in one of the following pathways: light-dependent, chemical, or biological (Martínez 2009). In the environment, a basic process of substance elimination is dependent on bacterial activity (particularly in sediment, soil and polluted water). Antibiotics entering the ecosystem are subject to transformations, such as photodegradation, hydrolysis (catalyzed by bacterial enzymes), decarboxylation, or hydroxylation (Thiele-Bruhn 2003; Kümmerer 2004). We should remember that the process of antimicrobial inactivation proceeds not always fully effectively, e.g., it is slowed down in low temperatures (Martínez 2009). A crucial process to neutralize all pharmaceuticals is their biodegradation and/or bacterial sorption. As a result of antibiotic degradation, the compound may be mineralized as carbon dioxide, converted to more hydrophobic derivative, or transformed into more hydrophilic substance (Kim and Aga 2007). Notably, products of antibiotic decomposition often differ only slightly from initial substrate. Another as important process is the sorption, which may promote resistant strains, as the drug maintains its properties.

Table 9.4 Degradation rates of pharmaceuticals in the environment

Class of pharmaceuticals	Concentration (μg/g)[a] Soil	Manure	Degradation (%)	Time (days)
Tetracyclines (chlorotetracycline, tetracycline, oxytetracycline)	4.7–300 μg/kg	5.6–100 μg/l	0–50	10–180
Sulfonamides (sulfabenzamide, sulfadiazine, trimethoprim)	1.0	250–1000 μg/l	0–50	14–64
Aminoglycosides (streptomycin)	5.6	5.6	0	30
β-lactams (penicillin, mecillinam)	5.6	500 μg/l	0–50	1–49
Macrolides[b] (erythromycin, spiramycin, tylosin)	5.6–100	25 mg/l	0–50	5–30
Fluoroquinolones (sarafloxacin, enrofloxacin, ciprofloxacin)	10	10	0–30	56–80
Imidazole (metronidazole)	10	–	50	14–75
Polypeptides (bacitracin, virginiamycin)	1.0	5.6–25	12–90	2–173
Polyethers (monensin)	–	–	30	70
Phospholipoglycosides (flavomycin, flavophospholipol)	5.6	10	0–100	6–119

Based on Thiele-Bruhn (2003)
[a]If not indicated otherwise
[b]This group does not include the modern macrolides with very long elimination half-lives

It is known that most antibiotics are not fully inactivated through sewage treatment; especially semisynthetic and synthetic compounds are very stable and have an excessive sorption capacity. This is why they accumulate easily and reach very high concentrations. It was proven that as a result of water chlorination, trimethoprim and β-lactams undergo degradation, likewise sewage utilization enables fluoroquinolones and tetracyclines elimination. Other antibiotics may be removed by carbon filtering, ionization, or coagulation. In the case of antimicrobials bound with soil, sediment, or clay particles, decomposition progress is slower, but they may be separated from water, which stops the long-distance dissemination (Martínez 2009).

9.4 Antibiotic-Resistant Bacteria

According to Her Majesty's Stationery Office, the definition of resistance stands as "the ability of the microorganism to oppose to antibiotic." It can be classified as intrinsic or acquired. Acquired resistance regards initially sensitive bacteria, which evolve into resistant either as a result of mutations or by acquisition of resistance-determining gene or set of genes. Intrinsic resistance is a natural feature of a specific strain or species (Markiewicz and Kwiatkowski 2006; Martínez et al. 2015).

To describe the influence of antibiotic on microorganism, minimal inhibitory concentration (MIC) and minimal bactericidal concentration (MBC) were established. These indexes specify the level of resistance to antibiotic or chemotherapeutic agent. MIC value describes a minimal drug concentration, which inhibits the growth of the microbe under given laboratory conditions. MBC states a minimal bactericidal drug concentration, on which the number of microorganisms capable of forming a colony decreases to zero (in practice the value of less than 0.1% is adopted). For clinically relevant bacteria, there are adequate guidelines for determining which strains are sensitive and which are resistant to a given group of antibiotics (EUCAS, The European Committee on Antimicrobial Susceptibility Testing—breakpoint tables for interpretation of MICs and zone diameters in EU or CLSI, The Clinical and Laboratory Standards Institute—Performance Standards for Antimicrobial Susceptibility Testing in the USA). Such guidelines for environmental bacteria do not exist.

The phenomenon of antimicrobial resistance is, in fact, not new. Studies from the past decade have shown that antibiotic resistance is an ancient biological mechanism whose proliferation has been strongly amplified through human activity. Metagenomes isolated from samples of up to 30,000-year-old permafrost show the presence of many genetic elements capable of conferring resistance to some beta-lactam, aminoglycoside, and glycopeptide antibiotics, among others. These genes are mostly identical to those functioning in contemporary pathogens (D'Costa et al. 2011; Perron et al. 2015). Studies involving species of bacteria isolated from a 4-million-year-old cave in Mexico, from soil samples from 30,000 years ago as well as from the deep ocean and deep below the Earth's surface showed that most of the isolates showed some level of resistance to one or more of the many antibiotics tested, including daptomycin, currently considered a treatment of last resort. In addition to well-known patterns of resistance, the studies revealed novel, not previously known, drug resistance mechanisms (Bhullar et al. 2012; Pawlowski et al. 2016). Research by Italian and American researchers on the microbiome of paleofeces in the colon of an eleventh century A.D. pre-Columbian Andean mummy using 16S rRNA gene high-throughput sequencing and metagenomics, which avoid the need for cultivation of bacteria, demonstrated that the microbiome profile of the paleofeces was unique when compared to previously characterized coprolites that did not undergo natural mummification. Unexpectedly, putative antibiotic resistance genes conferring resistance to penicillin, fosfomycin, chloramphenicol, aminoglycosides, macrolides, sulfa, quinolones, tetracycline, and vancomycin were found (Santiago-Rodriguez et al. 2015). It has been postulated that, in fact, some antibiotics may date back from more than 40 million to even 2 billion years (D'Costa et al. 2011). The examples mentioned above well document the fact that bacterial drug resistance was established long before the "discovery" of these compounds and their subsequent use in medicine, agriculture, and other areas.

A real breakthrough in resistance research was the discovery made in 1973, declaring that antibiotic resistance is not limited to pathogenic bacteria. It turned out that the source of resistance lies within non-pathogenic bacteria, inhabiting the natural environment and capable of producing the antibiotics themselves. It was an astounding finding, as opportunistic pathogens showed much higher resistance level of those infectious strains. What is more, it was shown that resistant bacteria were not restricted to the upper soil layer; at a depth of 173÷259 m, numerous multidrug-resistant strains were identified (Benveniste and Davies 1973). Originally, antibiotic presence within the natural environment was a natural selective factor contributing to genetic diversity of bacteria. In response to surrounding stimulus, complex defensive mechanisms, such as receptors, transporters, or enzymes introducing chemical modifications to the molecules, were developed (Wright 2010). Shortly after introduction of antibiotics into human treatment, bacteria had already been capable of developing resistance, not only via mutations, but mainly by acquisition of existing genes, encoding the resistance to antimicrobials. The reservoir of these genes and even resistance plasmids (R) was attributed to environmental microbiota. It a common knowledge now that those "original" antibiotic resistance genes were associated mainly with antibiotic-producing bacteria and usually encoded within the chromosomes, but moreover not always were they directly related to resistance, but rather with general physiological functions. Research demonstrates that bacteria, which do not produce antimicrobial compounds, carry resistance or even multiresistance determinants as well. *Efflux* pumps are great examples of such mechanism, as they remove all toxic compounds from the cell. Mutation leading to constitutive transporter expression ensures bacterial resistance. Pumps are omnipresent within all groups of microorganisms, with one cell usually encoding more than one type of the pump (Martínez 2008; Martínez and Baquero 2014). The original function of pumps, contributing to MDR phenotype (encoded mainly chromosomally), was proven to be connected with detoxification of metabolic pathways intermediates, homeostasis, and internal signal transduction and also with virulence (Aminov 2009). Another example is a group of enzymes hydrolyzing β-lactam ring—β-lactamases, at first being probably PBP proteins, involved in peptidoglycan synthesis. The ability to inhibit β-lactamases emerged as "side effect" of their basal, initial function (Martínez et al. 2009).

It is the result of a strong selective pressure in the course of the last decades, from the moment of introduction of antibiotics into therapy, that the genes changed their location and functions without alterations in nucleotide sequences to become resistance genes. Primary resistance genes present in the environment are termed as a "natural resistome." Natural response of the cell to antibiotics comprises the use of actual "machinery," i.e., engaging resistance genes encoded in antibiotic-producing bacteria, modification of existing (proactive response), or acquired (post-active response) genes. In most cases, bacteria from *Streptomyces* spp. isolated from soil are the natural source of resistance genes (Canton 2009). Basic mechanisms determining the resistance are distinguished as follows: *efflux* pumps systems, modifications within genes, which products serve as targets for antibiotics, and enzymatic inactivation of antimicrobials.

Environmental microbiota has been proven to possess considerably higher number of resistance genes than only those acquired by pathogenic bacteria, and not always are they limited to the producers of antibiotics (Allen et al. 2010). Moreover, different habitats may contain diverse resistance genes, which implies one's disability to even estimate the number of potential resistance genes, residing within the natural environments. What is more important, genes located on MGEs in human pathogens are found far and wide, including uncharted lands or wild animals with no record of antibiotic contact (Martínez 2009). This means that antibiotic resistance genes have always been present in the natural environment, and they persist irrespectively of selective pressure. However, it was not until the antibiotics were introduced into the environment because of human actions that they have spread among bacteria as an effect of omnipresent selective pressure and also have developed novel specialized resistance mechanisms and their rearrangements.

9.4.1 Antibiotic Resistance of Natural Bacteria

The dissemination of antibiotic resistance among pathogenic bacteria, especially resistance to multiple antibiotics, is a serious problem in the management and treatment of human diseases. The recent WHO's first global report on antibiotic resistance (WHO 2014, 2015) has shown how relevant the problem is worldwide, being a major threat to public health. This report has been followed up by a WHO Antimicrobial Resistance fact sheet (WHO 2016) and a list of "priority pathogens" for which new effective antibiotics are urgently needed (WHO 2017). These priority pathogens include some bacterial species commonly found in soils.

As mentioned above, most of the information we have regarding the production of antibiotics or the mechanisms of resistance to these compounds comes from laboratory or clinical studies of single or limited number of strains of microorganisms. It is only recently that research on drug resistance has focused on the soil environment from which antibiotics were originally obtained. The situation in the soil is very complex, and, in addition, most of the microorganisms living in it, anywhere up to 99% of them, are unculturable (Pham and Kim 2012).

However, it can safely be said that the vast majority of resistance mechanisms found in the soil are the same as those intensively studied under laboratory conditions. This is because the targets of the inhibitory action of antimicrobials are the same. There are approximately 200 conserved essential proteins in a bacterial cell, but the number of targets currently exploited by humans is very small. The most successful antibiotics used in medicine hit only one of a limited number of targets or pathways: the protein synthesis machinery, i.e., the 50S and 30S subunits of the ribosome, which are the target of aminoglycosides, chloramphenicol, macrolides, oxazolidinones (e.g., linezolid), or tetracyclines; nucleic acid synthesis, which is inhibited by fluoroquinolones and the rifamycins; cell wall synthesis that is disrupted by beta-lactams and glycopeptides (e.g., vancomycin) as well as by antibiotics like certain glycopeptides, targocil, and ticlopidine that interfere with teichoic acid synthesis; and, last but not the least, bacterial membranes which are disrupted by numerous polymyxins and lipopeptides (e.g., daptomycin) (Markiewicz and Kwiatkowski 2006, Davies and Davies 2010, Sengupta et al. 2013). Of course, the number of targets overall is larger since some antibiotics, such as the sulfonamides, can interfere with specific metabolic pathways (Markiewicz and Kwiatkowski 2006; Davies and Davies 2010; Blair et al. 2015; Munita and Arias 2016).

Some bacteria can be intrinsically resistant to certain antibiotics, which may result from the inherent absence of a target for a particular antibiotic or be more complex. Examples are rather numerous, from very simple to quite complex. *Klebsiella* spp. and several other species produce beta-lactamases that destroy ampicillin and/or other penicillins before they can reach their penicillin-binding protein (PBP) target. *Listeria monocytogenes* is intrinsically resistant to broad-spectrum cephalosporin antibiotics, but not to the other group of the beta-lactam antibiotics, the penicillins. This resistance is mediated by a complex cephalosporin resistome (Krawczyk-Balska and Markiewicz 2016). All Gram-negative bacteria are resistant to the action of vancomycin, which disrupts the synthesis of cell wall peptidoglycan because the presence of an outer membrane stops the antibiotic from reaching its site of action, the D-Ala-D-ala terminal dipeptide of the cell wall precursor. However, some Gram-positive bacteria, like *Lactobacilli* or *Leuconostoc* spp., are also resistant to the action of vancomycin because they lack the D-ala-D-ala moiety the antibiotic binds to (Swenson et al. 1990; Handwerger et al. 1994). Anaerobic bacteria are inherently resistant to aminoglycosides that bind to the 30S ribosomal subunit and cause a misreading of the genetic code, which leads to the disruption of normal bacterial protein synthesis. This resistance is caused by the lack of oxidative metabolism to drive the uptake of aminoglycosides. Similarly, insufficient oxidative metabolism is the reason for the intrinsic resistance of enterococci to this group of antibiotics. Gram-negative bacteria are resistant to the action of daptomycin, which kills Gram-positive bacteria. This resistance is brought about by differences in the composition of the cytoplasmic membrane (CM) of both these groups of bacteria. Gram-negative bacteria have a lower proportion of anionic phospholipids in the CM than Gram-positive bacteria, which reduces the efficiency of calcium-mediated insertion of the antibiotic into the CM that is a prerequisite of its antibacterial activity (Randall et al. 2013; Mueller et al. 2016).

In addition to intrinsic resistance, bacteria can acquire or develop resistance to various antibiotics, and this resistance can arise via different mechanisms which basically fall into one of three major categories: (1) those that minimize the intracellular concentration of an antibiotic as a result of its impeded penetration into the cell or its removal from the cell by efflux, (2) alteration or protection of the bacterial target sites as a result of genetic mutation or posttranslational modification of the target, (3) inactivation of the antibacterial drug by hydrolysis or its modification, and (4) modification of a metabolic pathway to bypass the antibiotic effect. From an evolutionary perspective, bacteria use two major genetic strategies to adapt to the antibiotic attack: mutations in genes often associated with the mechanism of action of the compound and the acquisition of extraneous DNA coding for resistance determinants through horizontal gene transfer.

9.4.2 Dissemination of Antibiotic-Resistant Bacteria to Soil

Over 80% of antibiotics in clinical use today originated from soil bacteria, either directly or as their semisynthetic derivatives. Knowledge of resistance mechanisms gained from laboratory and clinical research prompted numerous studies on the dissemination of antibiotic resistance genes in the soil environment. These studies have employed the polymerase chain reaction (PCR) amplicon surveys and, more recently, metagenomics to screen for resistance genes, i.e., resistance mechanisms, in the soil environment (Zhang et al. 2011; McGarvey et al. 2012; Popowska et al. 2012; Nesme et al. 2014; Nesme et al. 2016). Large-scale endeavors have also been launched, such as TerraGenome (Vogel et al. 2009) or the Earth Microbiome Project (Gilbert et al. 2014). The function of ARGs in the environment would seem, of course, to protect strains carrying them from the action of antibiotics, frequently produced by the ARG-harboring strains themselves, and also to regulate the responses induced by subinhibitory concentrations of antibiotics. Some ARGs, for instance, participate in the regulation of antibiotic biosynthesis.

One of the first similarities between the mechanisms of resistance of soil bacteria and pathogenic strains of the *Enterobacteriaceae* to antibiotics were described for the aminoglycosides kanamycin A and B, gentamicin, and neomycin (Benveniste and Davies 1973). The soil species *Streptomyces kanamyceticus* and *S. spectabilis* were found to contain aminoglycoside-modifying enzymes acetylating the 6′-amino group of kanamycin A and B, gentamicin, and neomycin and the 2′-amino group of the hexose ring of gentamicin. These acetylases catalyzed reactions identical to those carried out by enzymes found in plasmid-carrying Gram-negative bacteria. Other good examples of similarities are the presence of resistance genes coding for extended spectrum CTN-X beta-lactamase in soil bacteria and the plasmid-borne gene *qnr* mentioned above that encodes pentapeptide repeat proteins, which reduce susceptibility to quinolones. These resistance mechanisms were found in free-living *Kluyvera* and *Shewanella* isolates, respectively, and are identical to those of well-known pathogens (Farmer et al. 1981; Poirel et al. 2005).

There are many studies that prove that contamination of soil by either antibiotics or bacteria-carrying ARGs contributes to enhanced levels of antibiotic-resistant strains in the soil over the intrinsic levels and that ARB in livestock manure used as fertilizer and spread over agriculture land may pose a threat to the health of humans. Considering this, the concentrations of tetracycline, oxytetracycline, and sulfathiazole in the surface soil were quantified using liquid chromatography-mass spectrometry. These antibiotics were used in animal production and were found in fertilizer produced from livestock excretions. Species of ARB were identified using 16S rDNA. Soil samples were collected, and three compounds were detected: tetracycline, oxytetracycline, and sulfathiazole. The results of 16S rDNA gene analysis indicated that *Pseudomonas* spp., *Arthrobacter* spp., and *Rhodococcus* spp. showed persistent resistance to the three antibiotics tested. DNA quantification results revealed strong resistance of *Pseudomonas* spp. to sulfathiazole, whereas *Arthrobacter* spp. and *Rhodococcus* spp. were resistant to tetracycline and oxytetracycline (Yeom et al. 2017). An elegant study by Popowska et al. (2012) points to the widespread presence of high-level antibiotic-resistant bacteria in agricultural soils.

Bacterial drug resistance is being potentiated by human activity, creating "hotspots" in which antibiotic pressure leads to higher levels of resistance and the acquisition ad exchange of genetic material, also that from pathogenic bacteria introduced into the environment from medical settings, municipal wastewater systems, animal husbandry facilities, etc. (Berendonk et al. 2015).

9.5 Antibiotic Resistance Genes in Soil

Conventionally, the presence ARGs in natural environments is believed to stem from the release of animal and human bacteria from anthropogenic point sources such as sewage and animal husbandry, coupled to selective pressure from clinically and agriculturally derived antibiotic compounds. However, the realization that many clinically associated ARGs originated in terrestrial and aquatic microbiomes, and the fact that these genes predated antibiotic use by millions of years (Lebreton et al. 2017), necessitates a more holistic approach when assessing the distribution, abundance, and dynamics of ARGs in soil ecosystems. In the following sections, we explore current knowledge on native ARGs in soil; attempt to elucidate the impact of anthropogenic activities on the dissemination of ARGs in soil; and finally, present a holistic model that integrates native and anthropogenic factors, in attempt to better understand antibiotic resistance gene dynamics in soil.

9.5.1 Native Antibiotic Resistance Genes in Soil

A comprehensive study by D'Costa et al. (2006) first revealed the vast dimensions of antibiotic resistance in soil. The authors demonstrated that this resistance is associated with both novel and clinically characterized ARGs, postulating that the scope of antibiotic resistance in soil stems from both the high complexity of its microbial community and the profusion of antibiotic-producing bacteria in soil. The authors coined the "resistome," a collection of all known and novel ARGs in a specific environment, and indicated that the soil resistome may be a critical source for the emergence of novel ARGs in clinical pathogens (D'Costa et al. 2007). A metagenomic analysis of ancient DNA from 30,000-year-old permafrost soil strongly supported the resistome hypothesis by demonstrating that soil-associated ARGs predated clinical use of antibiotics. In this rigorously authenticated study, a highly diverse collection of ARGs encoding resistance to β-lactam, tetracycline, and glycopeptide antibiotics were detected that were highly similar to modern pathogen-associated ARGs (D'Costa et al. 2011). It has been suggested that under certain conditions, mobilization of ARGs from environmental microbiomes to clinical pathogens can significantly increase ARGs activity. This is believed to the case for certain β-lactamases that became significantly overexpressed when mobilized from soil bacterial chromosomes to pathogen-associated plasmids with strong promotors (Perry and Wright 2013).

To further elucidate the scope of the soil resistome (Allen et al. 2009), apply a novel functional metagenomics pipeline to pinpoint ARGs in highly pristine Alaskan soil. Functional metagenomics provides evidence that detected ARGs are actually functional in a specific host cell such as *Escherichia coli*, contrary to many other molecular-based methods that identify ARGs based on in silico comparisons. The authors characterized a β-lactam-resistant gene that was evolutionarily distant from clinically known β-lactamases. The fact that these genes conferred resistance in *E. coli* without manipulating the hosts gene expression machinery, demonstrated that soil ARGs have the potential to mobilize and function in human pathogens. A metagenomic analysis aimed at discovering streptomycin resistance in an apple orchard soil that had experienced repeated treatment with the antibiotic failed, but the authors discovered 13 antibiotic resistance genes that included β-lactamases, aminoglycoside acetyltransferases, multidrug efflux pump, and bifunctional protein containing a natural fusion of a β-lactamase and a sigma factor (predicted transcriptional regulator). This bifunctional protein conferred resistance to ceftazidime, a third-generation cephalosporin with broad-spectrum activity against Gram-positive and Gram-negative bacteria. The low sequence identities of the enzymes identified in this study highlight the novelty of these proteins. Many of the best matches in the NCBI nr database shared less than 60% amino acid identity with the enzymes found in the soil metagenomic library. Furthermore, nucleotide BLAST analysis of the NCBI nr database consistently failed to identify genes with significant matches (Donato et al. 2010). The pool of ARGs in the soil environment is little explored, and in addition many of these genes may be carried by bacteria that are not yet culturable. For pathogenic bacteria to acquire an ARG from the soil environment pool, it must be transferred by horizontal gene transfer. Functional metagenomics employed to search for ARGs in soil samples not subjected to antibiotic pressure revealed the presence of 11 novel resistance genes, four conferring resistance to trimethoprim, three to ampicillin, two to gentamicin and two to chloramphenicol. Shortly after the detection of this novel soil-derived β-lactamase, Wachino et al. (2011) described a carbapenem-resistant *Serratia marcescens* strain that was isolated from a Japanese hospital in 2010. Surprisingly, the carbapenem resistance of this strain was attributed to the production of a novel metallo-β-lactamase that was highly similar (75% identity) to the fused β-lactamase identified in the apple orchard soil analysis described above (Allen et al. 2009). In another study soil samples were collected from 20 prairie sites and screened for tetracycline-, sulfonamide-, β-lactamase-, and macrolide-resistance genes and characterized for soil physical and chemical parameters. All prairie sites contained tetracycline- and cefotaxime-resistant bacteria, and 48% of isolates collected were resistant to two or more antibiotics. Most (98%) of the soil samples and all 20 prairies sites had at least one tetracycline resistance gene. Sulfonamide genes, which are considered a marker of human or animal activity, were detected in 91% of the samples, despite the lack of human or animal inputs at these sites (Durso et al. 2016). In another paper were identified 45 clones conferring resistance to minocycline, tetracycline, streptomycin, gentamicin, kanamycin, amikacin, chloramphenicol, and rifampicin. The similarity of identified ARGs with the closest protein in GenBank ranged from 26 to 92%; more than 60% of identified ARGs had low similarity less than 60% at amino acid level. The identified ARGs include aminoglycoside acetyltransferase, aminoglycoside 6-adenyltransferase, ADP-ribosyltransferase, ribosome protection protein, transporters, and other antibiotic-resistant determinants (Su et al. 2014). These results suggest that functional metagenomic approach is powerful in discovering novel ARGs and resistant mechanisms.

Despite sporadic indications, very few studies have provided direct evidence of ARG mobilization from soil resistomes to clinical pathogens. Using a high-throughput functional metagenomic approach that targeted DNA from a pooled collection of multidrug-resistant soil bacterial isolates, Forsberg et al. (2012) identified ARGs that conferred resistance to five classes of antibiotics (β-lactams, aminoglycosides, amphenicols, sulfonamides, and tetracyclines) that had perfect nucleotide identity to genes from known human pathogens. Surprisingly, not only were the ARG sequences identical, but sequences encoding transposable elements that flanked both the soil strains and the pathogens were also 100% identical. However, most of the identified resistance genes had previously not been known. While still circumstantial, this study provided strong evidence that ARGs are mobilized from the soil microbiomes to clinical pathogens. However, a more comprehensive functional metagenomics-based study by the same research group that targeted 18 agricultural and grassland soils (Forsberg et al. 2014) indicated that the frequency of ARG transfer between soil bacteria and human pathogens is extremely rare. In this study, the authors identified close to 3000 ARGs that conferred resistance to 18 different antibiotics. Many of the ARGs identified here were novel, and most were only distantly related to clinically associated ARGs. The authors found that ARG composition was strongly linked to the bacterial community structure, indicating that phylogeny significantly dictates resistome composition. Furthermore, they determined that the association between ARGs and mobile elements was much lower in the soil resistome than in human-associated pathogens, providing a mechanistic explanation for the low mobility of ARGs from soil microbiomes to clinical environments. How can the discrepancies between these two studies be explained? Potentially, by the fact that the first study specifically targeted fast-growing "culturable" bacteria, while the second applied a culture-independent approach that incorporated ARGs from the entire soil microbiome, including a vast array of slow-growing bacteria that are commonly found in soil (Pham and Kim 2012). It may be suggested that significant differences in HGT frequency occur between fast- and slow-growing bacterial strains, and these may also be linked to phylogenetic constraints as described above. For example, certain fast-growing β- and γ-proteobacteria strains, including members of the *Burkholderia* and *Enterobacteriales* taxa, have hyper-potential for acquisition and dissemination of functional genes (primarily via conjugation), and therefore these taxa may represent "hubs" within gene transfer networks (Kloesges et al. 2011). Furthermore, certain *Burkholderia* and *Enterobacteriales* strains are common in both soil and clinical ecosystems; and these or similar taxa may bridge between the two environments. On the other hand, numerous studies have pointed to the presence of ARGs in sites subject to human activity, be it the release of antibiotics or of pathogenic bacteria to waters and soils. The abundance of pathogens that can survive in soils gives a potent mix that can result in the emergence of antibiotic resistance in the clinical setting.

9.5.2 Mechanisms of Resistance Dissemination

ARGs are a major platform for the spread of antibiotic resistance in both clinical and natural environments (Davies 1987). Dissemination of these genes within bacterial communities is facilitated by either vertical acquisition, where genes are transferred from parent bacteria to their offspring, or horizontal (or lateral) gene transfer, where ARGs are mobilized to a recipient bacterium (Ochman et al. 2000). Three modes of HGT are primarily intra- and interspecies transfer of ARGs: (a) transformation, where free extracellular DNA enters a recipient bacterium, and is integrated into the bacterial chromosome; (b) transduction, where bacteriophage DNA containing an ARG is injected into a recipient bacterium and is integrated into the bacterial chromosome (lysogenic stage); and (c) conjugation, where plasmids and MGEs such as conjugative transposons are transferred from a host to a recipient bacterium and are either autonomously replicated or are integrated into the chromosome of a recipient bacterium (de la Cruz and Davies 2000).

Bacteria acquire resistance through transfer of determinants by HGT and specific mutations. Still a vast majority of mechanisms are obtained through HGT from different, often distantly related species (Alekshun and Levy 2007; Aminov 2009). Among HGT-related mechanisms, they comprise drug modification, antibiotic target "protection," replacement of the sensitive target, and developing novel system of *efflux* pumps (Andersson and Hughes 2010). Mutation in genes encoding target proteins is a common mechanism of resistance to, e.g., quinolones, rifamycins, and phosphomycin.

The advancement in resistance is a complex and not fully understood process. But three facts stand undeniably as follows: (1) use of antibiotics promotes the increase in the resistance to various compounds, even those displaying distinct mechanisms of action (cross resistance); (2) resistance cannot always be foreseen, i.e., frequently, there is no correlation between the concentration of antibiotic or its metabolite and the effect caused by such pressure; and (3) it is hard to estimate the time of antibiotic persistence without selective pressure (Kümmerer 2004). Variations in organism tolerance resulting from exposure to certain compound are measured by pollution-inducted community tolerance index (PICT). PICT value is often analyzed in addition to other methods, e.g., PLFA (phospholipid fatty acid analysis), which indicates simultaneous changes in microbial community structure (Ding and He 2010).

It is also known that transfer and development of new resistance genes combinations occur more often in complex communities with high density of bacterial populations, i.e., in biofilms (Høiby et al. 2010). Biofilms are found in various areas related to medicine, industry, wastewater treatment tanks, sediments, soil, and water (Kümmerer 2004). Formation of such consortia is a method for survival in challenging and unstable environmental conditions. Literature data indicates stress factors in polluted environment as promoters of recombination and HGT, which additionally results in resistance genes dissemination (Martínez 2008). Plasmids may encode several genes providing the ability to survive in the presence of toxic compounds, such as petroleum products, detergents, heavy metals, and pesticides, which, regarding the increasing pollution, favor the survival of microbes carrying such plasmids (Martínez et al. 2009, 2015). It was proven that contamination with heavy metals promotes the selection of resistant strains (Martínez 2008). It was also shown that cross resistance to heavy metals and antibiotics linked with plasmid resistance genes aids resistance maintenance even in the case of absence of the antibiotic but with the presence of metal (Martínez 2009).

Studies demonstrate the potential of diverse, previously unknown resistance mechanisms lurking within the soil ecosystem (Riesenfeld et al. 2004). In one of the articles, the authors describe 480 strains of *Streptomyces* genus, capable of growth in the presence of 21 antibacterial compounds (natural, semisynthetic, and synthetic) of eight different groups, divided by cellular target (D'Costa et al. 2006). All strains displayed resistance to 7 or 8 compound at once; two strains were resistant to 15 and 21 antibiotics. Genetic analysis revealed almost 200 distinct resistance profiles, and all strains were resistant to phosphomycin, trimethoprim, and daptomycin—a novel drug in streptococcal infection treatment. The resistance to vancomycin and macrolides: erythromycin and telitromycin, was also observed, the latter being a drug of last resort in eradication of pathogens resistant to other macrolides (D'Costa et al. 2006). Another papers brought the discovery of resistance mechanisms connected with enzymatic drug modification or *efflux* pumps, encoded within genes unrecognized before (Riesenfeld et al. 2004; Dantas et al. 2008). Furthermore, a gene encoding broad-spectrum β-lactamase (*CTX-M*) was detected before it became a serious clinical concern (Knapp et al. 2010). Regarding the range of the research and our capability to culture only a small percentage of all bacteria, it seems that nature bears considerably higher level of resistance and mechanism diversity than it was previously believed.

What is crucial for predicting the fate of resistance genes in natural environment is a thorough understanding of their influence on bacterial physiology. As was mentioned, antibiotics target the essential life functions in bacteria. Therefore, it seems reasonable that resistant strains "suffer" from reduced fitness, i.e., metabolic ability to survive and reproduce (Andersson and Hughes 2010). Transfer of resistance genes involves the decrease in metabolism, as it is energetically costly process for the cell. It is demanded that reduction of antibacterial drugs intake is to be accomplished to raise the sensitivity to antibiotics in bacteria and replace resistant bacteria with sensitive ones. In a long-time perspective, such approach would enable sensitive bacteria to win the competition with resistant strains. However, many examples illustrate that in spite of the potential, reversion of such phenotype proceeds very slowly (Andersson and Hughes 2010). In addition, a part of the population remains resistant, as resistance genes are difficult to eradicate despite the lack of antibiotic selective pressure (see above, heavy metals). It was proven that if a small number of resistant strains survive in the habitat, in the case of reintroduction of antibiotic, resistance spreads and transforms much more rapidly than originally. Taking this into account, a full and stable reversion is unfortunately improbable.

9.5.3 Dissemination of Anthropogenically Associated Antibiotic Resistance Genes to Soil

Section 9.5.1 provides solid evidence that soil microbiomes encompass a diverse array of novel- and clinically characterized ARGs. Nonetheless, the transient and constant influx of anthropogenically derived factors to soil ecosystems is believed to result in proliferation of the soil resistome (Perry and Wright 2013; Perry et al. 2014). Based on the "wisdom of the crowd," this phenomenon is believed to be associated with dissemination and persistence of ARGs from anthropogenic sources such as wastewater, animal manure, or municipal biosolids, coupled to residual concentrations of selective elements such as antibiotics, heavy metals, and detergents linked to these activities. Anthropogenic-derived ARGs can be horizontally transferred to native soil communities, or, alternatively, selective elements can stimulate native soil resistome constituents. Although several studies supporting both of these concepts have been published, much of the data is circumstantial and does provide direct evidence linking specific factors to individual ARGs; while other studies show clear evidence that factors previously unconsidered may also be important. Collectively, it is becoming increasingly clear that the dynamics of the soil resistome is much more complex than originally assumed. In the following section, we explore how anthropogenic activities impact the diversity and abundance of ARGs in soil, provide suggestions for differentiating between background ARG levels and anthropogenic effects, and make suggestions for risk assessment of ARG in soils.

The realization that clinically associated ARGs are naturally present in soil microbiomes necessitates understanding of background levels in order to differentiate between native soil- and anthropogenically derived ARGs. Various studies that compared ARG levels in wastewater- and biosolid-amended vs. treated soils, have determined that in many cases background levels can be higher than the actual treatments (Munir and Xagoraraki 2011; Negreanu et al. 2012). This not only necessitates identifying ARGs that are abundant in anthropogenic point sources that can be used from source tracking but also requires identification of prominent "background" ARGs that are profuse in specific soils. A recent study that applied a high-throughput qPCR-based approach to identify prominent ARGs and MGEs in DNA extracted from soils in Antarctica (Wang et al. 2016) identified seven ARGs (blaTEM, blaSFO, blaFOX, cphA, mexF, oprD, and oprJ) that were ubiquitous to all of the soils regardless of anthropogenic activities, and these were regarded to be background genes. In contrast, several other genes were found to be associated with anthropogenic activities, as determined by the distance of these soils to research stations; however, specific biotic and abiotic factors that facilitated the propagation of these "anthropogenic" genes could not be determined.

Soil archives provide a unique opportunity to evaluate ARG diversity and abundance in soil. A groundbreaking study by Knapp et al. (2010) that assessed the relative abundance of 18 clinically relevant ARGs in an archived soil collection found that the relative abundances of all of the targeted ARG families significantly increased in the soils since the 1940s when antibiotic compounds were first used. This was especially evident for specific tetracycline resistance genes, whose abundance increased by 15-fold by the 1970s. The study strongly suggests that anthropogenic activities associated with antibiotic use may be responsible for the increase in ARG abundance but fails to characterize specific factors associated with the observed increase in antibiotic resistance. A subsequent study by the same authors (Graham et al. 2016) applied a similar qPCR approach to assess ARG levels in archived Danish soils documented from 1923. In this study, the authors specifically targeted four broad-spectrum β-lactamases bla(TEM), bla(SHV), bla(OXA), and bla(CTX-M) as well as class-1 integron genes (int1) in manure and inorganically fertilized soils. Two interesting trends were revealed from this study: (a) composite assessment of all of the targeted β-lactamase abundances were significantly higher in manure-fertilized vs. inorganically fertilized soils suggesting potential transfer of manure-associated ARGs to the soil; and (b) from a temporal perspective, the abundance of individual β-lactamases increased approximately parallel to the initial detection of these ARGs in clinical isolates. While this study provides broader evidence of anthropogenic stimulation of the soil resistome, it does not specifically highlight individual factors (i.e., residual concentrations of antibiotic or other contaminants, influx of antibiotic-resistant bacteria, or MGE-harboring ARGs) that facilitate this stimulation. Although specific links between antibiotic exposure and increased abundance of ARGs in archived soils has not been shown, there is evidence indicating that heavy metals, especially copper, stimulate ARG abundance in soil. This may be linked to profusion of multidrug efflux pumps in bacterial communities exposed to heavy metals or to co-resistance whereby linkage between metal resistance genes and specific ARGs result in increased ARG abundance in metal-contaminated environments (Knapp et al. 2011).

Several antibiotic compounds are frequently used in animal husbandry for clinical and prophylactic treatment and as growth promotors; and these compounds may stimulate antibiotic resistance in animal microbiomes. Animal manure is a rich source of nitrogen and phosphorus and, therefore, is often applied to soil as organic fertilizer with or without stabilization. However, application of manure to soil transfers fecal-derived ARGs from animals as well as residual antibiotic compounds in the manure that can select for native soil ARGs (Heuer et al. 2011). In order to specifically assess the effects of antibiotic residues in manure on ARG dynamics, Jechalke et al. (2013) monitored the mobility of sulfonamide resistance genes (*sul1* and *sul2*) in the rhizosphere of maize and grass and in bulk soil, in response to sulfadiazine-spiked vs. antibiotic-free manure. The authors found that the abundances of both *sul* genes were significantly higher in the sulfadiazine-spiked manure-amended plots than in plots that received manure without antibiotics. Furthermore, they found that plasmid transfer frequencies were higher in the antibiotic manure amendments suggesting that the increased abundance of *sul* in the soil may be partially associated with HGT. Interestingly, contrary to the bulk soil where the difference in *sul* abundance between the two treatments decreased over time, the rhizosphere samples continued to show significantly higher *sul* abundance in the antibiotic treatment even 6 weeks after manure application, suggesting that organic matter in the form of root exudates or mucilage may also impact ARG dynamics in the soil.

Temporal factors are crucial when evaluating the impact of manure and biosolid application on ARG abundance, because these amendments are generally performed only once or twice in a growing season. A comprehensive study that applied a qPCR array that detected over 300 ARGs and MGEs determined that the relative abundance of many of these genes rapidly decreased in soil following application of manure of untreated animals (Muurinen et al. 2017), suggesting that proper environmental stewardship can significantly reduce the risks of ARG propagation due to agricultural practices. Comprehensive understanding of temporal ARG dynamics in anthropogenically impacted soils necessitates long-term monitoring to assess accumulative effects on the soil resistome. A recent study applied a high-throughput qPCR ARG and MGE array-based approach that monitored the long-term effects of sewage sludge soil application (Xie et al. 2016). Initially, the authors characterized a wide array of multidrug efflux pumps, as well as specific β-lactam, macrolide-lincosamide-streptogramin B (MLSB), tetracycline, vancomycin, and aminoglycoside resistance genes that were present in both amended and non-amended soils, as background ARGs. Subsequently, they identified specific aminoglycoside and tetracycline resistance genes that were attenuated shortly after sludge application. Interestingly, the authors found that while annual applications of sludge impacted the soil resistome, this was primarily associated with stimulation of the intrinsic soil resistome and not with the influx of sludge associated ARGs. Evidence of limited dissemination of ARGs from anthropogenic sources to soil is also supported by previous studies that assessed impact of wastewater irrigation on the soil microbiome (Gatica et al. 2016; Gatica et al. 2015). Collectively, the failure of anthropogenically associated bacteria and

ARGs to survive in soil may be explained by the fact that the vast diversity and highly competitive nature of the soil microbiome significantly limits influx of invaders (van Elsas et al. 2012). An additional long-term experiment that monitored the abundance of specific ARGs and MGEs in Mexican soils irrigated with raw sewage for different periods of time found increased abundance of several of the genes in the sewage-irrigated vs. nonirrigated soils but failed to find a correlation between abundances of these genes and the duration of the irrigation. This indicates that while soil resistome are influenced by anthropogenic impacts, there appears to be a certain resilience that regulates the scope of this impact.

As described in Sect. 9.5.1, functional metagenomics is a groundbreaking method because it sheds light on soil-associated ARGs that are active and can be transcribed by model human pathogens such as *E. coli*. In attempt to pinpoint ARGs and elucidate factors that impact the soil resistome, Udikovic-Kolic et al. (2014) applied a functional metagenomics approach to soils with and without manure amendment from cows that were not treated with antibiotics. It revealed that soil treated with manure contained a higher abundance of β-lactam-resistant bacteria, which were attributed to enrichment of resident soil-associated β-lactamases. The study presents two interesting findings. First, it supports previous observations that anthropogenic perturbations primarily impact the soil resistome through activation/modification of soil-associated components and not as a result of ARG influx; and, second, based on the notion that applied manure did not contain residual antibiotic concentrations, the study indicates that the stimulation of the soil resistome by anthropogenic perturbations may often be associated with biotic or abiotic factors such as nutrients and organic matter and not necessarily with antibiotics, detergents, or other contaminant believed to induce antibiotic resistance.

9.5.4 *Integrating the Native and Anthropogenic: A Holistic Model for Understanding Antibiotic Resistance Gene Dynamics in Soil*

As comprehensively explored above, the scope and dynamics of ARGs in soil are determined by a myriad of factors. These may include the abundance and diversity of ARGs within the native soil microbiome and their capacity to replicate (vertical transfer) and mobilize (horizontal transfer) to other bacteria, the abundance and diversity of ARGs that are transmitted to the soil through natural and anthropogenic inputs (i.e., dust, runoff, applied manure or biosolids, irrigation) and their capacity to successfully mobilize into native soil bacterial genomes (horizontal transfer), the ability of ARG-harboring bacteria from the external environments described above to survive and replicate in soil ecosystems (vertical transfer), the scope and intensity of selective pressure (i.e., residual antibiotic compounds, heavy metals, or detergents) exerted on the soil microbiome, and by the level of competition within the soil microbiome, which can be facilitated by a variety of biotic (i.e., predation,

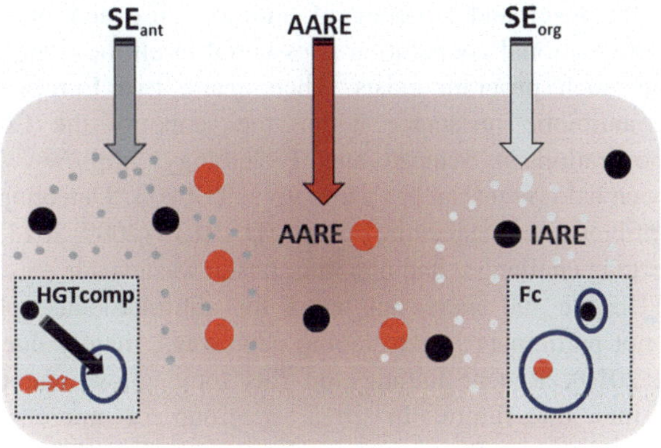

Fig. 9.2 Conceptual model depicting the effect of various native and anthropogenic factors on the soil resistome. Legend: IARG—indigenous soil-associated antibiotic resistance elements; EARG—external (anthropogenically derived) antibiotic resistance elements that enter and persist in the soil (by vertical or horizontal gene transfer); SE_{ant}—level of selective pressure propagated by antibiotic compounds or co-selection propagated by heavy metals, pesticides, etc.; SE_{comp}—selective pressure propagated by increased competition associated with external organic inputs and moisture; HGT_{comp}—horizontal gene transfer frequency of individual indigenous and external ARGs; Fc—fitness cost for maintaining individual ARGs; Re—random founder effects—*defined as IRE containing resistant bacteria proliferate following antibiotic amendment*

vegetation) and abiotic (i.e., soil organic matter, nutrients, moisture, temperature) factors. A conceptual model that summarizes the various factors that impact of the scope and abundance of ARGs in soil microbiomes is shown in Fig. 9.2. It is well known that soil is highly heterogeneous, and therefore the soil resistome is dictated by both spatial and temporal dynamics. Future studies need to produce empirical data that can be used to feed this and similar models in order to better understand the dynamics of the soil resistome.

9.6 Summary and Perspectives

Regarding the position of environmental strains of bacteria as a source of examined, as well as novel resistance mechanisms, it appears that the problem of resistance dissemination is unequivocally linked with omnipresent pressure resulting from antibiotic inflow into the environment as a consequence of human activity.

In 1998, the European Union banned the administration of clinical antimicrobials to livestock as growth promoters. The goal was to diminish the impact of antibiotics used in agriculture on selection of resistant strains among human pathogens. In 2006, the ban was extended to all antibiotics. What is more, in line with the European Commission's regulations, all EU Member States must establish and involve cross-sectoral teams to establish rules for the monitoring of antibiotic-resistant strains, relevant etiologic agents, in human and veterinary medicine, and assessment of the level and structure of antibiotic use and make appropriate interventions. International cooperation is essential in all these areas. For natural environment no such program exists. Then again, two European actions on environmental antibiotic resistance within the scope of the COST program (European Cooperation in Science and Technology, http://www.cost.esf.org/) have already been held on researcher's initiative: TD0803 "Detecting evolutionary hot spots of antibiotic resistances in Europe (DARE)" (2009–2013) and ES1403 "New and emerging challenges and opportunities in wastewater reuse (NEREUS)" (2014–2018). Despite the increasing need for antimicrobials, pharmaceutical industry does not participate in developing new drugs, mainly due to enormous costs (Wright 2010). Recent findings on this topic allow one to look more optimistically into the future. In 2015, a group of microbiologists from Northeastern University in Boston, Massachusetts, published the discovery of a novel antibiotic, teixobactin, synthesized by soil bacterium *Eleftheria terrae* (Ling et al. 2015). Research showed that teixobactin damages the cell wall of Gram-positive bacteria. It is effective in treatment of many prevalent bacterial infections, including tuberculosis, sepsis, and infection caused by *Staphylococcus aureus*, with strain resistant to methicillin (MRSA) among them (Smith et al. 2009). It has been the first "new" antibiotic for 30 years, but before it can be introduced into the so-called market, long-term clinical trials are necessary. This discovery demonstrates how big a potential is available for new potential bacteriostatic or bactericidal agents in the soil.

The question of the direction of antibiotic resistance transmission remains open. Initially many scientists and physicians were of opinion that the route is as follows: medicine-veterinary medicine-environment. After thorough analyses of collected data, it seems that it may be the other way round. The multitude of bacteria and their physiological and genetical determinants evolve in response to selective pressure, generating improved mechanisms. In the era of routine administration of antibiotics, all routes of dissemination of antibiotics, ARB, and ARGs seem to somehow connect. Then above all, what matters is the limitation of antibiotic abuse in medicine (particularly in nonhospital treatment), veterinary medicine, and agriculture as well to eradicate the mechanisms associated with selective pressure.

References

Aga DS, Lenczewski M, Snow D, Sallach JB, Muurinen J, Wallace JS (2016) Challenges in the measurement of antibiotics and in evaluating their impacts in the agroecosystems: a critical review. J Environ Qual 45(2):407–419. doi:10.2134/jeq2015.07.0393

Alekshun MN, Levy SB (2007) Molecular mechanisms of antibacterial multidrug resistance. Cell 128:1037–1050

Allen HK, Moe LA, Rodbumrer J, Gaarder A, Handelsman J (2009) Functional metagenomics reveals diverse beta-lactamases in a remote Alaskan soil. ISME J 3:243–251

Allen HK, Donato J, Wang HH, Cloud-Hansen KA, Davies J, Handelsman J (2010) Call of the wild: antibiotic resistance genes in natural environments. Nat Rev Microbiol 8:251–259

Allen CA, Babakhani F, Sears P, Nguyen L, Sorg JA (2013) Both fidaxomicin and vancomycin inhibit outgrowth of Clostridium difficile spores. Antimicrob Agents Chemother 57(1):664–667. doi:10.1128/AAC.01611-12

Aminov RI (2009) The role of antibiotics and antibiotic resistance in nature. Environ Microbiol 11:2970–2988

Aminov R, Mackie RI (2007) Evolution and ecology of antibiotic resistance genes. FEMS Microbiol Lett 271:147–161

Anderson DI, Hughes D (2012) Evolution of antibiotic resistance at non-lethal drug concentrations. Drug Resist Updat 15(3):162–72. doi:10.1016/j.drup.2012.03.005

Andersson DI, Hughes D (2010) Antibiotic resistance and its cost: is it possible to reverse resistance? Nat Rev Microbiol 2319:260–271

Anjum R, Krakat N (2015) A review: improper antibiotic utilization evokes the dissemination of resistances in biotic environments—a high risk of health hazards. Pharm Anal Acta 6:454. doi:10.4172/2153-2435.1000454

Auerbach EA, Seyfried EE, McMahon KD (2007) Tetracycline resistance genes in activated sludge wastewater treatment plants. Water Res 41:1143–1151

Babakhani F, Gomez A, Robert N, Sears P (2011) Killing kinetics of fidaxomicin and its major metabolite OP-1118 against *Clostridium difficile*. J Med Microbiol 60:1213–1217

Baltz RH (2007) Antimicrobials from actinomycetes: back to the future. Microbe 2:125–131

Baquero F, Tedim AP, Coque TM (2013) Antibiotic resistance shaping multi-level population biology of bacteria. Front Microbiol 4:15. doi:10.3389/fmicb.2013.00015

Bean TG, Boxall AB, Lane J, Herborn KA, Pietravalle S, Arnold KE (2014) Behavioural and physiological responses of birds to environmentally relevant concentrations of an antidepressant. Philos Trans R Soc Lond Ser B Biol Sci 369:20130575. doi:10.1098/rstb.2013.0575

Benveniste R, Davies J (1973) Aminoglycoside antibiotic-inactivating enzymes in actinomycetes similar to those present in clinical isolates of antibiotic-resistant bacteria. Proc Natl Acad Sci U S A 70(8):2276–2280

Berendonk TU, Manaia CM, Merlin C, Fatta-Kassinos D, Cytryn E, Walsh F, Bürgmann H, Sørum H, Norström M, Pons MN, Kreuzinger N, Huovinen P, Stefani S, Schwartz T, Kisand V, Baquero F, Martinez JL (2015) Tackling antibiotic resistance: the environmental framework. Nature 13:310–317

Berger CN, Sodha SV, Shaw RK, Griffin PM, Pink D, Hand P, Frankel G (2010) Fresh fruit and vegetables as vehicles for the transmission of human pathogens. Environ Microbiol 12:2385–2397

Berglund B (2015) Environmental dissemination of antibiotic resistance genes and correlation to anthropogenic contamination with antibiotics. Infect Ecol Epidemiol 5:28564. doi:10.3402/iee.v5.28564

Bernier SP, Surette MG (2013) Concentration-dependent activity of antibiotics in natural environments. Front Microbiol 4:20. doi:10.3389/fmicb.2013.00020

Bhullar K, Waglechner N, Pawlowski A, Koteva K, Banks ED, Johnston MD, Barton HA, Wright GD (2012) Antibiotic resistance is prevalent in an isolated cave microbiome. PLoS One 7: e34953. doi:10.1371/journal.pone.0034953

Blair JM, Webber MA, Baylay AJ, Ogbolu DO, Piddock LJ (2015) Molecular mechanisms of antibiotic resistance. Nat Rev Microbiol 13(1):42–51. doi:10.1038/nrmicro3380

Blazquez J, Couce A, Rodriguez-Beltran J, Rodriguez-Rojas A (2012) Antimicrobials as promoters of genetic variation. Curr Opin Microbiol 15:561–569

Brameyer S, Kresovic D, Bode HB, Heerman R (2015) Dialkylresorcinols as bacterial signaling molecules. PNAS 112(2):572–577. doi:10.1073/pnas.1417685112

Brötz-Oesterhelt H, Brunner NA (2008) How many modes of action should an antibiotic have? Curr Opin Pharmacol 8:564–573

Buscot F, Varma A (eds) (2005) Microorganisms in soils: roles in genesis and functions. Springer, Berlin

Cafaro MJ, Currie CR (2005) Phylogenetic analysis of mutualistic filamentous bacteria associated with fungus-growing ants. Can J Microbiol 51(6):441–446

Cantas L, Shah SQ, Cavaco LM, Manaia CM, Walsh F, Popowska M, Garelick H, Bürgmann H, Sørum H (2013) A brief multi-disciplinary review on antimicrobial resistance in medicine and its linkage to the global environmental microbiota. Front Microbiol 4:96. doi:10.3389/fmicb.2013.00096

Canton R (2009) Antibiotic resistance genes from the environment: a perspective through newly identified antibiotic resistance mechanisms in the clinical setting. Clin Microbiol Infect 15:20–25

Cevheri C (2012) Maize growth promotion of bacteria isolates from semi-arid region of turkey. J Food Agric Environ 10:1269–1272

Chopra I, Brennan P (1998) Molecular action of antimycobacterial agents. Tuber Lung Dis 78:89–98

D'Costa VM, Mcgrann KM, Hughes DW, Wright GD (2006) Sampling the antibiotic resistome. Science 311:374–377

D'Costa VM, Griffiths E, Wright GD (2007) Expanding the soil antibiotic resistome: exploring environmental diversity. Curr Opin Microbiol 10:481–489

D'Costa VM, King CE, Kalan L, Morar M, Sung WW, Schwarz C, Froese D, Zazula G, Calmels F, Debruyne R, Golding GB, Poinar HN, Wright GD (2011) Antibiotic resistance is ancient. Nature 477:457–461

Dantas G, Sommer MOA, Oluwasegun RD, Church GM (2008) Bacteria subsisting on antibiotics. Science 320:100–103

Davies J (1987) Antibiotic-resistance genes—ecology, transfer, and expression—Levy, SB, Novick, RP. Nature 329:212–212

Davies J (2011) How to discover new antibiotics: harvesting the parvome. Curr Opin Chem Biol 15(1):5–10. doi:10.1016/j.cbpa.2010.11.001

Davies J, Davies D (2010) Origins and evolution of antibiotic resistance. Microb Mol Biol Rev 74:417–533

Davies J, Spiegelman GB, Yim G (2006) The world of subinhibitory antibiotic concentrations. Curr Opin Microbiol 9(5):445–453. doi:10.1016/j.mib.2006.08.006

De La Cruz F, Davies J (2000) Horizontal gene transfer and the origin of species: lessons from bacteria. Trends Microbiol 8:128–133

De La Torre A, Iglesias I, Carballo M, Ramírez P, Muñoz MJ (2012) An approach for mapping the vulnerability of European Union Soils to Antibiotic Contamination. Sci Total Environ 414:672–679

Derewacz DK, Goodwin CR, McNees CR, McLean JA, Bachmann BO (2013) Antimicrobial drug resistance affects broad changes in metabolomic phenotype in addition to secondary metabolism. Proc Natl Acad Sci USA 110:2336–2341

Ding CH, He J (2010) Effect of antibiotics in the environment on microbial populations. Appl Microbiol Biotechnol 87:925–941

Donato JJ, Moe LA, Converse BJ, Smart KD, Berklein FC, Mcmanus PS, Handelsman J (2010) Metagenomic analysis of apple orchard soil reveals antibiotic resistance genes encoding predicted bifunctional proteins. Appl Environ Microbiol 76:4396–4401

Dungait JA, Cardenas LM, Blackwell MS, Wu L, Withers PJ, Chadwick DR, Bol R, Murray PJ, Macdonald AJ, Whitmore AP, Goulding KW (2012) Advances in the understanding of nutrient dynamics and management in UK agriculture. Sci Total Environ 434:39–50. doi:10.1016/j.scitotenv.2012.04.029

Durso LM, Wedin DA, Gilley JE, Miller DN, Marx DB (2016) Assessment of selected antibiotic resistances in ungrazed native Nebraska prairie soils. J Environ Qual 45(2):454–462

Farmer JJ, Fanning GR, Huntley-Carter GP, Holmes B, Hickman FW, Richard C, Brenner DJ (1981) Kluyvera, a new (redefined) genus in the family *Enterobacteriaceae*: identification of *Kluyvera ascorbata* sp. nov. and *Kluyvera cryocrescens* sp. nov. in clinical specimens. J Clin Microbiol 13(5):919–933

FDA (2015) 2013 summary report on antimicrobials sold or distributed for use in food-producing animals. FDA, Department of Health and Human Services, Center for Veterinary Medicine, Silver Springs, MD

Fevre C, Jbel M, Passet V, Weill FX, Grimont PA, Brisse S (2005) Six groups of the OXY β-lactamase evolved over millions of years in *Klebsiella oxytoca*. Antimicrob Agents Chemother 49:3453–3464

Fick H, Soderstrom H, Lindberg RH, Phan C, Tysklind M, Larsson DGJ (2009) Contamination of surface, ground, and drinking water from pharmaceutical production. Environ Toxicol Chem 28:2522–2527

Forsberg KJ, Reyes A, Wang B, Selleck EM, Sommer MO, Dantas G (2012) The shared antibiotic resistome of soil bacteria and human pathogens. Science 337:1107–1111

Forsberg KJ, Patel S, Gibson MK, Lauber CL, Knight R, Fierer N, Dantas G (2014) Bacterial phylogeny structures soil resistomes across habitats. Nature 509(7502):612–616. doi:10.1038/nature13377

Frost LS, Leplae R, Summers AO, Toussaint A (2005) Mobile genetic elements: the agents of open source evolution. Nat Rev Microbiol 3:722–732

Garau G, Di Guilmi AM, Hall BG (2005) Structure-based phylogeny of the metallo-beta-lactamases. Antimicrob Agents Chemother 49:2778–2784

Gatica J, Cytryn E (2013) Impact of treated wastewater irrigation on antibiotic resistance in the soil microbiome. Environ Sci Pollut Res Int 20(6):3529–3538. doi:10.1007/s11356-013-1505-4

Gatica J, Yang K, Pagaling E, Jurkevitch E, Yan T, Cytryn E (2015) Resistance of undisturbed soil microbiomes to ceftriaxone indicates extended spectrum β-lactamase activity. Front Microbiol 6:1233. doi:10.3389/fmicb.2015.01233

Gatica J, Tripathi V, Green SJ, Manaia CM, Berendonk T, Cacace D, Merlin C, Kreuzinger N, Schwartz T, Fatta-Kassinos D (2016) High throughput analysis of integron gene cassettes in wastewater environments. Environ Sci Technol 50(21):11825–11836

Gilbert JA, Jansson JK, Knight R (2014) The Earth microbiome project: successes and aspirations. BMC Biol. doi:10.1196/s12915-014-0069-1

Gillings MR (2013) Evolutionary consequences of antibiotic use for the resistome, mobilome and microbial pangenome. Front Microbiol 4:4. doi:10.3389/fmicb.2013.00004

Goh E-B, Yim G, Tsui W, McClure J, Surette MG, Davies J (2002) Transcriptional modulation of bacterial gene expression by subinhibitory concentrations of antibiotics. Proc Natl Acad Sci U S A 99:17025–17030

Gootz TD (2010) The global problem of antibiotic resistance. Crit Rev Immunol 30:79–93

Graham DW, Knapp CW, Christensen BT, Mccluskey S, Dolfing J (2016) Appearance of β-lactam resistance genes in agricultural soils and clinical isolates over the 20th century. Sci Rep 6:21550. doi:10.1038/srep21550lk

Hall BG, Barlow M (2004) Evolution of the serine beta-lactamases: past, present and future. Drug Resist Updat 7:111–123

Handwerger D, Pucci MJ, Volk KJ, Liu J, Lee MS (1994) Vancomycin-resistant *Leuconostoc mesenteroides* and *Lactobacillus casei* synthesize cytoplasmic precursors that terminate in lactate. J Bacteriol 176(1):260–264

Heeb S, Fletcher MP, Chhabra SR, Diggie SP, Williams P, Camaara M (2011) Quinolones: from antibiotics to autoinducers. FEMS Microbiol Rev 35:247–274

Herridge DF, Peoples MB, Boddey RM (2008) Global inputs of biological nitrogen fixation in agricultural systems. Plant Soil 311:1–18. doi:10.1007/s11104-008-9668-3

Heuer H, Schmitt H, Smalla K (2011) Antibiotic resistance gene spread due to manure application on agricultural fields. Curr Opin Microbiol 14:236–243

Høiby N, Bjarnsholt T, Givskov M, Molin S, Ciofu O (2010) Antibiotic resistance of bacterial biofilms. Int J Antimicrob Agents 35:322–332

Humeniuk C, Arlet G, Gautier V, Grimont P, Labia R, Philippon A (2002) Beta-lactamases of Kluyvera ascorbata, probable progenitors of some plasmid-encoded CTX-M types. Antimicrob Agents Chemother 46(9):3045–3049

Jechalke S, Kopmann C, Rosendahl I, Groeneweg J, Weichelt V, Krogerrecklenfort E, Brandes N, Nordwig M, Ding GC, Siemens J, Heuer H, Smalla K (2013) Increased abundance and transferability of resistance genes after field application of manure from sulfadiazine-treated pigs. Appl Environ Microbiol 79:1704–1711

Kato S, Shindo K, Kawai H, Matsuoka M, Mochizuki J (1993) Studies on free radical scavenging substances from microorganisms. III. Isolation and structural elucidation of a novel free radical scavenger resorstatin. J Antibiot (Tokyo) 46:1024–1026

Kim S, Aga DS (2007) Potential ecological and human health impacts of antibiotics and antibiotic-resistant bacteria from wastewater treatment plants. J Toxicol Environ Health B Crit Rev 10:559–573

Kleineidam K, Sharma S, Kotzerke A, Heuer H, Thiele-Bruhn S, Smalla K, Wilke BM, Schloter M (2010) Effect of sulfadiazine on abundance and diversity of denitrifying bacteria by determining nirK and nirS genes in two arable soils. Microb Ecol 60:703–707. doi:10.1007/s00248-010-9691-9

Kloesges T, Popa O, Martin W, Dagan T (2011) Networks of gene sharing among 329 proteobacterial genomes reveal differences in lateral gene transfer frequency at different phylogenetic depths. Mol Biol Evol 28(2):1057–1074. doi:10.1093/molbev/msq297

Knapp CW, Dolfing J, Ehlert PA, Graham DW (2010) Evidence of increasing antibiotic resistance gene abundances in archived soils since 1940. Environ Sci Technol 44:580–587

Knapp CW, Mccluskey SM, Singh BK, Campbell CD, Hudson G, Graham DW (2011) Antibiotic resistance gene abundances correlate with metal and geochemical conditions in archived Scottish soils. PLoS One 6:e27300

Krawczyk-Balska A, Markiewicz Z (2016) The intrinsic cephalosporin resistome of *Listeria monocytogenes* in the context of stress response, gene regulation, pathogenesis and therapeutics. J Appl Microbiol 120(2):251–265

Kümmerer K (2004) Resistance in the environment. J Antimicrob Chemother 54:311–320

Kümmerer K (2009) Antibiotics in the aquatic environment. A review Part I. Chemosphere 75:417–434

LaPara TM, Burch TR, McNamara PJ, Tan DT, Yan M, Eichmiller JJ (2011) Tertiary-treated municipal wastewater is a significant point source of antibiotic resistance genes into Duluth-Superior harbor. Environ Sci Technol 45:9543–9549. doi:10.1021/es202775r

Lathers CM (2001) Role of veterinary medicine in public health: antibiotic use in food animals and humans and the effect on evolution of antibacterial resistance. J Clin Pharmacol 41:595–599

Lebreton F, Manson AL, Saavedra JT, Straub TJ, Earl AM, Gilmore MS (2017) Tracing the enterococci from paleozoic origins to the hospital. Cell 169(5):849–861. doi:10.1016/j.cell.2017.04.027

Levy SB, Marshall B (2004) Antibacterial resistance worldwide: causes, challenges and responses. Nat Med 12:S122–S129

Ling LL, Schneider T, Peoples AJ, Spoering AL, Engels I, Conlon BP, Mueller A, Schäberle TF, Hughes DE, Epstein S, Jones M, Lazarides L, Steadman VA, Cohen DR, Felix CR, Fetterman KA, Millett WP, Nitti AG, Zullo AM, Chen C, Lewis K (2015) A new antibiotic kills pathogens without detectable resistance. Nature 517:455–459

Liu B, Pop M (2009) ARDB—Antibiotic Resistance Genes Database. Nucleic Acids Res 37:D443–D447

Majewsky M, Wagner D, Delay M, Brase S, Yargeaua V, Horn H (2014) Antibacterial activity of sulfamethoxazole transformation products (TPs): general relevance for sulfonamide TPs modified at the parA position. Chem Res Toxicol 27:1821–1828. doi:10.1021/tx500267x

Manaia CM, Novo A, Coelho B, Nunes OC (2010) Ciprofloxacin resistance in domestic wastewater treatment plants. Water Air Soil Pollut 208:335–343

Mannanov RN, Sattarova RK (2001) Antibiotics produced by *Bacillus* bacteria. Chem Nat Compd 37:117

Markiewicz Z, Kwiatkowski Z (2006) Bacteria, antibiotics, drug resistance. Polish Scientific Publishers, PWN, Warsaw

Marshall BM, Levy SB (2011) Food animals and antimicrobials: impacts on human health. Clin Microbiol Rev 24(4):718–733. doi:10.1128/CMR.00002-11

Martínez JL (2008) Antibiotics and antibiotic resistance genes in natural environments. Science 321:365–367

Martínez JL (2009) Environmental pollution by antibiotics and by antibiotic resistance determinants. Environ Pollut 157:2893–2902
Martínez JL, Baquero F (2014) Emergence and spread of antibiotic resistance: setting a parameter space. Ups J Med Sci 119:68–77
Martínez JL, Fajardo A, Garmendia L, Hernandez A, Linares JF, Martinez-Solano L, Sanchez MB (2009) A global view of antibiotic resistance. FEMS Microbiol Rev 33:44–65
Martínez JL, Coque TM, Baquero F (2015) What is a resistance gene? Ranking risk in resistomes. Nat Rev Microbiol 13:116–123
McGarvey KM, Queitsch K, Fields S (2012) Wide variation in antibiotic resistance proteins identified by functional metagenomic screening of a soil DNA library. Appl Environ Microbiol 78:1708–1714
McLain JE, Williams CF (2014) Sustainability of water reclamations: long-term recharge with reclaimed wastewater does not enhance antibiotic resistance in sediment bacteria. Sustainability 6:1313–1327. doi:10.3390/su6031313
McManus PS, Stockwell VO, Sundin GW, Jones AL (2002) Antibiotic use in plant agriculture. Annu Rev Phytopathol 40:443–465
Miller JH, Novak JT, Knocke WR, Pruden A (2016) Survival of antibiotic resistant bacteria and horizontal gene transfer control antibiotic resistance gene content in anaerobic digesters. Front Microbiol 7:263. doi:10.3389/fmicb.2016.00263
Mueller A, Wenzel M, Strahl H, Saaki TN, Kohl B, Siersma T, Bandow JE, Sahi HG, Schneider T, Hamoen LW (2016) Daptomycin inhibits cell envelope synthesis by interfering with fluid membrane microdomains. Proc Natl Acad Sci USA pii:201611173
Munir M, Xagoraraki I (2011) Levels of antibiotic resistance genes in manure, biosolids, and fertilized soil. J Environ Qual 40:248–255. doi:10.2134/jeq2010.0209
Munir M, Wong K, Xagoraraki I (2011) Release of antibiotic resistant bacteria and genes in the effluent and biosolids of five wastewater utilities in Michigan. Water Res 45:681–693. doi:10.1016/j.watres.2010.08.033
Munita JM, Arias CA (2016) Mechanisms of antibiotic resistance. Microbiol Spectr 4(2). doi:10.1128/microbiolspec.VMBF-0016-2015
Muurinen J, Stedtfeld RD, Karkman A, Pärnänen K, Tiedje JM, Virta MP (2017) Influence of manure application on the environmental resistome under Finnish agricultural practice with restricted antibiotic use. Environ Sci Technol. doi:10.1021/acs.est.7b00551
Nannipieri P, Ascher J, Ceccherini MT, Landi L, Pietramellara G, Renella G (2003) Microbial diversity and soil functions. Eur J Soil Sci 54:655–670
National Research Council (1980) National Research Council (US) Committee to Study the Human Health Effects of Subtherapeutic Antibiotic Use in Animal Feeds. Washington (DC): National Academies Press (US).
Neeno-Eckwall EC, Kinkel LL, Schottel JL (2001) Competition and antibiosis in the biological control of potato scab. Can J Microbiol 47(4):332–340
Negreanu Y, Pasternak Z, Jurkevitch E, Cytryn E (2012) Impact of treated wastewater irrigation on antibiotic resistance in agricultural soils. Environ Sci Technol 46:4800–4808
Nesme J, Cecillon S, Delmont TO, Monier JM, Vogel TM, Simonet P (2014) Large-scale metagenomic-based study of antibiotic resistance in the environment. Curr Biol 24:1096–1100
Nesme J, Achouak W, Agathos SN, Bailey M, Baldrian P et al (2016) Back to the future of soil metagenomics. Front Microbiol. doi:10.3389/fmicb.2016.00073
Ochman H, Lawrence JG, Groisman EA (2000) Lateral gene transfer and the nature of bacterial innovation. Nature 405:299–304
Pawlowski AC, Wang W, Koteva K, Barton HA, McArthur AG, Wright GD (2016) A diverse intrinsic antibiotic resistome from a cave bacterium. Nat Comm. doi:10.1038/ncomms13803
Perron GG, Whyte L, Turnbaugh PJ, Goordial J, Hanage WP, Dantas G, Desai MM (2015) Functional characterization of bacteria isolated from ancient arctic soil exposes diverse resistance mechanisms to modern antibiotics. PLoS One. doi:10.1371/journal.pone.0069533

Perry JA, Wright GD (2013) The antibiotic resistance "mobilome": searching for the link between environment and clinic. Front Microbiol 4:138. doi:10.3389/fmicb.2013.00138

Perry JA, Westman EL, Wright GD (2014) The antibiotic resistome: what's new? Curr Opin Microbiol. doi:10.1016/j.mib.2014.09.002

Pham VH, Kim J (2012) Cultivation of unculturable soil bacteria. Trends Biotechnol 30:475–484

Piotrowska M, Popowska M (2014) The prevalence of antibiotic resistance genes among *Aeromonas* species in aquatic environments. Ann Microbiol 64:921–934

Piotrowska M, Popowska M (2015) Insight into the mobilome of *Aeromonas* strains. Front Microbiol 6:494. doi:10.3389/fmicb.2015.00494

Piotrowska M, Przygodzińska D, Matyjewicz K, Popowska M (2017) Occurrence and variety of β-lactamase genes among *Aeromonas* spp. isolated from urban wastewater treatment plant. Front Microbiol 8:863. doi:10.3389/fmicb.2017.00863

Poirel L, Rodrigues-Martinez J-M, Mammeri H, Liard A, Nordmann P (2005) Origin of plasmid-mediated quinolone resistant determinant QnrA. Antimicrob Agents Chemother 49 (8):3523–3525

Popowska M, Krawczyk-Balska A (2013) Broad-host-range IncP-1 plasmids and their resistance potential. Front Microbiol 4:44. doi:10.3389/fmicb.2013.00044

Popowska M, Miernik A, Rzeczycka M, Łopaciuk A (2010) The impact of environmental contamination with antibiotics on levels of resistance in soil bacteria. J Environ Qual 39:1679–1687

Popowska M, Rzeczycka M, Miernik A, Krawczyk-Balska A, Walsh F, Duffy B (2012) Influence of soil use on prevalence of tetracycline, streptomycin, and erythromycin resistance and associated resistance genes. Antimicrob Agents Chemother 56:1434–1443

Procópio RE, Silva IR, Martins MK, Azevedo JL, Araújo JM (2012) Antibiotics produced by *Streptomyces*. Braz J Infect Dis 16:466–471

Rampioni G, Falcone M, Heeb S, Frangipani E, Fletcher MP, Dubern JF, Visca P, Leoni L, Cámara M, Williams P (2016) Unravelling the genome wide contributions of specific 2-alkyl-4-quinolones and PqsE to quorum sensing in *Pseudomonas aeruginosa*. PLoS Pathog 2(11): e1006029. doi:10.1371/journal.ppat.1006029

Randall CP, Mariner KR, Chopra I, O'Neill AJ (2013) The target of daptomycin is absent from *Escherichia coli* and other gram-negative pathogens. Antimicrob Agents Chemother 57 (1):637–639

Riesenfeld CS, Goodman RM, Handelsman J (2004) Uncultured soil bacteria are a reservoir of new antibiotic resistance genes. Environ Microbiol 6:981–989

Sales of veterinary antimicrobial agents in 29 European countries in 2014. Sixth ESVAC report. www.ema.europa.eu/ema/pages/includes/document/open_document.jsp?webContentId=WC500214217

Santiago-Rodriguez TM, Fornaciari G, Luciani S, Dowd SE, Toranzos GA, Marota I, Cano RJ (2015) Gut microbiome of an 11th century A.D. Pre-Columbian Andean Mummy. PLoS One. doi:10.1371/journal.pone.0138135

Sarmah AK, Meyer MT, Boxall AB (2006) A global perspective on the use, sales, exposure pathways, occurrence, fate and effects of veterinary antibiotics (VAs) in the environment. Chemosphere 65:725–759

Seiler C, Berendonk TU (2012) Heavy metal driven co-selection of antibiotic resistance in soil and water bodies impacted by agriculture and aquaculture. Front Microbiol 3:399. doi:10.3389/fmicb.2012.00399

Sen D, Van Der Auwera G, Rogers L, Thomas CM, Brown CJ, Top EM (2011) Broad-host-range plasmids from agricultural soils have IncP-1backbones with diverse accessory genes. Appl Environ Microbiol 77:7975–7983

Sengupta S, Chattopadhyay MK, Grossart HP (2013) The multifaceted roles of antibiotics and antibiotic resistance in nature. Front Microbiol. doi:10.3389/fmicb.2013.00047

Shtienberg D, Zilberstaine M, Oppenheim D, Herzog Z, Manulis S, Shwartz H, Kritzman G (2001) Efficacy of oxolinic acid and other bactericides in suppression of *Erwinia amylovora* in pear orchards in Israel. Phytoparasitica 29:143–154

Smith TC, Male MJ, Harper AL, Kroeger JS, Tinkler GP, Moritz ED, Capuano AW, Herwaldt LA, Diekema DJ (2009) Methicillin-resistant *Staphylococcus aureus* (MRSA) strain ST398 is present in Midwestern U.S. swine and swine workers. PLoS One 4:e4258

Song JS, Jeon JH, Lee JH, Jeong SH, Jeong BC, Kim SJ, Lee JH, Lee SH (2005) Molecular characterization of TEM-type beta-lactamases identified in cold-seep sediments of Edison Seamount (south of Lihir Island, Papua New Guinea). J Microbiol 43:172–178

Stockwell VO, Duffy B (2012) Use of antibiotics in plant agriculture. Rev Sci Tech 31:199–210

Stokes HW, Gillings MR (2011) Gene flow, mobile genetic elements and the recruitment of antibiotic resistance genes into Gram-negative pathogens. FEMS Microbiol Rev 35:790–819

Su JQ, Wei B, Xu CY, Qiao M, Zhu YG (2014) Functional metagenomic characterization of antibiotic resistance genes in agricultural soils from China. Environ Int. doi:10.1016/j.envint.2013.12.010

Swenson JM, Facklam RR, Thornsberry C (1990) Antimicrobial susceptibility of vancomycin-resistant *Leuconostoc, Pediococcus* and *Lactobacillus* species. Antimicrob Agents Chemother 34(4):543–549

Thaker M, Spanogiannopoulos P, Wright GD (2010) The tetracycline resistome. Cell Mol Life Sci 67:419–431

Thanner S, Drissner D, Walsh F (2016) Antimicrobial resistance in agriculture. mBio 7. doi:10.1128/mBio.02227-15

Thiele-Bruhn S (2003) Pharmaceutical antibiotic compounds in soils—a review. J Plant Nutr Soil Sci 166:145–167

Tsui WHW, Yim G, Wang HH, McClure JE, Surette MG, Davies J (2004) Dual effects of MLS antibiotics: transcriptional modulation and interactions on the ribosome. Chem Biol 11:1307–1316

Udikovic-Kolic N, Wichmann F, Broderick NA, Handelsman J (2014) Bloom of resident antibiotic-resistant bacteria in soil following manure fertilization. Proc Natl Acad Sci USA 111:15202–15207

Van Elsas JD, Chiurazzi M, Mallon CA, Elhottovā D, Krištůfek V, Salles JF (2012) Microbial diversity determines the invasion of soil by a bacterial pathogen. Proc Natl Acad Sci USA 109:1159–1164

Vogel TM, Simonet P, Jansson JK, Hirsch PR, Tiedje JM, van Elsas JD, Bailey MJ, Renaud N, Philippot L (2009) TerraGenome: a consortium for the sequencing of a soil metagenome. Nat Rev Microbiol 7:252–253

Wachino J, Yoshida H, Yamane K, Suzuki S, Matsui M, Yamagishi T, Tsutsui A, Konda T, Shibayama K, Arakawa Y (2011) SMB-1, a novel subclass B3 metallo-β-lactamase, associated with ISCR1 and a class 1 integron, from a carbapenem-resistant *Serratia marcescens* clinical isolate. Antimicrob Agents Chemother 55:5143–5149. doi:10.1128/AAC.05045-11

Walsh F (2013) The multiple role of antibiotics an antibiotic resistance in nature. Front Microbiol 4:255–259

Wang F, Stedtfeld RD, Kim O-S, Chai B, Yang L, Stedtfeld TM, Hong SG, Kim D, Lim HS, Hashsham SA (2016) Influence of soil characteristics and proximity to Antarctic research stations on abundance of antibiotic resistance genes in soils. Environ Sci Technol 50:12621–12629

Weber T, Charusanti P, Musiol-Kroll EM, Jiang X, Tong Y, Kim HU, Lee SY (2015) Metabolic engineering of antibiotic factories: new tools for antibiotic production in actinomycetes. Trends Biotechnol 33:15–23

Wholey WY, Kochan TJ, Storck DN, Dawid S (2016) Coordinated bacteriocin expression and competence in *Streptococcus pneumoniae* contributes to genetic adaptation through neighbor predation. PLoS Pathog. doi:10.1371/journal.ppat.1005413

Williams-Nguyen J, Sallach JB, Bartelt-Hunt S, Boxall AB, Durso LM, McLain JE, Singer RS, Snow DD, Zilles JL (2016) Antibiotics and antibiotic resistance in agroecosystems: state of the science. J Environ Qual 45(2):394–406. doi:10.2134/jeq2015.07.0336

World Health Organization (2014) Antimicrobial resistance: global report on surveillance. http://www.who.int/antimicrobial-resistance/publications/en/

World Health Organization (2015) Global action plan on antimicrobial resistance. http://www.wpro.who.int/entity/drug_resistance/resources/global_action_plan_eng.pdf

World Health Organization (2016) Antimicrobial resistance fact sheet. http://www.who.int/mediacentre/factsheets/fs194/en/

World Health Organization (2017) Global priority list of antibiotic-resistant bacteria to guide research, discovery, and development of new antibiotics. http://www.who.int/medicines/publications/global-priority-list-antibiotic-resistant-bacteria/en/

Wright GD (2010) Antibiotic resistance in environment: a link to the clinic? Curr Opin Microbiol 13:589–594

Xie F, Murray JD, Kim J, Heckmann AB, Edwards A, Oldroyd GED, Downie JA (2012) Legume pectate lyase required for root infection by rhizobia. Proc Natl Acad Sci U S A 109:633–638. doi:10.1073/pnas.1113992109

Xie W-Y, Mcgrath SP, Su JQ, Hirsch PR, Clark IM, Shen Q, Zhu Y-G, Zhao F-J (2016) Long-term impact of field applications of sewage sludge on soil antibiotic resistome. Environ Sci Technol 50:12602–12611

Yeom JR, Su Y, Kim CG (2017) Quantification of residual antibiotics in cow manure being spread over agricultural land and assessment of their behavioral effects on antibiotic resistant bacteria. Chemosphere 182:771–780

Yim G, Wang H, Davies J (2006) The truth about antibiotics. Int J Med Microbiol 296:163–170

Yim G, Wang H, Davies J (2007) Antibiotics as signalling molecules. Philos Trans R Soc Lond Ser B Biol Sci 362:1195–1200

Yim G, McClure J, Surette MG, Davies JE (2011) Modulation of Salmonella gene expression by subinhibitory concentrations of quinolones. J Antibiot (Tokyo) 64:73–78

Zhang XX, Zhang T, Fang HHP (2009) Antibiotic resistance genes in water environment. Appl Microbiol Biotechnol 82:397–414

Zhang T, Zhang XX, Ye L (2011) Plasmid metagenome reveals high levels of antibiotic resistance genes and mobile genetic elements in activated sludge. PLoS One 6:e26041

Chapter 10
Dissemination Mechanism of Antibiotic Resistance Genes in Environment

Manoj Kumar, Ram Prasad, Shivesh Sharma, Ajit Varma, and Vivek Kumar

10.1 Introduction

Antimicrobials or antibiotics, and their resistance genes do appear naturally in the environment owing to the protective race between microbial communities competing for their territory and survival. Nearly 50% of actinomycetes species isolated from ecosystem have capabilities of antibiotics synthesis, which results in a natural antibiotic deposit in the ecosystem. But the application of antibiotics to encourage the livestock growth also increases the resistance to the whole new level, as established by the variances in the resistance level in forest and agricultural soils. Chen et al. (2007) and D'Costa et al. (2011) have demonstrated that manure from antibiotic fed animals aggravates the spread of resistance, as established by the high levels in the manure applied vegetable garden soils. The increase of antibiotic resistance and multiresistant strains of microbial pathogens and opportunistic bacterial strains that can infect humans and animals is helped and supported by the fact that they are regularly carried on mobile genetic elements, such as transposons and plasmids, that can be transferred not only among bacteria of the same species, but also among the different species, as reported by Davies (2006).

The ever growing degree of antimicrobial resistance (AMR) that comes across human pathogens is a great apprehension for public health all over the world. On the other hand, treatment limitations options for bacterial infections and thereby

The original version of this chapter was revised. An erratum to this chapter can be found at DOI 10.1007/978-3-319-66260-2_22.

M. Kumar • R. Prasad • A. Varma • V. Kumar (✉)
Amity Institute of Microbial Technology, Amity University, Noida, Uttar Pradesh, India
e-mail: vivekbps@gmail.com

S. Sharma
Department of Biotechnology, MLN National Institute of Technology, Allahabad, Uttar Pradesh, India

M.Z. Hashmi et al. (eds.), *Antibiotics and Antibiotics Resistance Genes in Soils*, Soil Biology 51, DOI 10.1007/978-3-319-66260-2_10

lessening clinical effectiveness while enhancing the treatment prices and mortality rate. With a lack in progress of new antibiotics, and ever increasing resistance even to last resort antibiotics, there is an urgent need to preserve the ones available with us. The natural antibiotics have been existed for billions of years (Aminov 2009; Sengupta et al. 2013), which provides a choosy benefit for the antibiotic producing microbial strains by constraining or eliminating other bacteria competing for the resources (Bernier and Surette 2013). Furthermore, their function as cell-to-cell signaling molecules involved has also been described by Davies and Davies (2010). We understand that as the presence of antibiotics in the environment are ancient, so are the antibiotic resistance genes (ARGs) in the environment, as demonstrated by studies which identify numerous antibiotic resistance genes in early permafrost samples (D'Costa et al. 2011; Wright 2010) and on the other hand (Allen et al. 2010) isolated microbiomes from cave ecosystem. Resistance or challenge to antibiotics can happen either by acquirement of resistance conferring genes via horizontal gene transfer method (HGT) or by mutations, of which the earlier is deliberated to be the most significant factor in the existing epidemic of AMR as described by Kümmerer (2004).

Nowadays, antibiotic resistance genes among microbes is a well-known worldwide health problem. As per the report of WHO (2014), many antibiotics are nevertheless very effective at treating several bacterial infections, but some strains are very problematic to treat, and therefore, therapeutical options are getting rarer. This is further exacerbated by the point that the short expected time of valuableness of a novel antibiotic compound before resistance arises means that limited companies are attracted in developing the new antibiotics (Martínez 2008). It has been assessed that in the European Union, in the USA, and in Thailand, antibiotic-resistant bacteria (ARB) are responsible for more than 27,000, 25,000, and 39,000 deaths per annum, correspondingly (Wright 2010; WHO 2014). In a nutshell, the antibiotic resistance is getting more predominant and widespread while few new antibiotics are in progress. It has been presumed that, as the situation of antibiotic resistance stands today, there is a clear-cut danger that manhood will return to medical conditions like those before the therapeutical beginning of the antibiotics (Bernier and Surette 2013). The overuse of antibiotics "should be regulated to hazardous bacterial infections, and to severe medical administration to prevent its widespread in the environment and also not to allow the sensitive bacteria as well as pathogenic bacteria resistant to antibiotics (Davies 2006)".

10.2 Initiation of Antibiotics in the Environment

The antibiotics were originally discovered as chemical compounds manufactured by the environmental fungi and bacteria capable of killing other microbes. Subsequently, these chemical compounds were effective at eliminating bacteria, and it was commonly presumed that the reason of antibiotic production was to fend off competing microbes (Levy and Marshall 2004). On the same note, when antibiotic resistance genes were discovered, it has been believed that they have evolved in bacteria producing the antibiotics which aim bacteria in order to protect against the antibiotics effects (Aminov 2009; Allen et al. 2010). Although this view point is not essentially wrong, newly other aspects of the ecological functions of antibiotics have begun to be searched. The concentration levels of antibiotic manufactured by the environmental bacteria are normally far below the minimum inhibitory concentrations (MICs), which advocates that antibiotics may have primarily some other important function (Cabello 2006; Aminov 2009). Research proof suggests that subinhibitory doses of antibiotics excreted by bacteria play numerous roles in the ecosystem as a regulatory or controlling substance and as a signaling molecule in interbacterial interaction (Martínez 2009; Andersson and Hughes 2012). Strangely, subinhibitory concentrations (SIC) of various antibiotic compounds have been shown to persuade different states in bacteria, chiefly including the biofilm formation, SOS response, and changes in key metabolism. These states can upsurge tolerance towards antibiotics (Wright 2010; Bernier and Surette 2013). Antibiotic resistance genes have also likely to be evolved to fulfill other reasons than being defensive bacteria from antibiotics. One probability is that the main functions of antibiotic resistance genes in the ecosystem are to act as rheostat to manage the responses generated from (SICs) of antibiotic compounds. Some antibiotic resistance genes may play controlling roles in the biological synthesis of the antibiotics (Allen et al. 2010). It has been advocated that β-lactamases are the enzymes which at one point were implicated in peptidoglycan biosynthesis (Martínez 2009). It is important to point out that even though antibiotic compounds and antibiotic resistance genes seem to have functions distinct to antibiosis in the natural ecosystem, it has been established that subinhibitory concentrations of antibiotic compounds, about 200 times below minimum inhibitory concentration values, can select for antibiotic-resistant bacteria (ARB) (Cabello 2006; Sengupta et al. 2013).

10.3 Antibiotics Functions and Resistance Genes in the Ecosystem

Antibiotic compounds were initially discovered as chemical compounds produced by fungi and bacteria present in ecosystem and capable of killing other microbes. Since these chemical compounds were effective at eliminating bacteria, it was commonly presumed that the reason of antibiotic compounds production was to stave off the competing microbes. On the same note, when antibiotic resistance genes were discovered, they were supposed to have evolved in bacteria manufacturing the antibiotics and aim bacteria in order to defend against the effects of antibiotic compounds (Wright 2010). While this view point is not essentially wrong, in recent times other features of the environmental functions of antibiotic compounds have also begun to be investigated. The concentration levels of antibiotic produced by the environmental bacteria are usually far below the minimum inhibitory concentrations (MICs), which advocates that antibiotic compounds may mainly serve some other function also (Aminov 2009).

The antibiotic resistance genes have also likely progressed to fulfill other points rather than protecting the bacteria from antibiotic compounds. One option is that the key functions of antibiotic resistance genes in environment are to control the responses persuaded from subinhibitory antibiotics concentrations. Some antibiotic resistance genes may play controlling roles in the biological synthesis of antibiotics (Martínez 2009; Allen et al. 2010; Andersson and Hughes 2012).

10.4 Role of Gene Transfer in Ecosystem and ARG's Propagation

In an environment, for a pathogenic bacterium to get an antibiotic resistance genes from the existing environmental gene pool, it need to be transferred via one of the three processes of horizontal gene transfer (HGT) methods such as conjugation, transformation, and transduction.

10.4.1 Conjugation Process

Conventionally, the conjugation method has been regarded as the chief helper of antibiotic resistance genes transfer between bacteria. This process of antibiotic resistance transferable by bacterial conjugation means was discovered by Davies and Davies (2010) in the 1950s. Since then, antibiotic resistance genes transferable by the process of conjugation have been recorded numerous times. The advantages of this mode of gene transfer include the possibility to transfer DNA among the extensive host range of species (Smillie et al. 2010). The conjugation process has even been established from bacterial cells to the eukaryotic cells also (Calero-Caceres et al. 2014) and has been observed in numerous diverse environments, which includes soil, aquatic sediment, marine water, activated sludge and sewage wastewater (Brown-Jaque et al. 2015). The most significant genetic elements capable of being transferred by conjugation process are by the plasmids and integrative conjugative elements (ICEs) (Calero-Caceres et al. 2014).

10.4.2 Transformation Process

The transformation is one of three processes of gene transfer in bacterial cell in which the exogenous genetic material can be introduced into the bacterial cell. Not long ago, the significance of transformation process in facilitating environmental transfer of antibiotic resistance genes has begun to be reconsidered. Transformation process in environment may instinctively sound like an infrequent event considering the sensitivity of DNA to degradation by enzyme nucleases and by the dilution effects in aquatic environments. Nonetheless, DNA strand may be stabilized by linkage to particles from sediment and the soil. The dilution effects may also be a lesser amount of imperative if transformation process occurs in the biofilms where newly dead bacteria lyse and allow their adjacent bacteria to uptake their freed DNA (Davison 1999). The natural transformation process has also been established in many dissimilar environments, such as groundwater, marine water, rivers, wastewater, and soil (Davison 1999), and it has been connected as accountable for the distribution of penicillin resistance genes in *Streptococcus* spp. (Johnsborg and Håvarstein 2009). In one study conducted by Mao et al. (2014), concentrations of extracellular DNA were compared to intracellular DNA in a river basin in China. It was observed that extracellular DNA (including antibiotic resistance genes) was more plentiful than intracellular amount of DNA, implying that extracellular DNA is a significant ecological pool for genes availability via transformation process.

10.4.3 Transduction Process

In transduction process, transfer of DNA takes place between bacteria via bacteriophages or phage; this process may also be more significant in environmental gene transfer rather than formerly thought (Muniesa et al. 2013). The bacteriophage particles are well suitable for facilitating DNA transfer in the ecosystem. Opposing to the naked DNA, they are comparatively resilient to the environmental degradation and their compacted size further make simpler their distribution (Johnsborg and Håvarstein 2009). Additionally, some phages are also known to have very broad host ranges, some phages even are capable of infecting dissimilar bacterial classes (Jensen et al. 1998). These possessions make a phage ideal for transporting genes between spatially indistinct bacterial groups, for example, from ecological communities to the human microbiomes (Muniesa et al. 2013). The transduction process has also been shown to be very common in the marine environments (Jiang and Paul 1998). Moreover, the evolutionary studies have established that significant parts of bacterial genomes have prophage origins, which implies the prominence and degree of viral alterations in the bacterial chromosome (Brüssow and Hendrix 2002). Using viral metagenome analyses technique, the β-lactamase genes have been spotted in the activated sludge and urban sewage also (Rolain et al. 2012). Gene conferring methicillin resistance in the methicillin-resistant *S. aureus* bacterium (MRSA), mecA, has also been found in phage DNA from a wastewater treatment plant and from the receiving water (Colomer-Lluch et al. 2011).

10.4.4 Integrons Process

A particularly well-studied genetic transfer element in environmental contexts is the integron. Integrons are genetic assembly platforms capable of capturing and expressing gene cassettes, which can encode antibiotic resistance determinants. A defining feature of all types of integrons is a gene coding for a site-specific tyrosine recombinase called an integrase which can excise and integrate gene cassettes into the integron. It can also reshuffle the order of gene cassettes which affect the relative rate of expression of the individual cassettes (Mazel 2006; Cambray et al. 2010). The gene encoding the integrase (intI) is induced by the SOS response. Interestingly, antibiotics such as trimethoprim, quinolones, and β-lactams are known to induce the SOS response. Exposure to any of these antibiotics will thus induce excision, integration, and changes in expression rates of gene cassettes, increasing the likelihood that at least some bacteria in the population carrying the integron will have a high expression rate of a relevant ARG (Guerin et al. 2009). The SOS response has also been shown to be inducible by conjugative DNA transfer, which means that integrons transferred to another host cell on a conjugative plasmid are likely to reshuffle their cassettes and thus increase the phenotype variability in a population (Baharoglu et al. 2010).

Integrons are typically divided into mobile integrons and chromosomal integrons. While chromosomal integrons are usually stationary in the bacteria, mobile integrons are readily disseminated between bacteria. While mobile integrons cannot mobilize and transfer themselves per se, they are often associated with genetic elements which can, such as plasmids (Guerin et al. 2009; Domingues et al. 2012a). Recent studies have also indicated that natural transformation may be important in the dissemination of integrons (Domingues et al. 2012b). Mobile integrons are often capable of changing genetic locations within the host cell as well since they are commonly associated with transposable genetic elements such as insertion sequences (ISs) and transposons (Domingues et al. 2012a). Class 1 integrons, a class of mobile integrons commonly found among clinical isolates, are associated with transposons derived from Tn402, which in turn can be carried by larger transposons, such as Tn21. Although mobile integrons usually only have a few gene cassettes in their cassette arrays, they often encode antibiotic resistance functionality and other phenotypes which give the host bacteria an adaptive advantage (Mazel 2006, Cambray et al. 2010).

10.5 Propagation of ARGs in Ecosystem

The environmental presence of antibiotic resistance genes is well recognized for numerous and different types of environments. The extent of antibiotic resistance genes dissemination is likely to be dependent upon the genetic context of particular gene.

10.5.1 Sulphonamide Resistance

Sulphonamide resistance genes sulI and sulII are examples of antibiotic resistance genes which are prevalent in environment. As sulI is normally found as a conserved part of class 1 integrons, as described by Johnsborg and Håvarstein (2009), it can also be expected to be found wherever these extensive mobile genetic elements are omnipresent. Gene sulII is most usually found on plasmids of the incQ group (Huovinen et al. 1995; Sköld 2001). Gene sulI and sulII have also been found in the river water from Colorado region, USA (Pei et al. 2006), from Danish pigs (Wu et al. 2010), from Australian and German surface waters (Stoll et al. 2012), and from the freshwater and marine water in Philippines (Suzuki et al. 2013). Gene sulI has also been found to be reported in wastewater (Gao et al. 2012).

10.5.2 Quinolone Resistance

The first quinolone resistance gene was discovered, qnrA, located on the plasmid pMG252. Since then, many qnr genes have been found to be associated with plasmids and other related mobile genetic elements. Genes qnrA and qnrB are often found on class 1 integrons, making quinolone resistance a trait which is often associated with other resistance determinants which is co-carried on the integrin part (Robicsek et al. 2006). Numerous studies have been reported the isolation of qnr genes from the environmental sources. Gene qnrS has also been isolated from numerous dissimilar sources, including the activated sludge of a wastewater treatment plant in Germany (Bönemann et al. 2006), also from the Seine river in France (Cattoir et al. 2008), from a lake in the Switzerland (Picão et al. 2008) and from a river water in Turkey (Özgumus et al. 2009). Gene qnrB has also been found in wastewater effluent from a wastewater treatment plant in Italy (Forcella et al. 2010), genes qnrB and qnrS have been reported in Mexican soils which has been irrigated with wastewater (Dalkmann et al. 2012), and gene qnrA, qnrB, and qnrS have also been found in an urban coastal wetland which close to the US-Mexican border (Cummings et al. 2011).

10.5.3 Trimethoprim Resistance

The trimethoprim resistance genes of the dfr family normally appear to be found on the integrons as cassettes (Alekshun and Levy 2007). For example, dfrA1 has been found as a gene cassette on both class 1 and class 2 of integrons. The dfr genes tendency for being carried on integrons is likely to have helped their widespread distribution in the ecosystem (Alekshun and Levy 2007). Genes dfrA1, dfrA5, dfrA6, dfrA12, and dfrA17 have also been detected as cassettes in the class

1 integrons in a river from India (Mukherjee and Chakraborty 2006). In Portugal, genes dfrA1, dfrA7, dfrA12, and dfrA17 were also found as integron cassettes in a polluted lagoon (Henriques et al. 2006), and genes dfrA1 and dfrA12 were detected in a wastewater treatment plant which is connected to a slaughterhouse (Moura et al. 2007). Gene drfA1 has been detected in the surface waters from areas of Germany and Australia.

10.5.4 Vancomycin Resistance

The vanA operon, which encodes vancomycin resistance, is characteristically found to be carried on Tn1546 or Tn1546 like elements. While the former transposable element is non-conjugative, the latter is frequently found to be associated with conjugative plasmids. Distribution of the vanB operon is believed to be chiefly owing to the spread of Tn916 like integrative and conjugative elements (ICEs) and related elements carrying the gene cluster (Courvalin 2006). Both vanA and vanB gene have been found in wastewater in England (Caplin et al. 2008) and in Sweden (Iversen et al. 2002). Gene vanB has also been reported in receiving river and wastewater effluent in Sweden (Berglund et al. 2013). Furthermore, gene vanA and vanB have also been found in meat from bovine and swine sources (Messi et al. 2006), and in oceanic water in the USA (Roberts et al. 2009). Gene vanA has also been found from Portugal wastewater (Araújo et al. 2010), and drinking water and wastewater from Germany (Schwartz et al. 2003). Remarkably, a variant of the gene vanA operon has also been found in 30,000-year-old Beringian permafrost (D'Costa et al. 2011), which advocates that vancomycin resistance gene is both prehistoric and also prevalent in the ecosystem.

10.5.5 Tetracycline Resistance

There are many tetracycline resistance determinants which are chromosomally encoded, and the majority of tet genes are found on the plasmids, transposons, and integrative and conjugative elements. Many of the mobile genetic elements which carry tet genes are conjugative in nature and also carry genes encoding resistance to the other antibiotic compounds. For instance, tetM gene can be found on the Integrative and conjugative element Tn2009 which also carries the macrolide-lincosamide-streptogramin B resistance gene ermB, and macrolide efflux genes mefA and mfrD. It is very likely that great range of tet genes and the variety and mobility of the genetic elements in which they live have provided appreciably to their distribution among many different bacterial genera (Roberts 2005). Genes tetA, tetB, tetC, tetD, tetE, tetG, tetM, tetO, tetS, and tetQ have been found in the wastewater from two wastewater plants in Wisconsin region, USA (Auerbach et al. 2007), gene tetO and tetW from river water in Colorado, USA (Stoll et al. 2012), genes tetB, tetL, tetM, tetO, tetQ, and tetW from archived soil collected from the Netherlands (Knapp et al. 2010). Genes tetA, tetC, tetG, tetM, tetS, and tetX have also been discovered in the activated sludge from 15 diverse sewage treatment plants in China (Zhang and Zhang 2011), and genes tetA and tetB from the surface water from Germany and Australia (Stoll et al. 2012).

10.5.6 Macrolide Resistance

The ermB is the most prevalent of the macrolide resistance genes, and it is connected with a variety of dissimilar mobile genetic elements including integrative and conjugative elements located on both the chromosomes and plasmids as well as on to the non-conjugative transposons as reported by Roberts (2008). The integrative and conjugative elements among which ermB gene have been found to be carried include transposable elements Tn1545, Tn2010, and others of the Tn916 family, representing that ermB gene is often associated with other antibiotic resistance determinants on the conjugative platform (Roberts 2008). The erm genes are predominant in environment, and they have also been found in a variety of dissimilar ecosystems. Genes ermA and ermB have also been found in milk from Brazil cows (Duarte et al. 2004) and from poultry production environments in the eastern seaboard of the USA (Hayes et al. 2005).Furthermore, gene ermB have also been found in the poultry samples (Novais et al. 2005), isolated from wastewater in Portugal (Araújo et al. 2010), and also from the German and Australian surface waters (Stoll et al. 2012). Genes ermA, ermB, ermC, ermF, ermT, and ermX have also been found in swine and bovine manure as well as from the swine waste lagoon (Chen et al. 2007).

10.6 Distribution Ways of Antibiotics and ARGs

Antibiotics of human origin can enter the environment through a number of different routes. Antibiotics and their metabolites are released from hospitals with urine and feces from patients as hospital wastewater effluent. Similarly, antibiotics are released into the wastewater treatment system via people taking antibiotics from home. From the WWTPs, the antibiotics can end up in sludge dispersed on fields as fertilizer, or released as run-off directly into the receiving surface waters (Halling-Sørensen et al. 1998). Wastewater can also be treated by releasing it into wetlands (Scholz and Lee 2005). In such cases, the wetlands will be exposed to antibiotic contaminants in the wastewater. Antibiotics are also used therapeutically or as growth promoters in livestock and poultry. Antibiotics and their metabolites will spread through animal excrements and end up in fields and groundwater, or in the case of antibiotic use in fish farms, directly into the aquatic environment (Halling-Sørensen et al. 1998). It is also worth noting that wherever antibiotics are spread, it is also likely that resistant bacteria follow the same routes of dispersal (Baquero et al. 2008). This results in environments where antibiotics, ARGs, resistant bacteria, and the environmental bacterial flora (which may also harbor ARGs and potential ARGs) are mixed. These types of environments are likely resistance hotspots where ARGs proliferate and new resistant strains are created by HGT. The routes by which humans may come into contact with these bacteria are numerous. They include consumption of crops grown by contaminated sludge used as fertilizer, drinking water drawn from contaminated ground or surface water, and frolicking in marine water linked to contaminated surface water. When these resistant bacteria enter humans, they have the opportunity to spread their ARGs to the human microbiome (Wellington et al. 2013).

10.7 The Gene Transferring Agents

The gene transfer agents (GTAs) were first identified in bacteria *Rhodobacter capsulatus* (RcGTA) in 1974 by Marrs (1974). These are host cell or by bacterial cell produced small particles that look like or resemble a phage and are capable of transferring the genetic content of a bacterial cell. The gene transfer agents have several typical and distinctive features, such as: (1) the gene transfer agents carry random pieces of the producing cell's genome rather than carrying the DNA encoding for their own cell machinery (as happens with the self-propagating phages), (Marrs 1974; Hynes et al. 2012); (2) the amount of DNA packaged by the gene transfer agents is not sufficient to encode all of their protein constituents, therefore, making them unable to self-replicate or self-propagate (Lang and Beatty 2001, 2007); (3) the gene transfer agents production is controlled by the cell regulatory tools (Leung et al. 2012; Brimacombe et al. 2014); (4) the gene transfer agent particles are discharged through the bacterial cell lysis (Hynes et al. 2012) though the bacterial cultures do not display noticeable lysis on Petri plates (Marrs 1974). This is because that only a small subpopulation of the gene transfer agents producing cultures (~3%) is responsible for ~95% of the gene transfer agents discharge (Fogg et al. 2012; Hynes et al. 2012); (5) not long ago, it has been anticipated that the gene transfer agents combine key aspects of mechanisms like transduction and transformation for bacterial cell entry, which require proteins involved in natural process of transformation (Brimacombe et al. 2015).

Though the gene transfer agents particles do not essentially carry any GTA-encoding genes (Lang et al. 2012), the RcGTA-like gene clusters are prevalent in alpha-proteobacteria, particularly in the order Rhodobacterales. Lang and Beatty (2007) and Lang et al. (2012) have demonstrated that a comprehensive set of RcGTA-like structural genes has been demonstrated in nearly every sequenced member of the Rhodobacterales. Furthermore, in the order of Rhodobacterales, there are two species viz. *Roseovarius nubinhibens* and *Ruegeria mobilis*, which are known to manufacture the gene transfer agents, and there is evidence of GTA production in bacterium *Ruegeria pomeroyi*

(McDaniel et al. 2010; Lang et al. 2012). Other known gene transfer agents have also been reported such as in the spirochaete *Brachyspira hyodysenteriae* (VSH-1), in the delta-proteobacterium *Desulfovibrio desulfuricans* (Dd1), and in the archaeon *Methanococcus voltae* (VTA) (Lang and Beatty 2007; Lang et al. 2012). The genes required for gene transfer agents manufacture are contained within the bacterial host genome and seem to have been proliferated through the process of vertical transmission (Lang and Beatty 2007).

It has been proposed that gene transfer agents have numerous benefits over the mechanisms like horizontal gene transfer (HGT) (Stanton 2007). The gene transfer agents' particles also afford DNA defense and protection by damaging environmental factors. This is opposed to the naked DNA involved in the process of natural transformation and compared to the process of conjugation; the transfer ability of gene transfer agents is likely to be maintained after conditions such as killing the bacterial cell. Moreover, this is not inhibited by cell-to-cell contact; and lastly, compared to the process of transduction, the gene transfer agents' particles chiefly carry arbitrary pieces of the host or bacterial genome, rather than mostly phage DNA. In the marine ecosystem, the gene transfer agents facilitated transfer events which have been reported to be unusually high, which up to the several million times greater than earlier estimates of horizontal gene transfer in the marine environments, and which exceeds formerly pronounced process like transformation and transduction (McDaniel et al. 2010). Furthermore, the genes can be swapped between bacterial phyla (McDaniel et al. 2010; Lang et al. 2012), which suggests the possible prevalent involvement of gene transfer agents in determining and driving acclimatization of the natural environmental factors.

10.8 Concluding Remarks

The enhancement in the levels of antibiotics in the environment, which has been driven by medical and agricultural need, is unparalleled and has disturbed and interrupted the natural equilibrium and stability between microorganisms and antimicrobial products. The adverse effect of antibiotics on the microbial communities is increasing and wide-ranging, which result in a progressively perceptible risk to the healthcare, as resistance to all known antibiotics spreads quickly around the sphere. Our information and acquaintance of the communications between antibiotics or antimicrobials and challenge against it has been observed not only in the clinical field but also throughout diverse ecosystems around the globe. This situation is rapidly increasing and has provided valuable insights also to make scientists and clinical practitioners aware. Though, it is imperative that we continue to unknot the extent of and spreading between microbial resistomes and their ecosystems. Since we understand that any effort at coming to terms with the antimicrobial resistance problem will have to account for the vast pools of antibiotic resistance genes.

References

Alekshun MN, Levy SB (2007) Molecular mechanisms of antibacterial multidrug resistance. Cell 128:1037–1050

Allen HK, Donato J, Wang HH, Cloud-Hansen KA, Davies J, Handelsman J (2010) Call of the wild: antibiotic resistance genes in natural environments. Nat Rev Microbiol 8:251–259

Aminov RI (2009) The role of antibiotics and antibiotic resistance in nature. Environ Microbiol 11:2970–2988

Andersson DI, Hughes D (2012) Evolution of antibiotic resistance at non-lethal drug concentrations. Drug Resist Updat 15:162–172

Araújo C, Torres C, Silva N, Carneiro C, Gonçalves A, Radhouani H (2010) Vancomycin-resistant enterococci from Portuguese wastewater treatment plants. J Basic Microbiol 50:605–609

Auerbach EA, Seyfried EE, McMahon KD (2007) Tetracycline resistance genes in activated sludge wastewater treatment plants. Water Res 41:1143–1151

Baharoglu Z, Bikard D, Mazel D (2010) Conjugative DNA transfer induces the bacterial SOS response and promotes antibiotic resistance development through integron activation. PLoS Genet 6:e1001165

Baquero F, Martínez JL, Cantón R (2008) Antibiotics and antibiotic resistance in water environments. Curr Opin Biotechnol 9:260–265

Berglund B, Fick J, Lindgren PE (2013) Urban wastewater effluent increases antibiotic resistance gene concentrations in a receiving northern European river. Environ Toxicol Chem 34:192–196

Bernier SP, Surette MG (2013) Concentration-dependent activity of antibiotics in natural environments. Front Microbiol 4:20–25

Bönemann G, Stiens M, Pühler A, Schlüter A (2006) Mobilizable IncQ-related plasmid carrying a new quinolone resistance gene, qnrS2, isolated from the bacterial community of a wastewater treatment plant. Antimicrob Agents Chemother 50:3075–3080

Brimacombe CA, Ding H, Beatty JT (2014) *Rhodobacter capsulatus* DprA is essential for RecA-mediated gene transfer agent (RcGTA) recipient capability regulated by quorum-sensing and the CtrA response regulator. Mol Microbiol 92:1260–1278

Brimacombe CA, Ding H, Johnson JA, Beatty JT (2015) Homologues of genetic transformation DNA import genes are required for *Rhodobacter capsulatus* gene transfer agent recipient capability regulated by the response regulator CtrA. J Bacteriol 197:2653–2663

Brown-Jaque M, Calero-Caceres W, Muniesa M (2015) Transfer of antibiotic-resistance genes via phage-related mobile elements. Plasmid 79:1–7

Brüssow H, Hendrix RW (2002) Phage genomics: small is beautiful. Cell 108:13–26

Cabello FC (2006) Heavy use of prophylactic antibiotics in aquaculture: a growing problem for human and animal health and for the environment. Environ Microbiol 8:1137–1144

Calero-Caceres W, Melgarejo A, Colomer-Lluch M, Stoll C, Lucena F, Jofre J (2014) Sludge as a potential important source of antibiotic resistance genes in both the bacterial and bacteriophage fractions. Environ Sci Technol 48:7602–7611

Cambray G, Guerout AM, Mazel D (2010) Integrons. Annu Rev Genet 44:141–166

Caplin JL, Hanlon GW, Taylor HD (2008) Presence of vancomycin and ampicillin-resistant *Enterococcus faecium* of epidemic clonal complex-17 in wastewaters from the south coast of England. Environ Microbiol 10:885–892

Cattoir V, Poirel L, Aubert C, Soussy CJ, Nordmann P (2008) Unexpected occurrence of plasmid-mediated quinolone resistance determinants in environmental *Aeromonas* spp. Emerg Infect Dis 14:231–237

Chen J, Yu Z, Michel FC Jr, Wittum T, Morrison M (2007) Development and application of realtime PCR assays for quantification of germ genes conferring resistance to macrolides-lincosamides-Streptogramin B in livestock manure and manure management systems. Appl Environ Microbiol 73:4407–4416

Colomer-Lluch M, Jofre J, Muniesa M (2011) Antibiotic resistance genes in the bacteriophage DNA fraction of environmental samples. PLoS One 6:e17549

Courvalin P (2006) Vancomycin resistance in Gram-positive cocci. Clin Infect Dis 42(Suppl 1): S25–S34

Cummings DE, Archer KF, Arriola DJ, Baker PA, Faucett KG, Laroya JB (2011) Broad dissemination of plasmid-mediated quinolone resistance genes in sediments of two urban coastal wetlands. Environ Sci Technol 45:447–454

Dalkmann P, Broszat M, Siebe C, Willaschek E, Sakinc T, Huebner J (2012) Accumulation of pharmaceuticals, Enterococcus, and resistance genes in soils irrigated with wastewater from zero to 100 years in central Mexico. PLoS One 7:e45397

Davies J (2006) Are antibiotics naturally antibiotics? J Ind Microbiol Biotechnol 33:496–499

Davies J, Davies D (2010) Origins and evolution of antibiotic resistance. Microbiol Mol Biol Rev 74:417–433

Davison J (1999) Genetic exchange between bacteria in the environment. Plasmid 42:73–91

D'Costa VM, King CE, Kalan L, Morar M, Sung WW, Schwarz C (2011) Antibiotic resistance is ancient. Nature 477:457–461

Domingues S, da Silva GJ, Nielsen KM (2012a) Integrons: vehicles and pathways for horizontal dissemination in bacteria. Mob Genet Elements 2:211–223

Domingues S, Harms K, Fricke WF, Johnsen PJ, da Silva GJ, Nielsen KM (2012b) Natural transformation facilitates transfer of transposons, integrons and gene cassettes between bacterial species. PLoS Pathog 8:e1002837

Duarte RS, Miranda OP, Bellei BC, Brito MA, Teixeira LM (2004) Phenotypic and molecular characteristics of *Streptococcus agalactiae* isolates recovered from milk of dairy cows in Brazil. J Clin Microbiol 42:4214–4222

Fogg PC, Westbye AB, Beatty JT (2012) One for all or all for one: heterogeneous expression and host cell lysis are key to gene transfer agent activity in *Rhodobacter capsulatus*. PLoS One 7: e43772

Forcella C, Pellegrini C, Celenza G, Segatore B, Calabrese R, Tavio MM (2010) QnrB9 in association with TEM-116 extended-spectrum beta-lactamase in Citrobacter freundii isolated from sewage effluent: first report from Italy. J Chemother 22:243–245

Gao P, Munir M, Xagoraraki I (2012) Correlation of tetracycline and sulfonamide antibiotics with corresponding resistance genes and resistant bacteria in a conventional municipal wastewater treatment plant. Sci Total Environ 421–422:173–183

Guerin E, Cambray G, Sanchez-Alberola N, Campoy S, Erill I, Da Re S (2009) The SOS response controls integron recombination. Science 324:1034

Halling-Sørensen B, Nors Nielsen S, Lanzky PF, Ingerslev F, Holten Lützhøft HC, Jørgensen SE (1998) Occurrence, fate and effects of pharmaceutical substances in the environment—a review. Chemosphere 36:357–393

Hayes JR, Wagner DD, English LL, Carr LE, Joseph SW (2005) Distribution of streptogramin resistance determinants among *Enterococcus faecium* from a poultry production environment of the USA. J Antimicrob Chemother 55:123–126

Henriques IS, Fonseca F, Alves A, Saavedra MJ, Correia A (2006) Occurrence and diversity of integrons and beta-lactamase genes among ampicillin-resistant isolates from estuarine waters. Res Microbiol 57:938–947

Huovinen P, Sundström L, Swedberg G, Sköld O (1995) Trimethoprim and sulfonamide resistance. Antimicrob Agents Chemother 39:279–289

Hynes AP, Mercer RG, Watton DE, Buckley CB, Lang AS (2012) DNA packaging bias and differential expression of gene transfer agent genes within a population during production and release of the *Rhodobacter capsulatus* gene transfer agent, RcGTA. Mol Microbiol 85:314–325

Iversen A, Kühn I, Franklin A, Möllby R (2002) High prevalence of vancomycin-resistant enterococci in Swedish sewage. Appl Environ Microbiol 68:2838–2842

Jensen EC, Schrader HS, Rieland B, Thompson TL, Lee KW, Nickerson KW (1998) Prevalence of broad-host-range lytic bacteriophages of *Sphaerotilus natans*, *Escherichia coli*, and *Pseudomonas aeruginosa*. Appl Environ Microbiol 64:575–580

Jiang SC, Paul JH (1998) Gene transfer by transduction in the marine environment. Appl Environ Microbiol 64:2780–2787

Johnsborg O, Håvarstein LS (2009) Regulation of natural genetic transformation and acquisition of transforming DNA in *Streptococcus pneumonia*. FEMS Microbiol Rev 33:627–642

Knapp CW, Dolfing J, Ehlert PA, Graham DW (2010) Evidence of increasing antibiotic resistance gene abundances in archived soils since 1940. Environ Sci Technol 44:580–587

Kümmerer K (2004) Resistance in the environment. J Antimicrob Chemother 54:311–320

Lang AS, Beatty JT (2001) The gene transfer agent of *Rhodobacter capsulatus* and "constitutive transduction" in prokaryotes. Arch Microbiol 175:241–249

Lang AS, Beatty JT (2007) Importance of widespread gene transfer agent genes in alpha-proteobacteria. Trends Microbiol 15:54–62

Lang AS, Zhaxybayeva O, Beatty JT (2012) Gene transfer agents: phagelike elements of genetic exchange. Nat Rev Microbiol 10:472–482

Leung MM, Brimacombe CA, Spiegelman GB, Beatty JT (2012) The GtaR protein negatively regulates transcription of the gtaRI operon and modulates gene transfer agent (RcGTA) expression in *Rhodobacter capsulatus*. Mol Microbiol 83:759–774

Levy SB, Marshall B (2004) Antibacterial resistance worldwide: causes, challenges and responses. Nat Med 10(Suppl):S122–S129

Mao D, Luo Y, Mathieu J, Wang Q, Feng L, Mu Q (2014) Persistence of extracellular DNA in river sediment facilitates antibiotic resistance gene propagation. Environ Sci Technol 48:71–78

Marrs B (1974) Genetic recombination in *Rhodopseudomonas capsulata*. Proc Natl Acad Sci USA 71:971–973

Martínez JL (2008) Antibiotics and antibiotic resistance genes in natural environments. Science 321:365–367

Martínez JL (2009) The role of natural environments in the evolution of resistance traits in pathogenic bacteria. Proc Biol Sci 276:2521–2530

Mazel D (2006) Integrons: agents of bacterial evolution. Nat Rev Microbiol 4:608–620

McDaniel LD, Young E, Delaney J, Ruhnau F, Ritchie KB, Paul JH (2010) High frequency of horizontal gene transfer in the oceans. Science 330:50

Messi P, Guerrieri E, de Niederhäusern S, Sabia C, Bondi M (2006) Vancomycin-resistant enterococci (VRE) in meat and environmental samples. Int J Food Microbiol 107:218–222

Moura A, Henriques I, Ribeiro R, Correia A (2007) Prevalence and characterization of integrons from bacteria isolated from a slaughterhouse wastewater treatment plant. J Antimicrob Chemother 60:1243–1250

Mukherjee S, Chakraborty R (2006) Incidence of class 1 integrons in multiple antibiotic-resistant Gram-negative copiotrophic bacteria from the River Torsa in India. Res Microbiol 157:220–226

Muniesa M, Colomer-Lluch M, Jofre J (2013) Could bacteriophages transfer antibiotic resistance genes from environmental bacteria to human-body associated bacterial populations? Mob Genet Elements 3:e25847

Novais C, Coque TM, Costa MJ, Sousa JC, Baquero F, Peixe LV (2005) High occurrence and persistence of antibioticresistant enterococci in poultry food samples in Portugal. J Antimicrob Chemother 56:1139–1143

Ozgumus OB, Sandalli C, Sevim A, Celik-Sevim E, Sivri N (2009) Class 1 and class 2 integrons and plasmid-mediated antibiotic resistance in coliforms isolated from ten rivers in northern Turkey. J Microbiol 47:19–27

Pei R, Kim SC, Carlson KH, Pruden A (2006) Effect of river landscape on the sediment concentrations of antibiotics and corresponding antibiotic resistance genes (ARG). Water Res 40:2427–2435

Picão RC, Poirel L, Demarta A, Silva CS, Corvaglia AR, Petrini O (2008) Plasmid-mediated quinolone resistance in *Aeromonas allosaccharophila* recovered from a Swiss lake. J Antimicrob Chemother 62:948–950

Roberts MC (2005) Update on acquired tetracycline resistance genes. FEMS Microbiol Lett 245:195–203

Roberts MC (2008) Update on macrolide-lincosamide-streptogramin, ketolide and oxazolidinone resistance genes. FEMS Microbiol Lett 282:147–159

Roberts MC, Soge OO, Giardino MA, Mazengia E, Ma G, Meschke JS (2009) Vancomycin-resistant *Enterococcus* spp. in marine environments from the West Coast of the USA. J Appl Microbiol 107:300–307

Robicsek A, Jacoby GA, Hooper DC (2006) The worldwide emergence of plasmid mediated quinolone resistance. Lancet Infect Dis 6:629–640

Rolain JM, Canton R, Cornaglia G (2012) Emergence of antibiotic resistance: need for a new paradigm. Clin Microbiol Infect 8:615–616

Scholz M, Lee BH (2005) Constructed wetlands: a review. Int J Environ Stud 62:421–447

Schwartz T, Kohnen W, Jansen B, Obst U (2003) Detection of antibiotic-resistant bacteria and their resistance genes in wastewater, surface water, and drinking water biofilms. FEMS Microbiol Ecol 43:325–335

Sengupta S, Chattopadhyay MK, Grossart HP (2013) The multifaceted roles of antibiotics and antibiotic resistance in nature. Front Microbiol 4:47–51

Sköld O (2001) Resistance to trimethoprim and sulfonamides. Vet Res 32:261–273

Smillie C, Garcillán-Barcia MP, Francia MV, Rocha EP, de la Cruz F (2010) Mobility of plasmids. Microbiol Mol Biol Rev 74:434–452

Stanton TB (2007) Prophage-like gene transfer agents-novel mechanisms of gene exchange for *Methanococcus*, *Desulfovibrio*, *Brachyspira*, and *Rhodobacter* species. Anaerobe 13:43–49

Stoll C, Sidhu JP, Tiehm A, Toze S (2012) Prevalence of clinically relevant antibiotic resistance genes in surface water samples collected from Germany and Australia. Environ Sci Technol 46:9716–9726

Suzuki S, Ogo M, Miller TW, Shimizu A, Takada H, Siringan MA (2013) Who possesses drug resistance genes in aquatic environment?: sulfamethoxazole (SMX) resistance genes among the bacterial community in water environment of Metro-Manila, Philippines. Front Microbiol 4:102–110

Wellington EM, Boxall AB, Cross P, Feil EJ, Gaze WH, Hawkey PM (2013) The role of the natural environment in the emergence of antibiotic resistance in Gram-negative bacteria. Lancet Infect Dis 13:155–165

World Health Organization (2014) Antimicrobial resistance: global report on surveillance 2014. WHO, Geneva

Wright GD (2010) Antibiotic resistance in the environment: a link to the clinic? Curr Opin Microbiol 13:589–594

Wu S, Dalgaard A, Hammerum AM, Porsbo LJ, Jensen LB (2010) Prevalence and characterization of plasmids carrying sulfonamide resistance genes among *Escherichia coli* from pigs, pig carcasses and human. Acta Vet Scand 52:47–51

Zhang XX, Zhang T (2011) Occurrence, abundance, and diversity of tetracycline resistance genes in 15 sewage treatment plants across China and other global locations. Environ Sci Technol 45:2598–2604

Chapter 11
Fate of Antibiotics in Soil

Rida Akram, Asad Amin, Muhammad Zaffar Hashmi, Abdul Wahid, Muhammad Mubeen, Hafiz Mohkum Hammad, Shah Fahad, and Wajid Nasim

11.1 Introduction

Antibiotics exist and are being used for the betterment of public health; use of antibiotic starts in the beginning of twentieth century in cattle feed (Knapp et al. 2010). A beneficial aspect of antibiotic addition in animal feed was reported by American Cyanamid publication in 1950 (Ogle 2013). Since then, the practices of antibiotics addition in animal feed have become more frequent and fired as global trend. Approximately 150 antibiotics are used of which about 90% is obtained from the natural compounds like bacteria, fungi, and semisynthetic modifications and taken as

The original version of this chapter was revised. An erratum to this chapter can be found at DOI 10.1007/978-3-319-66260-2_22.

R. Akram • A. Amin • M. Mubeen • H.M. Hammad
Department of Environmental Sciences, COMSATS Institute of Information Technology (CIIT), Vehari, Pakistan

M.Z. Hashmi
Department of Meteorology, COMSATS Institute of Information Technology (CIIT), Islamabad, Pakistan

A. Wahid
Department of Environmental Sciences, Bahauddin Zakarya University, Multan, Pakistan

S. Fahad
College of Plant Science and Technology, Huazhong Agricultural University, Wuhan, China

W. Nasim (✉)
Department of Environmental Sciences, COMSATS Institute of Information Technology (CIIT), Vehari, Pakistan

CIHEAM-Institut Agronomique Mediterraneen de Montpellier (IAMM), Montpellier, France

National Research Flagship, CSIRO Sustainable Ecosystems, Toowoomba, QLD, 4350, Australia
e-mail: wajidnaseem2001@gmail.com

M.Z. Hashmi et al. (eds.), *Antibiotics and Antibiotics Resistance Genes in Soils*, Soil Biology 51, DOI 10.1007/978-3-319-66260-2_11

"natural products," and some are totally synthetic (von Nussbaum et al. 2006). The total amount of annual use of antibiotic is about 100,000–200,000 tons worldwide including medical and veterinary antibiotic (Wang and Tang 2010).

About 400 active chemical constituents are synthetically manufactured and used to make 2000 veterinary drugs to treat animals (including various species like pigs, cattle, horses, sheep, goats, birds, fish, deer, cats, and dogs) (FDA 2012). The use of these veterinary drugs is crucially important for animal production. But, animals don't have the ability to utilize all these applied antibiotics, depending on the animal species and chemical constituents of veterinary drugs, 10–90% of the supplemented antibiotics is excreted in the form of animal urine or feces and behave as integral parent complexes or bioactive metabolites (Kumar et al. 2005).

Most antibiotics are organic in nature with a wide range of functional groups and can behave as ionic, amphiphilic, or amphoteric and have the capability of absorption on the soil surface (Tasho and Cho 2016). The absorption and fixation rate of different antibiotics on the soil particles surfaces mainly depends on the soil pH (HoltenLutzhoft et al. 2000), physico-chemical characteristics, climatic conditions, soil type, composition, and quality of organic matter and many other environmental factors (Doretto and Rath 2013). Antibiotics with different sorption affinity at the solid phase depends upon their Kd value that controls their mobility in the environment and considered as an indicator of the potential of antibiotics mobility through soil into different environmental sources like groundwater and surface runoff. There is another important point that dissolved organic matter minimized the sorption of antibiotics to clay with maximize the mobility rate (Kulshrestha et al. 2004).

With land application of farmyard manure (FYM) (containing animal waste) as an organic fertilizer these residual veterinary antibiotic and antibiotic-resistant microorganisms may enter into soil and influence ecosystem (Solomon et al. 2010; Carlsson et al. 2013). Environmental features like climate and soil characteristics also affect the fate of antibiotics (Boxall 2004; Topp et al. 2008).

To assess the fate and probable risks associated with antibiotics land application of FYM, the existence of animal antibiotic compounds in animal manure and their fate and transport in agricultural ecosystems are essential to be addressed. There are several procedures that are involved including chemical nature, transport, leaching and runoff, sorption, plant uptake, and biodegradation that determine the fate of antibiotic in soil (Meena et al. 2015).

11.2 Chemical Nature of Antibiotics in Soil

Biodegradation/transformation into metabolites forms a sink for antibiotics in soil. Therefore, high quantities of antibiotics reach to agricultural fields by manure application or directly through animal excretion (via urine and faces), depending on their structure and chemical nature they can persist for years (Du and Liu 2012).

Sequential extraction is used to analyze soil (manure-amended) samples to explain the sequestration behavior of SDZ and its main metabolites further

(Forster et al. 2009). However, this residual fraction is apparently not covalently bound. So it can be remobilized to a possible long-term risk in soil. Several data sets are available on the fate of SDZ after application of manure to soil (Hammesfahr et al. 2008; Heuer et al. 2008).

Sulfadiazine, widely used antibiotic, has two main metabolites that are:

(a) 4-hydroxy-SDZ
(b) *N*-acetyl-SDZ

These two metabolites are formed during the passage through the animal (Lamshoft et al. 2007). Soil amended with manure having 14C-labeled SDZ and with its main metabolites, the only substantial transformation process evaluated was de-acetylation of biologically inactive *N*-acetyl-SDZ to the parent compound SDZ (Forster et al. 2009) (Fig. 11.1).

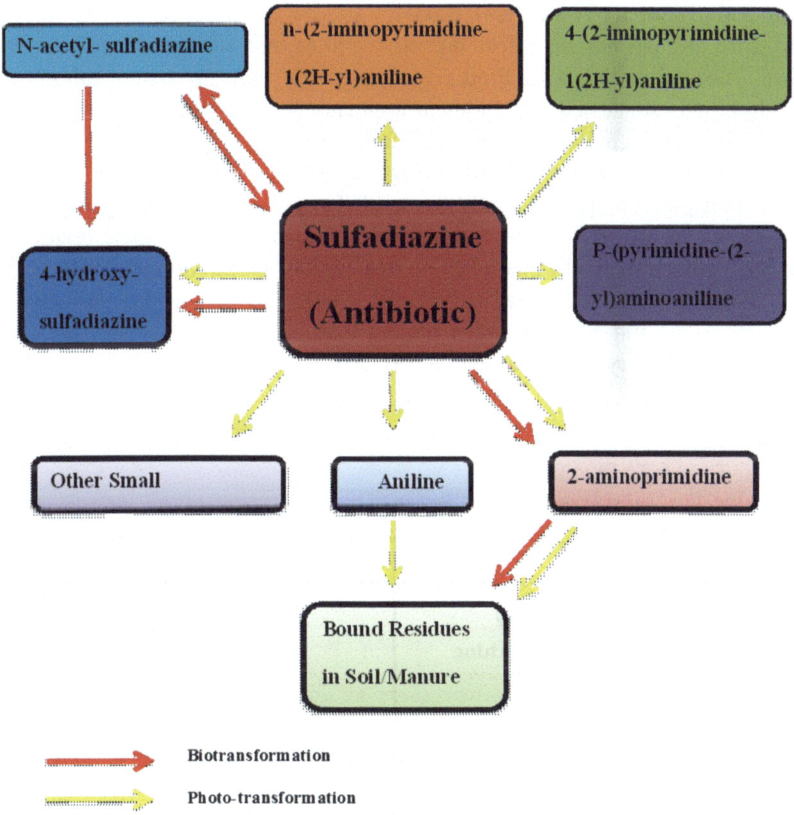

Fig. 11.1 Chemical behavior of sulfadiazine modified after (Sukul et al. 2008)

11.3 Effect of Transport on Fate of Antibiotics in Soil

Antibiotics transport in the environment is associated with its physico-chemical properties. The transport of antibiotics in soils is mainly controlled by the sorptivity, life time, soil solution pH, and ionic strength (Chen et al. 2012). In addition, sorptive ability of antibiotics it also forms strong associations with the colloids and dissolved organic matter and migrates in soil profile via preferential flow channels (Ding et al. 2014).

The antibiotics like ciprofloxacin, tetracycline, doxycycline, and clindamycin are strongly sorbed on the surface of aerobically digested biosolids. But other antibiotics such as sulfamethazine and sulfamethoxazole are weakly sorbed on particles surface. The affinity of certain antibiotics to adsorb on the particles surface may reduce their bioavailability (Wu et al. 2009) and results in decreased degradation rates (Dolliver et al. 2007). Mostly, antibiotics have short half-lives (days to weeks), but at high concentration some antibiotics persist for months to years within soil (Teeter and Meyerhoff 2003).

Manure storage has no effect on tetracyclines and sulfadiazine (Chen et al. 2012). The metabolites of sulfadiazine reversibly converted into sulfadiazine. Therefore, it is suggested that frequent application of manure contaminated with sulfadiazine and its metabolites may accumulate in soil and causes environmental contamination (Lamshoft et al. 2010) (Fig. 11.2).

The transport of approximately seven different antibiotics that are used for animal production, during the rainfall event, also determines their relationship with the sediment/aqueous phase. The percentage of partitioning is different from these seven antibiotics into aqueous/solid phases. Consequently, sulfathiazole, sulfamethazine, and monensin are mostly connected with the aqueous phase, while tylosin and erythromycin are closely associated with the solid phase (Davis et al. 2006).

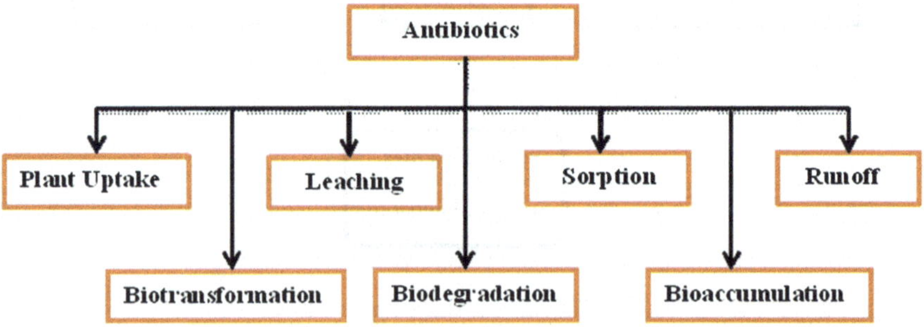

Fig. 11.2 Possible pathways of antibiotics transport in soil

11.4 Runoff and Leaching from Soil

Antibiotic transport within soils and to ground/surface water may occur by both leaching and runoff. Surface transport of sulfonamides via runoff was attributed to delayed infiltration of water into the soil because of surface sealing through manure and particle bound transport (Burkhardt et al. 2005; Kreuzig and Holtge 2005).

But some antibiotic like tylosin was not noticed in soil or leachates (2–120 days) after application of large amounts of tylosin when dissolved in slurry (Kay et al. 2005). The mobility rate of antibiotics in the soil depends on several factors, including chemical properties, temperature, moisture content, the timing of manure application, and weather conditions (Sarmah et al. 2006). The rapid movement of antibiotics, largely depends on soil macropores, while smaller macropores have less significant effect on antibiotic leaching (Kay et al. 2004).

To decrease runoff losses of antibiotic chemicals, instantaneous soil integration of land applied animal manure becomes important. Runoff losses of antibiotic (sulfonamides) might be one to two orders of level higher from the grassland than from the cultivated land that receives surface use of manure slurry/waste (Kreuzig and Holtge 2005). Furthermore, surface application of animal manure can considerably increase the rate of runoff water from the treated field, due to the surface sealing effect of manure particles (Burkhardt et al. 2005). Surface runoff of antibiotics from animal waste spreads the chemicals to the general water environment.

It is also reported that the erosion control practice has the ability to control the leaching and runoff of antibiotics like tetracyclines because it has extremely low aqueous concentration with lowest absolute losses, and this method proved to be beneficent to reduce the transport of these antibiotics in soil (Davis et al. 2006). On the other hand, sulfonamides are easily water soluble (Hu et al. 2010) so can transfer easily through leaching and runoff.

11.5 Sorption of Antibiotics in Soil

In soils, antibiotics interact with soil organic matter, clay, and mineral particles that result in sorption, binding/attachment, and fixation of the chemicals on the soil matrix. The strength of this interaction totally depends on the chemical nature of species and the soil characteristics and also influenced by temperature, humidity, and the soil solution chemistry (Kumar et al. 2005).

Antibiotics with increased aromaticity and electropolarity show higher sorption values and strongly bind on soils surface (Thiele-Bruhn et al. 2004). The sorption is usually rapid when antibiotics spiked (400–12,000 mg kg^{-1}) soil slurry under agitation. In this condition, more than 95% of the chlortetracycline adsorption occurred within 10 min and 95% of the tylosin adsorption occurred within 3 h to a sandy loam and a clay soil (Allaire et al. 2006).

Many antibiotics have functional groups like amines, carboxyls, and hydroxyls. So, protonation or deprotonation of these functional groups in pH-specific media produces positive or negative charges, positively charged antibiotics bind to soil particles through the electrostatic attraction or cation exchange (Gao and Pedersen 2005; Wang et al. 2012).

Anionic antibiotic molecules form complex compound with cations that are adsorbed on the surface of negatively charged soil particles. The cation linking ability enables the antibiotics to being retained in soils (Tolls 2001). In case of alkaline solutions, antibiotics may form complexes with particles of clay minerals through the anion exchange processes and show strong sorption when carboxylic groups of antibiotics directly substitute the OH-groups on mineral surfaces (Sassman and Lee 2007).

Furthermore, sorption rate of antibiotics to soil minerals is also affected by pH, ionic strength, and natures of exchangeable cations (Pils and Laird 2007; Wang et al. 2012). The media pH effects antibiotic–soil interactions by changing the charges of antibiotics and the cation exchange capacity (CEC) of soils. At pH 5.0, oxytetracycline has zero charges and interacts with organic matter of soil generally via hydrophobic partitioning. But at lower pH, the antibiotic becomes positively charged and sorbed on the surface of mineral soil via cation exchange and higher pH becomes negatively charged and sorbed to soil mineral particles mainly via cation bridging (Kulshrestha et al. 2004). The presence of Cu^{2+} improves the sorption rate of tetracycline to montmorillonite in an extensive pH range (Wang et al. 2008). Polar antibiotic compounds are also sorbed to soil particles through the interactions (via van der Waals force, electric attraction, cation bridging, and anion exchange) with organic matter (MacKay and Canterbury 2005; Gu et al. 2007) (Fig. 11.3).

Most of the antibiotics contain both hydrophobic and hydrophilic moieties (chemistry), it can be concluded that sorption of antibiotics in soils is a result of interactions of the chemicals with soil clay minerals and soil organic matter chiefly through the hydrophobic separating, electric attraction, and cation binding. The interactions are determined by the physico-chemical nature of the antibiotics and the soils, e.g., the quantity and the type of soil clay and soil organic matter and are stuck by the soil solution chemistry (Song and Guo 2014).

The sorption rate of antibiotics on different surfaces like (1) sterile manure, (2) compost, and (3) humic acid was affected by contact time and high pH for sorption of sulfathiazole and sulfamethazine to sand particles and loamy soils (Kurwadkar et al. 2007). Electrostatic forces are responsible for the sorption of these antibiotic by-products to charged surfaces of mineral and organic exchange sites in soil (HoltenLutzhoft et al. 2000).

Presence of sulfonamides in the environment brings changes in microbial population which is also hazardous to human health as well (Baran et al. 2011). Noxiousness due to the degraded antibiotic composites has been reported in microorganisms (Ge et al. 2010). Jones et al. (2005) reported that soil texture, CEC, and iron oxide content as the most important factors that determined the sorption values of antibiotic (oxytetracycline) in soils. Sorption of sulfamethazines in different

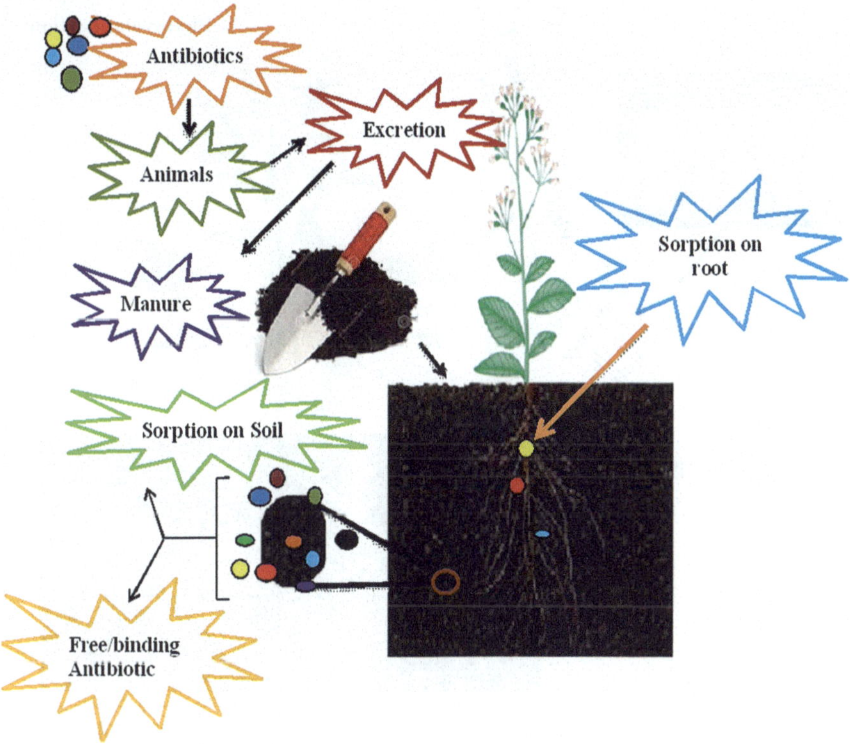

Fig. 11.3 Sorption of antibiotic in soil

mineral soils was influenced by soil organic content, total soil surface area, and also soil solution pH (Lertpaitoonpan et al. 2009).

The toxicity associated with antibiotics can range from damaging the vital microbes that are needed for supplying nutrients to the plants, difference in microbial population due to the resistant selection by altering soil microbial composition and function, and increased occurrence in different soil bacteria (Knapp et al. 2010).

11.6 Uptake of Antibiotics by Plants

An extensive use of antibiotics promotes the growth and reduces the diseases in animals. So the continuous application of antibiotics in soil through repetitive manure use may ultimately build up concentration high enough to enter into the terrestrial environment as a potent hazard (Bassil et al. 2013) (Fig. 11.4).

Different plant organ and tissues respond differently towards antibiotics depending upon its concentration and exposure time. Mostly roots, cotyledons, and cotyledon petioles exhibited a toxic effect, while other parts like internodes and

Fig. 11.4 Phytoremediation to uptake/accumulate antibiotic

leaf length showed an increased growth at lower antibiotic concentrations and toxicity at a higher level (Migliore et al. 2010a).

Recently, new research is also conceded out on the phytoremediation potential of plants against the different antibiotics. In general, plants are used for the phytoremediation of toxic materials (like heavy metals, PAH) from soil in the past. But nowadays, phytoremediation (phytostabilization, phytotoextraction, phytovolatilization, and phytoaccumulation) emerging as a new technique was considered to be effective in elimination of antibiotic (tetracycline and sulfadimethoxine) from planted soil (Lee et al. 2009; Michelini et al. 2014).

Vegetables like corn, potatoes, and lettuce have the ability to absorb antibiotics at different rates when grown in soil that is fertilized with animal manure. The vegetables that are grown on the soil improved with liquid manure containing antibiotic mainly sulfamethazine in a greenhouse. All plant leaves and potato tubers showed the presence of sulfamethazine; thus, root crops (potatoes, carrots, and radishes) that have direct contact with soil are more vulnerable to antibiotic contamination. Some of the staple foods that are enjoyed worldwide are mostly root vegetables, and the uncontrolled application of antibiotics in the agriculture can jeopardize food security (Dolliver et al. 2007). The antibiotics bound tightly to the soil particles and moved deeper/further into the field soil (Hu et al. 2010). Small concentrations of chlortetracycline accumulated in different plant tissues (Kumar et al. 2005).

Plants uptake antibiotic (chlortetracycline) from loamy sand and sandy loam soils when mixed with antibiotic having raw hog manure. Thus, surface/groundwater and agricultural soil become reservoirs to antibiotics due to the current manure managing practices (Hu et al. 2010; Chang et al. 2010). Different antibiotics have effects on normal plant growth (Migliore et al. 2003) (Tables 11.1 and 11.2).

Table 11.1 Effect of different antibiotics on plant growth

Plants	Antibodies used (via animal manure)	Effects on plants	Reference
Cucumis sativus L., *Lactuca sativa* L., *Phaseolus vulgaris* L., *Raphanus sativus* L.	Enrofloxacin	Length of primary root, hypocotyl, cotyledons, and the number/length of leaves modified	Migliore et al. (2003)
Medicago sativa	Oxytetracycline	Stem and root growth inhibited	Kong et al. (2007)
Oryza sativa, *Cucumis sativus*, *Avena sativa*	Chlortetracycline, tetracycline, tylosin, sulfamethazine, trimethoprim	Seed germination inhibited	Liu et al. (2009)
Brassica rapa sub. *pekinensis*, *Solanum lycopersicum*	Sulfadiazine, sulfamonomethoxine, enrofloxacin	Root elongation and shoot elongation affected	Jin et al. (2009)
Lolium (ryegrass)	Tetracycline	Plant biomass, especially the roots reduced; plant P assimilation decreased	Wei et al. (2009)
Lythrum salicaria	Sulfadimethoxine	Phytotoxic effect varied from organ to organ. Internodes and leaf length showed an increased growth at lower drug concentrations	Migliore et al. (2010a)
Zea mays	Oxytetracycline, chlortetracycline	Hormetic growth in the hypogeal system. Antibiotic absorption varies in field and pot cultures	Migliore et al. (2010b)
Arabidopsis thaliana mutant	Sulfamethazine	Decreased plant folate pool size which causes methyl deficiency and reduction in DNA methylation and the repressive histone mark	Zhang et al. (2012)
Salix fragilis	Sulfadimethoxine, Sulfonamide antibiotics	Sulfonamide antibiotics, tolerance to antibiotic increased with the exposure duration, probably due to the onset of acclimation mechanisms	Michelini et al. (2014)

Table 11.2 Accumulation of antibiotics in dry weight (DW) and fresh weight (FW) in plants

Antibiotics	Plants	Accumulation rate	Reference
Chlortetracycline, tylosin	*Brassica oleracea* var. *capitata*, *Allium fistulosum*	0.002–0.017 mg kg^{-1} (FW) chlortetracycline detected	Kumar et al. (2005)
Sulfamethazine	*Zea mays*, *Lactuca sativa* L., *Solanum tuberosum*	0.1–1.2 mg kg^{-1} (DW) sulfamethazine in all three plants	Dolliver et al. (2007)
Oxytetracycline, tetracycline	*Capsicum annum*, *Solanum tuberosum*, *Ipomoea batatas*, *Ipomoea aquatica*, Chinese flowering cabbage, *Lactuca sativa* L., *Daucus carota*, *Momordica charantia*, *Benincasa hispida*	Oxytetracycline and tetracycline concentrations in the range of 0.041–0.174 and 0–0.048 mg kg^{-1} (DW) detected, respectively	Yao et al. (2010)
Chlortetracycline, monensin, sulfamethazine, tylosin, and virginiamycin	Vegetables	10 μg kg^{-1} (FW) concentration of all five antibiotics detected from the test crops	Kang et al. (2013)

11.7 Biodegradation of Antibiotics

Bioavailability of antibiotics depends on the chemical compound's hydrophobicity, which determines their degradation rate (Ingerslev and Halling-Sorensen 2000). Therefore, chemical properties antibiotics and manure-related medium characteristics transform the antibiotics reluctance to biodegradation and also play a major role in removal of antibiotic from the soil (Storteboom et al. 2007). The specific adsorption characteristics are different for the same antibiotic in different types of animal manure matrices (Motoyama et al. 2011). The physico-chemical characteristics of different antibiotics associated with the degradation profiles (Thiele-Bruhn 2003).

The physico-chemical properties, the chemical structure of antibiotics and their degradation determines whether they degrade during the biological treatment (Ben et al. 2008). Degradation of antibiotics in different mediums like compost, soil, manure, and sediments have same metabolic mechanisms. However, between liquid and solid phases differences among media matrices bring changes in fractioning of antibiotics (Buyuksonmez et al. 2000). The removal of antibiotic (oxytetracycline) varied from as low as in soil (55–70%) to anaerobic digestion (55–75%) to composting (85–99%). The high rate of degradation during composting is due to the presence of the extra aerobic bio-activity as compared to anaerobic digestion alone. Both soil and composting have same aerobic-anoxic situations but composting shows that higher degradation rate is just because of the existence of good inoculum as compared to soil conditions (Masse et al. 2014).

The degradation kinetics of antibiotic (sulfadimethoxine) is affected by initial concentration because the activity of microorganisms is inhibited at high antibiotic

Table 11.3 Biodegradation of antibiotics in soil

Treatment	Antibiotic	Concentration	Observed reduction	Reference
Manure amended soil				
Soil	Tetracycline	5–300 μg kg^{-1}	0%	Hamscher et al. (2002)
	Chlortetracycline	4.7 μg kg^{-1}	0%	
	Sulfanilamide	0.25–1.0 mg L^{-1}	0%	
	Tylosin	5.6 μg L^{-1}	0%	
	Erythromycin	5.6 μg L^{-1}	25%	
Storage	Sulfadiazine	156 mg L^{-1}	0% (10 and 20°C)	Lamshoft et al. (2010)
	Difloxacin	17.6 mg L^{-1}	7% (10 and 20°C)	

concentrations (Wang et al. 2006). It is also reported that tetracycline and sulfamethoxydiazine initial concentrations is decreased to approximately 50% within 12 h after continuous anaerobic digestion and just traces of antibiotics are detected after 2–3 days (Shi et al. 2011) (Table 11.3).

11.8 Conclusion

It is concluded that if the ability of antibiotics utilization by the animals is enhanced it can decrease the exposure rate of these antibiotics to the soil via manure application. In addition, phytoremediation as new emerging techniques can be helpful in removal of antibiotics from the soil and these plants with accumulation of antibiotics in different plant parts can be discarded in a proper way.

References

Allaire SE, Del Castillo J, Juneau V (2006) Sorption kinetics of chlortetracycline and tylosin on sandy loam and heavy clay soils. J Environ Qual 35:969–972

Baran E, Larinier M, Ziv G, Marmulla G (2011) Review of the fish and fisheries aspects in the feasibility study and the environmental impact assessment of the proposed Xayaburi dam on the Mekong mainstream. Report prepared for the WWF Greater Mekong

Bassil RJ, Bashour II, Sleiman FT, Abou-Jawdeh YA (2013) Antibiotic uptake by plants from manure-amended soils. J Environ Sci Health B 48:570–574

Ben W, Qiang Z, Adams C, Zhang H, Chen L (2008) Simultaneous determination of sulfonamides, tetracyclines and tiamulin in swine wastewater by solid-phase extraction and liquid chromatography-mass spectrometry. J Chromatogr 1202:173–180

Boxall A (2004) The environmental side effects of medication. EMBO J 5:1110–1116

Burkhardt M, Stamm C, Waul C, Singer H, Muller S (2005) Surface runoff and transport of sulfonamide antibiotics and tracers on manured grassland. J Environ Qual 34:1363–1371

Buyuksonmez F, Rynk R, Hess TF, Bechinski E (2000) Occurrence, degradation and fate of pesticides during composting: Part II: Occurrence and fate of pesticides in compost and composting systems. Compost Sci Util 8:61–81

Carlsson G, Patring J, Kreuger J, Norrgren L, Oskarsson A (2013) Toxicity of 15 veterinary pharmaceuticals in zebrafish (Daniorerio) embryos. Aquat Toxicol 126:30–41

Chang X, Meyer MT, Liu X, Zhao Q, Chen H, Chen JA, Qiu Z, Yang L, Cao J, Shu W (2010) Determination of antibiotics in sewage from hospitals, nursery and slaughter house, wastewater treatment plant and source water in Chongqing region of Three Gorge Reservoir in China. Environ Pollut 158:1444–1450

Chen Y, Zhang H, Luo Y, Song J (2012) Occurrence and assessment of veterinary antibiotics in swine manures: a case study in east china. Chin Sci Bull 57:606–614

Davis JG, Truman CC, Kim SC, Ascough JC, Carlson K (2006) Antibiotic transport via runoff and soil loss. J Environ Qual 35:2250–2260

Ding GC, Radl V, Schloter-Hai B, Jechalke S, Heuer H, Smalla K, Schloter M (2014) Dynamics of soil bacterial communities in response to repeated application of manure containing sulfadiazine. PLoS One 9(3):e92958. doi:10.1371/journal.pone.0092958

Dolliver H, Kumar K, Gupta S (2007) Sulfamethazine uptake by plants from manure-amended soil. J Environ Qual 36:1224–1230

Doretto KM, Rath S (2013) Sorption of sulfadiazine on Brazilian soils. Chemosphere 90:2027–2034

Du LF, Liu WK (2012) Occurrence, fate, and ecotoxicity of antibiotics in agroecosystems. A review. Agron Sustain Dev 32:309–327

FDA (2012) The Green Book—FDA approved products. Food and Drug Administration, Rockville

Forster M, Laabs V, Lamshoft M, Groneweg J, Zuhlke S, Spiteller M, Krauss M, Kaupenjohann M, Amelung W (2009) Sequestration of manure-applied sulfadiazine residues in soils. Environ Sci Technol 43:1824–1830

Gao J, Pedersen JA (2005) Adsorption of sulfonamide antimicrobial agents to clay minerals. Environ Sci Technol 39:9509–9516

Ge LK, Chen JW, Zhang SY, Cai XY, Wang Z, Wang CL (2010) Photodegradation of fluoroquinolone antibiotic gatifloxacin in aqueous solutions. Chin Sci Bull 55:1495–1500

Gu C, Karthikeyan KG, Sibley SD, Pedersen JA (2007) Complexation of the antibiotic tetracycline with humic acid. Chemosphere 66:1494–1501

Hammesfahr U, Heuer H, Manzke B, Smalla K, Thiele-Bruhn S (2008) Impact of the antibiotic sulfadiazine and pig manure on the microbial community structure in agricultural soils. Soil Biol Biochem 40:1583–1591

Hamscher G, Sczesny S, Hoper H, Nau H (2002) Determination of persistent tetracycline residues in soil fertilized with liquid manure by high-performance liquid chromatography with electrospray ionization tandem mass spectrometry. Anal Chem 74:1509–1518

Heuer H, Focks A, Lamshoft M, Smalla K, Matthies M, Spiteller M (2008) Fate of sulfadiazine administered to pigs and its quantitative effect on the dynamics of bacterial resistance genes in manure and manured soil. Soil Biol Biochem 40:1892–1900

HoltenLutzhoft HC, Vaes WH, Freidig AP, Halling-Sorensen B, Hermens JL (2000) 1-octanol/water distribution coefficient of oxolinic acid: influence of pH and its relation to the interaction with dissolved organic carbon. Chemosphere 40:711–714

Hu X, Zhou Q, Luo Y (2010) Occurrence and source analysis of typical veterinary antibiotics in manure, soil, vegetables and groundwater from organic vegetable bases, Northern China. Environ Pollut 158:2992–2998

Ingerslev F, Halling-Sorensen B (2000) Biodegradability properties of sulfonamides in activated sludge. Environ Toxicol Chem 19:2467–2473

Jin C, Chen Q, Sun R, Liu J (2009) Eco-toxic effects of sulfadiazine sodium, sulfamonomethoxine sodium and enrofloxacin on wheat, Chinese cabbage and tomato. Ecotoxicology 18:878–885

Jones AD, Bruland GL, Agrawal SG, Vasudevan D (2005) Factors influencing the sorption of oxytetracycline to soils. Environ Toxicol Chem 24:761–770

Kang DH, Gupta S, Rosen C, Fritz V, Singh A, Chander Y, Murray H, Rohwer C (2013) Antibiotic uptake by vegetable crops from manure-applied soils. J Agric Food Chem 9:9992–10001

Kay P, Blackwell PA, Boxall ABA (2004) Fate of veterinary antibiotics in a macroporous tile drained clay soil. Environ Toxicol Chem 23:1136–1144

Kay P, Blackwell PA, Boxall ABA (2005) Column studies to investigate the fate of veterinary antibiotics in clay soils following slurry application to agricultural land. Chemosphere 60:497–507

Knapp CW, Dolfing J, Ehlert PAI, Graham DW (2010) Evidence of increasing antibiotic resistance gene abundances in archived soils since 1940. Environ Sci Technol 44:580–587

Kong W, Zhu YG, Liang Y, Zhang J, Smith FA, Yang M (2007) Uptake of oxytetracycline and its phytotoxicity to alfalfa (Medicago sativa L.) Environ Pollut 147:187–193

Kreuzig R, Holtge S (2005) Investigations on the fate of sulfadiazine in manured soil: laboratory experiments and test plot studies. Environ Toxicol Chem 24:771–776

Kulshrestha P, Giese RF, Aga DS (2004) Investigating the molecular interactions of oxytetracycline in clay and organic matter: insights on factors affecting its mobility in soil. Environ Sci Technol 38:4097–4105

Kumar K, Gupta SC, Baidoo SK, Chander Y, Rosen CJ (2005) Antibiotic uptake by plants from soil fertilized with animal manure. J Environ Qual 34:2082–2085

Kurwadkar ST, Adams CD, Meyer MT, Kolpin DW (2007) Effects of sorbate speciation on sorption of selected sulfonamides in three loamy soils. J Agric Food Chem 55:1370–1376

Lamshoft M, Sukul P, Zuhlke S, Spiteller M (2007) Metabolism of 14C-labelled and non-labelled sulfadiazine after administration to pigs. Anal Bioanal Chem 388:1733–1745

Lamshoft M, Sukul P, Zuhlke S, Spiteller M (2010) Behaviour of (14)c-sulfadiazine and (14)c-difloxacin during manure storage. Sci Total Environ 408:1563–1568

Lee HY, Lim JE, Kwon KK, Seo YH, Yang JE, Ok YS (2009) Removal of antibiotics in soil and water: a literature review. J Agric Life Sci 20:45–54

Lertpaitoonpan W, Ong SK, Moorman TB (2009) Effect of organic carbon and pH on soil sorption of sulfamethazine. Chemosphere 76:558–564

Liu F, Ying GG, Tao R, Zhao JL, Yang JF, Zhao LF (2009) Effects of six selected antibiotics on plant growth and soil microbial and enzymatic activities. Environ Pollut 157:1636–1642

MacKay AA, Canterbury B (2005) Oxytetracycline sorption to organic matter by metal bridging. J Environ Qual 34:1964–1971

Masse DI, Saady NMC, Gilbert Y (2014) Potential of biological processes to eliminate antibiotics in livestock manure: an overview. Animals 4:146–163

Meena VD, Dotaniya ML, Shah JK, Patra AK (2015) Antibiotics and antibiotic resistant bacteria in wastewater: impact on environment, soil microbial activity and human health. Afr J Microbial Res 9:965–978

Michelini L, Gallina G, Capolongo F, Ghisi R (2014) Accumulation and response of willow plants exposed to environmental relevant sulphonamide concentration. Int J Phytoremediation 16:947–961

Migliore L, Cozzolino S, Fiori M (2003) Phytotoxicity to and uptake of enrofloxacinin crop plants. Chemosphere 52:1233–1244

Migliore L, Rotini A, Cerioli NL, Cozzolino S, Fiori M (2010a) Phytotoxic sulfadimethoxine elicits a complex hormetic response in the weed Lythrum salicaria L. Dose Response 8:414–427

Migliore L, Godeas F, De Filippis SP, Mantovi P, Bonazzi G, Barchi D, Testa C, Rubattu N, Brambilla G (2010b) Hormetic effect(s) of tetracyclines as environmental contaminant on Zea mays. Environ Pollut 158:129–134

Motoyama M, Nakagawa S, Tanoue R, Sato Y, Nomiyama K, Shinohara R (2011) Residues of pharmaceutical products in recycled organic manure produced from sewage sludge and solid waste from livestock and relationship to their fermentation level. Chemosphere 84:432–438

Ogle M (2013) In meat we trust: an unexpected history of carnivore. America Houghton Mifflin Harcourt, Boston. ISBN: 978-0151013401

Pils JRV, Laird DA (2007) Sorption of tetracycline and chlortetracycline on K- and Ca-saturated soil clays, humic substances, and clay-humic complexes. Environ Sci Technol 41:1928–1933

Sarmah AK, Meyer MT, Boxall ABA (2006) A global perspective on the use, sales, exposure pathways, occurrence, fate and effects of veterinary antibiotics (VAs) in the environment. Chemosphere 65:725–759

Sassman SA, Lee LS (2007) Sorption and degradation in soils of veterinary ionophore antibiotics: monensin and lasalocid. Environ Toxicol Chem 26:1614–1621

Shi JC, Liao XD, Wu YB, Liang JB (2011) Effect of antibiotics on methane arising from anaerobic digestion of pig manure. Anim Feed Sci Technol 166–167:457–463

Solomon KR, Hillis DG, Lissemore L, Sibley PK (2010) Risks of agricultural pharmaceuticals in surface water systems and soils. In: Henderson K et al (eds) Veterinary pharmaceuticals in the environment, ACS symposium series. American Chemical Society, Washington, DC, pp 191–204

Song M, Guo M (2014) Residual veterinary pharmaceuticals in animal manures and their environmental behaviors in soils. He Z, Zhang H (eds) Applied manure and nutrient chemistry for sustainable agriculture and environment. Springer, Dordrecht., Chapter 2, pp 23-52

Storteboom HN, Kim SC, Doesken KC, Carlson KH, Davis JG, Pruden A (2007) Response of antibiotics and resistance genes to high-intensity and low-intensity manure management. J Environ Qual 36:1695–1703

Sukul P, Lamshoft M, Zuhlke S, Spiteller M (2008) Photolysis of 14C SDZ in water and manure. Chemosphere 71:717–725

Tasho RP, Cho JY (2016) Veterinary antibiotics in animal waste, its distribution in soil and uptake by plants: A review. Sci Total Environ 563–564:366–376

Teeter JS, Meyerhoff RD (2003) Aerobic degradation of tylosin in cattle, chicken, and swine excreta. Environ Res 93:45–51

Thiele-Bruhn S (2003) Pharmaceutical antibiotic compounds in soils—a review. J Plant Nutr Soil Sci 166:145–167

Thiele-Bruhn S, Seibicke T, Schulten HR, Leinweber P (2004) Sorption of sulfonamide pharmaceutical antibiotics on whole soils and particle-size fractions. J Environ Qual 33:1331–1342

Tolls J (2001) Sorption of veterinary pharmaceuticals in soils: a review. Environ Sci Technol 35:3397–3406

Topp E, Hendel J, Lapen D, Chapman R (2008) Fate of the non-steroidal anti-inflammatory drug naproxen in agricultural soil receiving liquid municipal biosolids. Environ Toxicol Chem 27:2005–2010

von Nussbaum F, Brands M, Hinzen B (2006) Medicinal chemistry of antibacterial natural products exodus or revival? Angew Chem Int Ed 45:5072–5129

Wang M, Tang JC (2010) Research of antibiotics pollution in soil environments and its ecological toxicity. J Agro Environ Sci 29:261–266

Wang QQ, Bradford SA, Zheng W, Yates SR (2006) Sulfadimethoxine degradation kinetics in manure as affected by initial concentration, moisture, and temperature. J Environ Qual 35:2162–2169

Wang YJ, Jia DA, Sun RJ, Zhu HW, Zhou DM (2008) Adsorption and cosorption of tetracycline and copper (II) on montmorillonite as affected by solution pH. Environ Sci Technol 42:3254–3259

Wang C, Teppen BJ, Boyd SA, Li H (2012) Sorption of lincomycin at low concentrations from water by soils. Soil Sci Soc Am J 76:1222–1228

Wei X, Wu S, Nie XP, Wong MH (2009) The effects of residual tetracycline on soil enzymatic activities and plant growth. J Environ Sci Health B 44:461–471

Wu C, Spongberg AL, Witter JD (2009) Sorption and biodegradation of selected antibiotics inbiosolids. J Environ Sci Health A 44:454–461

Yao JH, Niu DK, Li ZJ, Liang YC, Zhang SQ (2010) Effects of antibiotics oxytetracycline on soil enzyme activities and microbial biomass in wheat rhizosphere. Sci Agric Sin 43:721–728

Zhang H, Deng X, Miki D, Cutler S, La H, Hou YH (2012) Sulfamethazine suppresses epigenetic silencing in Arabidopsis by impairing folate synthesis. Plant Cell 24:1230–1241

Chapter 12
Uptake of Antibiotics by Plants

Sandra A. Youssef and Isam I. Bashour

12.1 Introduction

Antibiotics are often used to maintain the health of humans and animals. Those used in human medicine belong to the same general classes as those used in animals, and even if they are not exactly the same compounds their mode of action is similar (Phillips et al. 2004).

Their regular use in animals, especially food-producing animals, has raised several apprehensions about not only the development of antibiotic resistant bacteria in the environment but also the appearance of antibiotics in food and water supplies. According to Kang et al. (2013), the main cause behind these effects is when manure holding antibiotic is applied to land. As proven by research, up to 90% of an administered dose of antibiotics may be excreted through urine and feces (Phillips et al. 2004; Kumar et al. 2005a) ending up in manure. Therefore, crops become exposed to antibiotics because antibiotics tend to persist in soils from a few to several hundred days depending on the antibiotic compound, sorption interactions with soil, and environmental conditions (Dolliver et al. 2007).

The continuous use of antibiotics in animal production increases the chances of bacteria developing resistance to antibitoics used in humans. The accumulation of antibiotics may or may not disrupt the growth and development of plants; however, the uptake into plants may indicate a notable exposure pathway of these compounds to humans and other biota. Thus, as stated by Kong et al. (2007) there is a potential risk that plants are capable of spreading antibiotics from the soil into the food chain.

The original version of this chapter was revised. An erratum to this chapter can be found at DOI 10.1007/978-3-319-66260-2_22.

S.A. Youssef • I.I. Bashour (✉)
Faculty of Agricultural and Food Sciences, American University of Beirut, Beirut, Lebanon
e-mail: sandrayoussef4@gmail.com; sy27@aub.edu.lb; iibashour@gmail.com; ib02@aub.edu.lb

M.Z. Hashmi et al. (eds.), *Antibiotics and Antibiotics Resistance Genes in Soils*,
Soil Biology 51, DOI 10.1007/978-3-319-66260-2_12

Therefore, it is important to understand the potential impact of veterinary antibiotics in the environment and their uptake and accumulation in different plant tissues.

12.2 Classification of Antibiotics

Antibiotics can mainly be classified according to their spectrum of activity and their mechanism of action.

Broad-spectrum antibiotics refer to an antibiotic that function against a wide variety of disease-causing bacteria. In other words, it acts against both Gram-negative and Gram-positive bacteria. Narrow-spectrum antibiotics tend to have a limited action against bacteria. They are effective against selective families of bacteria. In other words, they are either effective against Gram-negative or Gram-positive bacteria. Antibiotics can also be separated according to their mechanism of action and their target sites in the bacterium (Kohanski et al. 2010). Mechanism of action varies with the varying classes of antibiotics (Table 12.1).

12.3 Veterinary Antibiotics

Veterinary antibiotics are antibiotics having disease-fighting and growth-promoting capabilities. Approximately half of all antibiotics manufactured are for human consumption, and the other 50% are administered to livestock either to treat sick animals or used as growth promoters. In general, antimicrobials are used in everything from apples to aquaculture. As mentioned by Henderson and Coats (2010), veterinary antibiotics are feed additives of poultry, swine, cattle, equine, and aquaculture. The problem ascends when such practice results in the development of bacterial resistance. Livestock production is one of the fastest growing agricultural subsectors in the developing countries (Thornton 2010). Therefore, there is a growing demand for livestock products driven mainly by the population growth leading to a continuous and upsurge usage of veterinary antibiotics.

Table 12.1 List of some antibiotic classes with different mechanism of action

Antibiotic	Mechanism of action
Tetracycline, macrolides, aminoglycosides	Protein synthesis inhibitors
Beta-lactams	Cell wall synthesis inhibitors
Fluoroquinolones	Nucleic acid inhibitors
Isoniazid	Mycolic acid synthesis inhibitors
Sulfonamides	Folic acid synthesis inhibitors

Source: Kohanski et al. (2010)

12.3.1 Usage in Animal Production

With the rapid development of stockbreeding and aquaculture an extraordinary augmentation in the used amount of veterinary antibiotics is undergoing (Du and Liu 2012).

Veterinary antibiotics serve a wide purpose of administration. They may be used therapeutically in animals as an integrated disease management approach for treating bacterial diseases, or used non-therapeutically as growth promoters, prophylaxis, and metaphylaxis treatments (Song and Guo 2014). The latter is the main cause of the problem. The non-therapeutic administration to animals is a long-term process given through feed and water but at low dosage and is fed to whole flock or herds. According to McEwen and Fedorka-Cray (2002), growth-promoting antibiotics are often administered in relatively low concentrations, ranging from 2.5 to 125 mg/kg (ppm), depending on the drug and species treated. Prophylactic practice is a preventive measure, where antibiotics are admitted to healthy animals before disease exposure. The metaphylactic practice is mass medication of a group of animals that is administered after an exposure to an infectious agent to prevent spread of disease. The growth promotion treatment of antibiotic applied to feed creates the ideal situation for the selection of antibiotic resistant bacteria (Barton 2014). All antibiotics acting as growth promoters were banned in the European Union in 2006 (Thornton 2010). This issue is under discussion among many countries in the world.

Tables 12.2 and 12.3 are adopted from Regassa et al. (2009) showing a partial list of FDA-approved antibiotics used in the production of beef, cow-calf, and poultry.

12.3.2 Usage Around the World

It was estimated that the total amounts of annual use of antibiotics had reached 100,000–200,000 tons worldwide including veterinary antibiotics and medical

Table 12.2 FDA-approved commonly used antibiotics for therapeutic and subtherapeutic purposes in beef and cow-calf production

Drug	Level in feed (mg/head/day)	Treatment objective
Bacitracin zinc	35–70	Feed efficiency and growth
Bambermycin	1–5	Feed efficiency and growth
Chlortetracycline	350	Disease control
Monensin	25–400	Intensive feeding and weight gain
Oxytetracycline	75	Feed efficiency and growth
Oxytetracycline	75	Disease control
Oxytetracycline	0.1–5 mg/lb. of body weight	Disease control
Tylosin	8–10	Disease control

Table 12.3 FDA-approved commonly used antibiotics for the therapeutic and subtherapeutic use in poultry production

Drug	Level in feed (g/ton)	Treatment objective
Arsanilic acid	75–120	Feed efficiency, growth, and pigmentation
Bacitracin	4–50	Feed efficiency and growth
Bambermycin	1–20	Feed efficiency and growth
Chlortetracycline	10–100	Feed efficiency, growth, and disease control
Oxytetracycline	5–50	Feed efficiency, growth, and disease control
Tylosin (Banned in EU)	4–50	Feed efficiency and growth

antibiotics (Wang and Tang 2010). In the year 2000, it has been stated that 897 tons of antibiotics were applied to animal production in the United Kingdom (Thiele-Bruhn 2003). In Turkey; antimicrobial usage has been reported to be 33% of the total veterinary pharmaceutical consumptions (Karcı and Balcıoğlu 2009). Kumar et al. (2005a) stated that by the year 2005 the annual EU consumption of veterinary antibiotics was approximately 5000 tons.

Intensive animal farming implies considerable drug use. It is vital to stress on the fact that most antibiotics used in animal production are more or less comparable to those used in humans. The World Health Organization (2011) estimated that the top three classes by global sales for animal use in 2009 were macrolides ($600 million), penicillins ($600 million), and tetracyclines ($500 million) all of which are considered as critically important in human medicine.

Nonetheless, as stated by Van Boeckel et al. (2015), worldwide in 2010, at least 63,200 tons of antibiotics were mainly used up by livestock, an amount much likely to be matched by human consumption. This digit is expected to escalate by two-third reaching 105,600 tons to meet the demand of a projected 8.5 billion human population in year 2030. The two-third upsurge is contributed to the increase in the number of food-producing animals and to the shift from small scale to industrial scale production system.

It is evident that the uncontrolled consumption and usage of veterinary antibiotics disturb not only the environment and ecosystems but also the human health. In many countries, unfortunately, there is no adequate data or statistics displaying an assessment of the total amount of veterinary pharmaceuticals utilized and this is because livestock production is barely monitored or surveyed and farmers may tend to use more than the recommended dosage. In Lebanon, a survey done to assess antibiotic usage across several Lebanese farms stated that the top five mostly used antibiotics by dairy farms are streptomycin, gentamicin, penicillin, oxytetracycline, and tylosin. (Choueiri 2008).

12.4 Antibiotics in the Environment

The occurrence of antibiotics in the environment is caused by unmonitored excretion done by humans and animals. Antimicrobials can be present in the environment through several different ways, these include, the drug manufacturing process, disposal of unused drug containers, medical waste, and through the use and application of waste material containing the drugs.

Animal agriculture is only one source of entry of drug residues in the environment. The problem is that livestock manure holds elevated levels of veterinary antibiotics that stay active even after normal digestive procedure (Kim et al. 2011). Once in the environment, antibiotics can be transported either in dissolved phase or adsorbed to colloids or soil particles into surface and groundwater (Chee-Sanford et al. 2009).

Consequently, the persistence of antibiotic in the environment will lead to microbial resistance. Therefore, drugs have the properties they need to accumulate in organisms and cause change in water and soil ecosystems (Lillenberg et al. 2010). All characteristics of antibiotics in soils are interrelated and determined by crop, soil microorganisms, water, and anthropogenic activities, which will eventually decide the spatial temporal distribution and environmental impacts of antibiotics (Du and Liu 2012).

12.5 Antibiotic Levels in Manure and Soil

Manure containing antibiotic residue is being used as a source of fertilizer to enhance soil quality, consequently affecting the soil flora and accumulating in plants. Drugs and their metabolites found in soil are either mineralized by soil organisms or enter the groundwater unaltered (Lillenberg et al. 2010). Most antibiotics fed to animals are poorly absorbed in the animal's gut and as much as 90% of them can be excreted as their parent compounds. Boxall et al. (2002) and Kumar et al. (2005a) illustrated further in the topic. They explained that the excretory organs eliminate polar compounds such as tetracycline and tylosin more efficiently than compounds that have high lipid solubility. Lipid soluble antibiotics are often not eliminated until they are metabolized to more polar compounds. A field study in Germany showed that a concentration of 15 μg/kg of sulfamethazine, member of the sulfonamide group, was measured in the soil after 7 months of manure fertilization on fields (Accinelli et al. 2006). Sarmah et al. (2006) stated that about 95% of the excreted antibiotics enter the environment in active forms. Namely, out of a dose of 70 mg/head/day of chlortetracycline, a growth promoter, 14 μg/g was found in fresh manure (Sarmah et al. 2006).

The destiny and persistence of an antibiotic in the environment rest upon several aspects such as binding to soil, biodegradation, chemical complexation or chelation, hydrolysis, and photolysis. Consequently, the antibiotic residues lead to serious environmental problems including ecological risk and human health damage.

Song and Guo (2014) mentioned that the worldwide heavy use of veterinary pharmaceuticals in confined animal feeding operations has resulted in annual discharge of 3000–27,000 tons of drug chemicals through livestock manure into the environment. Baguer et al. (2000) claimed that land application of antibiotic-laced manure seems to be the ruling pathway for the release of antibiotics in terrestrial environment and in fact is the main source of resistance.

With the advance of analytical techniques, many researchers estimated the level of antibiotics in manure. For example, antibiotics such as tetracycline, tylosin, monensin, sulfadimidine, and sulfathiazole have been detected in swine slurry, cattle manure, poultry litter, and fish farm sediment from different countries at a wide concentration ranging from traces to 200 mg/kg (Kumar et al. 2005b). In addition, in their study, Hamscher et al. (2002) estimated the level of antibiotics in manure and confirmed that the concentrations of tetracycline and chlortetracycline were 4.0 and 0.1 mg/kg, respectively. Song and Guo (2014) reported that more than 50 major antibiotics have been detected in poultry, swine, cattle, and horse manures at 0.01–765 mg/kg dry manure mass. Table 12.4 shows a number of reported concentrations of residual veterinary antibiotics in animal manures.

Some veterinary pharmaceuticals degrade rapidly through biochemical reactions, demonstrating a half-life of 2–30 days. Heuer et al. (2011) indicated that the macrolide class of antibiotics such as tylosin once excreted along with manure degrades quickly during storage with half-life in the order of days, yet many other antibiotic compounds are transferred to soil. Moreover, tetracycline class of antibiotic is the most persistent antibiotic and persists in water, manure, and soil, with half-lives approaching 100 days.

The physical and chemical properties, such as molecular structure, size, shape, solubility, and hydrophobicity of antibiotic vary with the compound and thus, the sorption and fixation of these substances in soils vary significantly. Some antibiotics seem to persist a long time in the environment, especially in soil, while others degrade very fast. Many aspects can possibly affect the distribution of antibiotics in soils. The dilution with soil, degradation, leaching, and uptake by plants are main reasons why the residue of antibiotic in soil is usually much lower than that in manure (Hu et al. 2010).

The persistence of antibiotic in soil poses an environmental, animal, and human risk making it a controversial research topic. To date, however, according to Larsson (2014), there is no clear evidence for interrupted ecosystem services in soil communities due to antibiotic exposure given the prevailing exposure levels documented in the field.

Table 12.4 Reported concentrations of residual veterinary antibiotics in animal manures

Manure type	Antibiotic	Concentration (mg/kg)	Country	References
Swine	Tetracycline	0.3–56.8	China	Li et al. (2013)
	Tylosin	0.2–1.9		
Poultry	Tetracycline	0.5–13.4	China	Li et al. (2013)
	Tylosin	0.2–0.4		
Poultry	Tetracycline	0.05–0.5	Turkey	Karcı and Balcıoğlu (2009)
	Enrofloxacin	0.01–0.08		
Dairy cow	Tetracycline	0.2–10.4	China	Li et al. (2013)
	Tylosin	0.2–0.3		
Fresh cattle	Oxytetracycline	872	Italy	De Liguoro et al. (2003)
	Tylosin	116		
Newly removed cattle bedding	Oxytetracycline	367	Italy	De Liguoro et al. (2003)
	Tylosin	32.8		
Aged cattle	Tetracycline	0.05–0.4	Turkey	Karcı and Balcıoğlu (2009)
Cattle (matured-5m)	Oxytetracycline	0.82	Italy	De Liguoro et al. (2003)
	Tylosin	0.1		
Cattle (day 30–day 135)	Oxytetracycline	2–19	Italy	De Liguoro et al. (2003)
	Tylosin	0.001–0.1		
Poultry	Chlortetracycline	23	Canada	Warman and Thomas (1981)
Liquid	Tetracycline	20	Germany	Winckler and Grafe (2001)
	Sulfadimidine	40		

Source: Song and Guo (2014) and Kumar et al. (2005a)

12.6 Fate of Veterinary Antibiotics in Soil

Antibiotics may become persistent in the environment once it is released from manure into soil (Wang and Yates 2008). It was estimated that the concentration of antibiotics found in soil ranged 0.1–2683 μg/kg (Hu et al. 2010). Several factors may possibly affect the dispersal of antibiotics in soils, the dilution with soil, degradation, leaching, and uptake by vegetables.

Depending on the antibiotic, excreted material may contain not only the original antibiotic but also a significant proportion of both active and inactive metabolites (Aust et al. 2008). Antibiotics are mostly excreted as their parent compounds or their metabolites. As stated by Kim et al. (2011), these metabolites are usually bioactive and even if they weren't they can be transformed back to the bioactive parent compound after excretion. The parent compound or its metabolites may reach the aquatic environment through surface run-off or leaching through the soil profile depending on the sorptive properties of the antibiotic and the rate of degradation (Rabølle and Spliid 2000).

The extent of antibiotic adsorption to soils rest on the antibiotic species and the soil properties including pH, organic matter content, and the concentration and type of divalent cations existing (Rabølle and Spliid 2000). A parameter known as distribution coefficient (Kd) is a parameter used to predict the transport and behavior of organic contaminants in the soil. The soil Kd values of animal antibiotics vary dramatically with the chemical species, from 0.3 to 6300 L/kg (Song and Guo 2014). Compounds with high Kd values are strongly bound to soil particles and less mobile. Compounds with less Kd value are less strongly bound and more mobile in the soil. The latter group of antibiotics can be easily transported to contaminate the ground as well as surface waters. Strongly bound antibiotics can however be transported mainly to surface waters with the sediments during run-off losses of soil (NAAS 2010). For example, it has been found that chlortetracycline strongly binds to soil components and may accumulate in the soil environment (Hamscher et al. 2002) and that a substantial amounts of oxytetracycline were bound to the soil irrespective of soil type (Rabølle and Spliid 2000). Nevertheless, the Kd varies with the soil type. For example, in sandy and sandy loam soils, oxytetracycline Kd were 420 and 1030 L/kg, respectively, while the Kd values for tylosin ranged from 8 to 11 L/kg for sandy soil and 62–128 L/kg for sandy loam soils (Rabølle and Spliid 2000). Rabølle and Spliid (2000) concluded that oxytetracycline and tylosin are strongly adsorbed to soils making them weakly mobile. In other studies, the Kd of 11 different soils varied from 950 to 7200 and 10 to 3707 L/kg for oxytetracycline and tylosin, respectively (TerLaak and Gebbink 2006), and the Kd for oxytetracycline and tylosin increased with increasing soil pH (TerLaak and Gebbink 2006). These outcomes implied that in general, oxytetracycline was more strongly adsorbed to soil than tylosin and antibiotic adsorption was mainly governed by soil pH. Table 12.5 presents the Kd for some of the veterinary antibiotics.

The pH plays a role in the interaction between the antibiotic and the soil by altering the charges of the pharmaceuticals and the cation exchange capacity of the soil. For example, at pH 5, oxytetracycline has zero charge and interacts with organic matter mainly through hydrophobic partitioning; at lower and higher pH, it becomes positively and negatively charged, respectively, and was sorbed to soil minerals mainly through cation exchange and cation bridging, respectively (Kulshrestha et al. 2004).

Degradation of veterinary pharmaceuticals in agricultural soils is a comprehensive result of microbial decomposition, organic transformation, oxidation,

Table 12.5 Distribution coefficient (Kd) of several antibiotics

Antibiotic	Kd, solid (L/kg)
Tetracycline	400–1620
Oxytetracycline	420–1030
Enrofloxacin	260–6310
Tylosin	8.3–128
Sulfamethazine	0.6–31

Source: modified from Tolls (2001) and Kumar et al. (2005a)

photolysis, and hydrolysis. Many aspects influence the degradation process, such as variation of veterinary chemicals, transformation rate, soil type, soil conditions, manure type, soil-manure ratio, pH, light, temperature, moisture, and oxygen status (Lin and Gan 2011). It is known that the antibiotics beta-lactams, macrolides, and sulfonamides are susceptible to hydrolysis (Huang et al. 2011). Compared to other reactions, photodegradation of antibiotics may be insignificant under field conditions due to limited light exposure (Beausse 2004). Adding to biodegradation, chemical processes other than hydrolysis and photolysis are similarly important for antibiotic transformation in soil such as temperature and oxygen availability. Sorption to soil minerals and soil organic matter preserves veterinary antibiotics and enhances their persistence in soils (Zitnick et al. 2011).

12.7 Antibiotics in Water

Recently, many countries have been investigating the occurrence and fate of antibiotics in the aquatic environment. In the USA, a nationwide survey of pharmaceutical compounds discovered that a number of antibiotics were detected in 27% of 139 rivers at concentrations up to 0.7 μg/L (Kolpin et al. 2002). According to Kemper (2008), veterinary antibiotics and their metabolites or their degradation products reach the aquatic environment through surface run-off, or leaching. Thus, soil act as an antibiotic reservoir gathering antibiotic contaminating the aquatic environment (Thiele-Bruhn 2003). Lillenberg et al. (2010) clarified that significant volume of drugs reaching the surface water can end up in drinking water.

In wastewater and sewage treatment plants, resistant and multi-resistant bacteria have been detected, possibly entering the food chain straight through sewage sludge used as fertilizer or wastewater serving for irrigation (Kümmerer 2004).

The movement of antibiotics into the aquatic environment varies with the antibiotic compound and its physiochemical properties. For example, penicillin and tetracycline are not typically anticipated to be established in aquatic environment. This is due to the easy hydrolysation of penicillin and the precipitation and accumulation of tetracycline (Myllyniemi et al. 2000). This coincides with Hamscher et al. (2002) where neither tetracycline nor tylosin was detected in any water sample.

A study was conducted, in northwest Germany, sampling a series of surface waters detected a wide range of antibiotics in all samples such as macrolides, sulphonamides, and lincosamides were examined regularly, but no traces of beta-lactams antibiotics were found. Moreover, tetracyclines were also not detected due to their strong adsorption to organic matter of the soil (Christian et al. 2003). On the other hand, in Germany, Hamscher et al. (2002) collected soil water samples and found concentrations of chlortetracycline, oxytetracycline, tetracycline, and tylosin ranging from 0.1 to 0.3 μg/L. Also, in another study residual oxytetracycline at concentrations ranging from 500 to 4000 μg/kg were observed in marine sediment following chemotherapy treatment in fish farms in the USA (Capone et al. 1996).

The transport of antibiotics to ground and surface water poses a risk of some antibiotics to enter the drinking water supply especially those that are highly mobile and do not easily degrade during water treatment process. On the other hand, less mobile antibiotics are potentially toxic to plants and soil organisms and provide an environment for antibiotic resistance to emerge in native soil bacteria (Tolls 2001).

12.8 Antibiotics in Plants

The misuse and overuse of antibiotics in food animals contribute to the emergence of resistant form of disease-causing bacteria. Such resistant bacteria can be communicated from food animals to humans, mainly through the food (WHO 2000). Any kind of antibiotic use in people, animals or plants can encourage the development and spread of antibiotic resistance. Many researchers provided evidence for animal to human spread of antibiotic resistance. The latter was either through direct acquisition from animal to human or through resistance transmission along the food chain. Resistance genes travel from a resistant bacterium in animals to a bacterium pathogenic to people. Resistance genes can willingly be transferred between bacteria from terrestrial animals, fish, and people. Further, such transfers can take place in various environments, such as kitchens, barns, and water sources (WHO 2011).

12.8.1 Uptake and Accumulation of Antibiotics in Plants

Many researchers studied the accumulation and uptake of veterinary antibiotics by various plants and its potential health risks. It is important to note that on a daily basis, an adult consumes 0.512 kg of plant material from crops grown above ground and 0.333 kg of plant material from crops grown below ground (Boxall et al. 2006).

A plant uptake study of ten antibiotics to lettuce and carrot from a sandy soil spiked at a soil antibiotic concentration of 1 mg/kg detected florfenicol, levamisole, and trimethoprim in lettuce leaves at concentrations ranging from 6 to 170 μg/kg, whereas enrofloxacin, florfenicol, and trimethoprim were detected in carrot at concentrations ranging from 2.8 to 13 μg/kg fresh weight (Boxall et al. 2006). Moreover, Lillenberg et al. (2010) suggests that when the vegetation period is longer, antibiotics accumulate in plants; it was highest in lettuce and lowest in cucumber.

Kumar et al. (2005b) conducted a greenhouse study to test whether or not plants take up antibiotics from manure-amended soil. The tested crops were corn, green onion, and cabbage. The study concluded that the three crops absorbed chlortetracycline at a rate of 2–17 ng/g fresh weight but did not absorb tylosin and that the more antibiotic present in the manure the higher the concentration of it in the plant tissue (Kumar et al. 2005b).

Sulfamethazine, which has a low molecular weight and is not strongly adsorbed to soil particles, was also taken up by plants such as corn, lettuce, and potato (Dolliver et al. 2007). A study made by Hu et al. (2010) reported that antibiotics in vegetables were apparent, and the range of antibiotics was 0.1–532 μg/kg in vegetables. Moreover, it has been stated that antibiotics from manure reach up plants by passive absorption (Hu et al. 2010). In a study conducted in Lebanon, Bassil et al. (2013) reported that carrot, lettuce, and radish absorbed relatively higher amounts of gentamicin than streptomycin. They also mentioned that the levels of antibiotics in plant tissue increased with increasing the antibiotic concentration in the manure (1 > 0.5 mg/kg). Willow and maize grown in greenhouse potting soils spiked with 10 mg/kg sulfadiazine for 40 days showed the presence of the chemical in the roots at 333 and 26.5 mg/kg dry weight, respectively, but not in the above ground tissues (Michelini et al. 2012).

A study was done by Youssef (2016) to test the accumulation of gentamicin, oxytetracycline, and tylosin by lettuce and radish plants in a greenhouse pot experiment at different antibiotic concentrations 0, 2.5, 5, and 10 mg/kg from two growth media (manure-amended soil and soil without manure). The results showed that gentamicin accumulated in lettuce roots (12.7 ng/g) and translocated to the leaves (17.7 ng/g) whereas in radish it accumulated in the roots (16.4 ng/g) and translocated to the leaves (31.51 ng/g). Tylosin, only at the highest concentration treatment (10 mg/kg), accumulated in lettuce roots (11.23 ng/g) and translocated to the leaves (3.58 ng/g) whereas in radish the average accumulation of tylosin in the roots is (56.6 ng/g) and in leaves (62.9 ng/g). Oxytetracycline was not absorbed by lettuce but it accumulated in radish roots (3.93 ng/g) and was translocated to the radish leaves of the highest concentration treatment only (10 mg/kg) at (6.69 ng/g). The addition of manure to the soil enhanced the uptake of the three antibiotics. The concentrations of the three antibiotics in the radish root and leaves were higher than their concentrations in lettuce. The obtained results indicated also that increasing the concentrations of the antibiotic in the growing media did not always lead to a significant increase in the accumulation levels of antibiotics in plant tissues and lettuce and radish responded differently to the presence of three antibiotics in the growing media indicating that the chemical and molecular formula of the antibiotic decides its behavior in the soil. Moreover, a study done by Shenker et al. (2011) on the uptake of carbamazepine by cucumber plants suggested that the antibiotic carbamazepine is mainly translocated by water mass flow and it highly accumulated in the older leaves, most carbamazepine were detected in the cucumber leaves relatively higher than in the roots and stems.

The bioaccumulation of veterinary antibiotics in food crops may be insignificant since the concentrations of residual antibiotics in soils receiving manure is much lower compared with the levels of antibiotics tested in laboratory of greenhouse research. Hence, it is still unclear whether or not the bioaccumulation of antibiotics in field crops poses health risks to consumers.

12.8.2 Antibiotics Effect on Plant Growth

The effect of antibiotics on plant growth was studied by many scientists. Kong et al. (2007) showed that oxytetracycline had a significant inhibitory effect on alfalfa growth. The effect was more obvious on root growth than on shoot indicating that the roots are the main accumulation site for antibiotics. As concentration of oxytetracycline increased, the leaves turned from light green to yellow. Oxytetracycline inhibited alfalfa shoot and root growth by up to 61% and 85%, respectively (Kong et al. 2007). Moreover, in a study on pinto beans grown in aerated nutrient media with chlortetracycline and oxytetracycline at 160 mg/L, top and root dry matter were reduced by 71–87% and 66–94%, respectively (Patten et al. 1980). Boxall et al. (2006) established that carrot and lettuce growth were repressed by spiking at a concentration of 1 mg antibiotic per kilogram soil. In china, oxytetracycline was found to inhibit the growth of lettuce (Cui et al. 2008) and repress root and shoot elongation of wheat (Bao et al. 2008). Also, Xie et al. (2009) found that when treating 63 wheat species with oxytetracycline the biomass and chlorophyll in leaves decreased. Phytotoxicity of enrofloxacin on cucumber, lettuce, common beans, and radish was determined in a laboratory experiment by determining the post-germinative growth of primary root, hypocotyl, cotyledons, and leaves. Concentrations between 50 and 5000 μg/L induced both toxic effect and hormesis in plants (Migliore et al. 2003). A study by Adomas et al. (2013) reported that with increasing enrofloxacin concentration the root growth was inhibited more severely and dry mass increased slightly but steadily. At the highest enrofloxacin concentration, the dry mass of both roots and stems did not exceed 15% fresh mass. While studying the effect of oxytetracycline on radish plant, Xu and Zhang (2014) reported that radish plant was capable of accumulating oxytetracycline from manure-amended soils and that the higher the concentration of oxytetracycline found in the soil the higher the concentration in plant tissue. Moreover, they reported that oxytetracycline did not have any negative effect on the growth of the radish plant when its concentration in the soil was less than 10 mg/kg, however, at higher concentrations of soil oxytetracycline (>25 mg/kg) the antibiotic not only stressed the plant but also reduced photosynthetic rate of leaves and biomass of both roots and shoot.

12.9 Antibiotic Resistance and Impact on Human Health

The misuse and overuse of antibiotics in food animals contribute to the emergence of resistant form of disease-causing bacteria. Such resistant bacteria can be communicated from food animals to humans, mainly through the food (WHO 2000). The rise in antibiotic resistance is now acknowledged worldwide as one of the greatest possible threat to human and animal health. The public has become increasingly alarmed about the connection between the overuse of antibiotics in both medicine and the agriculture agro-food industry and the emergence and spread of antibiotic resistant bacteria.

Marshall and Levy (2011) explained that the low-dose and prolonged courses of antibiotics among food animals create ideal selective pressures for the propagation of resistant strains that could facilitate the emergence and spread of resistant pathogens to humans.

Many researchers provided evidence for animal to human spread of antibiotic resistance. The latter was either through direct acquisition from animal to human or through resistance transmission along the food chain. Resistance genes travel from a resistant bacterium in animals to a bacterium pathogenic to people. Resistance genes can willingly be transferred between bacteria from terrestrial animals, fish, and people. Further, such transfers can take place in various environments, such as kitchens, barns, and water sources (WHO 2011). Farm workers are directly at risk of acquiring resistance since they are always in close contact with colonized or infected animals. Thus, this might provide a channel of spread of resistance genes into the environment wherever possible (Marshall and Levy 2011). Levy et al. (1976) reported the very first incidence of acquisition of resistance in human from direct contact with infected animals. It was proved with a study where they reported the existence of the same tetracycline-resistant *E. coli* strains in the gut flora of the chicken workers as in the chicken receiving tetracycline-rich feed. Gentamicin is an antibiotic mostly used in poultry as growth-promoting agent, it prevent early poultry mortality. A revelatory 2007 study established that the threat for carrying gentamicin-resistant *E. coli* was 32 times higher in poultry workers than in other members of the community: half of all poultry workers were colonized with gentamicin-resistant *E. coli*, compared to only 3% of non-poultry workers were colonized (Price et al. 2007). Several studies documented the transmission of resistance to humans through contact with infected animals. Marshall and Levy (2011) demonstrated several examples of bacterial species (*E. coli*, *Salmonella*, *Enterococcus faecalis*, *E. faecium*, and MRSA) and antibiotic resistance including poultry, pigs, and cattle and even resistance in humans to a range of antibiotics only used in animals.

The hypothesis is that the food chain is the main mean of transmission. But data on antibiotic resistance is limited and mainly gathered through research papers. For instance, Marshall and Levy (2011) stated that resistant *E. coli* have been found in beef carcasses that were stored for 24 h in a cooler and later made into ground beef in North American Feedlot. Work-related transmission of methicillin-resistant Staphylococcus aureus (MRSA) from food animals to humans is well documented and transfer of MRSA through the food chain has also been documented (Hanselman et al. 2006; Lewis et al. 2008).

Manure amendment of agricultural soils may add a considerable amount of bacteria carrying antibiotic resistance genes. Resistant bacteria attach to crops and are exposed to humans through antibiotic uptake by plants. Quantitatively, the massive input of resistance genes and selective agents with manure could well contribute to the resistance problem in human antibiotic therapy (Heuer et al. 2011).

Moreover, fish farming involves the use of antibiotics and fish as food may be contaminated with resistant bacteria (Phillips et al. 2004).

Alternatives to growth-promoting and prophylactic use of antimicrobials in agriculture include improved management practices, wider use of vaccines, and introduction of probiotics. Monitoring programs, prudent use guidelines, and educational campaigns provide approaches to minimize the further development of antimicrobial resistance. The existing information concerning the insinuations of veterinary antibiotics on the terrestrial environment and impacts on human health is still limited. Thus, a wide range of investigations to interpret the impact of antibiotics on humans and the environment is essential to launch safe management protocols for antibiotic usage and treatments.

12.10 Conclusion

In many countries, veterinary antibiotics are used by farmers not only to prevent diseases but to promote growth as well. Some of these antibiotics eventually find their way to contaminate the food chain by applying antibiotic-rich manures to agricultural land. Resistance bacteria attach to crops and are exposed to humans through uptake by plants. Thus, there is a need to investigate in specific what antibiotics are absorbed by plants, their accumulation rates in the edible parts, and their effect on human health and environment.

References

Accinelli C, Hashim M, Epifani R, Schneider RJ, Vicari A (2006) Effects of the antimicrobial agent sulfamethazine on metolachlor persistence and sorption in soil. Chemosphere 63:1539–1545

Adomas B, Antczak-Marecka J, Nalecz-Jawecki G, Piotrowicz-Cieslak AI (2013) Phytotoxicity of enrofloxacin soil pollutant to narrow-leaved lupin plant. Pol J Environ Stud 22(1):71–76

Aust MO, Godlinski F, Travis GR, Hao X, McAllister TA, Leinweber P, Thiele-Bruhn S (2008) Distribution of sulfamethazine, chlortetracycline and tylosin in manure and soil of Canadian feedlots after subtherapeutic use in cattle. Environ Pollut 156(3):1243–1251

Baguer AJ, Jensen J, Krogh PH (2000) Effects of the antibiotics oxytetracycline and tylosin on soil fauna. Chemosphere 40:751–757

Bao YY, Zhou QX, Xie XJ (2008) Influence of tetracycline kind antibiotics on the control of wheat germination and root elongation. China Environ Sci 6:024

Barton MD (2014) Impact of antibiotic use in the swine industry. Curr Opin Microbiol 19:9–15

Bassil R, Bashour I, Sleiman F, Abou-Jawdeh Y (2013) Antibiotic uptake by plants from manure-amended soils. J Environ Sci Health Part B 48(7):570–574

Beausse J (2004) Selected drugs in solid matrices: a review of environmental determination, occurrence and properties of principal substances. TrAC Trends Anal Chem 23(10):753–761

Boxall AB, Blackwell P, Cavallo R, Kay P, Tolls J (2002) The sorption and transport of a sulphonamide antibiotic in soil systems. Toxicol Lett 131:19–28

Boxall AB, Johnson P, Smith EJ, Sinclair CJ, Stutt E, Levy LS (2006) Uptake of veterinary medicines from soils into plants. J Agric Food Chem 54(6):2288–2297

Capone DG, Weston DP, Miller V, Shoemaker C (1996) Antibacterial residues in marine sediments and invertebrates following chemotherapy in aquaculture. Aquaculture 145(1):55–75

Chee-Sanford JC, Mackie RI, Koike S, Krapac IG, Lin Y-F, Yannarell AC, Aminov RI (2009) Fate and transport of antibiotic residues and antibiotic resistance genes. Environ Qual 38:1086–1102

Choueiri M (2008) Emerging contaminants in Lebanon: antibiotic residues in cow manure and soil. Ms-Thesis, Department of Land and Water Resources. American University of Beirut, Beirut-Lebanon

Christian T, Schneider RJ, Färber HA, Skutlarek D, Meyer MT, Goldbach HE (2003) Determination of antibiotic residues in manure, soil, and surface waters. Acta Hydrochimica et Hydrobiologica 31(1):36–44

Cui X, Qiao XL, Han CW, Wang Z (2008) Uptake of oxytetracycline and its phytotoxicity to lettuce. Agro Environ Sci 27(3):1038–1042

De Liguoro M, Cibin V, Capolongo F, Halling-Sørensen B, Montesissa C (2003) Use of oxytetracycline and tylosin in intensive calf farming: evaluation of transfer to manure and soil. Chemosphere 52(1):203–212

Dolliver H, Kumar K, Gupta S (2007) Sulfamethazine uptake by plants from manure-amended soil. J Environ Qual 36(4):1224–1230

Du L, Liu W (2012) Occurrence, fate, and ecotoxicity of antibiotics in agro-ecosystems. A review. Agron Sustain Dev 32(2):309–327

Hamscher G, Sczesny S, Höper H, Nau H (2002) Determination of persistent tetracycline residues in soil fertilized with liquid manure by high-performance liquid chromatography with electrospray ionization tandem mass spectrometry. Anal Chem 74(7):1509–1518

Hanselman BA, Kruth SA, Rousseau J, Low DE, Willey BM, McGeer A, Weese JS (2006) Methicillin-resistant Staphylococcus aureus colonization in veterinary personnel. Emerg Infect Dis 12(12):1933–1938

Henderson KL, Coats JR (2010) Veterinary pharmaceuticals in the environment: an introduction, ACS symposium series. American Chemical Society, Washington, DC, pp 3–7

Heuer H, Schmitt H, Smalla K (2011) Antibiotic resistance gene spread due to manure application on agricultural fields. Curr Opin Microbiol 14:236–243

Hu X, Zhou Q, Luo Y (2010) Occurrence and source analysis of typical veterinary antibiotics in manure, soil, vegetables and groundwater from organic vegetable bases, northern China. Environ Pollut 158(9):2992–2998

Huang C-H, Renew J, Smeby K, Pinkston K, Sedlak D (2011) Assessment of potential antibiotic contaminants in water and preliminary occurrence analysis. J Contemp Water Res Educ 120 (1):30–40

Kang DH, Gupta S, Rosen C, Fritz V, Singh A, Chander Y, Rohwer C (2013) Antibiotic uptake by vegetable crops from manure-applied soils. J Agric Food Chem 61(42):9992–10001

Karcı A, Balcıoğlu IA (2009) Investigation of the tetracycline, sulfonamide, and fluoroquinolone antimicrobial compounds in animal manure and agricultural soils in Turkey. Sci Total Environ 407(16):4652–4664. Elsevier

Kemper N (2008) Veterinary antibiotics in the aquatic and terrestrial environment. Ecol Indic 8:1–13

Kim KR, Owens G, Kwon SI, So KH, Lee DB, Ok YS (2011) Occurrence and environmental fate of veterinary antibiotics in the terrestrial environment. Water Air Soil Pollut 214:163–172. Springer

Kohanski MA, Dwyer DJ, Collins JJ (2010) How antibiotics kill bacteria: from target to networks. Nat Rev Microbiol 8(6):423–435. National Institute of Health

Kolpin DW, Furlong ET, Meyer MT, Thurman EM, Zaugg SD, Barber LB, Buxton HT (2002) Pharmaceuticals, hormones, and other organic wastewater contaminants in US streams, 1999-2000: a national reconnaissance. Environ Sci Technol 36(6):1202–1211

Kong WD, Zhu YG, Liang YC, Zhang J, Smith FA, Yang M (2007) Uptake of oxytetracycline and its phytotoxicity to alfalfa (Medicago sativa L.) Environ Pollut 147(1):187–193
Kulshrestha P, Giese JR, Diana S (2004) Investigating the molecular interactions of oxytetracycline in clay and organic matter: insights on factors affecting its mobility in soil. Environ Sci Technol 38(15):4097–4105
Kumar K, Gupta SC, Chander Y, Singh AK (2005a) Antibiotic use in agriculture and its impact on the terrestrial environment. Adv Agron 87:1–54. Elsevier
Kumar K, Gupta SC, Baidoo SK, Chander Y, Rosen CJ (2005b) Antibiotic uptake by plants from soil fertilized with animal manure. J Environ Qual 34:2082–2085
Kümmerer K (2004) Resistance in the environment. J Antimicrob Chemother 54(2):311–320
Larsson DJ (2014) Antibiotics in the environment. Ups J Med Sci 119(2):108–112
Lewis HC, Mølbak K, Reese C, Aarestrup FM, Selchau M, Sørum M, Skov RL (2008). Pigs as Source of Methicillin-Resistant Staphylococcus aureus CC398 Infections in Humans, Denmark. Emerging Infectious Diseases 14(9):1383–1389
Levy SB, FitzGerald GB, Macone AB (1976) Changes in intestinal flora of farm personnel after introduction of a tetracycline-supplemented feed on a farm. N Engl J Med 295:583–588
Li YX, Zhang XL, Li W, Lu XF, Liu B, Wang J (2013) The residues and environmental risks of multiple veterinary antibiotics in animal feces. Environ Monit Assess 185(3):2211–2220
Lillenberg M, Litvin SV, Nei L, Roasto M, Sepp K (2010) Enrofloxacin and Ciprofloxacin Uptake by Plants from Soil. Agron Res 8(1):807–814
Lin K, Gan J (2011) Sorption and degradation of wastewater-associated non-steroidal anti-inflammatory drugs and antibiotics in soils. Chemosphere 83(3):240–246
Marshall BM, Levy SB (2011) Food animals and antimicrobials: impacts on human health. Clin Microbiol Rev 24(4):718–733. American Society for Microbiology
McEwen SA, Fedorka-Cray PJ (2002) Antimicrobial use and resistance in animals. Clin Infect Dis 34(suppl 3):S93–S106. Oxford Journals
Michelini L, Reichel R, Werner W, Ghisi R, Thiele-Bruhn S (2012) Sulfadiazine uptake and effects on Salix fragilis L. and Zea mays L. plants. Water Air Soil Pollut 223(8):5243–5257
Migliore L, Cozzolino S, Fiori M (2003) Phytotoxicity to and uptake of enrofloxacin in crop plants. Chemosphere 52(7):1233–1244
Myllyniemi AL, Rannikko R, Lindfors E, Niemi A, Bäckman C (2000) Microbiological and chemical detection of incurred penicillin G, oxytetracycline, enrofloxacin and ciprofloxacin residues in bovine and porcine tissues. Food Addit Contam 17(12):991–1000
NAAS (2010) Antibiotics in manure and soil—a grave threat to human and animal health. National Academy of Agricultural Sciences, 1–20
Patten DK, Wolf DC, Kunkle WE, Douglass LW (1980) Effect of antibiotics in beef cattle feces on nitrogen and carbon mineralization in soil and on plant growth and composition. J Environ Qual 9(1):167–172
Phillips I, Casewell M, Cox T, De Groot B, Friis C, Jones R, Nightingale C, Preston R, Waddell J (2004) Does the use of antibiotics in food animals pose a risk to human health? A critical review of published data. J Antimicrob Chemother 53(1):28–52
Price LB, Graham JP, Lackey LG, Roess A, Vailes R, Silbergeld E (2007) Elevated risk of carrying gentamicin-resistant Escherichia coli among US poultry workers. Environ Health Perspect 115(12):1738–1742
Rabølle M, Spliid N (2000) Sorption and mobility of metronidazole, olaquindox, oxytetracycline and tylosin in soil. Chemosphere 40(7):715–722
Regassa TH, Koelsch RK, Wortmann CS, Randle RF, Abunyewa AA (2009) Antibiotic use in animal production: environmental concerns. University of Nebraska-Lincoln Extension RP196, pp 1–8. Retrieved from http://extensionpublications.unl.edu/assets/pdf/rp196.pdf
Sarmah AK, Meyer MT, Boxall AB (2006) A global perspective on the use, sales, exposure pathways, occurrence, fate and effects of veterinary antibiotics (VAs) in the environment. Chemosphere 65(5):725–759

Shenker M, Harush D, Ben-Ari J, Chefetz B (2011) Uptake of carbamazepine by cucumber plants—a case study related to irrigation with reclaimed wastewater. Chemosphere 82 (6):905–910

Song W, Guo M (2014) Residual veterinary pharmaceuticals in animal manures and their environmental behavior in soils. In: He Z, Zhang H (eds) Applied manure and nutrient chemistry for sustainable agriculture and environment. Springer, Dordrecht, pp 24–46

TerLaak TL, Gebbink WA (2006) Estimation of soil sorption coefficients of veterinary pharmaceuticals from soil properties. Environ Toxicol Chem 25(4):933–941

Thiele-Bruhn S (2003) Pharmaceutical antibiotic compounds in soils—a review. J Plant Nutr Soil Sci 166:145–148

Thornton PK (2010) Livestock production: recent trends, future prospects. Philos Trans R Soc Lond B Biol Sci 365(1554):2853–2867

Tolls J (2001) Sorption of veterinary pharmaceuticals in soils: a review. Environ Sci Technol 35 (17):3397–3406

Van Boeckel TP, Brower C, Gilbert M, Grenfell BT, Levin SA, Robinson TP, Laxminarayan R (2015) Global trends in antimicrobial use in food animals. PNAS 112(18):5649–5654

Wang M, Tang JC (2010) Research of antibiotics pollution in soil environments and its ecological toxicity. J Agro Environ Sci 29:261–266

Wang Q, Yates SR (2008) Laboratory study of oxytetracycline degradation kinetics in animal manure and soil. J Agric Food Chem 56(5):1683–1688

Warman PR, Thomas RL (1981) Chlortetracycline in soil amended with poultry manure. Can J Soil Sci 61(1):161–163

WHO (2000) Antimicrobial resistance: a global threat. Essential drugs monitor—WHO, pp 7–19. Retrieved from apps.who.int/medicinedocs/pdf/s2248e/s2248e.pdf

WHO (2011) Combat antimicrobial resistance. World Health Organization, Geneva. Retrieved from http://www.who.int/world-health-day/2011/en/

Winckler C, Grafe A (2001) Use of veterinary drugs in intensive animal production. J Soils Sediments 1(2):66

Xie XY, Zhang YQ, Li ZJ, Liang YC, Yao JH, Zhang SQ (2009) Cultivar differences in toxic effects of oxytetracycline on wheat (Triticum durum). Asian J Ecotoxicol 4(4):577–583

Xu Q, Zhang M (2014) Oxytetracycline uptake and growth of radish plants (Raphanussativus L.) in animal manure-amended soils. Agric Sci Technol 15(7):1229–1234

Youssef SA (2016) Uptake of gentamicin, tylosin, and oxytetracycline by lettuce and radish plants. MS-Thesis. Department of Agriculture, American University of Beirut, Beirut

Zitnick KK, Shappell NW, Hakk H, DeSutter TM, Khan E, Casey FX (2011) Effects of liquid swine manure on dissipation of 17β-estradiol in soil. J Hazard Mater 186(2):1111–1117

Chapter 13
Recent Advances in Methods for the Detection of Antibiotics and Antibiotics Resistance Genes in Soil

Hussani Mubarak, Nosheen Mirza, and Muhammad Zaffar Hashmi

13.1 Introduction

Revolutionized initiation of drug innovation and implementation in health and agriculture, during the last era has resulted in the discovery and development of hundreds of antibiotics (Xiao et al. 2016; Dasgupta and Sengupta 2015). Annually, about 200,000 tons of antibiotics commercially are being produced for the animal and human care (Rehman et al. 2015; Ok et al. 2011). According to one report in the USA alone, almost 4.5 and 22 million pounds is being invested on antibiotics in medical facilities and animal farming (McEachran et al. 2015; Phillips 2004). In

The original version of this chapter was revised. An erratum to this chapter can be found at DOI 10.1007/978-3-319-66260-2_22.

H. Mubarak
Department of Soil and Environmental Sciences, Ghazi University, Dera Ghazi Khan, Pakistan

School of Metallurgy and Environment, Central South University, Changsha 410083, China

Chinese National Engineering Research Center for Control and Treatment of Heavy Metal Pollution, Changsha 410083, China
e-mail: hussani.mubarak@yahoo.com

N. Mirza (✉)
School of Metallurgy and Environment, Central South University, Changsha 410083, China

Chinese National Engineering Research Center for Control and Treatment of Heavy Metal Pollution, Changsha 410083, China

Department of Environmental Sciences, COMSATS Institute of Information Technology, Abbottabad 22060, Pakistan
e-mail: nosheen_mirza@ciit.net.pk

M.Z. Hashmi
Department of Meteorology, COMSATS Institute of Information Technology, Islamabad, Pakistan
e-mail: hashmi_qau@yahoo.com

M.Z. Hashmi et al. (eds.), *Antibiotics and Antibiotics Resistance Genes in Soils*, Soil Biology 51, DOI 10.1007/978-3-319-66260-2_13

livestock industry to prevent losses and for weight-gaining strategies, 25 million pounds of antimicrobials are being administered (Dasgupta and Sengupta 2015).

Normally, antibiotics exist in the environment in the range of ng L^{-1} to μg L^{-1} but can result in chronic toxicity even in the concentration of ≤ 1 ng L^{-1} (Rehman et al. 2015; Christen et al. 2010).The sources of antimicrobial contamination differ from country to country, in accordance with their uses. In Germany, the European Union, the UK, and the USA, 25%, 5–20%, 70%, and 75% of the detected antibiotic concentrations were caused by hospitals. The other major source of contaminations is veterinary (disease control, breeding, and growth promoters of animals, i.e., fattening—methods to improve protein and fat contents in meat). Thus, the main contributors in the release of excessive antibiotics in the soil and water bodies are from urine and feces of livestock, runoff from agriculture fields, unused fodder, and inappropriate disposal of waste from livestock being utilized as fertilizers (Dasgupta and Sengupta 2015).

Unmonitored, overuse, and misuse of antibiotics in agriculture played crucial role in the aggregated resistance; as a result, accelerated development of newer antibiotics is overtaken by the pace of bacterial resistance (Rehman et al. 2015; Anjum 2015; McEachran et al. 2015; Edgar et al. 2011). Currently, due to manure application, a higher level of antibiotic resistance genes (ARGs) has been detected in agricultural lands (Zhang et al. 2015). It has become a global concern (Ahammad et al. 2014; Hersher 2012; Spellberg et al. 2013; WHO 2012, 2014; Sarmah et al. 2006; Kreuzig and Höltge 2005), and the situation is worsened in the developing countries (Kostic et al. 2015; Byarugaba 2004; Chan 2012; Okeke et al. 2005). ARG could be the result of mutation in the chromosome or entering of mobile genetic element (Anjum 2015). Antibiotic resistance genes (ARGs) are capable of growing in the presence of antibiotics. They develop this resistance via effluxing antibiotic, degrading antibiotic, and modifying the target point of antibiotic (Luby et al. 2016).

Previous reports are about the sorption and mobility of few classes of antibiotics, but little is known about the degradation and fate of antibiotics and ARG in the soil environment. The conventional technique of antibiotic screening involved culturing colonies (selective/nonselective agar plate culturing), purification later dilution (disc, broth, and gradient strip dilution), and then identification, although it was informative but was time-consuming for isolation (Anjum 2015). With the advent of sophisticated analytical methodologies, detection and quantification of antibiotics have drawn the needed attention (Dasgupta and Sengupta 2015; Richardson and Ternes 2014).The present chapter reviews the recent advancements of analytical, molecular, and other applied technologies for the detection, characterization, and quantification of antibiotics and ARG in the soil matrix.

13.2 Analytical Techniques for Assessing Antibiotics and Antibiotic Resistance Genes

Detection of pharmaceutical compounds, i.e., drugs, antibiotics, and beta-blockers in different water matrices using LC–MS–MS and GC–MS, began in the duration of the late 1970s–1990 (Garrison et al. 1976; Hignite and Azarnoff 1977; Ternes et al. 1998). Later Mitani and Kataoka (2006), using automated solid-phase microextraction (SPME) coupled with LC–MS–MS, reported the presence of numerous pharmaceutical compounds in the marine ecology (Dasgupta and Sengupta 2015).

13.2.1 Sampling and Extraction of Antibiotics

Soil is the substrate from where antibiotic synthesis originally evolved, and over 80% of antibiotics in clinical use today were extracted from soil; thus, soil can be a source of diverse antibiotic resistance (AR) determinant, i.e., antibiotic resistance gene (ARG) (Donato et al. 2010; Lang et al. 2010). The determination of antibiotics and ARGs depending upon their concentration, i.e., ng g^{-1} to μg g^{-1}, from soil, needs efficient extraction and sensitive analytical technology. Hence pretreatment of sample, extraction, chromatographic separation (cleanup), and finally identification by mass spectrometry are the usual practices of analysis (Pintado-Herrera et al. 2016; Albero et al. 2015; Szulejko et al. 2014). Sample preparation from soil is time-consuming as it needs multistep extractions and cleanup. The major concern for sampling and extraction of any antibiotic is its availability in low concentrations. Solid-phase extraction method (SPEM) is the most common method for sampling prior to LC–MS–MS analysis (Salvia et al. 2015). Previously SPE procedures were used offline, but with the recent advance of technology, automated online, miniaturized sample preparation, followed by LC–MS–MS analysis, has been reported (Mitani and Kataoka 2006; Pozo et al. 2006; Stoob et al. 2005). Traditional sample preparation methods from soil matrix include Soxhlet extraction, ultrasonic extraction (UAE), pressurized liquid extraction (PLE) (Salvia et al. 2015), microwave-assisted extraction (MAE), accelerated solvent extraction (ASE), and supercritical fluid extraction (SFE), where MAE combined with solvent bar has emerged as green and effective method (Kathi 2017).

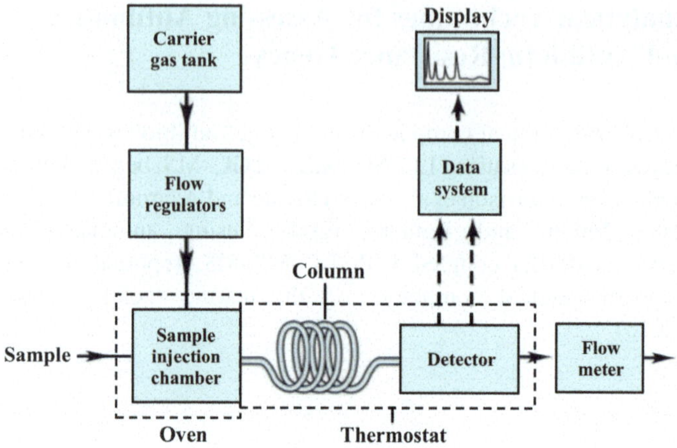

Fig. 13.1 Schematic diagram of gas chromatography (GC)

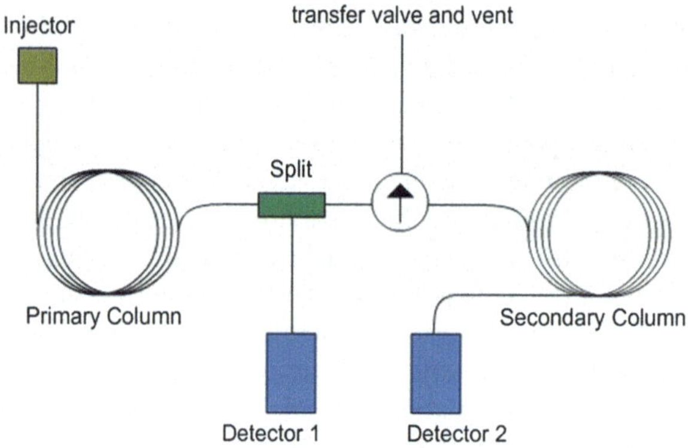

Fig. 13.2 Schematic diagram of 2D gas chromatography (GC)

13.2.2 Separation of Antibiotics by Chromatography

13.2.2.1 Gas Chromatography

GC is the most used separation technique for pesticide analysis since 1950s (Kathi 2017; Nolvachai et al. 2015) (Fig. 13.1).

For wide range of screening, 2D GC (Fig. 13.2) has proved more effective (Kathi 2017; Nolvachai et al. 2015; Tranchida et al. 2016). Shang et al. (2014) have compared the results of analysis of soil antibiotic extracts of compared GC/MS/MS pseudo multiple reaction monitoring mode (PMRM) and GC/MS/MS classic multiple reaction monitoring (CMRM) and reported improved sensitivity of PMRM. For the confident identification of individual compounds, on complex matrix and high resolution, MDGC techniques, i.e., 2D GC coupled with time-of-flight mass

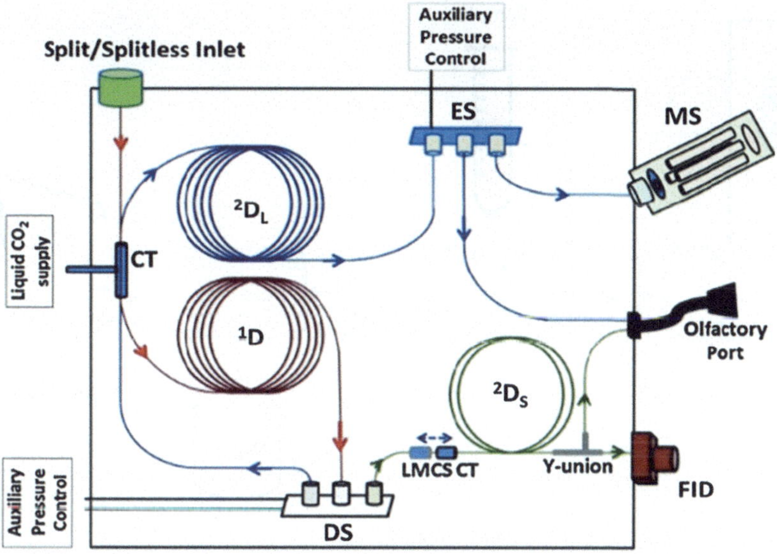

Fig. 13.3 Schematic diagram of MD gas chromatography (GC) (Chin and Marriott 2014)

spectrometry (GC×GC-TOF/MS), are reported (Kathi 2017; Jennerwein et al. 2014; Verma et al. 2015; Pena-Abaurrea et al. 2014) (Fig. 13.3).

13.2.2.2 Liquid Chromatography

The reversed-phase liquid chromatography (RPLC) technology is well reported for the separation, purification, and identification of organic compounds only with the hydrophobic interaction. The right column and solvent selection, i.e., water, acetonitrile, and methanol, are the crucial factors for the successful separation of compound in mixture. Recently, a new method is developed for hydrophilic interaction antibiotic compounds where water and acetonitrile were used as mobile phases in the HP column (Fig. 13.4), while for the increased efficiency in ionization and separation, organic reagents are employed. LC–MS (Fig. 13.5) analysis can be performed using several ionization modes (Salvia et al. 2015; Kathi 2017).

13.2.2.3 Detection of Antibiotics by Mass Spectrometry

With the use of mass spectrometry-based method, identification/quantification of organic compounds and detection of antibiotics became much easier and cost-effective. The method is an inevitable for unambiguous detection and identification of organic compounds in trace amounts and is quantified with reference to standard. The best tool for identification and quantification of antibiotics is the combination

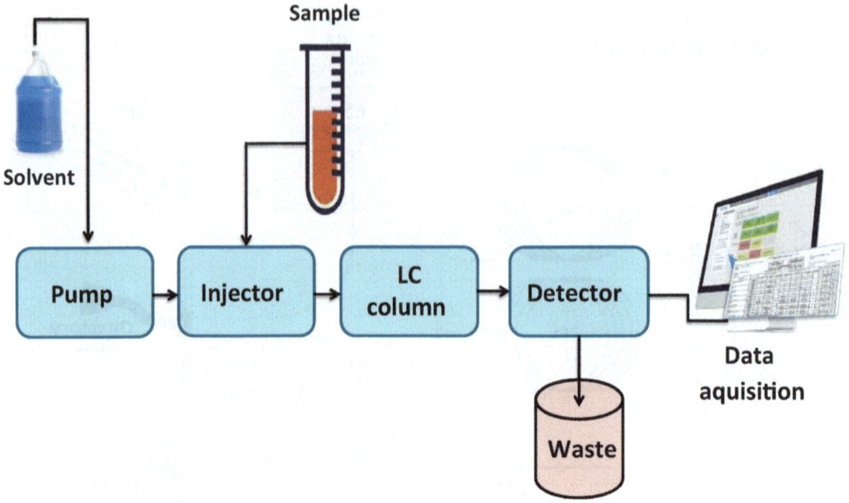

Fig. 13.4 Schematic diagram of liquid chromatography (LC)

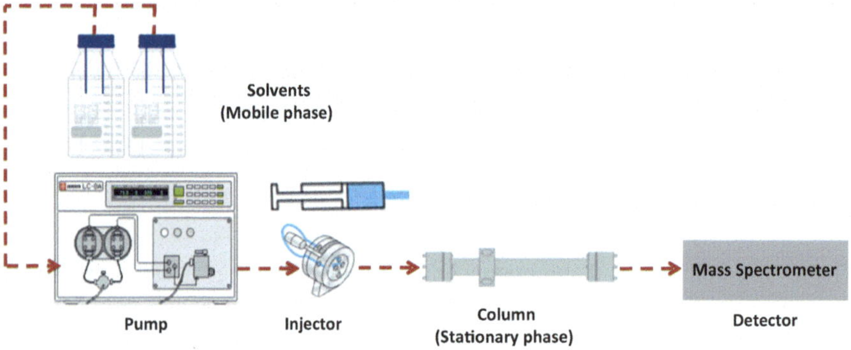

Fig. 13.5 Schematic diagram of liquid chromatography–mass spectrometry (LC–MS) (Ardrey 2003; Niessen 2010)

of liquid chromatography and mass spectrometry coupled with tandem mass spectrometry (MS–MS), i.e., LC–MS (Fig. 13.5) and/or LC–MS/MS (Salvia et al. 2015).

The crucial step in MS is the ionization of the compounds for separation by chromatography. Tributylamine (TBA) was first used to enhance the ionization efficiency of the molecular ions (Kathi 2017). The acidic nature antibiotics are analyzed by negative ionization mode, while neutral and basic antibiotics are analyzed in positive ionization mode (Planche et al. 2015). For successful detection of antibiotics, selection of instrument in MS analysis is crucial. Variations in the design, experimental conditions, and different instruments, i.e., Both 3D ion trap (IT) mass spectrometers, Thermo Finnigan IT, triple stage quadrupole (TSQ), Sciex TSQ, and Micromass TSQ in LC–MS–MS analysis are reported, for the detection of

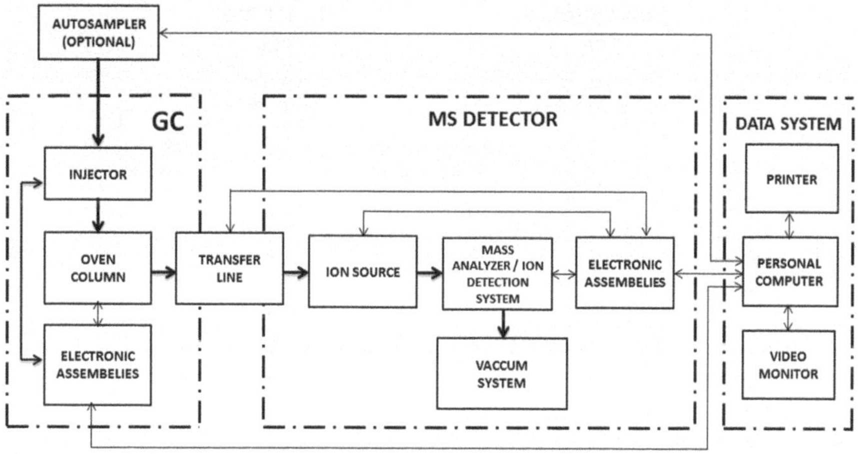

Fig. 13.6 Block diagram of gas chromatography–mass spectrometry (GC–MS)

antibiotics (Salvia et al. 2015; Regueiro et al. 2008; Richter et al. 2006). The Waters (Micromass) Q-TOF system and the Sciex QSTAR system are the commercially used mass spectrometer instruments (Kathi 2017).

Quantification from soil extracts can be achieved by employing gas chromatography–mass spectrometry (GC–MS) (Fig. 13.6), gas chromatography–tandem mass spectrometry (GC–MS/MS), comprehensive two-dimensional gas chromatography coupled to microelectron capture detection (GC × GC–μECD), liquid chromatography–mass spectrometry (LC–MS), or liquid chromatography–tandem mass spectrometry (LC–MS/MS) (Salvia et al. 2015; McEachran et al. 2015; Zhang et al. 2015; Dasgupta and Sengupta 2015; Kathi 2017; Albero et al. 2012).

13.3 Molecular Methodsof Analysis of Antibiotics and Antibiotic Resistance Genes

Molecular methods, i.e., DNA, RNA, plasmid, and/or proteins determination, are rapidly replacing the conventional technique, due to the advancement in technology, accuracy, precision, sophistication, less turnaround time, and increase in popularity. Molecular methods do not need culturing and access DNA, RNA, and/or protein after extraction (Microscopic vision), which later are compared with the publically available databases. Antibiotics and ARGs can be identified by PCR-based methods (PCR and/or qPCR) (Fig. 13.7), or with the recent advancement of technology, next-generation sequencing—metagenomic techniques proved more efficient for their identification (Luby et al. 2016).

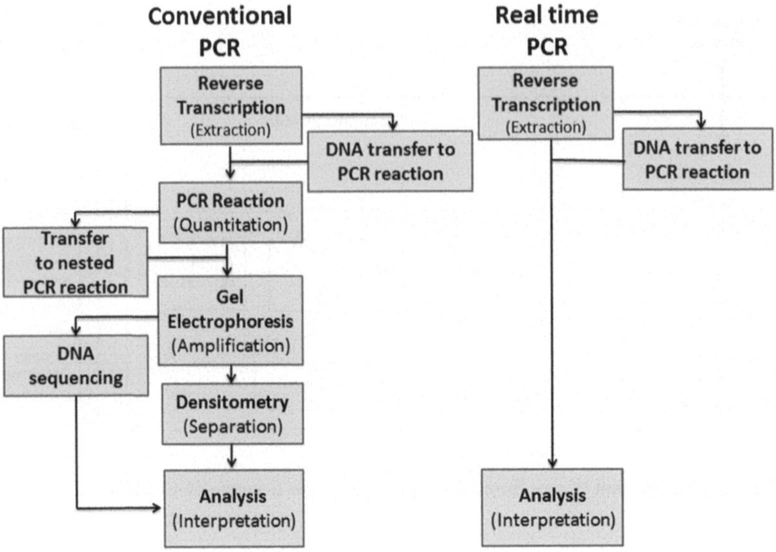

Fig. 13.7 Block diagram of conventional and real-time polymerase chain reaction (PCR)

13.3.1 Conventional Polymerase Chain Reaction

Polymerase chain reaction (PCR) has successfully and widely being applied to detect ARGs since at least 2001 (Chee-Sanford et al. 2001), and a wide range of primers for ARGs are widely available in the literature. PCR is a popular method for detecting ARGs due to its benefits, such as high sensitivity, provides relatively rapid results, and yields direct information. Conventional PCR depends upon the extracted DNA which varies in efficiency across different matrices; inhibitors could interfere with PCR functionality and could amplify the wrong target, thus resulting in false negative and false positive, respectively. Blank samples are consistently applied as a quality control measure and to false negative and/or positive. Conventional PCR only detect the presence or absence or detection below limit of ARGs. It does not quantify the concentration nor can predict the activity of ARGs (Seitz and Blokesch 2013).

PCR amplification is of crucial importance to sequence PCR for the intended target and to compare the genetic variability among ARGs databases are explored (Garder et al. 2014; Koike et al. 2007). Ideally, internal amplifications are employed to avoid false negatives (Hoorfar et al. 2004). DNA dilution is also considered as a mean diluting out inhibitors, and using dilution series technique, one can yield the required quantity of sample for optimal signal. Bead-beating appeared important step for high-quality extraction (Guo and Zhang 2013). To report any PCR-specific results, the used PCR primers, annealing temperature, and essential assay conditions with validated citations are important.

13.3.2 Real-Time Quantitative Polymerase Chain Reaction

Quantitative PCR (qPCR) detects and quantifies the abundance of the target ARGs (Fig. 13.7). In qPCR, the threshold cycle, or CT value, where the signal crosses the baseline, is compared against a standard curve to determine the gene copy number of an ARG. Quantitative PCR has also recently been adapted to manage longer templates (up to 6 kb) (Rodríguez et al. 2013; Luby et al. 2016; McKinney and Pruden 2012). During the last decade, qPCR has been widely applied to track both ARGs and markers of mobile genetic elements (Pei et al. 2006; Heuer and Smalla 2007; Koike et al. 2007; Nandi et al. 2004; Smith et al. 2004).

To report any qPCR-specific results, R_2 values of calibration curves to report limit of quantification, ideal value of 0.7 for applied assays, dilution factor to limit the interference of inhibitors, extraction, and other processing steps to ensure quantification standpoint of the actual sample are reported along. 16S rRNA genes aid to the extracting efficiency and provide proportion of total ARGs (Luby et al. 2016; Pruden et al. 2006; Heuer et al. 2011; Pei et al. 2006; Knapp et al. 2010). Inclusion of blank samples verifies the quality assurance of the process. The recent development of qPCR arrays made possible the simultaneous quantification of hundreds of ARGs (Wagner et al. 2007). Qiagen recently developed the Antibiotic Resistance Genes Microbial DNA qPCR Array, configured for quantification of 97 ARGs. Wafergen Bio-systems SmartChip Real-Time PCR, is configured for the quantification and track enrichment of 244 ARGs in the soil samples (Wang et al. 2014; Zhu et al. 2013). Limitations of using qPCR) are the optimization for individual qPCR uniformity in the annealing temperature throughout the array, inevitable higher detection limits, and volume of the reaction mixture. Further research could help in coping up with the mentioned limitations.

13.3.3 Next-Generation DNA Sequencing: Metagenomic Methods

Metagenomics is the application of genomic technologies and bioinformatics to directly access the total sum of genes and/or the genetic makeup of entire communities of target organism (Thomas et al. 2012). The last 5–10 years has big contribution in the molecular characterization (Donato et al. 2010; Nesme and Simonet 2015; Kristiansson et al. 2011; Bengtsson-Palme et al. 2014). Metagenomic studies revealed that soil possesses high diversity of ARGs, which is subject to continuous alterations, i.e., vertical and horizontal mutations are evident from analysis (Nesme et al. 2014; Fitzpatrick and Walsh 2016; Durso et al. 2012). The advantages of metagenomics are such as they can target specific ARGs, genomes get sequenced in a single step, i.e., 10–1000 Gb of DNA get sequenced in a single HiSeq 2500 Illumina lane, and later target ARGs can be detected and quantified through online databases (Luby et al. 2016). Publically available online

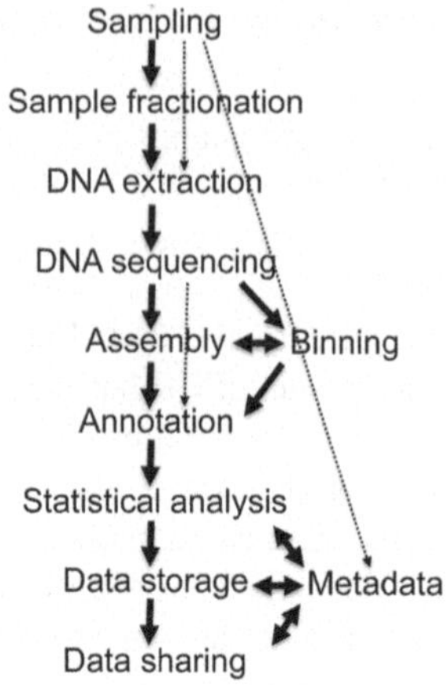

Fig. 13.8 Flow diagram of a typical metagenome projects. Note: *Dashed arrows* indicate steps that can be omitted (Thomas et al. 2012)

databases are the Integrated Microbial Genome (IMG) database, the Comprehensive Antibiotic Resistance Database (CARD) project, and MG-RAST (Meyer et al. 2008; Fitzpatrick and Walsh 2016; McArthur et al. 2013; Markowitz et al. 2012).

Metagenomics being capable of detection of total gene pool has been employed to compare the ARGs and markers of gene transfer between manure and cultivated soil (Durso et al. 2012; Nesme et al. 2014; Fang et al. 2015; Wichmann et al. 2014) (Fig. 13.8). In contrast to PCR techniques, metagenomics can be used to detect the host cells/organisms of target ARGs, but during this process quantification of ARGs can be lost (Henry et al. 2011; Thomas et al. 2012). Similar as PCR, metagenomics cannot confirm the functionality of target ARGs, but the purpose could be served by using functional metagenomics (de Castro et al. 2014; Su et al. 2014; Wichmann et al. 2014). For functional metagenomics, ligated plasmids are formed by inserting extracted fragmented DNA in the plasmids. The ligated plasmids later are transformed into the host and are plated onto the media of interest (Uchiyama and Miyazaki 2009; Zhou et al. 2012; Su et al. 2014). Functional metagenomics are reported to employ in the discovery of new antibiotics (Gillespie et al. 2002; Lim et al. 2005) and ARGs (Bengtsson-Palme et al. 2014). A total of 16 types of ARGs and 110 ARG subtypes are identified in the paddy soil of Southern China (Xiao et al. 2016).

The main disadvantages of functional metagenomics are the involved labor, gene expression biases on selection of specific host, and nonavailability of relevant databases in metagenomic libraries. In spite of these disadvantages currently, next-generation sequencing is the best applied molecular technology. To report metagenomic data, used extraction/amplification methods and ways and means to avoid exogenous DNA contamination are required to be indicated along.

13.4 Strategic Combinations

Molecular techniques are to detect and quantify the genetic potential of ARGs, while phenotypic reality is confirmed by cultural techniques. Molecular techniques are unambiguous, well defined, and less laborious as compared to the culturing techniques. But they alone cannot confirm the functionality of the target specimen. But with appropriate experimental design, the purpose can be served by combining molecular technologies with the other analytical technologies, i.e., GC–MS or LC–MS (Luby et al. 2016). The possible disconnection between the genetic potential and phenotypic reality could be bridged by employing cultural technique followed by molecular analysis. To understand the limitations associated with cultural and molecular methods and to obtain reliable results, researchers are combining the techniques together for ARGs detection (Sato et al. 2014; Zhou et al. 2009; Heuer et al. 2011; Jindal et al. 2006). Keeping in view the relevant constraints, multiple molecular methods, combination of molecular and analytical techniques, and cultural method followed by molecular and/or analytical methods are being applied, within a single study, i.e., combination of metagenomics and qPCR (Looft et al. 2012), combination of electrospray ionized (ESI) liquid chromatography and mass spectrometry coupled with tandem mass spectrometry (LC–MS/MS) and qPCR (McEachran et al. 2015), combination of liquid chromatography and mass spectrometry coupled with tandem mass spectrometry (LC–MS/MS) and DNA quantification by Fast DNA-Spin Kit for Soil (Bio-Rad, Carlsbad, CA, USA) (Zhang et al. 2015), and combination of culturing technique, DNA quantification using soil kit (PowerSoil DNA isolation kit; Mo Bio) and qPCR (Knappik et al. 2015).

13.5 Future Prospective of Molecular Methods of ARGs Analysis

Successful assessment of antibiotics and ARGs depends upon the address of appropriate research questions. Continued development of metagenomic technique, development of monitoring and standard protocols, and understanding the risk factor of misuse of antibiotics and ARGs to the human health are the dire needs for the safe survival and existence of life on Earth. Hence, appropriate risk assessment and advancement of molecular techniques are required.

13.5.1 Lab on a Chip: Using Smartphone (Point-of-Care Device)

For field, epidemic outspread and remote area monitoring of ARGs and with the advancement of information technology (IT) and globalization, a portable,

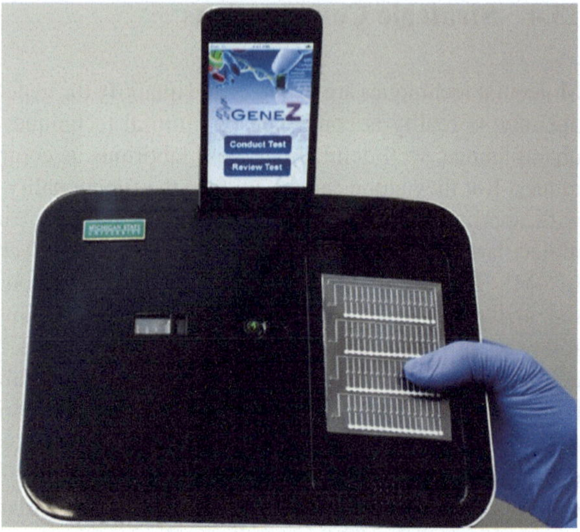

Fig. 13.9 Graphic presentation of point-of-care device (POC) (Stedtfeld et al. 2012)

inexpensive, and user-friendly diagnostic devices became dire need and dream of users. Keeping in view all these prospects, researchers have invented point of care (POC). Point of care (POC) is a compact, user-friendly, and inexpensive device, recently invented by a group of researchers (Stedtfeld et al. 2012, 2016; Kostic et al. 2015), with the characteristics of rapid quantitative analysis of multiple genetic markers. Multiple genetic testing for disease diagnosis and wireless connectivity are emerging key attributes of POC. It consists of disposable valveless polymer microfluidic chip (multiple reaction wells), i.e., Gene-Z and an iPod Touch application (Wi-Fi interface) (Fig. 13.9). It is capable of automatic analysis and reporting. POC is loaded with the testing of viral load with HIV (Shen et al. 2011), tuberculosis (ARGs of tuberculosis) (Lee et al. 2010), and microRNAs of cancer (Li et al. 2011).

13.5.2 Gene-Z

It is a disposable valve, polymer microfluidic chip containing four arrays of reaction wells each with dehydrated primers for isothermal amplification. To reduce the real time of detection, of analysis cost, and for quantification, loop-mediated amplification (LAMP) is incorporated in the device. It can analyze four samples simultaneous analysis, with a single step pipetting per sample (Stedtfeld et al. 2016). LAMP can provide highly specific and high yielding amplification at a single temperature (Mori et al. 2001). The optical unit of Gene-Z consists of light-emitting diodes (LEDs), per reaction well. For cross well optical inhibition, the chip is embossed, fabricated and is a thin-filmed microstructure (size, 22.5 (L) × 17.3(W) × 3.5 (H) in cm; weight,

930 g). As compared to conventional antibiotic resistance (AR) detection and molecular-based method (qPCR), Gene-Z gave multiplexed and comprehensive detection (69% detection, 98% in agreement with qPCR result), with turnaround time of less than 30 min (Kostic et al. 2015).

13.5.3 iPod Touch Application (Wi-Fi Interface)

Data transmission potential of Wi-Fi devices, i.e., smartphones, computational phone, and personal computer (Stedtfeld et al. 2012; Breslauer et al. 2009), and global positioning system (Google Maps and online databases) are used for automatic data transmission, analysis, and reporting. The barcode scanning capabilities of autofocus in smartphone cameras are combined with Wi-Fi and online databases to ensure specified quantitative genetic testing (ARGs).

The same groups of researchers (Stedtfeld et al. 2016) have initiated an online database for antibiotic resistance genes (ARGs) named antibiotic resistance (AR) dashboard application (beta stage of development). It is equipped to gather information about the occurrence and widespread distribution of ARGs and antibiotic resistance bacteria (ARB). The dashboard app is in sync with next-generation sequencing, qPCR, bioinformatics and metagenomics, etc. AR studies can be geospatially mapped by AR dashboard database (online available) and on integration can be utilized for AR studies (Stedtfeld et al. 2016).

13.6 Conclusions

Inappropriate and unmonitored disposal of drugs have contaminated the environment. The resultant development of antibiotic resistance genes (ARGs) in target bacteria is well documented. Soil is the source and sinks for antibiotics and ARGs. Antibiotics and ARGs extraction from soil is a complex and multistep task. In this chapter, advanced analytical, molecular, and other applied techniques were reviewed for antibiotics and ARGs detection in soil. It is found that strategic combination of traditional, advanced analytical, and/or molecular techniques are needed for the phenotypic, quantification, and functionality confirmation of antibiotics and ARGs. For reliable results of antibiotics and ARGs detection, genetic potential, and phenotypic reality, researchers should adopt combination of techniques. In case of field monitoring and/or epidemic outspread, POC could prove the best option for on-spot identification and quantification of antibiotics and ARGs. Overall, successful assessment of antibiotics and ARGs depends on the inclusions of relevant controls, focused research, and incorporation of appropriate combination of methods.

References

Ahammad ZS, Sreekrishnan TR, Hands CL, Knapp CW, Graham DW (2014) Increased waterborne *bla*NDM-1 resistance gene abundances associated with seasonal human pilgrimages to the upper Ganges River. Environ Sci Technol 48(5):3014–3020

Albero B, Sánchez-Brunete C, Miguel E, Pérez RA, Tadeo JL (2012) Determination of selected organic contaminants in soil by pressurized liquid extraction and gas chromatography tandem mass spectrometry with in situ derivatization. J Chromatogr A 1248:9–17

Albero B, Sanchez-Brunete C, Garcia-Valcarcel AI, Pérez RA, Tadeo JL (2015) Ultrasound-assisted extraction of emerging contaminants from environmental samples. TrAC Trends Anal Chem 71:110–118

Anjum MF (2015) Screening methods for the detection of antimicrobial resistance genes present in bacterial isolates and the microbiota. Future Microbiol 10(3):317–320

Ardrey RE (2003) Liquid chromatography—mass spectrometry: an introduction. Wiley, Chichester, pp 1–288. ISBNs: 0-471-49799-1 (HB); 0-47-49801-7 (PB)

Bengtsson-Palme J, Boulund F, Fick J, Kristiansson E, Larsson DGJ (2014) Shotgun metagenomics reveals a wide array of antibiotic resistance genes and mobile elements in a polluted lake in India. Front Microbiol 5:1–14

Breslauer DN, Maamari RN, Switz NA, Lam WA, Fletcher DA (2009) Mobile phone based clinical microscopy for global health applications. PLoS One 4:6320

Byarugaba DK (2004) A view on antimicrobial resistance in developing countries and responsible risk factors. Int J Antimicrob Agents 24:105–110

Chan M (2012) Antimicrobial resistance in the European Union and the world. WHO. http://www.who.int/dg/speeches/2012/amr_20120314/en/index.html

Chee-Sanford JC, Aminov RI, Krapac IJ, Garrigues-Jeanjean N, Mackie RI (2001) Occurrence and diversity of tetracycline resistance genes in lagoons and groundwater underlying two swine production facilities. Appl Environ Microbiol 67:1494–1502

Chin ST, Marriott PJ (2014) Multidimensional gas chromatography beyond simple volatiles separation. Chem Commun 50:8819–8833

Christen V, Hickmann S, Rechenberg B, Fent K (2010) Highly active human pharmaceuticals in aquatic systems: a concept for their identification based on their mode of action. Aquat Toxicol 96:167–181

Dasgupta D, Sengupta TK (2015) Techniques and methods: detection of antibiotics in environmental samples. In: Méndez-Vilas A (ed) The battle against microbial pathogens: basic science, technological advances and educational programs. FORMATEX, Badajoz, pp 1084–1090

de Castro AP, Fernandes GDR, Franco OL (2014) Insights into novel antimicrobial compounds and antibiotic resistance genes from soil metagenomes. Front Microbiol 5:1–9

Donato JJ, Moe LA, Converse BJ, Smart KD, Berklein FC, McManus PS, Handelsman J (2010) Metagenomic analysis of apple orchard soil reveals antibiotic resistance genes encoding predicted biofunctional proteins. Appl Environ Microbiol 76:4396–4401

Durso LM, Miller DN, Wienhold B (2012) Distribution and quantification of antibiotic resistant genes and bacteria across agricultural and non-agricultural metagenomes. PLoS One 7:48325

Edgar R, Friedman N, Molshanski-Mor S, Qimron U (2011) Reversing bacterial resistance to antibiotics by phage-mediated delivery of dominant sensitive genes. Appl Environ Microbiol 78(3):744–751

Fang H, Wang HF, Cai L, Yu YL (2015) Prevalence of antibiotic resistance genes and bacterial pathogens in long-term manured greenhouse soils as revealed by metagenomic survey. Environ Sci Technol 49:1095–1104

Fitzpatrick D, Walsh F (2016) Antibiotic resistance genes across a wide variety of metagenomes. FEMS Microbiol Ecol 92(2):1–8

Garder JL, Moorman TB, Soupir ML (2014) Transport and persistence of tylosin-resistant enterococci, genes, and tylosin in soil and drainage water from fields receiving swine manure. J Environ Qual 43:1484–1493

Garrison AW, Pope JD, Allen FR (1976) GC/MS Analysis of organic compounds in domestic wastewaters. In: Keith LH (ed) Identification and analysis of organic pollutants in water. Ann Arbor Science, Ann Arbor, MI, pp 517–556

Gillespie DE, Brady SF, Bettermann AD, Cianciotto NP, Liles MR, Rondon MR (2002) Isolation of antibiotics turbomycin A and B from a metagenomic library of soil microbial DNA. Appl Environ Microbiol 68:4301–4306

Guo F, Zhang T (2013) Biases during DNA extraction of activated sludge samples revealed by high throughput sequences. Appl Microbiol Biotechnol 97:4607–4616

Henry CS, Overbeek R, Xia FF, Best AA, Glass E, Gilbert J, Larsen P, Edwards R, Disz T, Meyer F et al (2011) Connecting genotype to phenotype in the era of high-throughput sequencing. Biochim Biophys Acta 1810(10):967–977

Hersher R (2012) Indian TB cases highlight need for drug-resistance tests. Nat Med 18(3):333

Heuer H, Smalla K (2007) Manure and sulfadiazine synergistically increased bacterial antibiotic resistance in soil over at least two months. Environ Microbiol 9:657–666. doi:10.1111/j.1462-2920.2006.01185.x

Heuer H, Solehati Q, Zimmerling U, Kleineidam K, Schloter M, Muller T, Focks A, Thiele-Bruhn S, Smalla K (2011) Accumulation of sulfonamide resistance genes in arable soils due to repeated application of manure containing sulfadiazine. Appl Environ Microbiol 77: 2527–2530

Hignite C, Azarnoff DL (1977) Drugs and drug metabolites as environmental contaminants: chlorophenoxyisobutyrate and salicylic acid in sewage water effluent. Life Sci 20(2):337–341

Hoorfar J, Malorney B, Abdulmawjood A, Cook N, Wagner M, Fach P (2004) Practical considerations in design of internal amplification controls for diagnostic PCR assays. J Clin Microbiol 42:1863–1868

Jennerwein MK, Eschner M, Gröger T, Wilharm T, Zimmermann R (2014) Complete group-type quantification of petroleum middle distillates based on comprehensive two-dimensional gas chromatography time-of-flight mass spectrometry (GC × GC-TOFMS) and visual basic scripting. Energy Fuels 28(9):5670–5681

Jindal A, Kocherginskaya S, Mehboob A, Robert M, Mackie RI, Raskin L, Zilles JL (2006) Antimicrobial use and resistance in swine waste treatment systems. Appl Environ Microbiol 72:7813–7820

Kathi S (2017) An overview of extraction, clean-up and instrumentation techniques for quantification of soil-bound xenobiotic compounds. Xenobiotics Soil Environ 49:101–118

Knapp CW, Dolfing J, Ehlert PAI, Graham DW (2010) Evidence of increasing antibiotic resistance gene abundances in archived soils since 1940. Environ Sci Technol 44:580–587. doi:10.1021/es901221x

Knappik M, Dance DAB, Rattanavong S, Pierret A, Ribolzi O, Davong V, Silisouk J, Vongsouvath M, Newton PN, Dittrich S (2015) Evaluation of molecular methods to improve the detection of *Burkholderia pseudomallei* in soil and water samples from Laos. Appl Environ Microbiol 81(11):3722–3727

Koike S, Krapac IG, Oliver HD, Yannarell AC, Chee-Sanford JC, Aminov RI, Mackie RI (2007) Monitoring and source tracking of tetracycline resistance genes in lagoons and groundwater adjacent to swine production facilities over a 3-year period. Appl Environ Microbiol 73: 4813–4823

Kostic T, Ellis M, Williams MR, Stedtfeld TM, Kaneene JB, Stedtfeld RD, Hashsham SA (2015) Thirty-minute screening of antibiotic resistance genes in bacterial isolates with minimal sample preparation in static self-dispensing 64 and 384 assay cards. Appl Microbiol Biotechnol 99:7711–7722

Kreuzig R, Höltge S (2005) Investigations on the fate of sulfadiazine in manured soil: laboratory experiments test plot studies. Environ Toxicol Chem 24:771–776

Kristiansson E, Fick J, Janzon A, Grabic R, Rutgersson C, Weijdegard B, Soderstrom H, Larsson DGJ (2011) Pyrosequencing of antibiotic-contaminated river sediments reveals high levels of resistance and gene transfer elements. PLoS One 6:e17038. doi:10.1371/journal.pone.0017038

Lang KS, Anderson JM, Schwarz S, Williamson L, Handelsman J, Singer RS (2010) Novel florfenicol and chloramphenicol resistance gene discovered in Alaskan soil by using functional metagenomics. Appl Environ Microbiol 76:5321–5326

Lee MF, Chen YH, Hsu HJ, Peng CF (2010) One-tube loop-mediated isothermal amplification combined with restriction endonuclease digestion and ELISA for colorimetric detection of resistance to isoniazid ethambutol and streptomycin in Mycobacterium tuberculosis isolates. J Microbiol Methods 83:53–58

Li C, Li Z, Jia H, Yan J (2011) One-step ultrasensitive detection of micro RNAs with loop-mediated isothermal amplification (LAMP). Chem Commun 47:2595–2597

Lim HK, Chung EJ, Kim JC, Choi GJ, Jang KS, Chung YR (2005) Characterization of a forest soil metagenome clone that confers indirubin and indigo production on *Escherichia coli*. Appl Environ Microbiol 71:7768–7777. doi:10.1128/AEM.71.12.7768-7777.2005

Looft T, Johnson TA, Allen HK, Bayles DO, Alt DP, Stedtfeld RD, Sul WJ, Stedtfeld TM, Chai B, Cole JR, Hashsham SA, Tiedje JM, Stanton TB (2012) In-feed antibiotic effects on the swine intestinal microbiome. Proc Natl Acad Sci USA 109(5, 1691):–1696

Luby E, Ibekwe AM, Zilles J, Pruden A (2016) Molecular methods for assessment of antibiotic resistance in agricultural ecosystems: prospects and challenges. J Environ Qual 45(2):441–453

Markowitz VM, Chen IMA, Palaniappan K, Chu K, Szeto E (2012) IMG: the integrated microbial genomes database and comparative analysis system. Nucleic Acids Res 40:115–122. doi:10.1093/nar/gkr1044

McArthur AG, Waglechner N, Nizam F, Yan A, Azad MA, Baylay AJ, Bhullar K, Canova MJ, De Pascale G, Ejim L, Kalan L, King AM, Koteva K, Morar M, Mulvey MR, O'Brien JS, Pawlowski AC, Piddock LJ, Spanogiannopoulos P, Sutherland AD, Tang I, Taylor PL, Thaker M, Wang W, Yan M, Yu T, Wright GD (2013) The comprehensive antibiotic resistance database. Antimicrob Agents Chemother 57:3348–3357. doi:10.1128/AAC.00419-13

McEachran AD, Blackwell BR, Hanson JD, Wooten KJ, Mayer GD, Cox SB, Smith PN (2015) antibiotics, bacteria, and antibiotic resistance genes: aerial transport from cattle feed yards via particulate matter. Environ Health Perspect 123(4):337–343

McKinney CW, Pruden A (2012) Ultraviolet disinfection of antibiotic resistant bacteria and their antibiotic resistance genes in water and wastewater. Environ Sci Technol 46(24): 13393–13400. doi:10.1021/es303652q

Meyer F, Paarmann D, D'Souza M, Olson R, Glass EM, Kubal M, Paczian T, Rodriguez A, Stevens R, Wilke A, Wilkening J, Edwards RA (2008) The metagenomics RAST server: a public resource for the automatic phylogenetic and functional analysis of metagenomes. BMC Bioinformatics 9:386. doi:10.1186/1471-2105-9-386

Mitani K, Kataoka H (2006) Determination of fluoroquinolones in environmental waters by in-tube solid-phase microextraction coupled with liquid chromatography–tandem mass spectrometry. Anal Chim Acta 562(1):16–22

Mori Y, Nagamine K, Tomita N, Notomi T (2001) Detection of loop-mediated isothermal amplification reaction by turbidity derived from magnesium pyrophosphate formation. Biochem Biophys Res Commun 289(1):150–154

Nandi S, Maurer JJ, Hofacre C, Summers AO (2004) Gram-positive bacteria are a major reservoir of Class 1 antibiotic resistance integrons in poultry litter. Proc Natl Acad Sci USA 101: 7118–7122

Nesme J, Simonet P (2015) The soil resistome: a critical review on antibiotic resistance origins, ecology and dissemination potential in telluric bacteria. Environ Microbiol 17:913–930. doi:10.1111/1462-2920.12631

Nesme J, Cécillon S, Delmont TO, Monier JM, Vogel TM, Simonet P (2014) Large-scale metagenomic-based study of antibiotic resistance in the environment. Curr Biol 24:1096–1100. doi:10.1016/j.cub.2014.03.036

Niessen WMA (2010) Group specific fragmentation of pesticides and related compounds in liquid chromatography–tandem mass spectrometry. J Chromatogr A 1217(25):4061–4070

Nolvachai Y, Kulsing C, Marriott PJ (2015) Pesticides analysis: advantages of increased dimensionality in gas chromatography and mass spectrometry. Crit Rev Environ Sci Technol 45(19): 2135–2173

Ok YS, Kim SC, Kim KR, Lee SS, Moon DH, Lim K, Sung JK, Hur SO, Yang JE (2011) Monitoring of selected veterinary antibiotics in environmental compartments near a composting facility in Gangwon Province, Korea. Environ Monit Assess 174(1):693–701

Okeke IN, Laxminarayan R, Bhutta ZA, Duse AG, Jenkins P, O'Brien T, Klugman KP (2005) Antimicrobial resistance in developing countries. Part I: Recent trends and current status. Lancet Infect Dis 5:481–493. doi:10.1016/S1473-3099(05)70189-4

Pei R, Kim SC, Carlson KH, Pruden A (2006) Effect of river landscape on the sediment concentrations of antibiotics and corresponding antibiotic resistance genes (ARG). Water Res 40: 2427–2435. doi:10.1016/j.watres.2006.04.017

Pena-Abaurrea M, Jobst KJ, Ruffolo R, Shen L, McCrindle R, Helm PA, Reiner EJ (2014) Identification of potential novel bioaccumulative and persistent chemicals in sediments from Ontario (Canada) using scripting approaches with GC× GC-TOF MS analysis. Environ Sci Technol 48(16):9591–9599

Phillips R (2004) Antibiotic usage in animals rise in 2004: life-saving compounds used to reduce animal pain, suffering and death. Animal Health Institute.

Pintado-Herrera MG, GonzálezMazo E, LaraMartín PA (2016) In-cell clean-up pressurized liquid extraction and gas chromatography–tandem mass spectrometry determination of hydrophobic persistent and emerging organic pollutants in coastal sediments. J Chromatogr A 1429: 107–118

Planche C, Ratel J, Mercier F, Blinet P, Debrauwer L, Engel E (2015) Assessment of comprehensive two-dimensional gas chromatography-time-of-flight mass spectrometry based methods for investigating 206 dioxin-like micropollutants in animal-derived food matrices. J Chromatogr A 1392:74–81

Pozo OJ, Guerrero C, Sancho JV, Ibáñez M, Pitarch E, Hogendoorn E, Hernández F (2006) Efficient approach for the reliable quantification and confirmation of antibiotics in water using on-line solid-phase extraction liquid chromatography/tandem mass spectrometry. J Chromatogr A 1103:83–93

Pruden A, Pei R, Storteboom H, Carlson KH (2006) Antibiotic resistance genes as emerging contaminants: studies in northern Colorado. Environ Sci Technol 40:7445–7450. doi:10.1021/es0604131

Regueiro J, Llompart M, Garcia-Jares C, GarciaMonteagudo JC, Cela R (2008) Ultrasound-assisted emulsification–microextraction of emergent contaminants and pesticides in environmental waters. J Chromatogr A 1190(1):27–38

Rehman MS, Rashid N, Ashfaq M, Saif A, Ahmad N, Han J (2015) Global risk of pharmaceutical contamination from highly populated developing countries. Chemosphere 138:1045–1055

Richardson SD, Ternes TA (2014) Water analysis: emerging contaminants and current issues. Anal Chem 86(6):2813–2848

Richter P, Jiménez M, Salazar R, Maricán A (2006) Ultrasound-assisted pressurized solvent extraction for aliphatic and polycyclic aromatic hydrocarbons from soils. J Chromatogr A 1132(1):15–20

Rodríguez RA, Bounty S, Linden KG (2013) Long-range quantitative PCR for determining inactivation of adenovirus 2 by ultraviolet light. J Appl Microbiol 114:1854–1865. doi:10.1111/jam.12169

Salvia MV, Fieu M, Vulliet E (2015) Determination of tetracycline and fluoroquinolone antibiotics at trace levels in sludge and soil. Appl Environ Soil Sci 2015:Article ID 435741. 10 pages

Sarmah AK, Meyer MT, Boxall ABA (2006) A global perspective on the use, sales, exposure pathways, occurrence, fate and effects of veterinary antibiotics (VAs). Chemosphere 65: 725–759

Sato T, Okubo T, Usui M, Yokota S, Izumiyama S, Tamura Y (2014) Association of veterinary third-generation cephalosporin use with the risk of emergence of extended-spectrum-cephalosporin resistance in *Escherichia coli* from dairy cattle in Japan. PLoS One 9(4):e96101. doi:10.1371/journal.pone.0096101

Seitz P, Blokesch M (2013) Cues and regulatory pathways involved in natural competence and transformation in pathogenic and environmental Gram-negative bacteria. FEMS Microbiol Rev 37:336–363. doi:10.1111/j.1574-6976.2012.00353.x

Shang D, Kim M, Haberl M (2014) Rapid and sensitive method for the determination of polycyclic aromatic hydrocarbons in soils using pseudo multiple reaction monitoring gas chromatography/tandem mass spectrometry. J Chromatogr A 1334:118–125

Shen F, Sun B, Kreutz JE, Davydova EK, Du W, Reddy PL, Joseph LJ, Ismagilov RF (2011) Multiplexed quantification of nucleic acids with large dynamic range using multivolume digital RT-PCR on a rotational Slip Chip tested with HIV and hepatitis C viral load. J Am Chem Soc 133(44):17705–17712

Smith MS, Yang RK, Knapp CW, Niu Y, Peak N, Hanfelt MM, Galland JC, Graham DW (2004) Quantification of tetracycline resistance genes in feedlot lagoons by real-time PCR. Appl Environ Microbiol 70:7372–7377

Spellberg B, Bartlett J, Gilbert D (2013) The future of antibiotics and resistance. N Engl J Med 368:299–302. doi:10.1056/NEJMp1215093

Stedtfeld RD, Tourlousse DM, Seyrig G, Stedtfeld TM, Kronlein M, Price S, Hashsham SA (2012) Gene-Z: a device for point of care genetic testing using a smartphone. Lab Chip 12:1454–1462. doi:10.1039/c2lc21226a

Stedtfeld RD, Williams MR, Fakher U, Johnson TA, Stedtfeld TM, Wang F, Khalife WT, Hughes M, Etchebarne BE, Tiedje JM, Hashsham SA (2016) Antimicrobial resistance dashboard application for mapping environmental occurrence and resistant pathogens. FEMS Microbiol Ecol 92(3):fiw020. doi:10.1093/femsec/fiw020

Stoob K, Singer HP, Goetz CW, Ruff M, Mueller SR (2005) Fully automated online solid phase extraction coupled directly to liquid chromatography–tandem mass spectrometry: quantification of sulfonamide antibiotics, neutral and acidic pesticides at low concentrations in surface waters. J Chromatogr A 1097:138–147

Su JQ, Wei B, Xu CY, Qiao ZYG (2014) Functional metagenomic characterization of antibiotic resistance genes in agricultural soils from China. Environ Int 65:9–15. doi:10.1016/j.envint.2013.12.010

Szulejko JE, Kim KH, Brown RJ, Bae MS (2014) Review of progress in solvent-extraction techniques for the determination of polyaromatic hydrocarbons as airborne pollutants. TrAC Trends Anal Chem 61:40–48

Ternes TA, Hirsch R, Mueller J, Haberer K (1998) Methods for the determination of neutral drugs as well as betablockers and β2-sympathomimetics in aqueous matrices using GC/MS and LC/MS/MS. Fresenius J Anal Chem 362:329–340

Thomas T, Gilbert J, Meyer F (2012) Metagenomics—a guide from sampling to data analysis Microbial Inform Exp 2(3): 12 pages.

Tranchida PQ, Franchina FA, Dugo P, Mondello L (2016) Comprehensive two-dimensional gas chromatography-mass spectrometry: recent evolution and current trends. Mass Spectrom Rev 35(4):524–534

Uchiyama T, Miyazaki K (2009) Functional metagenomics for enzyme discovery: challenges to efficient screening. Curr Opin Biotechnol 20:616–622

Verma B, Kumar P, Karthik L, Dhanasekaran D, Babalola OO, Banakar SP (2015) Gas chromatography—mass spectrometry analysis and antibacterial activity of bluish-green pigment from *Pseudomonas* sp. JJTBVK (KF836502). Braz Arch Biol Technol 58(4):628–635

Wagner M, Smidt H, Loy A, Zhou J (2007) Unravelling microbial communities with DNA-microarrays: challenges and future directions. Microb Ecol 53:498–506. doi:10.1007/s00248-006-9197-7

Wang FH, Qiao M, Su JQ, Chen Z, Zhou X, Zhu YG (2014) High throughput profiling of antibiotic resistance genes in urban park soils with reclaimed water irrigation. Environ Sci Technol 48:9079–9085

WHO (2012) The evolving threat of antimicrobial resistance: options for action. World Health Organization, Geneva

WHO (2014) Antimicrobial resistance: global report of surveillance. World Health Organization, Geneva

Wichmann F, Udikovic-Kolic N, Andrew S, Handelsman J (2014) Diverse antibiotic resistance genes in dairy cow manure. mBio 5(2):e01017-13. doi:10.1128/mBio.01017-13

Xiao KQ, Li B, Ma L, Bao P, Zhou X, Zhang T, Zhu YG (2016) Metagenomic profiles of antibiotic resistance genes in paddy soils from South China. FEMS Microbiol Ecol 92(3):1–6

Zhang S, Gu J, Wang C, Wang P, Jiao S, He Z, Han B (2015) Characterization of antibiotics and antibiotic resistance genes on an ecological farm system. J Chem, doi:10.1155/2015/526143, 8 pages

Zhou Z, Raskin L, Zilles JL (2009) Identification of macrolide resistant microorganisms on antimicrobial-free swine farms. Appl Environ Microbiol 75:5814–5820. doi:10.1128/AEM.00977-09

Zhou W, Wang Y, Lin J (2012) Functional cloning and characterization of antibiotic resistance genes from the chicken gut microbiome. Appl Environ Microbiol 78:3028–3032. doi:10.1128/AEM.06920-11

Zhu YG, Johnson TA, Su TQ, Qiao M, Guo GX, Stedtfeld RD, Hashsham SA, Tiedje JM (2013) Diverse and abundant antibiotic resistance genes in Chinese swine farms. Proc Natl Acad Sci USA 110:3435–3440. doi:10.1073/pnas.1222743110

Chapter 14
Elucidation of Emerging Nanomaterials Impacts on Antibiotic Resistance Against Soil and Aquatic Microflora

Toqeer Ahmed

14.1 Introduction

Microbial drug resistance in soil is increasing day by day by anthropogenic activities and becoming a serious challenge for the scientists especially in the field of soil biology. Fear of emerging new drug-resistant varieties and spread of new compound into the soil may affect plant's growth and normal soil microflora. Due to fast microbial growth, excessive application of antibiotics against bacteria and spread of antibiotics in soil have led to drug resistance and multiple drug resistance of different isolates (Brown et al. 2012). Antibiotic-resistant genes are spreading in the environment due to anthropogenic activity and multidrug resistance putting the environmentalists and soil microbiologists on more defensive side (Marti et al. 2014). Soil microflora is the largest reservoir of microbes on Earth which is unexplored (Mishra and Kumar 2009) especially about the impact of nanomaterials on soil microbes. Soil microflora perform numerous beneficial functions in soil which play important role in plant's growth. Soil habitat provides complex environment for their growth and interaction with contaminants. Microbes in rhizospheric soil are more functional than non-rhizospheric soils, and many factors like pH, oxygen, and water play significant role in plant's growth.

The evolution of nanotechnology and its vast applications in almost all the fields opened new ways to control the pathogens in different environment. Nanoparticles are synthesized by different methods, but with the evolution of green chemistry, biosynthesis of MNPs by using plants, bacteria, and fungi is used against pathogens,

---

The original version of this chapter was revised. An erratum to this chapter can be found at DOI 10.1007/978-3-319-66260-2_22.

T. Ahmed (✉)
Centre for Climate Research and Development (CCRD), COMSATS Institute of Information Technology, Islamabad Campus, Park Road Chak Shahzad, Islamabad 45550, Pakistan
e-mail: toqeer.ahmed@comsats.edu.pk

M.Z. Hashmi et al. (eds.), *Antibiotics and Antibiotics Resistance Genes in Soils*, Soil Biology 51, DOI 10.1007/978-3-319-66260-2_14

which is an eco-friendly, cheap, and effective method (Narayanan and Sakthivel 2010). Nanomaterials like single-walled and multi-walled carbon nanotubes (CNT) and metallic nanoparticles (MNPs) like silver (Ag), iron (Fe), titanium dioxide (TiO_2), and zinc oxide (ZnO) nanoparticle (NPs) have profound effect against microbes. Silver (Ag), iron (Fe), titanium dioxide (TiO_2), and zinc oxide (ZnO) nanoparticles (NPs) have antibacterial properties (Ahmed et al. 2012; Kandi and Kandi 2015), and NPs can stop drug resistance mechanisms and target microbes (Pelgrift and Friedman 2013). Nanotechnology plays an important role in targeted drug delivery, and nanomaterials enhanced application of other drugs and antibiotics when used in combination (Ahmed et al. 2014). Along with pathogens, soil beneficial microbes may be affected with the spread of antibiotics and nanomaterial in the soil. In this chapter, pros and cons of nanoparticles (NPs) and their effects on soil microbes, role of nanomaterials in combination with and without antibiotics, toxic effects of nanomaterial against microbes, and human and environmental management have been explored in detail. Impact of emerging nanoparticle on antibiotic resistance against soil microbes has been discussed in detail. Another factor like toxicity of NPs on plants and on human beings via food chain has also been discussed.

14.2 Nanoparticles as an Antibacterial Agent

Metallic nanoparticles are used as antibacterial agent against soil-, water-, and food-borne pathogens. Many previously published studies have shown the importance of metallic nanoparticles against environmental microflora. AgNPs were tested against ampicillin-resistant *Escherichia coli* O157:H7, multidrug-resilient *Pseudomonas aeruginosa*, and erythromycin-resistant *Streptococcus pyogenes* and were found very effective and recommended as broad-spectrum antibacterial agents in the community environment (Lara et al. 2010).

Metallic nanoparticles (MNPs) and engineered nanoparticles (ENPs) have great potential in almost all industries, and it is more impressing that their synthesis is green and economical (Pantidos and Horsfall 2014). Biosynthesis of metallic nanoparticles by using bacterial strains or soil bacteria is a common and cheap way. Scientists are using bacteria for biological synthesis of NPs these days and testing different strains. Ag and other metallic NPs are synthesized biologically as this method is bio-friendly which were tested against certain environmental bacteria. AgNPs have multi-bactericidal actions against bacteria (Fig. 14.1). Similarly, AgNPs were synthesized by soil bacteria *Bacillus* spp. and tested against multidrug resistance which showed excellent activity against *S. aureus*, *S. epidermidis*, *V. cholera*, *Salmonella typhi*, and *S. paratyphi*. Best action was found against *S. epidermidis*, and AgNPs showed synergistic effect with chloramphenicol when tested against *S. typhi* (Thomas et al. 2014). Similarly, Kumar et al. (2015) synthesized AgNPs (198–595 nm) by using *Streptomyces* spp. isolated from soil samples which were characterized by UV-vis spectroscopy, electron microscopy, and FTIR. The synthesized NPs had both cytotoxic and antibacterial properties. Eu^{3+} doped

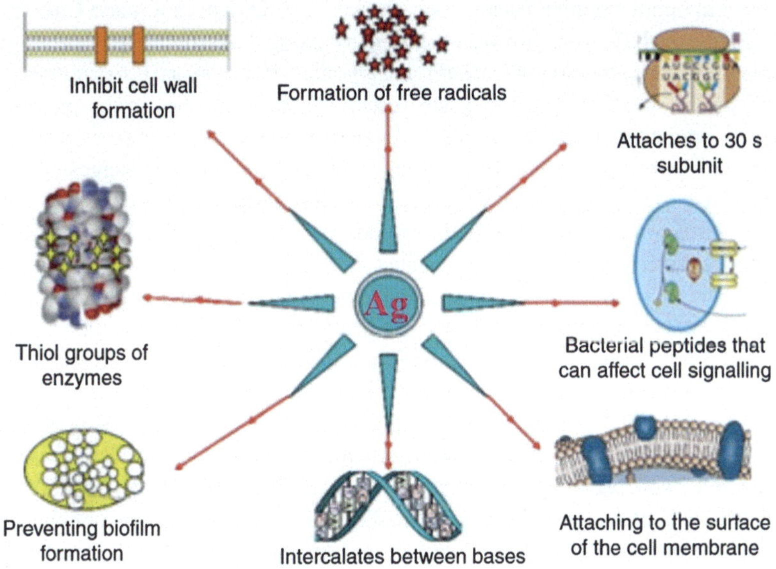

Fig. 14.1 Bactericidal actions of AgNPs (adopted from Rai et al. 2012)

lanthanum calcium manganate (LCMO) and lanthanum calcium manganate (LCM) NPs (50–200 nm) were compared against *Pseudomonas aeruginosa*; a common water, and soil bacteria and LCM showed more antibacterial activity than LECMO (De et al. 2010).

Similarly, ZnO, ZnS, and ZnNPs were prepared by using resistant *Pseudomonas stutzeri* grown on biofilm of Zn which were tested against and found active against Gram-positive and Gram-negative bacteria (Mirhendi et al. 2013). *Actinomycetes* and *Pilimelia columellifera* subsp. *pallida* were isolated from pine forest soils and used for the synthesis of AgNPs (12.7–15.9 nm) which were tested seperately and in amalgamation with antibiotics against *Bacillus subtilis*, *Staph. aureus*, *E. coli*, *Pseudomonas aeruginosa*, *Klebsiella pneumoniae*, and *Enterobacter*. The lowest inhibitory concentration (40 μg mL^{-1}) was observed against *E.coli* (Golinska et al. 2016). Similarly, *Weissella oryzae* DC6 isolated from mountain ginseng for the synthesis of AgNPs were tested against *Vibrio parahaemolyticus*, *Bacillus cereus*, *Bacillus anthracis*, *Staphylococcus aureus*, *Escherichia coli*, and *Candida albicans* and found active against *Staphylococcus aureus* and *Pseudomonas aeruginosa* (Singh et al. 2016). Biocompatible CuNPs were prepared by using Cu-resistant *B. cereus* isolated from soil, and the strains could tolerate >10 mM of Cu. CuNPs showed better antimicrobial activity against cell lines and was found safe when compared to $CuSO_4$ (Tiwari et al. 2016). They suggested the use of prepared CuNPs as antimicrobial agents, as effective delivery of copper, as biosensors, and for treatment of physiological disorders and cancer. ZnO, Fe_2O_3, and CuO NPs, which were prepared by a solgel combustion method and characterized by X-ray diffraction, were studied against Gram-positive and Gram-negative bacteria. The greatest antimicrobial activity against both Gram-positive and Gram-negative bacteria was shown by ZnO than Fe_2O_3 and CuO (Azam et al. 2012). AgNPs were mycosynthesized by using *Fusarium oxysporum* against resistant *Enterobacter* sp., *Klebsiella pneumonia*, *Pseudomonas aeruginosa*, and *E. coli*, and it was that found they are susceptible to antibiotics in the presence of AgNPs (Gopinath et al. 2015). Similarly, bioactive, single-step AgNps were prepared by using *Streptacidiphilus durhamensis* against various strains and suggested for use against resistant strains (Buszewski et al. 2016). Similarly, green synthesis of AgNPs by using extract of *Parkia speciosa Hassk* pods assisted by microwave irradiation was tested and showed enhanced activity against *P. aeruginosa*, *E. coli*, and *Staph. aureus* with microwave irradiation (Fatimah 2016). Antibacterial activity of phytopathogenic bacterium *Ralstonia solanacearum* was studied by using Tween-80 which stabilized AgNPs and was found to be the most perfect stabilizer. The synthesized AgNPs were found active against tobacco bacterial wilt and suggested as alternative for crop disease control (Chen et al. 2016). Antibacterial activity of AgNPs prepared by using *Satureja hortensis* extract showed significant activity against *Bacillus cereus* isolated from soil bacteria (Shirmohammadi et al. 2014). Effect of AgNPs against molds was studied, and it was reported that the environment can inhibit or enhance the antifungal properties of AgNPs (Pietrzak and Gutarowska 2015). Similarly, some more studies on ZnS NPs against fungi were studied and reported potential and emerging antifungal agent which caused irreversible damage to cell membrane (Ibrahim et al. 2016). Antibacterial properties of NPs depend on type of microbe, physicochemical properties of NPs as certain bacteria have self-defense mechanism, and they can become resistant to MNPs (Hajipour et al. 2012).

14.3 Antibiotic vs. Multidrug Resistance in Soil

Antibiotic resistance is an ancient phenomenon, but evolutionary drivers of resistant traits in natural settings remain unspoken (Hollowell et al. 2015). Extensive use of antibiotics both in human and animals especially in developing countries spreading antibiotic and multidrug resistance not only human and animals but also in soil microflora. Soil microflora can be both Gram-positive and Gram-negative bacteria along with both beneficial and non-beneficial microbes. Soil bacteria are divided into certain groups: (1) decomposers that use carbon as sole source for their growth; (2) mutualists that form association with plants; (3) pathogens like *Zymomonas*, *Agrobacterium*, and *Erwinia* species; (4) lithotrophs or chemoautotrophs, those that obtain their energy from compounds other than carbon like nitrogen, sulfur, iron, or hydrogen; and (5) among the beneficial actinomycetes, nitrifying and denitrifying bacteria that are common (Ingham 2016). Beneficial processes like nitrogen fixations, nitrification, and denitrification process in soil may be affected in the presence of AgNPs (Arnaout 2012). A study conducted by US agricultural research services department reported two commonly prescribed antibiotics, i.e., tetracycline and cefotaxime, from soil samples collected from where manure was deposited (USDA 2016). Rhizosphere soil microflora is not steadily more antibiotic resistant than non-rhizosphere soil microflora (Harris and Woodbine 1967). Manure can spread antibiotic resistance gene as reported that manure from treated animals augments the spread of antibiotic resistance genes in soil bacteria (Heuer and Smalla 2007). Similarly, antibiotic resistance in soil environment was analyzed, and it was found that dairy manure microbes are more resistant than in other sites. Tetracycline resistance was found predominant at 47–89% of total counts (Esibu et al. 2002). *Pseudomonas*, *Burkholderia*, and *Enterobacter* were found dominant at high drug level resistance (50–170 w g mL^{-1}). Similarly, by simple, fast, and eco-friendly method, NPs (30–70 nm) were synthesized by using newly identified *Cryphonectria* spp. of genus *Cryphonectria* and tested at a concentration of 5 μg mL^{-1} for antibacterial activity against *Staph. aureus*, *E. coli*, *S. typhi*, and *Candida albicans*, which showed best activity against *S. aureus* and *E. coli* while low activity against *S. typhi* and *C. albicans*. When used in combination with antibiotic, enhanced activity was observed against the resistant strains, and AgNPs showed more activity at a concentration of 5 μg mL^{-1} than AgNO3 and antibiotic, streptomycin (Dar et al. 2013). Another study biologically synthesized AgNPs by using fungus *Trichoderma viride*, and increased antibacterial activity was observed when using AgNPs with ampicillin, kanamycin, erythromycin, and chloramphenicol against test strains, and similar effect was observed against both Gram-positive and Gram-negative strains (Fayaz et al. 2010). A correlation between ARGs and metal resistance genes was studied. Animal waste was found as a major source of antibiotic resistance genes in Chinese dairy farms with high risk of environmental pollution and human health (Zhou et al. 2016). Another study reported that the use of antibiotics for agriculture application is the important factor for antibiotic resistance in soil (Ghosh and LaPara 2007). Anthropogenic activities play a significant role to disseminate antibiotic resistance genes (ARGs) in aquatic environments (Fig. 14.2).

Some bacteria in soil may also produce antibiotics and develop resistance against the produced antibiotics and their products or deposits. A study was conducted on isolation of 64 forest soil samples (no human activities) from family *Enterobacteriaceae*, and susceptibility tests were performed. In few isolates of *E. coli* and *Citrobacter koseri*, blaCTX-M, a gene responsible for resistance was found. Two of them showed resistance against amikacin (Upadhyay et al. 2016). This shows the natural phenomenon of antibiotic resistance in bacteria as samples were collected from forests where no anthropogenic activities exist. Upadhyay et al. (2016) also reported the cephalosporin resistance in soil, irrespective of human activities in soil, which provides a perception of environmental risk of antibiotic resistance and future loss of obtaining antibiotics from soil. Similarly, gentamicin is an important broad-spectrum antibiotic used against clinical isolates but now has been found in environmental samples. Im et al. (2016) conducted the screening of metagenome libraries obtained from soil samples; a fosmid clone (35–40 kb) was carefully chosen as it deliberated strong gentamicin resistance. They suggested soil metagenome, an important resource for the identification of antibiotic resistance to overcome antibiotic resistance. Antibiotic resistant was studied in *Bradyrhizobium* populations against all the tested antibiotics by Hollowell et al. (2015). They

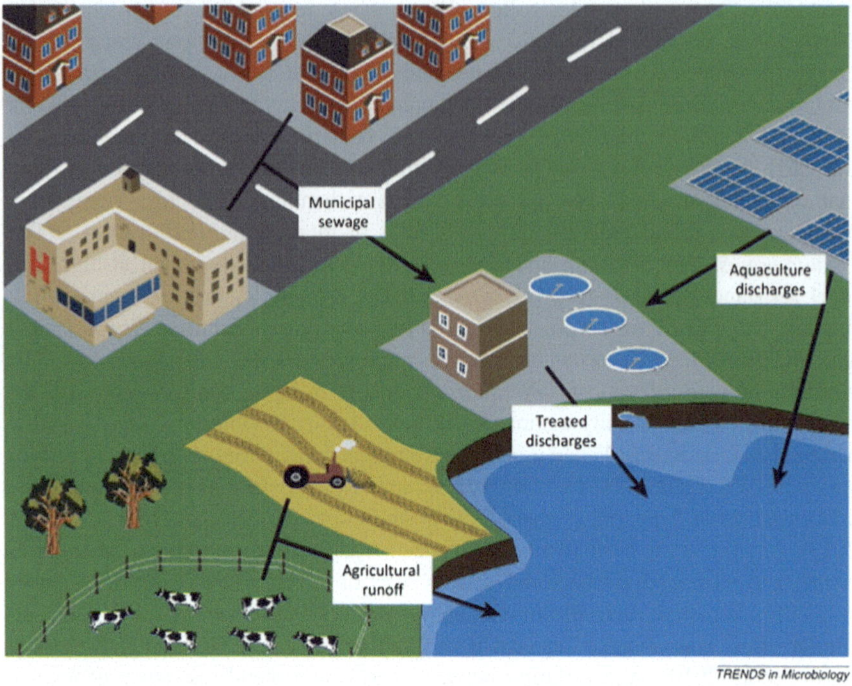

Fig. 14.2 Different anthropogenic activities that result in the dissemination of antibiotic resistance genes (ARGs) in aquatic environments (adopted from Marti et al. 2014)

discovered multidrug resistance, spatially structured subsets of resistance traits, and correlation between strain abundance and resistance traits. Zhang and Dick studied the soil bacterial isolates from *Proteobacteria* and *Bacteroidetes* treated with penicillin and neomycin antibiotics, which were not only resistant to the antibiotics but also using the chemicals as sole source of energy for their growth (Fig. 14.3). Some genes were not found in bacterial isolates but in soils. They reported that soil has large pool of resistant genes and suggested the need to explore the ecological and health aspects in the future.

Tyc et al. (2014) studied the impact of interactions on antimicrobial activity among 146 phylogenetically different soil bacterial isolates. Among them, 42% showed activity in interaction, and 323% showed activity in monoculture. Resistant bacteria belonging to family *Enterobacteriaceae* and non-fermenters (both Gram-positive and Gram-negative) were found prevalent in cattle manure and active against penicillin and levofloxacin while only resistant to ampicillin, chloramphenicol, and ampicillin-sulbactam. So, there is safe management of animal manure to stop contamination and spread of antibiotic resistance in soil along with protection of animal, human, and environmental health (Resende et al. 2014).

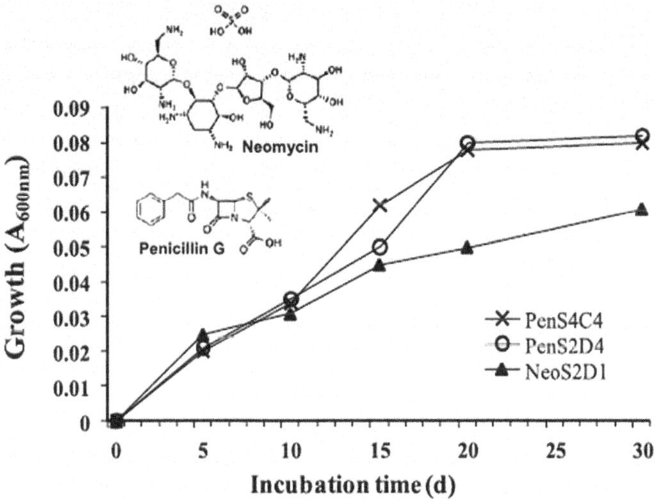

Fig. 14.3 Growth curves of isolates containing the antibiotics, penicillin, or neomycin, as sole carbon source at concentrations of 1000 mg L^{-1} (adopted from Zhang and Dick 2014)

Walsh and Duffy 2013 characterized the multidrug resistant profile in soil bacteria. They identified 412 antibiotic resistance in pristine, agriculture, and urban soils and found no involvement of ESBL, *bla*$_{NDM-1}$, and plasmid-mediated quinolone resistance genes. Among them, 80% were found resistant to 16–23 antibiotics. Similarly, enteric bacteria in dairy soil were found resistant to chloramphenicol, nalidixic acid, penicillin G, and tetracycline, and this may be due to poor absorption of antibiotics by animals that might persist on top soils (Burgos et al. 2005). Comprehensive analysis, impacts of antibiotic resistance on agroecosystem, and surveillance program for antibiotic resistant in soil are suggested by different studies (Burgos et al. 2005; Cantas et al. 2013; Rothrock et al. 2016). Some other studies reported the multidrug resistance in soil samples collected from premises of hospitals and damping sites of hospital wastes (Modi et al. 2013; Chandan et al. 2013).

14.4 Metallic Nanoparticles vs. Antibiotics Effects

It is important to deliberate that many studies have reported that MNPs not only have antibacterial activity against certain strains but also enhance the effects of certain antibiotics. Synergistic effect of AgNPs with doxycycline was studied against *Klebsiella pneumoniae*, and AgNPs enhance the effect of doxycycline against the tested strain (Kumar et al. 2016). Many other empirical studies supported the same evidence that MNPs enhance the activities of antibiotics against bacteria (Birla et al. 2009; Naqvi et al. 2013). Devi and Joshi (2012) screened the soil isolates from microhabitats for the synthesis of AgNPs and tested their effectiveness against microbes with and without combination of antibiotics. *Aspergillus terreus* SP5, *Fusarium* sp. MP5, and *Paecilomyces lilacinus* SF1 were isolated and identified by molecular technique (18S rRNA sequencing) which helped in synthesis of AgNPS. AgNPs synthesized by fungi were found active against *Staphylococcus aureus*, *Streptococcus pyogenes*, *Salmonella enterica*, and *Enterococcus faecalis* when used in combination with erythromycin, methicillin, chloramphenicol, and ciprofloxacin. In combination, they showed synergistic effects and higher activity as compared to AgNPS. Similarly, synergistic effects of AgNPs in combination with ampicillin, erythromycin, kanamycin, chloramphenicol, amoxicillin, penicillin G, erythromycin, clindamycin, and vancomycin against *E. coli*, *Pseudomonas aeruginosa*, *S. typhi*, *Micrococcus luteus*, and *S. aureus* have been reported by other studies (Shahverdi et al. 2007; Fayaz et al. 2010). Brown et al. (2012) worked on ampicillin-resistant strains like *E. aerogenes* and *P. aeruginosa* and methicillin-resistant *S. aureus* functionalized with Ag and Au nanoparticles (AgNPs and AuNPs). They found AgNPs as antibacterial agents, but AuNPs were only active in the presence of ampicillin. Both AgNPs and AuNPs functionalized with ampicillin were active against Gram-positive and Gram-negative bacteria, but there is a need to find the exact mechanism involved. Similarly, Burygin et al. (2009) reported the enhanced activity of gentamicin when mixed with AuNPs against *E. coli*. They further quoted that stable conjugates of AuNPs coated with antibiotic may enhance antibacterial properties. Similarly, another study on methicillin-resistant *S. aureus* has shown resistance to many antibiotics like nalidixic acid, ciprofloxacin, bacitracin, erythromycin, and vancomycin. Biosynthesis of Ag_2ONPs was performed by using *Aspergillus terreus*, which were tested and found active not only against methicillin-resistant *S. aureus* but also against other pathogenic bacteria like *Mycobacterium tuberculosis* (Sangappa and Thiagarajan 2015). Here, Ag_2ONPs showed similar activity like other antibiotics. Allahverdiyev et al. (2011) reviewed the combining effect of NPs and antibiotics and reported the enhanced effect of combining NPs and antibiotics at the spot of bacteria-antibiotic interface and decreased toxicity. Similarly, some other studies showed the enhanced activity of biologically synthesized MNPs when used in combination with antibiotics. AgNPs were prepared by using *Rhizophora apiculata* (15 nm in size) that showed enhanced activity of gentamicin and chloramphenicol in the presence of AgNPs against *E. coli*, *B. cereus*, *S. aureus*, and *P. mirabilis* strains (Dhas et al. 2013). Combining antibiotics with MNPs can prevent antibiotics-resistant development and can enhance the efficacy of antibiotics against the resistant strains (Raja and Singh 2013). Panacek et al. (2016) studied the synergistic effect of combined AgNPs (28 nm) and antibiotics against *P. aeruginosa*, *E. coli*, and *S. aureus* by using microdilution method and reported strong effect at very low concentration. They reported that very low concentration of Ag is required for broad-spectrum tested antibiotics to act against bacteria. Biofilm formation of gram-positive bacteria like *B. subtilis* and Gram-negative bacteria like *E. coli* and *Salmonella typhimurium* are affected at very low concentrations by AgNPs than MIC of antibiotics without or reducing cytotoxicity against mammalian cells (Sanyasi et al. 2016). AgNPs in

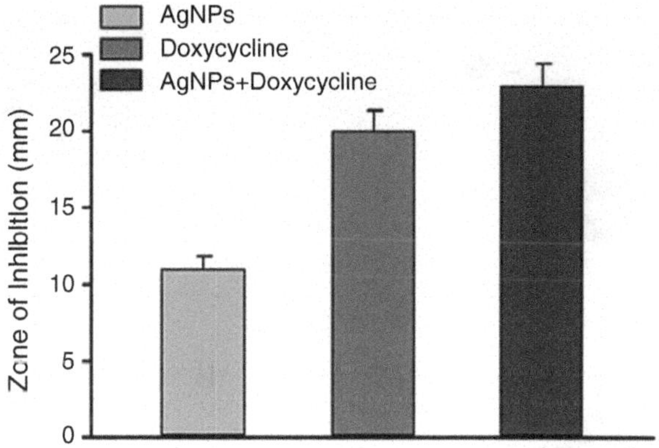

Fig. 14.4 Zone of inhibition (mm) of AgNPs, doxycycline, and both AgNPs + doxycycline (adopted from Kumar et al. 2016)

combination with doxycycline have more antibacterial activity than the individual impact of AgNPs and doxycycline (Fig. 14.4).

Li et al. (2014) studied the effect of AuNPs against multidrug-resistant pathogenic Gram-positive and Gram-negative bacteria which showed low toxicity to mammalian cells and no bacterial resistance after using AuNPs. Muhling et al. (2009) deliberated the impacts of AgNPs on antibiotic resistance bacteria in natural environment and reported that it depends on bioavailability and behavior of NPs once it is released into the environment. The enhanced activity of poly (*N*-vinyl-2-pyrrolidone) coated with AgNPs (Ag-PVP NPs) against *S. aureus*, *E. coli*, and gentamicin-resistant *E. coli* was studied. It was observed that gentamicin binds with dissolved Ag ions and affected the properties of silver ions (Wang et al. 2016). Collective approach of antibiosis of synthesized NPs and antibiotics is considered as more effective against microbes than the use of antibiotics alone, and this could be alternative cures to cope up with multidrug resistance. Ding et al. (2016) quoted that transfer of antibiotic resistance from dead to live bacteria is not known. However, they investigated that Al_2O_3 nanoparticles can facilitate the transfer of plasmid-facilitated resistance genes in Gram-positive (*Staph. aureus*) and Gram-negative (*E. coli*) bacteria by damaging cell membrane to facilitate the entry of plasmid. They suggested that nanomaterials are facilitating the microbes in obtaining resistance against drugs. Similarly, another study reported the transfer of resistance from *E. coli* to *Salmonella* through plasmid by nano-alumina (Qiu et al. 2012). This can be alarming and rather contradictory to previous studies that nanomaterials enhance the ability of antibiotics against resistant microbes. It can be concluded that nanomaterials not only enhance the properties of antibiotics against resistance but also support the weak microbes in obtaining resistance against antibiotics. AgNPs impact the bacterial cell in number of ways and have synergistic effects with antibiotics (Fig. 14.5).

Some other studies reported the green synthesis of MNPs, and enhanced activities of antibiotics against tested pathogens were observed when combined with AgNPs (Chauhan et al. 2013; Pantidos and Horsfall 2014; Bhosale et al. 2015).

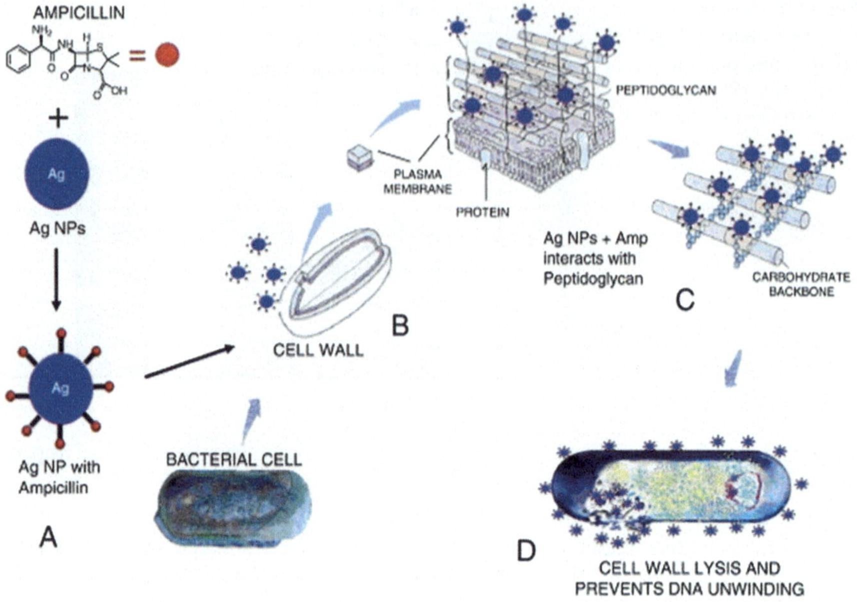

Fig. 14.5 Synergistic activity of AgNPs with ampicillin (Amp) against bacteria. (A) Formation of core silver nanoparticles with ampicillin. (B) Interaction of AgNPs-Amp complex over the cell wall of bacteria. (C) AgNPs-Amp complex inhibits the formation of cross-links in the peptidoglycan layer (which provides rigidity to the cell wall), leading to cell wall lysis. (D) AgNPs-Amp complex prevents the DNA unwinding (adopted from Fayaz et al. 2010)

14.5 Nanoparticle's Impacts on Soil Microflora, Enzyme Activity, and Plant's Growth

Empirical studies have been reported and have observed that plant beneficial bacteria play an important role in plant's growth by performing nitrogen fixation, nitrification, and denitrification processes (Ibekwe et al. 2003), and as reported in previous studies Ag has toxic effects against bacteria, but toxicity against denitrifying bacteria which convert nitrates into N_2 in aquatic and other environments can be harmful. Similarly, excess nitrogen is harmful and toxic for plants and can cause eutrophication in water bodies as well. Frenk et al. (2013) studied the effect of engineered NPs against soil bacteria which were assessed by determining bacterial community actions, size, and composition by subsequent exposure to CuO and magnetite (<50 nm) NPs in two dissimilar soils (sandy loam and clay loam). Hydrolytic activity, community composition, oxidative potential, and extent were strongly affected by CuO, while only hydrolytic activity and bacterial communal composition were affected by Fe_3O_4. Other soil groups like *Rhizobiales* and *Sphingobacteriaceae* were not affected by addition of CuO to soil. They suggested that both types of NPs are potentially damaging to soil surroundings, and both organic substance and clay fraction in diverse soils interact with NPs and reduced their toxicity. Concha-Guerrero et al. (2014) studied the impact of CuNPs against 56 bacterial strains which were isolated from soil, including 36 *Bacilli* strains, 2 *Flavobacteria* strains, and 18 *Gammaproteobacteria* strains. Among them, 21 were used for cytotoxicity and 11 showed high susceptibility. The effect of engineered NPs (Au, Al, fullerenes, etc.) on *Rhizobium*, a plant growth-promoting strain (PGRP), was studied, which showed significant cytotoxic effects on soil bacteria (Mishra and Kumar 2009). Hsueh et al. (2015) studied the impact of ZnO NPs against *B. subtilis* (which is a plant beneficial bacterium universally found in soil for ring formation, biofilm formation, and protein activity) and reported adverse effects on cell growth, biofilm formation capabilities of *B. subtilis*. Toxicity of MNPs is still under investigation and not clearly understood especially in soil environment. There is a dire need to investigate the effect of MNPs against resistant strains under natural settings. To promote green technologies, health and environmental impacts should be studied well before spreading the MNPs into the environment. Simonin and Richaume (2015) studied the impact of MNPs on soil microflora (Fig. 14.6) and reported the damaging effects of NPs against soil microbes even at <1 mg kg^{-1} concentration and at >250 mg kg^{-1}, and negative effect of CNPs has been reported. Functionalized MWCNTs were tested against soil bacterial communities (*Proteobacteria*, *Actinobacteria*, *Chloroflexi*, and *Bacteroidetes*), and they observed no profound effects on tested soil bacterial communities even at high concentrations; however, they suggested further experimental work for the change in soil nutrient cycling processes (Chung et al. 2011). Chai et al. (2015) found positive correlation between soil enzyme activity and total microbial activity and the presence of more functional bacteria. They further reported that ZnO and CeO_2 NPs were found to delay thermogenic metabolism and ultimately diminish numbers of soil *Azotobacter*. According to this study, ZnO, CeO_2, and TiO_2 NPs significantly decreased the numbers of P- and K-solubilizing bacteria and *Azotobacter* ($p < 0.05$).

Jin et al. (2013) studied the effect of SWCNTs on soil enzyme activity and bacterial biomass with different concentrations (0–1000 μg g^{-1} of soil) of SWCNTs. SWCNTs (300–1000 μg g^{-1} soil) considerably dropped activities of most enzymes and microbial biomass and had negative effect. It is important to note that SWCNTs showed similar effects to MWCNTs but at five times lower concentration. They also reported that it may be due to high surface area-to-volume ratio of SWCNTs than MWCNTs. It is important for policy making and spread of nanomaterials into the soil environment. Similarly, the effect of C60 fullerenes (50 nm to micron in size) was studied on soil microbes and protozoans, and they also reported the effect on microbial biomass and suggested not to spread the C60 fullerenes and other nanomaterials into the environment (Johansen et al. 2008), but other study showed little impact on functions and structure on soil microbes and processes (Tong et al. 2007). Similarly, another study reported the MWCNT's effect on plant's growth and soil microflora. They reported the decrease of *Proteobacteria* and *Verrucomicrobia* with more concentration and profusions of *Bacteroidetes* and *Firmicutes* (Khodakovskaya et al. 2013). In contrast to previous studies, the effect of FeO

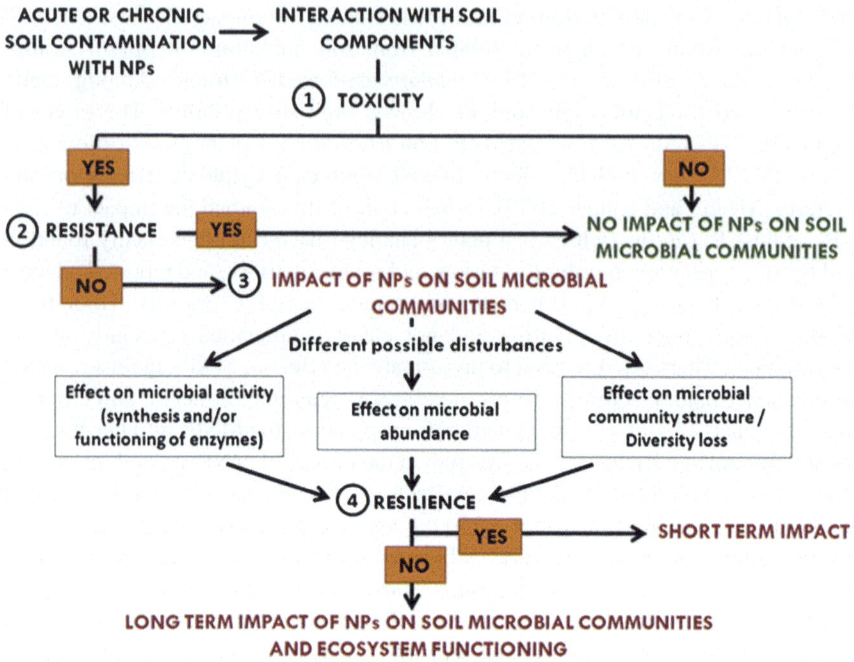

Fig. 14.6 The response of soil microbial communities to NP contamination (adopted from Simonin and Richaume 2015)

magnetic NPs was studied against soil microflora, and they reported positive effect on C and N of soil and its influence of related soil bacteria using C and N as sole source of energy but not the whole soil microflora (He et al. 2011). This indicates the specific and controlled use may help in increasing C and N of soil and beneficial microflora of soil. Another study reported that bacteria involved in C and P biogeochemical cycles are negatively affected by SWCNTs and have varying effects on applying higher concentrations (Rodrigues et al. 2013). Toxic effects of CuO and ZnO NPs were assessed against soil bacteria, and they reported all forms of Zn toxic to soil bacteria. Assessment of bioavailable metal concentration rather than general ecotoxicity has been suggested (Rousk et al. 2012). Effects of silica, palladium, Au, and Cu NPs against soil microbes and the germination of lettuce seeds were investigated and reported nonsignificant influence of the NPs in the soil on the no. of CFU, while NPs affected the lettuce seeds growth (Shah and Belozerova 2009). MNPs like TiO_2 and ZnO may affect plant growth as enzymes like protease, catalase, and peroxidase are inhibited, while urease remains undisturbed by these MNPs. Titanium dioxide adheres to the cell wall of wheat plants, while ZnO is dissolved and taken up by wheat plants (Du et al. 2011). Similarly, Ge et al. (2012) studied the effects of ZnO and TiO_2 on soil bacteria and reported genus *Bradyrhizobium* (symbiotic N_2-fixing bacteria) is negatively affected by these NPs, while genus *Streptomyces* is positively affected in the presence of these MNPs. Similarly, family *Sphingomonadaceae* (decomposers) is also affected positively. The effect of EMNPs like Ni, Fe, Co, and Ag was studied in winter field conditions, and they observed that Ag was distributed and Co and Ni remained on the top layer of soil when exposed to high precipitation for 50 days. However, *Flavobacterium* and *Niastella* were found increase in number, when analysis of individual genera was carried out. They found general migration pattern of NPs in soil, and their effect on microbial diversity was found reliant on environmental conditions (Shah et al. 2014). Similarly, Kumar et al. 2012 reported toxic behavior of MNPs on arctic bacterial communities even when applied at modest concentration. Thul et al. (2013) reported that worked done on toxicity is insufficient and does not convey any clear evidence on the effect on plant's growth, and this needs more solid work on plant beneficial bacteria, but there are some studies that reported negative effects on plant growth. Similarly, Xing et al. (2016) reported that little is known on effect of MNPs on plant microbe symbiosis in soil and regarding soil fertility, food security, and cultivation. A study was conducted to prove that plants absorb MNPs; butterfly (*Atrophaneura alcinous*) eggs were hatched after feeding on host plant's (*Aristolochia debilis*) leaves. The roots were dipped in 10 μg mL^{-1} in 100 mL TiO_2-NPs suspension, and TiO_2-NPs were absorbed and found in leave's veins which were later confirmed by X-ray analytical microscopy. TiO_2-NPs were transmitted from the plant to the larvae indicating absorption of MNPs by plants through roots (Kubo-Irie et al. 2016). Riahi-Madvar et al. (2012) reported the retarding effect of Al NPs on plant roots and quoted that antioxidant enzymes decrease the level of free radicals which diminish the toxic effect of NPs on plants. Similarly, another study reported the effect of engineered NPs on seed germination. They further quoted that plants not only directly absorb the MNPs but also absorb after mixing with water which remain in the soil for a long time, if not absorbed (Fig. 14.7), but contradictory to this study, some other studies reported no effect of MWCNTs on plant seedlings and germination and reported positive effects (Flores et al. 2014). Many studies reviewed absorption of nanomaterials in different plants with different concentrations (Aslani et al. 2014; Chichiricco and Poma 2014). So, it is clear from the literature that nanomaterials are absorbed by the plants, but effect of nanomaterials on plant's growth varies depending on the type of non-material used along with the concentration on type of plants species (Fig. 14.7).

Many empirical studies studied on the effect of carbon and other MNPs and their oxides on soil microbial communities. The data were collected between 1975 and 2016 from ISI web of knowledge by using "impact of MNPs on soil microbial communities" and by changing different metals and carbon NPs keywords. More work has been conducted on the impact of Ag, Fe, CuO, and fullerenes against soil microbial communities than other nanomaterials (Fig. 14.8). Simonin and Richaume (2015) performed the study till July 2014, and the same result has been reported but till December 2016 showed that not much work has been done to depict the clear results. More work in future is required to investigate the effect of nanomaterials again soil microbes.

Vesicular-arbuscular mycorrhizae (VAM) form symbiotic association with plant roots and fungi. TiO_2NPs effect was reported on VAM association (Burke et al. 2014). MNPs not only impact and contribute to phytotoxicity but also have effects on secondary process like metal nutrition. They further quoted that CuO NPs colonized by *Pseudomonas chlororaphis*, a root bacterium, had more significant

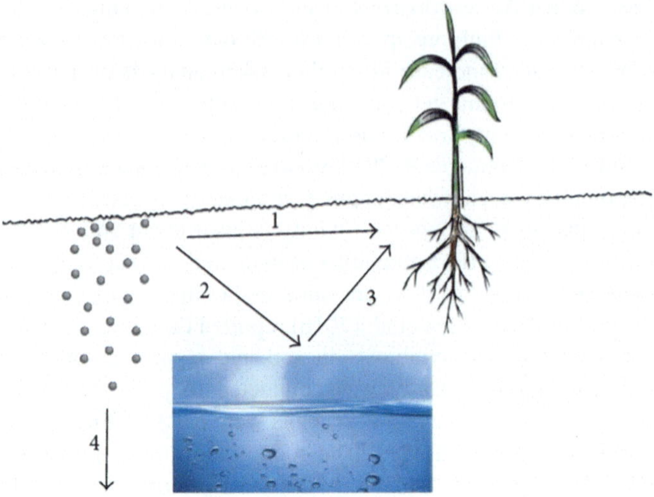

Fig. 14.7 Metallic NPs interaction with plants in the soil (adopted from Aslani et al. 2014)

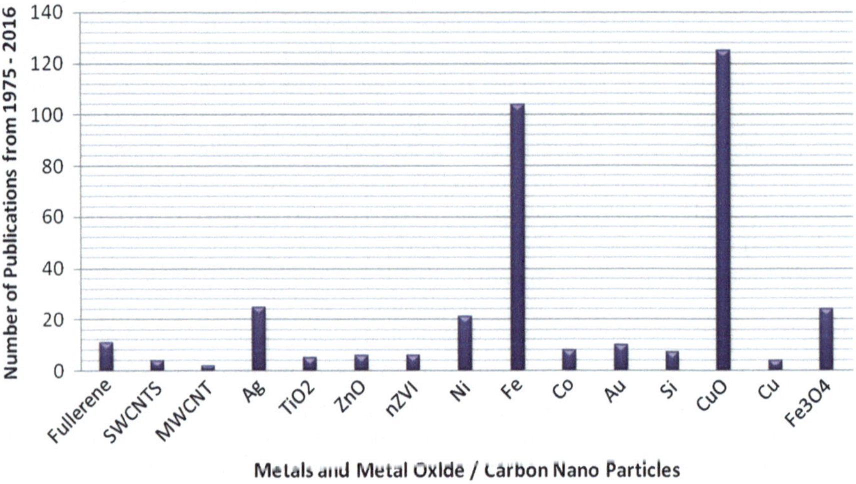

Fig. 14.8 Article published on the effect of MNPs and oxides on soil microbial flora between 1975 and 2016 (adapted from Simonin and Richaume 2015)

inhibition effect on roots than shoots of the plants when treated with 250 and 500 mg Cu kg^{-1}, and reduction in root length was observed when treated with CuO/ZnO mixture at both concentrations and depends on the exposure level (Dimkpa et al. 2015). A study on impact of NPs on soil bacterial communities with different pH ranges 4.5–7.2 reported less susceptibility with highest pH and pH dependent impact of all forms of zinc on bacterial communities. This indicates soil pH is an important factor in assessing the nanotoxicology against soil bacteria.

A missing link has been found among NPs, soil living organisms like bacterial communities, and properties of the soil (Read et al. 2016). Similarly, another study reported that Ag, CuO, and ZnO NPs have toxic effects against soil microbes which may be bactericidal or bacteriostatic (Gajjar et al. 2009).

14.6 Toxicity and Health Impacts of Nanoparticles and Antibiotics

Antibacterial activity of Ag is known for years, and AgNPs are now well studied for antimicrobial activities with different particle sizes and shapes with variable properties, but mechanism and toxicity of AgNPs are not well studied (Li et al. 2005; Mishra and Kumar 2009). It is important to assess the toxicity of Ag and other metallic NPs against beneficial and harmful microbes in soil as many available studies fail to identify the significant effects of MNPs on microbes in more complex systems (Neal 2008). To assess the toxicity of MNPs against soil microbes, it is important to analyze soil properties, organic contents, pH of soil, texture, and bioavailability of nutrients and which NPs are tested (EU 2016). Similarly, Simonin and Richaume (2015) reported that the influence of properties of soil on nanoparticle's toxicity is not well known. According to Zeliadt (2010), antimicrobial AgNPs in soil may harm plant growth as AgNPs not only kill the soil microflora but negatively affect the plant growth. Many other studies quoted in nanomaterial's impacts on soil microflora, enzyme activity, and plant's growth showed the toxic effect of nanomaterials, and all the previous studies suggested to stop the spread of nanoparticles and CNTs into the environment (Jin et al. 2013; Khodakovskaya et al. 2013; Johansen et al. 2008). They are inhibiting pathogens, but along with the benefits, they have significantly negative effect on plant's growth and reducing enzyme activity. These may be used under controlled environment and specific in nature for increasing beneficial microflora (He et al. 2011) against pathogens which may be helpful in reducing plant's pathogens. Due to the small size of MNPs and other nanomaterials, they easily enter soil, absorb, and may enter the food web through plant's absorptions. Pan and Xing (2012) suggested the modeling for MNP's fate and incusing for regulatory framework for their risk assessment. Studies proposed the need of more work on the interaction of NPs with soil microbes under natural settings (Dinesh et al. 2012). Maurer-Hones et al. (2013) studied the toxicity of nanoparticles in the environment and reported that toxicity is complex in the ecosystem and fate and transport of nanomaterial in the environment is unknown. They suggested the need of more analytical work on toxicity due to the behavior and toxicity of MNPs. This is important for the sustainable development of nanotechnology and its applications in all the fields especially in soil. Similarly, Panyala et al. (2008) reviewed the toxicity of AgNPs against soil beneficial bacteria and reported that little has been done on the effects of MNPs on soil communities, and they suggested the need of more work on this aspect. They reported that whether micro- or nano-Ag may cause problem both in human and the environment. Aslani et al. (2014) reported the toxicity of MNPs on plants, and it depends on size, composition, and plant's species. Similarly, Rana and Kalaichelvan (2013) reviewed the effect of MNPs on microbes, human, plants, and animals. They reported that toxicity of MNPs depends on the influence of environmental factor on bioavailability and natural uptake mechanism. It also varies and depends on the types of NPs as they behave differently in different environment. They further quoted that NPs do not degrade as they accumulate in the environment, while some mineral-based NPs like ZnO dissolve with the passage of time. Toxicity in microbes by NPs may be due to membrane disorganization or oxidative stress or generation of reactive oxygen (Niazi and Gu 2009). However, in plants, respiration, transpiration, and photosynthesis are affected by MNPs after absorption (Rana and Kalaichelvan 2013). A study on the absorption of AuNPs by tomato plants and genotoxicity reported that *Lycopersicon esculentum* can absorb AuNPs and can cause cellular toxicity (Agtuca 2014). Another study reported that MNPs are most widely used because of their applications and they can enter the food chain via terrestrial organisms and have shown adverse effects on gene expression in earthworms, but it depends on type and size of the MNPs (Unrine et al. 2008). According to another study, very limited studies are present on biotransformation of NMs in food crops (Rico et al. 2011). Plant protection mechanism actively guards the plants against toxic substances like NPs. A study reported that components of wheat plant roots change the activity of ZnO and CuO NPs and reported that CuO NP toxicity could be negated by plant metabolites (Martineau et al. 2014). Similarly, another study reported the toxicity of ZnO NPs in wheat plants by inhibition of plant roots by increasing the level of Zn taken from ZnO NPs as compared to control (Watson et al. 2015). However, 50–90% variations in Zn and Cu toxicity have been observed by linear regression model, and normal relationship has been recommended in assessing ecological risks (Warne et al. 2008). MNPs after aggregation in the soil may have less toxicity as bulk material have less antibacterial effect (Gajjar et al. 2009). Furthermore, aggregation status and MNP properties can be influenced by different chemical, physical, and biological processes (Pan and Xing 2012). Although green synthesis of MNPs has significant impacts against drug-resistant microbes, their mode of synthesis and social and environmental implications should be considered before use, and risk assessment must be addressed thoroughly (Rudramurthy et al. 2016; Pantidos and Horsfall 2014). Many other empirical studies warned the use of nanomaterials in the environment before assessing the potential impacts of NPs, and they emphasized the conduction of more research work and training in the field of agriculture (Kardos et al. 2015; Mukhopadhyay 2014). Environmental health risk using Al_2O_3 in obtaining antibiotic resistance has been highlighted (Ding et al. 2016).

14.7 Conclusion

It can be concluded from the above review study that MNPs or other nanomaterials like fullerenes and CNTs have certain impacts against bacteria. Nanomaterials are synthesized naturally or synthetically in the labs. Biosynthesis is common and used for the green synthesis of MNPs and used against pathogens. Empirical studies have shown the toxic effects of MNPs against pathogenic bacteria both in human and environmental samples, but studies against plant beneficial bacteria is still lacking and need more work on this important aspect. The role of MNPs in plant's growth and the effect of MNPs on enzymes are clear and reported by certain studies that MNPs have great impact on enzymes which have important role in plant's growth, such as protease, catalase, and peroxidase which are inhibited by applying MNPs. From the studies, it has been observed that effect of nanomaterials on plant's growth varies depending on the type of nanomaterial used along with the concentration on the type of plants species. Impacts of MNPs on plant beneficial bacteria were found negative and cytotoxic, but more work is required to clarify the situation in more complex system and under natural settings. MNPs showed synergistic effects and enhanced the effect of antibiotics against pathogens and resistant strains, but some studies indicated nano-alumina promotes resistance through plasmids in certain genera. MNPs may be more active when used in combination with antibiotics and may have fruitful results when used against resistant strains. Further studies on elucidation of impacts of MNPs and other nanomaterials on plant growth-promoting bacteria and sensitive microbes in obtaining resistance or transformation of antibiotic resistance genes in the presence of certain NPs are essential and recommended in future studies.

Acknowledgment The author gratefully acknowledges the support received from the Centre for Climate Research and Development (CCRD), COMSATS IIT, Islamabad. The CCRD has no financial and material contribution in writing and submission of this work.

References

Agtuca BJ (2014) Gold nanoparticles in the environment: studying the genetic toxicity and bioavailability in soils and hydroponic exposures with *Lycopersicon esculentum* (tomato 'brandywine'). Honors theses, Paper 37

Ahmed T, Imdad S, Ashraf S, Butt NM (2012) Effect of size and surface ligands of silver (ag) nanoparticles on waterborne bacteria. Int J Theor Appl Nanotechnol 1(1):111–116. doi:10.11159/ijtan.2012.017

Ahmed T, Yaldram K, Imdad S, Butt NM (2014) Emerging nanotechnology based methods for water purification—a review. Desalin Water Treat 52:4089–4101

Allahverdiyev AM, Kon KV, Abamor ES, Bagirova M, Rafailovich M (2011) Coping with antibiotic resistance: combining nanoparticles with antibiotics and other antimicrobial agents. Expert Rev Anti-Infect Ther 9(11):1035–1052

Arnaout CL (2012) Assessing the impacts of silver nanoparticles on the growth, diversity, and function of wastewater bacteria. Ph.D. thesis, Department of Civil and Environmental Engineering, Duke University

Aslani F, Bagheri S, Julkapli NM, Juraimi AS, Hashemi FSG, Baghdadi A (2014) Effects of engineered nanomaterials on plants growth: an overview. ScientificWorldJournal 1–28. doi:10.1155/2014/641759, 2014

Azam A, Ahmed AS, Oves M, Khan MS, Habib SS, Memic A (2012) Antimicrobial activity of metal oxide nanoparticles against gram-positive and gram-negative bacteria: a comparative study. Int J Nanomedicine 7:6003–6009

Bhosale RS, Hajare KY, Mulay B, Mujumdar S, Kothawade M (2015) Biosynthesis, characterization and study of antimicrobial effect of silver nanoparticles by *Actinomycetes spp*. Int J Curr Microbiol Appl Sci 2:144–151

Birla SS, Tiwari VV, Gade AK, Ingle AP, Yadav AP, Rai MK (2009) Fabrication of silver nanoparticles by Phoma Glomerata and its combined effect against *Escherichia coli*, *Pseudomonas aeruginosa* and *Staphylococcus aureus*. Lett Appl Microbiol 48:173–179

Brown AN, Smith K, Samuels TA, Lu J, Obare SO, Scott ME (2012) Nanoparticles functionalized with ampicillin destroy multiple-antibiotic-resistant isolates of *Pseudomonas aeruginosa* and *Enterobacter aerogenes* and methicillin-resistant *Staphylococcus aureus*. Appl Environ Microbiol 78(8):2768–2774

Burgos JM, Ellinggton BA, Varela MF (2005) Presence of multidrug resistant enteric bacteria in dairy farm topsoil. J Dairy Sci 88(4):1391–1398

Burke DJ, Zhu S, Pablico-Lansigan MP, Hewins CR, Samia ACS (2014) Titanium oxide nanoparticle effects on composition of soil microbial communities and plant performance. Biol Fertil Soils 50:1169–1173

Burygin GL, Khlebtsov BN, Shantrokha AN, Dykman LA, Bogatyrev VA, Khlebtsov NG (2009) On the enhanced antibacterial activity of antibiotics mixed with gold nanoparticles. Nanoscale Res Lett 4:794–801

Buszewski B, Railean-Plugaru V, Pomastowski P, Rafinska K, Szultka-Mlynska M, Wypij M, Laskowski D, Dahm H (2016) Antimicrobial activity of biosilver nanoparticles produced by a novel *Streptacidiphilus durhamensis* strain. J Microbiol Immunol Infect. doi:10.1016/j.jmii.2016.03.002

Cantas L, Shah SQA, Cavaco LM, Manaia CM, Walsh F, Popowska M, Garelick H, Burgmann H, Sorum H (2013) A brief multi-disciplinary review on antimicrobial resistance in medicine and its linkage to the global environmental microbiota. Front Microbiol 4:96. doi:10.3389/fmicb.2013.00096

Chai H, Yao J, Sun J, Zhang C, Liu W, Zhu M, Ceccanti B (2015) The effect of metal oxide nanoparticles on functional bacteria and metabolic profiles in agricultural soil. Bull Environ Contam Toxicol 94:490. doi:10.1007/s00128-015-1485-9

Chandan P, Mishra RP, Asif A, Gangwar VS, Shweta C (2013) Isolation and characterization of multi drug resistant super pathogens from soil samples collected from hospitals. Res J Recent Sci 2:124–129

Chauhan R, Kumar A, Abraham J (2013) A biological approach to the synthesis of silver nanoparticles with Streptomyces sp JAR1 and its antimicrobial activity. Sci Pharm 81:607–621

Chen J, Li S, Luo J, Wang R, Ding W (2016) Enhancement of the antibacterial activity of silver nanoparticles against phytopathogenic bacterium *Ralstonia solanacearum* by stabilization. J Nanomater 2016:1. doi:10.1155/2016/7135852

Chichiricco G, Poma A (2014) Penetration and toxicity of nanomaterials in higher plants. Nano 5: 851–873. doi:10.3390/nano5020851

Chung H, Son Y, Yoon TK, Kim S, Kim W (2011) The effect of multi-walled carbon nanotubes on soil microbial activity. Ecotoxicol Environ Saf 74:569

Concha-Guerrero SI, Brito ES, Piñón-Castillo HA, Tarango-Rivero SH, Caretta CA, Luna-Velasco A, Duran R, Orrantia-Borunda E (2014) Effect of CuO nanoparticles over isolated bacterial strains from agricultural soil. J Nanomater 2014:1. doi:10.1155/2014/148743

Dar MA, Ingle A, Rai M (2013) Enhanced antimicrobial activity of silver nanoparticles synthesized by *Cryphonectria* sp. evaluated singly and in combination with antibiotics. Nanomedicine 9(1):105–110

De D, Mandal SM, Gauri SS, Bhattacharya R, Ram S, Roy SK (2010) Antibacterial effect of lanthanum calcium manganate ($La_{0.67}Ca_{0.33}MnO_3$) nanoparticles against *Pseudomonas aeruginosa* ATCC 27853. J Biomed Nanotechnol 6(2):138–144

Devi LS, Joshi SR (2012) Antimicrobial and synergistic effects of silver nanoparticles synthesized using soil fungi of high altitudes of eastern Himalaya. Mycobiology 40(1):27–34

Dhas SP, Mukherjee A, Chandrasekar N (2013) Synergistic effect of biogenic silver nanocolloid in combination with antibiotics: a potent therapeutic agent. Int J Pharm Pharm Sci 5(1):292–295

Dimkpa CO, McLean JE, Britt DW, Anderson AJ (2015) Nano-CuO and interaction with nano-ZnO or soil bacterium provide evidence for the interference of nanoparticles in metal nutrition of plants. Ecotoxicology 24:119–129

Dinesh R, Anandaraj M, Srinivasan V, Hamza S (2012) Engineered nano-particles in the soil and their potential implications to microbial activity. Geoderma 173:19

Ding C, Pan J, Jin M, Yang D, Shen Z, Wang J, Zhang B, Liu W, Fu J, Guo X, Wang D, Chen Z, Yin J, Qiu Z, Li J (2016) Enhanced uptake of antibiotic resistance genes in the presence of nanoalumina. Nanotoxicology 10(8):1051–1060. doi:10.3109/17435390.2016.1161856

Du WC, Sun YY, Ji R, Zhu JG, Wu JC, Guo HY (2011) TiO2 and ZnO nanoparticles negatively affect wheat growth and soil enzyme activities in agricultural soil. J Environ Monit 13: 8220–8828

Esibu N, Armenta L, Ike J (2002) Antibiotic resistance in soil and water environments. Int J Environ Health Res 12(2):133–144

EU (2016) Science for environmental policy: nanoparticles' ecological risks: effects on soil micro-organisms. 463: European Commission DG, Environment News Alert Service, edited by SCU, The University of the West of England, Bristol

Fatimah I (2016) Green synthesis of silver nanoparticles using extract of *Parkia speciosa Hassk* pods assisted by microwave irradiation. J Adv Res 7(6):961–969

Fayaz AM, Balaji K, Girilal M, Yadav R, Kalaichelvan PT, Venketesan R (2010) Biogenic synthesis of silver nanoparticles and their synergistic effect with antibiotics: a study against gram-positive and gram-negative bacteria. Nanomedicine 6:103–109

Flores D, Chacon R, Alvarado L, Schmidt A, Alvarado C, Chaves J (2014) Effect of using two different types of carbon nanotubes for Blackberry (rubus adenotrichos) in vitro plant rooting Growth and Histology. Am J Plant Sci 5:3510–3518. https://doi.org/10.4236/ajps.2014.524367

Frenk S, Ben-Moshe T, Dror I, Berkowitz B, Minz D (2013) Effect of metal oxide nanoparticles on microbial community structure and function in two different soil types. PLoS One 8(12):84441

Gajjar P, Pettee B, Britt DW, Huang W, Johnson WP, Anderson AJ (2009) Antimicrobial activities of commercial nanoparticles against an environmental soil microbe, *Pseudomonas putida* KT2440. J Biol Eng 3:9. doi:10.1186/1754-1611-3-9

Ge YG, Schimel JP, Holden PA (2012) Identification of soil bacteria susceptible to TiO_2 and ZnO nanoparticles. Appl Environ Microbiol 78:6749–6758

Ghosh S, LaPara TM (2007) The effects of subtherapeutic antibiotic use in farm animals on the proliferation and persistence of antibiotic resistance among soil bacteria. ISME J 1:191–203. doi:10.1038/ismej.2007.31

Golinska P, Wypij M, Rathod D, Tikar S, Dahm H, Rai M (2016) Synthesis of silver nanoparticles from two acidophilic strains of Pilimelia columellifera subsp. pallida and their antibacterial activities. J Basic Microbiol 56(5):541–556

Gopinath PM, Narchonai G, Ranjani DD, Thanjuddin N (2015) Mycosynthesis, characterization and antibacterial properties of AgNPs against multidrug resistant (MDR) bacterial pathogens of female infertility cases. Asian J Pharm Sci 10:138–145

Hajipour MJ, Fromm KM, Ashkarran AA, Aberasturi DJ, Larramendi IRD, Rojo T, Serpooshan V, Parak WJ, Mahmoudi M (2012) Antibacterial properties of nanoparticles. Trends 30(10): 499–511

Harris PJ, Woodbine M (1967) Antibiotic resistance of bacteria from rhizosphere and non-rhizosphere soils. Plant Soil 27:167–171. doi:10.1007/BF01373387

He S, Feng Y, Ren H, Zhang Y, Gu N, Lin X (2011) The impact of iron oxide magnetic nanoparticles on the soil bacterial community. J Soils Sediment 11:1408–1417

Heuer H, Smalla K (2007) Manure and sulfadiazine synergistically increased bacterial antibiotic resistance in soil over at least two months. Environ Microbiol 9:657–666. doi:10.1111/j.1462-2920.2006.01185

Hollowell AC, Gano KA, Lopez G, Shahin K, Regus JU, Gleason N, Graeter S, Pahua V, Sachs JL (2015) Native California soils are selective reservoirs for multidrug-resistant bacteria. Environ Microbiol Rep 3:442–449

Hsueh Y, Ke W, Hsieh C, Lin K, Tzou D, Chiang C (2015) ZnO nanoparticles affect *Bacillus subtilis* cell growth and biofilm formation. PLoS One 10(6):e0128457. http://dukespace.lib.duke.edu/dspace/bitstream/handle/10161/5852/Arnaout_duke_0066D_11593.pdf. Last accessed 28 Dec 2016

Ibekwe AM, Grieve CM, Lyon SR (2003) Characterization of microbial communities and composition in constructed dairy wetland wastewater effluent. Appl Environ Microbiol 69:5060–5069

Ibrahim IS, Ali IM, Dheeb BI, Abbas QA, Ramizy MA, Elsa MH, Aljameel AI (2016) Antifungal activity of wide band gap Thioglycolic acid capped ZnS:Mn semiconductor nanoparticles against some pathogenic fungi. Mater Sci Eng 73:665. doi:10.1016/j.msec.2016.12.121

Im H, Kim KM, Lee S, Ryu C (2016) Functional metagenome mining of soil for a novel gentamicin resistance gene. J Microbiol Biotechnol 26(3):521–529

Ingham ER (2016) Soil bacteria. Natural resources conservation service soils. USDA. https://www.nrcs.usda.gov/wps/portal/nrcs/detailfull/soils/health/biology/?cid=nrcs142p2_053862. Retrieved 17 Nov 2016

Jin L, Son Y, Yoon TK, Kang YJ, Kim W, Chung H (2013) High concentrations of single-walled carbon nanotubes lower soil enzyme activity and microbial biomass. Ecotoxicol Environ Saf 88:9–15

Johansen A, Pedersen AL, Jensen KA, Karlson U, Hansen BM, Scott-Fordsmand JJ, Winding A (2008) Effects of C60 fullerene nanoparticles on soil bacteria and protozoans. Environ Toxicol Chem 27:1895–1903

Kandi V, Kandi S (2015) Antimicrobial properties of nanomolecules: potential candidates as antibiotics in the era of multi-drug resistance. Epidemiol Health 37:e2015020

Kardos J, Jemnitz K, Jablonkai I, Bota A, Varga Z, Visy J, Heja L (2015) The janus facet of nanomaterials. Biomed Res Int 2015:1. doi:10.1155/2015/317184

Khodakovskaya MV, Kim BS, Kim JN, Alimohammadi M, Dervishi E, Mustafa T, Cernigla CE (2013) Carbon nanotubes as plant growth regulators: effects on tomato growth, reproductive system, and soil microbial community. Small 9:115–123

Kubo-Irie M, Yokoyama M, Shinkai Y, Niki R, Takeda K, Irie M (2016) The transfer of titanium dioxide nanoparticles from the host plant to butterfly larvae through a food chain. Sci Rep 6:article number 23819. doi:10.1038/srep23819

Kumar N, Shah V, Walker VK (2012) Influence of a nanoparticle mixture on an arctic soil community. Environ Toxicol Chem 31:131–135

Kumar S, Balachandran C, Duraipandiyan V, Ramasamy D, Ignacimuthu S, Abdullah Al-Dhabi N (2015) Extracellular biosynthesis of silver nanoparticle using *Streptomyces sp.* 09 PBT 005 and its antibacterial and cytotoxic properties. Appl Nanosci 5:169. doi:10.1007/s13204-014-0304-7

Kumar N, Das S, Jyoti A, Kaushik S (2016) Synergistic effect of silver nanoparticles with doxycycline against *Klebsiella pneumonia*. Int J Pharm Pharm Sci 8(7):183–186

Lara HH, Ayala-Nunez NV, Turrent LCI, Padilla CR (2010) Bactericidal effect of silver nanoparticles against multidrug-resistant bacteria. World J Microbiol Biotechnol 26(4):615–621

Li P, Li J, Wu C, Wu Q, Li J (2005) Synergistic antibacterial effects of beta-lactam antibiotic combined with silver nanoparticles. Nanotechnology 16:1912–1917

Li X, Robinson SM, Gupta A, Saha K, Jiang Z, Moyano DF, Sahar A, Riley MA, Rotello VM (2014) Functional gold nanoparticles as potent antimicrobial agents against multi-drug-resistant bacteria. ACS Nano 8(10):10682–10686

Marti E, Variatzi E, Balcazar JL (2014) The role of aquatic ecosystems as reservoirs of antibiotic resistance. Trends Microbiol 22(1):36–41

Martineau N, McLean JE, Dimkpa CO, Britt DW, Anderson AJ (2014) Components from wheat roots modify the bioactivity of ZnO and CuO nanoparticles in a soil bacterium. Environ Pollut 187:65–72

Maurer-Hones MA, Gunsolus IL, Murphy CJ, Haynes CL (2013) Toxicity of engineered nanoparticles in the environment. Anal Chem 85(6):3036–3049

Mirhendi M, Emtiazi G, Roghanian R (2013) Antibacterial activities of nano magnetite ZnO produced in aerobic and anaerobic condition by *Pseudomonas stutzeri*. Jundishapur J Microbiol 6(10):10254. doi:10.5812/jjm.10254

Mishra VK, Kumar A (2009) Impact of metal nanoparticles on the plant growth promoting Rhizobacteria. Digest J Nanomater Biostruct 4(3):587–592

Modi G, Mishra SK, Modi BS, Modi D (2013) Production and characterization of multiple drug resistant cultures isolated from hospital premises. Indian J Life Sci 3(1):7–14

Muhling M, Bradford A, Readman JW, Somereld PJ, Handy RD (2009) An investigation into the effects of silver nanoparticles on antibiotic resistance of naturally occurring bacteria in an estuarine sediment. Mar Environ Res 68(5):278. doi:10.1016/j.marenvres.2009.07.001

Mukhopadhyay SS (2014) Nanotechnology in agriculture: prospects and constraints. J Nanotechnol Sci Appl 7:63–71

Naqvi SZ, Kiran U, Ali MI, Jamal A, Hameed A, Ahmed S (2013) Combined efficacy of biologically synthesized silver nanoparticles and different antibiotics against multidrug-resistant bacteria. Int J Nanomedicine 8:3187–3195

Narayanan KB, Sakthivel N (2010) Biological synthesis of metal nanoparticles by microbes. Adv Colloid Interf Sci 156(1-2):1–13

Neal AL (2008) What can be inferred from bacterium–nanoparticle interactions about the potential consequences of environmental exposure to nanoparticles? Ecotoxicology 17:362. doi:10.1007/s10646-008-0217-x

Niazi JH, Gu MB (2009) Toxicity of metallic nanoparticles in microorganism—a review. In: Kim YJ, Platt U, Gu MB, Iwahashi H (eds) Atmospheric and biological environmental monitoring. Springer, Berlin, pp 193–206

Pan B, Xing B (2012) Applications and implications of manufactured nanoparticles in soils: a review. Eur J Soil Sci 63:437–456

Panacek A, Smekalova M, Kilianova M, Prucek R, Bogdanova K, Vecerova R, Kolar M, Havrdova M, Płaza GA, Chojniak J, Zboril R, Kvitek L (2016) Strong and nonspecific synergistic antibacterial efficiency of antibiotics combined with silver nanoparticles at very low concentrations showing no cytotoxic effect. Molecules 21:26

Pantidos N, Horsfall LE (2014) Biological synthesis of metallic nanoparticles by bacteria, fungi and plants. J Nanomed Nanotechnol 05:233. doi:10.4172/2157-7439.1000233

Panyala NR, Pena-Mendez EM, Havel J (2008) Silver or silver nanoparticles: a hazardous threat to the environment and human health? J Appl Biomed 6:117–129

Pelgrift RY, Friedman AJ (2013) Nanotechnology as a therapeutic tool to combat microbial resistance. Adv Drug Deliv Rev 65(13-14):1803–1815

Pietrzak K, Gutarowska B (2015) Influence of the silver nanoparticles on microbial community in different environments. Acta Biochim Pol 62(4):721–724

Qiu Z, Yu Y, Chen Z, Jin M, Yang D, Zhao Z, Wang J, Shen Z, Wang X, Qian D, Huang A, Zhang B, Li J (2012) Nanoalumina promotes the horizontal transfer of multiresistance genes mediated by plasmids across genera. PNAS 109(13):4944–4949

Rai MK, Deshmukh AP, Ingle AP, Gade AK (2012) Silver nanoparticles: the powerful nanoweapon against multidrug-resistant bacteria. J Appl Microbiol 112:841–852

Raja RB, Singh P (2013) Synergistic effect of silver nanoparticles with the cephalexin antibiotic against the test strains. Biores Bull 2(4):1–8

Rana S, Kalaichelvan PT (2013) Ecotoxicity of nanoparticles. ISRN Toxicol 2013:1–11. doi:10.1155/2013/574648

Read DS, Matzke M, Gweon HS, Newbold LK, Heggelund L, Ortiz MD, Lahive E, Spurgeon D, Svendsen C (2016) Soil pH effects on the interactions between dissolved zinc, non-nano- and nano-ZnO with soil bacterial communities. Environ Sci Pollut Res 23:4120–4128

Resende JA, Silva VL, Oliveira TLR, Fortunato SO, Carneiro JC, Otenio MH, Diniz CG (2014) Prevalence and persistence of potentially pathogenic and antibiotic resistant bacteria during anaerobic digestion treatment of cattle manure. Bioresour Technol 153:284–291

Riahi-Madvar A, Rezaee F, Jalili V (2012) Effects of alumina nanoparticles on morphological properties and antioxidant system of *Triticum aestivum*. Iran J Plant Physiol 3(1):595–603

Rico CM, Majumdar S, Duarte-Gardea M, Peralta-Videa JR, Gardea-Torresdey JL (2011) Interaction of nanoparticles with edible plants and their possible implications in the food chain. J Agric Food Chem 59(8):3485–3498

Rodrigues DF, Jaisi DP, Elimelech M (2013) Toxicity of functionalized single-walled carbon nanotubes on soil microbial communities: implications for nutrient cycling in soil. Environ Sci Technol 47:625–633

Rothrock MJ, Keen PL, Cook KL, Durso LM, Franklin AM, Dungan RS (2016) How should we be determining background and baseline antibiotic resistance levels in agroecosystem research? In: Antibiotics in agroecosystems: state of the science. J Environ Qual 45(2):420–431

Rousk J, Ackermann K, Curling SF, Jones DL (2012) Comparative toxicity of nanoparticulate CuO and ZnO to soil bacterial communities. PLoS One 7:e34197

Rudramurthy GR, Swamy MK, Sinniah UR, Ghasemzadeh A (2016) Nanoparticles: alternatives against drug-resistant pathogenic microbes. Molecules 21:836. doi:10.3390/molecules21070836

Sangappa M, Thiagarajan P (2015) Combating drug resistant pathogenic bacteria isolated from clinical infections, with silver oxide nanoparticles. Indian J Pham Sci 77(2):151–155

Sanyasi S, Majhi RK, Kumar S, Mishra M, Ghosh A, Suar M, Satyam PV, Mohapatra H, Goswami C, Goswami L (2016) Polysaccharide-capped silver nanoparticles inhibit biofilm formation and eliminate multi-drug-resistant bacteria by disrupting bacterial cytoskeleton with reduced cytotoxicity towards mammalian cells. Sci Rep 6:24929. doi:10.1038/srep24929

Shah V, Belozerova I (2009) Influence of metal nanoparticles on the soil microbial community and germination of lettuce seeds. Water Air Soil Pollut 197:143–148

Shah V, Collins D, Walker VK, Shah S (2014) The impact of engineered cobalt, iron, nickel and silver nanoparticles on soil bacterial diversity under field conditions. Environ Res Lett 9:024001

Shahverdi AR, Fakhimi A, Shahverdi HR, Minaian S (2007) Synthesis and effect of silver nanoparticles on the antibacterial activity of different antibiotics against *Staphylococcus aureus* and *Escherichia coli*. Nanomedicine 3:168–171

Shirmohammadi E, Saeidi S, Mohasseli T, Boogar AR (2014) Antibacterial effects of silver nanoparticles produced by *Satureja hortensis* extract on isolated *Bacillus cereus* from soil of Sistan plain. Int J Inf Secur 1(3):e21944

Simonin M, Richaume A (2015) Impact of engineered nanoparticles on the activity, abundance, and diversity of soil microbial communities: a review. Environ Sci Pollut Res 22:13710

Singh P, Kim YJ, Wang C, Mathiyalagan R, Yang DC (2016) *Weissella oryzae* DC6 facilitated green synthesis of AgNPs and their antimicrobial potential. Artif Cells Nanomed Biotechnol 44(6):1569–1575

Thomas R, Nair AP, Kr S, Methew J, Ek R (2014) Antibacterial activity and synergistic effect of biosynthesized AgNPs with antibiotics against multidrug-resistant biofilm-forming coagulase-negative staphylococci isolated from clinical samples. Appl Biochem Biotechnol 173:449–460. doi:10.1007/s12010-014-0852-z

Thul ST, Sarangi BK, Pandey RA (2013) Nanotechnology in agroecosystem: implications on plant productivity and its soil environment. Expert Opin Environ Biol 2:1

Tiwari M, Jain P, Hariharapura RC, Narayanan K, Bhat UK, Udupa N, Rao JV (2016) Biosynthesis of copper nanoparticles using copper-resistant *Bacillus cereus*, a soil isolate. Process Biochem 51:1348–1356

Tong Z, Bischoff M, Nies L, Applegate B, Turco RF (2007) Impact of fullerene (C60) on a soil microbial community. Environ Sci Technol 41:2985–2991

Tyc O, Berg M, Gerards S, vanVeen JA, Raaijmakers JM, deBoer W, Garbeva P (2014) Impact of inter specific interactions on antimicrobial activity among soil bacteria. Front Microbiol 5: 1–10

Unrine J, Bertsch P, Hunyadi S (2008) Bioavailability, trophic transfer, and toxicity of manufactured metal and metal oxide nanoparticles in terrestrial environments. In: Grassian VH (ed) Nanoscience and nanotechnology: environmental and health impacts. Wiley, Hoboken, NJ. doi:10.1002/9780470396612.ch14

Upadhyay S, Mustafa M, Joshi SR (2016) Naturally evolving extended spectrum cephalosporin resistance in soil borne isolates of *Enterobacteriaceae*. Natl Acad Sci Lett 39(3):181–184

USDA/Agricultural Research Service (2016) Antibiotic resistance can occur naturally in soil bacteria. Science Daily, September 27. Retrieved 17 Nov 2016 from www.sciencedaily.com/releases/2016/09/160927151524.htm

Walsh F, Duffy B (2013) The culturable soil antibiotic resistome: a community of multi-drug resistant bacteria. PLoS One 8(6):e65567

Wang Y, Tang H, Wu D, Liu D, Liu Y, Cao A, Wang H (2016) Enhanced bactericidal toxicity of silver nanoparticles by the antibiotic gentamicin. Environ Sci Nano 3:788–798

Warne MS, Heemsbergen D, Stevens D, McLaughlin M, Cozens G, Whatmuff M, Broos K, Barry G, Bell M, Nash D, Pritchard D, Penney N (2008) Modeling the toxicity of copper and zinc salts to wheat in 14 soils. Environ Toxicol Chem 27:786–792

Watson JL, Fang T, Dimkpa CO, Britt DW, McLean JE, Jacobson A, Anderson AJ (2015) The phytotoxicity of ZnO nanoparticles on wheat varies with soil properties. Biometals 28:101–112

Xing B, Senesi N, Vecitis CD (2016) Engineered nanoparticles and the environment. biophysicochemical processes and toxicity. Wiley, Hoboken, NJ. ISBN:978–1–119-27582-4. 1119275822

Zeliadt N (2010) Silver beware: antimicrobial nanoparticles in soil may harm plant life. https://www.scientificamerican.com/article/silver-beware-antimicrobial-nanoparticles-in-soil-may-harm-plant-life. Last accessed 30 Nov 2016

Zhang Q, Dick WA (2014) Growth of soil bacteria, on penicillin and neomycin, not previously exposed to these antibiotics. Sci Total Environ 493:445–453

Zhou B, Wang C, Zhao Q, WangY HM, Wang J, Wang S (2016) Prevalence and dissemination of antibiotic resistance genes and coselection of heavy metals in Chinese dairy farms. J Hazard Mater 320:10–17

Chapter 15
The Effects of Antibiotics on the Structure, Diversity, and Function of a Soil Microbial Community

Anna Piotrowska-Długosz

15.1 Introduction

A wide variety of antibiotics are extensively applied worldwide as drugs for preventing or treating human, animal, and plant infections or as feed additives for animal to prevent diseases as well as growth promotion (Kumar et al. 2005; Aust et al. 2008; Martinez 2009; Du and Liu 2012). As not all antibiotics are metabolized completely in the human and animal digestive system, between 30 and 90% of these compounds, either unaltered or as metabolites as well as degradation products, some of which are still bioactive, are excreted from a body together with urine and feces and subsequently spread into the environment together with different forms of urban wastes, biosolids, and manures (Sarmah et al. 2006; Fang et al. 2014). In fact, a significant amount of antibiotics and their metabolites are introduced into the soil and water through fertilization and irrigation with antibiotic-polluted manures, biosolids, sewage sludge, sediments, and wastewaters (Du and Liu 2012). The periods for the detection of various antibiotics in soils range from days to months after an application (Hamscher et al. 2002), depending on the antibiotic being used and the corresponding degradation kinetics (Müller et al. 2002; De Liguoro et al. 2003). The most slowly degrading antibiotics can accumulate in the soil due to the repeated applications of manure. As a result, the amounts of antibiotic that are found in soils often far exceed the recommended value of 100 μg kg^{-1} (EMEA 2008) and often reach values in the range of a *mg* per *kg*. Subsequently, the

The original version of this chapter was revised. An erratum to this chapter can be found at DOI 10.1007/978-3-319-66260-2_22.

A. Piotrowska-Długosz (✉)
Division of Biochemistry, Department of Soil Science and Soil Protection, Faculty of Agriculture and Biotechnology, UTP University of Science and Technology, 6 Bernardyńska St, 85-029 Bydgoszcz, Poland
e-mail: ap03@wp.pl

M.Z. Hashmi et al. (eds.), *Antibiotics and Antibiotics Resistance Genes in Soils*, Soil Biology 51, DOI 10.1007/978-3-319-66260-2_15

accumulation of antibiotics in soil makes them potentially hazardous to nontarget bacteria and other soil organisms (Baguer et al. 2000). Besides unaltered antibiotics, considerable amounts of antibiotic-resistant bacteria and resistance genes (ARGs) are also introduced into soil via the organic amendments that are used as fertilizers (Heuer et al. 2011; Zhu et al. 2013). Bacteria have been shown to readily exchange genetic information in nature, thus permitting the transfer of different resistance mechanisms already present in the environment from one bacterium to another (Salyers and Amábile-Cuevas 1997). Antibiotic resistance has received considerable attention due to the problem of the emergence and rapid expansion of antibiotic-resistant pathogenic bacteria. Therefore, it is necessary to understand the environmental impact of the antibiotics and ARGs that are associated with the application of animal manures or other organic amendments on agricultural land. Although these substances have been considered harmless, until recently, their potential as contaminants is now in focus, and there is a growing interest in their fate and long-term effects on soils and waters (Vaclavik et al. 2004). The environmental risks of antibiotics have been studied less frequently compared to pesticides and biocides.

Soil microorganisms play an important role in many ecosystem processes such as the biogeochemical cycling of nutrients, soil structural and hydrological properties, and energy flow (Schulz et al. 2013). Thus, maintenance of the biological activity in the soil is generally regarded as a key feature of sustainable production in order to ensure the functions of an ecosystem (Swift 1994) and soil microbial properties are often used as indicators of soil quality (Schloter et al. 2003; Bastida et al. 2008; Navas et al. 2011). Although antibiotics in soil can potentially affect natural microbiological systems, their effects on the structure and function on a soil microbial community are not well understood. It is believed that antibiotics disturb the complex soil microbial system, even though its environmental concentration is below the clinically relevant minimum inhibitory concentration (Jechalke et al. 2014), although other authors have indicated that primarily higher doses of antibiotics, those that exceed environmentally relevant concentrations, negatively affect soil microorganisms (Kotzerke et al. 2008). It is worth stressing that many previous studies were conducted with doses of antibiotics in the range of *mg* per *kg* soil, whereas the antibiotics that are typically found in soil treated with animal manures are in the range of *μg* per *kg* (Hamscher et al. 2005; Blackwell et al. 2007; Aust et al. 2008; Song et al. 2010).

Antibiotics may affect microorganisms by reducing their number, biochemical activity, and diversity and by changing the microbial community structure (Bansal and Srivastava 2014). In fact, antibiotics applied with manure can lead to the enrichment of particular bacterial taxa in soil while suppressing others as was recently shown for the sulfadiazine antibiotic SDZ (Ding et al. 2014). A number of studies have reported the impact of antibiotics on various microbial activity indices such as respiration (Vaclavik et al. 2004; Kotzerke et al. 2008), nitrification (Gomez et al. 1996; Toth et al. 2011), iron reduction (Thiele-Bruhn 2005; Toth et al. 2011), and enzyme activities (Liu et al. 2009; Gutiérrez et al. 2010; Chen et al. 2013; Liu et al. 2014). The effects of antibiotics on the micro-ecosystems also included changes in microbial diversity and structure (Liu et al. 2012a, b; Ding et al. 2014) and an increase in antibiotic resistance (Heuer and Smalla 2007). The induction of the possible resistance in bacterial strains in the environment is the main current concern connected with these compounds (Vaclavik et al. 2004). It was suggested that manure-derived bacteria might not be well adapted to the soil conditions, and this would lead to a decrease in the abundance of resistant bacteria that are applied with manure over time (Hammesfahr et al. 2008; Heuer et al. 2011). However, the resistance genes from manure bacteria are often located on translocative elements such as integrons, transposons, and the insertion sequence common region (ISCR) elements that can be efficiently transferred into soil bacteria by conjugative elements such as broad host-range plasmids (Heuer et al. 2011). In microcosm experiments, manure spiked with SDZ increased the abundance of *sul1*- and *sul2*-resistant genes in soil over a period of at least 2 months (Heuer and Smalla 2007), and repeated application of manure with SDZ even led to an accumulation of *sul1* and *sul2* genes in the soil bacterial community, as compared with manure without SDZ and untreated soil (Heuer et al. 2011). Additionally, the application of manure containing SDZ significantly increased the proportion of SDZ-resistant cultivable bacteria and the frequencies of SDZ-resistance-plasmid capture in *Escherichia coli* (Heuer and Smalla 2007). The application of nutrients such as manure to the oligotrophic soil environment likely contributes to the effect of bacteriostatic antibiotics such as SDZ by stimulating horizontal gene transfer, as well as bacterial growth, because these antibiotics only affect growing bacteria (Brandt et al. 2009). In field experiments, it was shown that pig manure containing SDZ increased the abundance of *sul1* and *sul2* in soil relative to 16S rRNA genes and the transferability of sulfonamide resistance compared to manure without SDZ (Kopmann et al. 2013; Jechalke et al. 2013).

15.2 Factors Influencing the Effects of Antibiotics on Soil Microbial Communities

Many factors could affect the distribution of antibiotics in soil and their impact on microbial communities (Akimenko et al. 2015). The soil texture, absorption on soil particles, degradation, leaching, and uptake by plants and microorganisms are of the most importance (Hu et al. 2010). The various concentrations and lengths of exposure as well as differences in the chemical properties of antibiotics are also major factors (Jechalke et al. 2014).

Sorption-desorption is one of the major key processes that regulate the concentration of antibiotics in soil and hence their bioavailability to microorganisms (Tolls 2001). The antibiotics may be adsorbed on soil depending on the physical and chemical characteristics of the specific antibiotic and the particular soil. Antibiotics interact with soil surfaces through a variety of processes (e.g., hydrogen bonding, van der Waals forces, hydrophobic bonding, ion exchange, etc.) (Thiele-Bruhn et al. 2004). The extent of these processes depends on the characteristics of (1) the soil solid phase (e.g., organic matter and types of clay minerals) (Thiele-Bruhn 2003),

(2) the solution phase (e.g., pH and ion composition) (Boxall et al. 2003), and (3) the antibiotic (water solubility and functional groups) (Boxall et al. 2003). Sorption is an essential process because it controls the amount of chemicals that can be mobilized into surface water and groundwater and the amount that can be degraded by a variety of chemical and biological processes. Sorption can be fast and complete, sometimes taking only a few hours, as in the case of sulfonamides (Jones et al. 2005). The extent of sorption is commonly described by the distribution coefficient Kd of a compound, which is commonly determined in sorption isotherm experiments. Distribution coefficients for many antibiotics range from 0.2 to 6000 ml g^{-1} (Tolls 2001). Antibiotics with low distribution coefficients tend to be highly mobile and bioavailable compared to antibiotics with high coefficients. Tetracyclines have high Kd values (417–1026 ml g^{-1}) and a strong combination with soil particles. Thus, they were relatively stable and not easily migrated in soils (Lunestad and Goksøyr 1990). By contrast, the Kd values of sulfonamides were 0.9–18.1 ml g^{-1}, which suggested that sulfonamides had a strong water solubility, and therefore it was easy for them to move down from the surface soil (Boxall et al. 2003). Antibiotic adsorption in soil is facilitated by soil organic matter (SOM). For example, sulfapyridine adsorption increases on manure with a greater SOM content because this manure contains many lipids and lignin dimmers (de la Torre et al. 2012). pH is one of the most important regulators of antibiotic adsorption on mineral and organic surfaces because of the electrostatic forces involved in the process (Li et al. 2010). Adsorption is greater when the soil pH is near the dissociation pKa of antibiotics. The cationic forms of antibiotics are attracted by negatively charged surfaces, e.g., tylosin A, which at a cationic form at pH 7.8 binds negatively charged organic compounds through electrostatic interaction (Thiele-Bruhn 2003). Du and Liu (2012) reported a similar pH dependency for sulfathiazole and sulfamethazine adsorption on soil.

Antibiotics may also be degraded into simpler compounds by abiotic (e.g., hydrolysis) or biotic (e.g., enzymatic degradation) processes. Degradation of antibiotics is important because once broken down they often pose less of an adverse affect on microorganisms (Gavalchin and Katz 1994). The rate of degradation of an antibiotic is mainly described by its half-life, which is defined as the amount of time it takes to reduce the concentration of the compound by one half of its original amount. For many antibiotics, half-lives can range from less than a day (e.g., penicillin) to more than a year (e.g., tetracycline) (Zuccato et al. 2001). For antibiotics with long half-lives, adverse affects on soil microorganisms may persist for long periods after soil amendments (Halling-Sørensen et al. 2005; Furtula et al. 2010). Halling-Sørensen et al. (2005) found chlortetracycline's half-life in soil to be in the range of 20–42 days. Relatively mobile sulfonamides persisted at low concentrations (15% of the applied concentration) for at least 3 months (Stoob et al. 2006), while monensin was not detectable in soil within 1 month of its application (Donoho 1984). Many antibiotics can be degraded into simpler compounds by microbial enzymes, whose mechanisms of inactivation were proposed by Wright (2005) (Table 15.1). A classic example is the hydrolytic deactivation of the β-lactam ring in penicillins and cephalosporins by the bacterial enzyme called β-lactamase. The inactivated penicilloic acid will then be ineffective in binding to

Table 15.1 Enzymatic strategies of antibiotic inactivation

Strategy	Type	Antibiotics affected
Hydrolysis		β-lactam Macrolides
Group transfer	Acyl	Aminoglycoside Chloramphenicol Type A streptogramin
	Phosphoryl	Aminoglycoside Macrolide Rifamycin Peptide
	Thiol	Fosfomycin
	Nucleotidyl	Aminoglycoside Lincosamide
	ADP-ribosyl	Rifamycin
	Glycosyl	Macrolide Rifamycin
Other	Redox	Tetracycline Rifamycin Type A streptogramin
	Lyase	Type B streptogramin

Adopted from Wright (2005)

the penicillin-binding proteins (PBPs), thereby protecting the process of cell wall synthesis (Byarugaba 2009). Using glutathione S-transferases (GSTs), which catalyze the conjugation of reduced glutathione with a variety of hydrophobic chemicals containing electrophilic centers, tetracycline, sulfathiazole, and ampicillin were transformed into components that were nontoxic to microorganisms (Park and Choung 2007). The initial concentrations of tetracycline, sulfathiazole, and ampicillin were 100 mg L^{-1}, 100 mg L^{-1}, and 50 mg L^{-1}, respectively. They were 60–70% transformed by GSTs at the end of the degradation reaction. This lowered their inhibitory strength against microorganisms.

Antibiotic efficacy is also dependent on its concentration and on the amount of time that microorganisms are exposed to these compounds. Antibiotic bioavailability is directly dependent on its affinity with the soil and as was seen above, on soil texture and pH. In some cases, concentrations higher than those that are commonly found in soils can not only exert a bacteriostatic action but can even kill microorganisms that are sensitive to antibiotics. By contrast, lower concentrations can have only inhibitory effects or have no affect microorganisms at all (Ding and He 2010).

Some authors assumed a reduction in the effects of antibiotics on soil microorganisms over time, which was due to the degradation and binding of the chemical compounds, on the one hand, and the adaptation of resistant populations, on the other hand (Kotzerke et al. 2008). Liu et al. (2014) found that microbial properties were more affected by the incubation time than they were by the DOM and/or CTC treatments. This finding was consistent with several previously reported studies (Bundy et al. 2004; Bohme et al. 2005), but disagreed with the results of Hammesfahr et al. (2008), who indicated that the factor of incubation time was less important than treatment. After 1–4 days of incubation, the recovery of SDZ from soil that had been treated with 10 mg SDZ kg^{-1} ranged from 1.2 to 3.4 mg SDZ kg^{-1}, whereas after 32 days of incubation, almost no SDZ was detectable (Kotzerke et al. 2008). In the study of Hund-Rinke et al. (2004), at the beginning of the experiment, about 65–96% of the applied concentrations of TC were determined, decreasing to 16–28% at the end of the study (after 16 weeks). Thiele-Bruhn and Beck (2005) noticed that there was a reduction in substrate-induced respiration (SIR) in sandy soil after a 24 h exposure to sulfapyridine and oxytetracycline. Incubation for another 24 h resulted in an SIR reduction in the loamy soil. The authors attributed the reduction in SIR in the sandy soil to a decrease in the bioavailable antibiotic fraction and the microbial community's adaptation to soil spiked with antibiotics and the development of resistance.

15.3 Antibiotics as Agents that Affect the Structure and Diversity of a Soil Microbial Community

Some earlier studies indicated the significant influence of antibiotics on soil microorganisms. Thus, sulfonamides have commonly been reported to affect or inhibit soil microorganisms (Ma et al. 2014; Wang et al. 2016). Westergaard et al. (2001) found that tylosin, which was amended to agricultural soils at a rate of 3000 mg kg^{-1}, influenced the abundance of bacteria, fungi, and protozoa. Colinas et al. (1994) found that the antibiotics oxytetracycline and penicillin at concentrations of 10 mg kg^{-1} in forest soil decreased the total and active microbial cell counts by approximately 80%. In turn, the total bacterial and fungal biomass of soils receiving low and moderate oxytetracycline (OTC) inputs of less than 15 mg kg^{-1} were enhanced, while those in soils that had an extremely high OTC dose of 200 mg kg^{-1} were significantly depressed (Chen et al. 2013). The authors suggested that the manure in which the OTC was incorporated diminished the toxic effects of the antibiotics. Additionally, the aboveground vegetative covers enhanced the total microbial biomass in the soil receiving low and moderate OTC doses. The study of Bansal and Srivastava (2014) indicated that the total bacterial population was affected by three antibiotics—tetracycline (TC), chlortetracycline (CTC), and oxytetracycline (OTC), which were used at three different concentrations (50, 100, and 200 mg kg^{-1} soil). The bacterial population decreased significantly with an increase of antibiotics concentration in all of the studied soil depths (0–15, 15–30, 30–45, and 45–60 cm) along with an increase in the length of application. An increase of a bacterial population after antibiotics application over time might be due to dissipation of antibiotics and/or due to the development of antibiotic-resistant bacteria (Byarugaba 2009). These phenomena could be expected as a result of the utilization of antibiotics that are used as the source of carbon and other nutrient elements by soil bacteria (Zhang and Dick 2014). Different microbial properties and some laboratory techniques have been used to assess the influence of antibiotics and resistant genes on soil microbial communities. The most often studied soil microbial properties and the appropriate techniques are presented in Table 15.2.

Recently, the overall microbial community structure in soil under antibiotic pressure has often been studied by monitoring changes in the phospholipid fatty acid patterns or PCR-denaturing gradient gel electrophoresis (DGGE) of 16S rDNA (e.g., Thiele-Bruhn and Beck 2005; Zielezny et al. 2006; Gutiérrez et al. 2010). The last molecular method, which is based on recovery of community DNA from soil, offers a great potential for investigating the nonculturable part of complex microbial communities. DGGE has previously detected an altered genetic structure of a bacterial community due to environmental disturbances such as heavy metals or pesticide applications (Westergaard et al. 2001).

In the study of Gutiérrez et al. (2010), the effect of three commonly used and simultaneously applied sulfonamide antibiotics on the profiles of phospholipid fatty acids (PLFAs) was determined. Soil samples were applied with either mineral water only (W-treatments), liquid manure (M-treatments), or with glucose only (G-treatments), and with each of these were treated with a cocktail of three sulfonamides—sulfadimethoxine (SDT), sulfamethoxazole (SMX), and sulfamethazine (SMZ) at five total concentration levels ranging from 0 (control) to 900 mg kg_{dm}^{-1} (G-0–G-900) (Table 15.3). The lower $PLFA_{tot}$ concentrations in the treatments G-90 and G-900 compared to G-0 were consistent with the findings of Thiele-Bruhn and Beck (2005). Under similar conditions (glucose addition, incubation time of 14 days), they reported a reduction in the microbial biomass at 1000 mg kg^{-1} of sulfapyridine by 55% compared to the control, while a sulfapyridine concentration of 100 mg kg^{-1} decreased the microbial biomass by only 10%. The addition of sulfonamides caused a relative shift in the bacterial community toward gram-negative bacteria and increased the proportion of the fungal biomass compared to the bacterial biomass. This shift of the microbial community structure toward fungi is in line with the results presented by Thiele-Bruhn and Beck (2005), who amended a sandy soil with maize straw, glucose, and sulfapyridine. For a concentration of 1000 mg kg^{-1}, they reported an increased concentration of fungal ergosterol, while the total microbial biomass decreased. A smaller increase of the gram-positive compared to the gram-negative indicator PLFAs was also found in the test units containing manure and additionally 500 mg kg^{-1} TC per kg dry mass after 8 weeks of the experiment (Hund-Rinke et al. 2004).

None of the most commonly used antibiotics in poultry feed (bacitracin, roxarsone, virginiamycin) applied in the environmentally relevant doses (up to 1 mg kg^{-1}) significantly influenced either the concentration of the total FAMEs or the content of any specific FAMEs (59 types) compared to the control soils (Banerjee and D'Angelo 2013). Soil microbial community composition was only affected when antibiotics (expect of bacitracin) were used in a dose of 100 mg kg^{-1} with 1.2–1.6-fold increase in the relative amount of 16:1ω7 and 18:1ω7, which are biomarkers of aerobic gram-negative bacteria. This could be explained by the fact that this group of bacteria has gained competitive advantages over other microbial groups in the soil, whose growth was inhibited by antibiotics.

Table 15.2 Most often studied soil microbial properties under the influence of antibiotics and the techniques used

Antibiotic group	Antibiotic	Microbial property affected	Technique used	References
Macrolides	Tylosin (TYL)	1. Bacterial, protozoan, and fungal populations 2. Community structure 3. Community-level physiological profiling	1. Plate count method, MPN, chitinase activity 2. PCR-DGGE of 16S rDNA 3. Biolog Ecoplates®	Westergaard et al. (2001)
	Tylosin (TYL)	1. Soil microbial respiration (CO_2 evolution) 2. Phosphatase activity	1. Titration with 0.15 M NaOH 2. Spectrophotometry method (quantification of NP)	Liu et al. (2009)
Fluoroquinolones	Difloxacin (DIF)	1. Phospholipid-derived fatty acids (PLFAs) 2. Community structure	1. Gas chromatography (GC-MS) 2. PCR-DGGE of 16S rDNA	Reichel et al. (2013)
	Norfloxacin (NOR)	1. Soil-induced respiration	1. MicroResp™ (colorimetric method based on the color change of a pH indicator dye caused by the release of CO_2 by heterotrophic communities)	Wang et al. (2016)
β-lactam	Benzylpenicillin	1. Catalase activity 2. Population density of microorganisms	1. Gasometrical method 2. Method of deep inoculation from corresponding dilution onto solid nutrient media	Akimenko et al. (2015)
Tetracyclines	Chlortetracycline (CTC) Tetracycline (TC)	1. Soil microbial respiration (CO_2 evolution) 2. Phosphatase activity	1. Titration with 0.15 M NaOH 2. Spectrophotometry method	Liu et al. (2009)
	Chlortetracycline (CTC)	1. Community-level physiological profiling 2. Soil enzyme activities (dehydrogenases, acid and alkaline phosphatase, urease)	1.Biolog EcoPlates® (BIOLOG Inc. Hayward, CA, USA) 2. Spectrophotometry method (quantification of TPF, NP, and NH_4)	Liu et al. (2014)
	Chlortetracycline (CTC)	1. Bacterial community tolerance by measuring the metabolic activity 2. Soil enzyme activities (dehydrogenases, urease)	1. Biolog Eco microplate™ 2. Spectrophotometry method (quantification of TPF and NH_4)	Fang et al. (2014)

	Tetracycline (TC)	1. Phospholipid-derived fatty acids (PLFAs) 2. Soil microbial respiration 3. Enzymatic activity (dehydrogenases, phosphatase, urease) 4. Catalase activity 5. Bacteria population	1. Gas chromatography (GC-MS) 2. ISO/DIS 17155 (2001) using a "Sapromat" (Voith, Heidenheim) 3. Spectrophotometry method (quantification of TPF, NP, and NH_4) 4. Back-titratying residual H_2O_2 with $KMnO_4$ 5. Dilution plate technique	Hund-Rinke et al. (2004) Bansal (2015) Bansal and Srivastava (2014)
	Oxytetracycline (OTC)	1. Community-level physiological profiling	1. Biolog EcoPlates® (BIOLOG Inc. Hayward, CA, USA)	Liu et al. (2012a, b) Kong et al. (2006)
	Doxycycline (DOX)	1. Soil-induced respiration	1. MicroResp™ system	Wang et al. (2016)
	Oxytetracycline (OTC)	1. Phosphatase activity 2. Dehydrogenase activity 3. Substrate-induced soil microbial respiration 4. Alkaline phosphatase and arylsulfatase activities	1. Spectrofluorometrically at $\lambda ex = 360$ nm and $\lambda em = 465$ nm 2. Colorimetric quantification of TPF 3. Using BacTrac 4300 (SY-Lab, GmbH, Purkersdorf, Austria) 4. Colorimetric quantification of NP	Boleas et al. (2005) Chen et al. (2013)
Sulfonamides	Sulfamethazine (SMZ)	1. Soil microbial respiration (CO_2 evolution) 2. Enzyme activities (phosphatase, dehydrogenases, urease) 3. Biolog community-level physiological profile (CLPP)	1. Titration with 0.15 M NaOH 2. Spectrophotometry method (quantification of NP, TPF, and NH_4) 3. Biolog EcoPlates® (BIOLOG Inc. Hayward, CA, USA)	Liu et al. (2009) Pinna et al. (2012)
	Sulfadiazine (SDZ)	1. Phospholipid fatty acids (PLFAs) 2. Community structure	1. Gas chromatography (GC-MS) 2. PCR-DGGE of 16S rDNA	Reichel et al. (2013) Hammesfahr et al. (2008)

(continued)

Table 15.2 (continued)

Antibiotic group	Antibiotic	Microbial property affected	Technique used	References
	Sulfadiazine (SDZ)	1. Bacterial community tolerance by measuring the metabolic activity 2. Dehydrogenase and urease activities	1. Biolog Eco microplate™ 2. Colorimetric quantification of TPF and NH_4	Fang et al. (2014)
	Cocktail of sulfadimethoxine (SDT), sulfamethoxazole (SMX), sulfamethazine (SMZ)	1. Dehydrogenase and urease activities 2. Phospholipid fatty acids (PLFAs)	1. Colorimetric quantification of TPF and NH_4 2. Gas chromatography (GC-MS)	Gutiérrez et al. (2010)

DGGE denaturing gradient gel electrophoresis, *MPN* most probable number estimation model, *TPF* trifenyloformazan, *TTC* 2,3,5-triphenyltetrazolium chloride, *NP* nitrophenol

Table 15.3 PLFA concentrations (nmol g_{dm}^{-1}; indices—g+, gram-positive; g−, gram-negative; bact, sum of gram-positive and gram-negative bacteria; fungi, fungal markers; tot, sum of all analyzed PLFA) of selected microbial groups in different treatments at $t = 168$ and $t = 504$ h

	$PLFA_{g+}$		$PLFA_{g-}$		$PLFA_{bact}$		$PLFA_{fungi}$		$PLFA_{tot}$	
	168 h	504 h	168 h	504 h	168 h	504 h	168 h	504 h	168 h	504 h
G 0	246	n/a	281	n/a	527	n/a	36	n/a	856	n/a
G 90	152	269	265	420	417	689	28	42	773	1259
G 900	99	114	177	173	276	287	34	24	524	529
M 0	244	244	365	350	609	594	28	32	1117	1144
W 0	270	n/a	278	n/a	548	n/a	18	n/a	904	n/a

n/a, not analyzed; G 0, G 90, G 900, sulfonamide concentration (μg g_{dm}^{-1}); M 0, manure treatment only; W 0, water treatment only
Adopted from Gutiérrez et al. (2010)

No effect of sulfadiazine (SDZ) and chlortetracycline (CTC) at three different doses (1, 10, 50 mg kg^{-1} soil) on the bacterial community structure was found by Zielezny et al. (2006) when changes were visualized using PCR denaturing gradient gel electrophoresis (DGGE) of 16S rDNA that had been derived from soil samples after 1, 7, 11, and 48 days. In the presence of glucose (5 g kg^{-1}), SDZ affected the bacterial community structure, and a clear relationship between SDZ concentrations and changes in DGGE patterns became visible. Additional bands appeared, and some bands that were already visible at the beginning of incubations increased in intensity. In the study of Westergaard et al. (2001), a different number of bands compared to the control were found in the soil treated with tylosin. At day 15 in the tylosin-treated soil, the number of bands on the DGGE profiles decreased, and two bands had a high intensity, thus indicating that a few types of bacteria were responsible for growth. The results of DGGE profiles showed a shift in the bacterial population from cells that were hard to lyse in the control soil to more easily lysable cells in the tylosin-treated soil. Tylosin acts mainly on gram-positive bacteria, which are generally more resistant to lyse than gram-negative bacteria (McGuire et al. 1961). Müller et al. (2002), who investigated the effect of antibiotic tylosin on a soil bacterial community using denaturing gradient gel electrophoresis (DGGE) analysis, found a small difference in the diversity in the 16S rDNA of the samples compared to the control soil.

15.4 Influence of Antibiotics on Microbial Growth and Activity

Antibiotics not only modify the structure of the soil microbial community but also its function. The microbial community function influenced by antibiotic treatment was characterized by investigating some parameters such as the community-level physiological profile (CLPP), soil respiration, microbial nitrogen turnover, nitrification, iron reduction, and the activities of several soil enzymes (e.g., Hund-Rinke et al. 2004;

Thiele-Bruhn and Beck 2005; Boleas et al. 2005; Zielezny et al. 2006; Pinna et al. 2010; Toth et al. 2011; Fang et al. 2014; Bansal 2015; Cao et al. 2015).

15.4.1 Soil Respiration

Respiratory activity in the soil was measured in order to study the effect of sulfadiazine (SDZ) and chlortetracycline (CTC) at three different concentrations (1, 10, 50 mg kg^{-1} soil) on soil microbial activity (Zielezny et al. 2006). According to the authors, neither SDZ nor CTC significantly influenced respiration during 20 days of incubation. CTC probably had no effect on respiration because of its strong inactivation in the presence of the orthic luvisol soil that the authors used in the experiment. Similarly, tetracycline (TC) had no influence on the soil respiratory activity at concentrations up to 50 mg kg^{-1} soil in the study of Hund-Rinke et al. (2004). What is more, soil respiration was not affected regardless of the type of antibiotics or their environmentally relevant concentrations (up to 200 μg kg^{-1} for sulfadimethoxine, no more than 30 μg kg^{-1} for chlortetracycline and 100 μg kg^{-1} for monensin) during 50 days of the experiment that was carried out by Toth et al. (2011). The lack of the effects of antibiotic treatment on soil respiration could be explained by the fact that this is a universal process that is carried out by all of the types of organisms that inhabit soil. The data obtained by the above-mentioned authors are in contrast with the data presented by Vaclavik et al. (2004), who found a 1.3–1.7 times increase in respiration compared to the background respiration with different tetracyclines (including CTC), sulfonamides, and a sulfachloropyridazines, at initial concentrations of 60 and 600 mg kg^{-1} soil. Because of the absence of a lag phase and the fact that the antibiotics themselves could be quantitatively re-extracted at the end of the incubation, the authors concluded that the antibiotics did not serve as substrates but that the reason for the increased respiration remained unclear.

Visible inhibiting effects of antibiotics on soil respiration have been noted in some studies, whereat the doses of antibiotics were usually high. Boleas et al. (2005) observed inhibitions of soil respiration in the range of 16–25% and 28–38% in soils spiked with 100 and 1000 mg OTC kg^{-1} soil (Fig. 15.1). Earlier, Fründ et al. (2000) found an inhibiting effect on the microbial respiration activity at 133 mg kg^{-1} of tetracycline. In the study of Hund-Rinke et al. (2004), the inhibiting effect of TC on substrate induced respiration was only detected at the highest concentration (500 mg kg^{-1}) in the presence of pig manure rich in TC resistance genes. The low effects of TC on soil microbial activity can be explained by the strong sorption of the substance to soil resulting in a low bioavailability. According to Tolls (2001), TC displays very high Kd values (1100–2000 ml g^{-1}). For OTC distribution, quotients of 417 ml g^{-1} in sandy soil and 1026 ml g^{-1} in a sandy loam were determined, and the antibiotic could not be detected in the leachate of the soil column (Rabølle and Spliid 2000).

Similarly, in the study of Liu et al. (2009) little effects of tetracyclines and tylosin on soil microbial respiration were noted, with statistically significant variations observed only at the higher concentration levels. As was mentioned by the

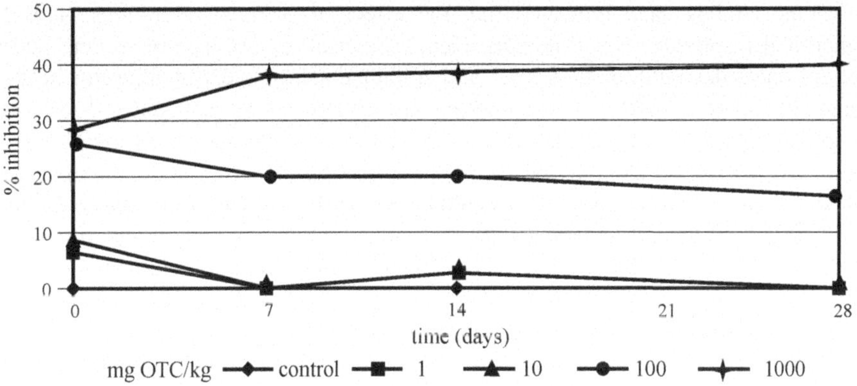

Fig. 15.1 Effects of oxytetracycline on the microbial respiration (measured as CO_2 production) of the soil used for setting the system (reproduced from Boleas et al. 2005)

authors, the sorption and degradation processes played certain roles in reducing the effects of these antibiotics. These three compounds exhibited strong adsorption into soil and were therefore less bioavailable (Sarmah et al. 2006). Earlier studies found that tylosin was not persistent in soil, and its DT50 (dissipation half-lives) was no more than 1 week (Hu and Coats 2007). In the study of Liu et al. (2009), tylosin had a DT50 of 8 days in the soil and that is why it does not accumulate in soil and poses very little risk to the soil microbial respiration process (Blackwell et al., 2007). Tetracyclines had DT50 values of more than 20 days in the soil used in the same study. Additionally, as was stated by Kemper (2008), Pils and Laird (2007), and Zielezny et al. (2006), tetracyclines show strong adsorption and can form complexes with cations such as the calcium in soil, which could significantly reduce the bioavailability and effects of tetracyclines on soil microbial respiration. In the recent study of Wang et al. (2016), doxycycline (DOX) application showed a generally positive effect on soil-induced respiration that was investigated using a MicroResp™ system.

Since no effects could be observed on bacterial population activity in the samples with soil and antibiotics, further incubations were conducted in the presence of an additional assimilable carbon source (5 g glucose kg^{-1} soil) (Zielezny et al. 2006). With substrate-induced respiration (SIR) concentrations of SDZ and CTC up to 50 mg kg^{-1}, soil showed no differences in respiration rates during the first 6–8 h compared to the addition of glucose alone. This was expected because the antibiotic activity of SDZ and CTC is based on the inhibition of folic acid and protein synthesis, respectively, and thus they should mainly have a growth inhibitory effect. For that reason, in order to determine any antimicrobial effects with SIR, the incubation time should be long enough to enable microorganisms to grow. That is why, Thiele-Bruhn and Beck (2005) extended the incubation time in their experiments to 48 h, and the respiratory activity in the experiments of Zielezny et al. (2006) showed a clear delay of 36 h at the beginning of exponentially increasing oxygen consumption.

Some studies have indicated that the effect of selected antibiotics on soil microbial respiration was time dependent (Liu et al. 2009). The increasing rates of sulfonamides (sulfamethoxazole and sulfamethazine) and trimethoprim in the study of Liu et al. (2009) were found to cause significant decreases in soil respiration within the first 4 days, while the values of this parameter increased significantly after this period. As was explained by Thiele-Bruhn and Beck (2005), the increase in soil respiration after a few days could be due to the decrease in the bioavailable antibiotic fraction and an increasing adaptation and resistance of the soil microorganisms. The DT50 values for three compounds (sulfamethoxazole, sulfamethazine, and trimethoprim) ranged from 2 to 5 days. That is why the recovery of soil respiration after the first 4 days was partially due to the significant loss of these antibiotics in the soil. Fang et al. (2014) studied the effects of antibiotics (SDZ and CTC) on soil respiration with five different treatments (control, manure, manure + SDZ, manure + CTC, and manure + SDZ + CTC) (Fig. 15.2). Soil respiration was measured 5 times, at 60-day intervals. Soil

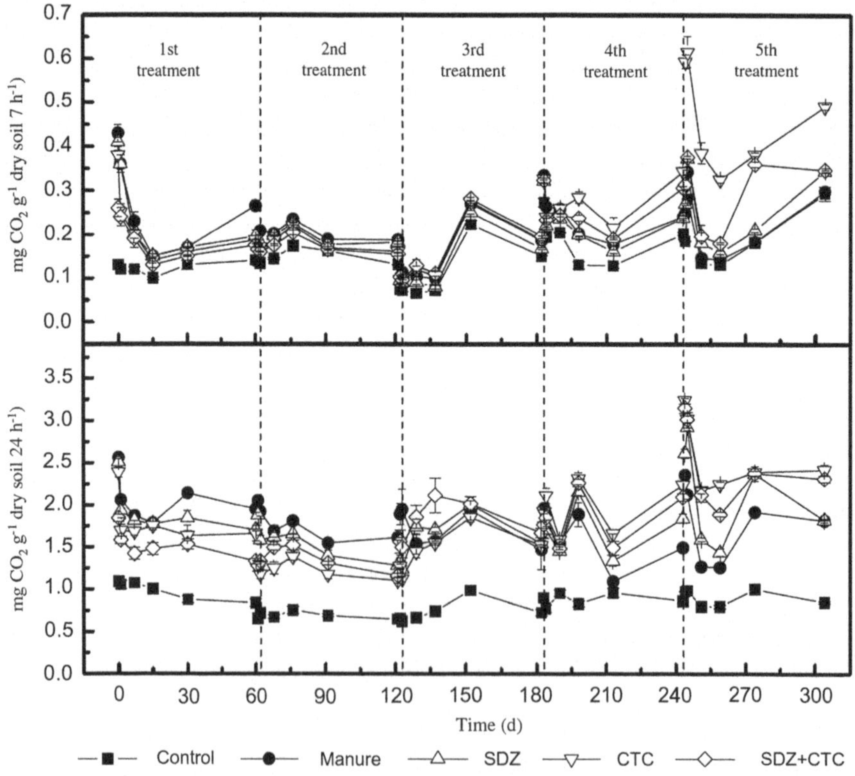

Fig. 15.2 Effects of SDZ and CTC, alone and in combination, together with manure on soil respiration. Respiration was measured as the cumulative CO_2 generated during 0–7 h and 7–24 h within different incubation periods during five repeated treatments. The first, second, third, fourth, and fifth treatments are designated by first, second, third, fourth, and fifth, respectively. The *error bars* represent the standard deviation of three replicates (reproduced from Fang et al. 2014)

respiration activity with increasing SDZ or CTC treatment was inhibited during the initial two treatments, and then it gradually recovered or reached the levels that were observed in an individual manure treatment. This trend may be due to the initial inhibitory effect of antibiotics on microbial populations followed by the formation and substantive proliferation of tolerant or resistant microbial populations, thus leading to a shift in the microbial community structure. An initial (up to 4 days) reduction in the CO_2 flux and its subsequent recovery were also found by Kotzerke et al. (2008) after 32 days in two arable soils after the application of manures containing 10 or 100 mg kg^{-1} SDZ.

15.4.2 Community-Level Physiological Profile

Community-level physiological profile (CLPP) assessed using BIOLOG Microplate have been widely used to investigate the functional diversity of soil microbial communities (Garland 1997; Mäder et al. 2002; Li et al. 2004). Despite a number of limitations, e.g., reflecting only a part of microbial community in environmental samples because of the focus on bacterial species that are able to respond rapidly to the substrates, domination of fast-growing species in substrate utilization, and changes in community composition during growth (Smalla et al. 1998; Preston-Mafham et al. 2002), the method can provide insights into the effect of a disturbance on microorganisms and was widely used to investigate the functional diversity of soil microbial communities in different environments (Gomez et al. 2006; Andersen et al. 2010; Zhang et al. 2010).

A 2-month-long study was undertaken to determine the effect of the antibiotic tylosin on the soil microbial community structure and function (Westergaard et al. 2001). The investigation of the community substrate utilization pattern that was obtained using Ecoplates® showed no differences in the number of substrates that were utilized between tylosin-treated and untreated soil. Thirty-one Ecoplates® substrates were selected according to the separation power of the environmental samples (Insam 1997). When using 23 of the substrates, a better differentiation of the diversity was found compared to the 95 substrates from Biolog GN plates (Derry et al. 1998), but the resolution of the microbial communities appeared to be reduced. In the study of Westergaard et al. (2001), the Ecoplates® were able to detect differences in the community structure but not in diversity. The parameters of CLPP were not inhibited by sulfadimethoxine, monensin, and chlortetracycline when spiked into manure and mixed with soil at environmentally relevant doses (up to 200 μg kg^{-1} soil) in a laboratory study (Toth et al. 2011).

In the study of Liu et al. (2012a), the influence of two antibiotics (sulfamethoxazole (SMX) and chlortetracycline (CTC)) at doses of 0, 1, 10, 40, and 100 mg kg^{-1} (additionally 300 mg CTC kg^{-1}) on CLPP in soil samples were analyzed at 7 and 21 days of incubation. When these broad-spectrum antibiotics reach the environment, the antibiotic parent compounds may exert an antibacterial potency toward bacteria (Halling-Sørensen et al. 2002; Hammesfahr et al. 2010). At day 7, SMX addition showed an obvious inhibition effect on AWCD as compared to the control. In particular, the highest SMX concentration of 100 mg kg^{-1} had the lowest utilization rate of the carbon substances and obtained a significant inhibition effect ($p < 0.05$) on AWCD (average well color development) compared to the other treatments (Liu et al. 2012a). On the other hand, the treatments with higher SMX concentrations increased AWCD at day 21. The CTC had less of an effect on soil microbial community function during the entire incubation period. At day 7, substrate utilization was similar between the control and CTC treatments. Even the dose of 300 mg CTC kg^{-1} had no remarkable effect on microbial activity at day 7, and it significantly increased AWCD at day 21. The improvement of soil microbial community function that was caused by antibiotic addition in the late incubation period of the study compared with the control (Liu et al. 2012a) could indicate that antibiotics may act as nutrients for soil microbial growth (Schmitt et al. 2005; Thiele-Bruhn and Beck 2005; Westergaard et al. 2001). Similar results of the effects of tetracycline on a soil microbial community were also found earlier by Hund-Rinke et al. (2004) and Zielezny et al. (2006). The differences in the influence of SMX and CTC were probably caused by their different adsorption tendency to soil and CTC showed a strong adsorption to soil organic matter and clay minerals (Kemper 2008; Zielezny et al. 2006). The strong adsorption of tetracyclines with soil resulted in low bioavailability for soil bacteria and mitigated the antibacterial effects (Sarmah et al. 2006; Schmitt et al. 2005; Thiele-Bruhn 2003).

The findings of Liu et al. (2012b) suggested that the addition of OTC stimulates soil microbial activities at a certain content. Their results showed that along with an increase of OTC in a 7-week greenhouse pot experiment, AWCD values increased with a peak at 200 mg kg^{-1} OTC and the utilization of sugar and its derivatives were enhanced. The higher AWCD of the soil microbial composition, which indicates high microbial activity, could be partly due to the stress effect of OTC on soil microorganisms, or due to higher energy demands for their survival and finally for changes in the microbial community composition (Liu et al. 2012b).

15.4.3 Soil Enzymatic Activity and Biochemical Processes

Enzymatic Activity

Enzymatic activity could reflect the activities of the entire soil microbial community, since microorganisms are the main source of enzymes in soil (Nannipieri et al. 2002). The impact of antibiotics on soil microorganisms can modify enzyme activities and biochemical processes. The effects of antibiotics on soil enzymes are different and depend on many factors, such as antibiotic properties, dose and length of influence, as well as soil properties (e.g., Thiele-Bruhn and Beck 2005; Pinna et al. 2010; Fang et al. 2014; Bansal 2015; Cao et al. 2015). Soil enzymatic activity is often reduced by an antibiotic that is applied at higher doses, while lower concentrations do not inhibit and even stimulate the enzyme activities, especially when combined together with organic amendments (Boleas et al. 2005; Thiele-Bruhn and Beck 2005; Liu et al. 2009, 2014;

Gutiérrez et al. 2010; Cao et al. 2015). Pinna et al. (2012) showed more than a 50% decrease of dehydrogenases and urease activities in soils after a 7-day-long incubation with SMZ compared to the same soils that had additionally been treated with farmyard manure (Fig. 15.3). The addition of manure to soils suppressed the antibiotic effect, especially after 1 week of incubation, and no inhibition effect of the manure containing an antibiotic was observed on urease activity as was reported by Gutiérrez et al. (2010). Similarly in the study of Chen et al. (2013), small and moderate inputs of OTC stimulated the enzyme activities (15 mg kg^{-1} in the case of urease and 1 mg kg^{-1} OTC for arylsulfatase), whereas an increasing concentration of OTC antibiotics (up to 200 mg kg^{-1}) lowered all of the tested enzyme activities (dehydrogenase, urease, alkaline phosphatase, arylsulfatase). In another study, OTC was able to reduce alkaline phosphatase activity by about 41–80% but did not affect acidic phosphatase, urease, and dehydrogenase activities (Ding and He 2010). Oxytetracyclines are antibiotics that have a broad spectrum of antimicrobial properties including blocking the transfer of RNA and preventing protein synthesis (Chopra and Roberts 2001). Similarly, Wei et al. (2009) showed that the presence of TC significantly disturbed the structure of microbial communities and inhibited soil microbial activities of urease, acid phosphatase, and dehydrogenase. The combined effects of chlortetracycline (CTC) and dissolved organic matter (DOM) that had been extracted from pig manure on the functional diversity of a

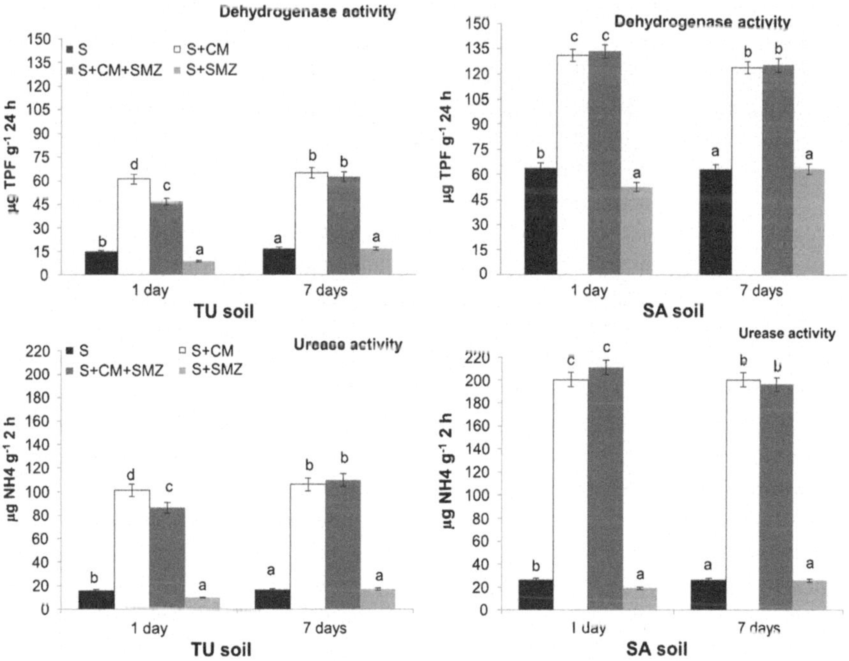

Fig. 15.3 Dehydrogenase and urease activities in TU and SA soils. For each enzyme activity, time point, and soil type, mean values ± standard deviations (*error bars*) sharing the same letter do not differ significantly at the 5% level ($p < 0.05$) according to the Fisher's LSD test (reproduced from Pinna et al. 2012)

soil microbial community determined by enzymatic activity were determined as well (Liu et al. 2014). DOM was applied to soils in combination with chlortetracycline (CTC) at three levels (0, 10, 100 mg kg^{-1} soil), and all of the treatments were incubated for 1, 6, 12, and 45 days. DOM and CTC exhibited opposite influences on the activities of dehydrogenases and urease; the DOM-treated soil showed an enhanced enzyme activity, which was decreased when spiked with CTC. This effect indicated that the enzyme activities were inhibited by the antibiotic disturbance as Boleas et al. (2005) and Liu et al. (2009) reported earlier. The study of Liu et al. (2009) suggested that the addition of antibiotics, including CTC, TC, TYL, SMX, SMZ, and trimethoprim at the concentrations that were used (1–300 mg kg^{-1}), can significantly inhibit soil phosphatase activity during a 22-day-long incubation ($p < 0.05$), although the inhibition rates were quite variable during the incubation period, which could have been caused by the heterogeneous nature of the soil. The EC10 values that were calculated for the six antibiotics ranged from 1 mg kg^{-1} for sulfamethazine to 406 mg kg^{-1} for tetracycline. Boleas et al. (2005) also observed significant effects of OTC on the soil microbial enzymatic activities (phosphatase and dehydrogenase) in a multi-species soil system (MS·3). Oxytetracycline concentrations of 0.01, 1, and 100 mg kg^{-1} soil were added to a 20 cm top arable soil layer with and without the addition of manure. At the beginning, a dose-response curve with an EC50 value of around 100 mg OTC kg^{-1} soil was observed for the dehydrogenase activity, while a non-dose-related inhibition was observed for phosphatase. In the system without manure, an initial induction of dehydrogenase activity at the low and medium concentrations was observed on day 7 and a clear inhibition at the highest concentration on day 21. Phosphatase activity showed a potentially dose-related reduction trend, but no significant differences were observed at any of the tested concentrations.
Similarly, in the study of Bansal (2015), both time- and dose-related relationships between soil enzyme activities (dehydrogenases, acid and alkaline phosphatases, urease, and catalase) and the tetracyclines (TC, OTC, and CTC) were determined. The activity of all of the studied enzymes was significantly inhibited for up to 14–21 days and the inhibition increased with an increase in tetracycline content. After 2–3 weeks, the inhibition got weaker or most of the activities almost reached the initial values.

In turn, Thiele-Bruhn and Beck (2005) found no effects on dehydrogenase activity even at a concentration of 1000 mg kg^{-1} of sulfapyridine and oxytetracycline, which may be related to the biomass and dormant state of most of the soil microorganisms and that these antibiotics can exert a temporary selective pressure on soil microbes. No clear trends in dehydrogenases activity as affected by three sulfonamides (sulfadimethoxine (SDT), sulfamethoxazole (SMX), and sulfamethazine (SMZ)) were observed in the study of Gutiérrez et al. (2010). Generally, dehydrogenases were substantially reduced when sulfonamides were present, but it appears that dehydrogenases inhibition was highest at the lowest sulfonamide level and decreased with an increasing concentration of sulfonamides. The inconsistent results on dehydrogenase activity could have been caused by various factors, and they may not be specific for the antibiotic effects in soil.

Antibiotic persistence following five successive treatments (at 60-day intervals) of sulfadiazine (SDZ) and chlortetracycline (CTC) (control, manure, manure + SDZ, manure + CTC, and manure + SDZ + CTC) in soil under laboratory conditions was studied by Fang et al. (2014). The activities of soil urease and dehydrogenases were initially unaffected, but after the fifth treatment, these activities were significantly stimulated in the CTC individual treatments and combined treatments compared to their activities in the individual manure treatment. The behavior of the enzymatic activity in this study was probably due to the proliferation of tolerant or resistant bacteria and nitrifying bacteria.

Cao et al. (2015) observed that soil enzyme activities had a significantly negative correlation with the residual OTC content. In their study, OTC contained one amide ($CONH_2$) group, and there was a significantly negative correlation between the urease activity and the residual OTC concentration. The authors suggested that urease breaks the C-N bonds in the amide group of OTC, which could enhance the degradation of OTC. It was found earlier that higher urease activity is required for enhancing antibiotic biodegradation (Chen et al. 2014). Similarly, dehydrogenase activity also had a significantly negative correlation with the residual OTC concentration, which indicates that dehydrogenase activity can also enhance the degradation of OTC. The reason for this may be that dehydrogenase catalyzes the dehydrogenation and biochemical transformation of OTC (Xie et al. 2012).

N-Transformation Processes

Nitrification is a process that is carried out primarily by nitrifying bacteria (Gram-negative) *Nitrosomonas* and *Nitrobacter*. The autotrophic-nitrifying bacteria oxidize ammonia into nitrite and then into nitrate in two steps (Maliszewska-Kordybach et al. 2007). Nitrification is an important process because it converts ammonium to nitrite and nitrate, which are the most bioavailable forms of N for plants and denitrifiers in soils. Generally, it has been found that broad-spectrum antibiotics such as tetracyclines, aminoglycosides, and sulfonamides are expected to inhibit the nitrification process (Halling-Sørensen 2001), while the narrow-spectrum antibiotics such as sulfadiazine, oxolinic acids, olaquindox, and tylosin stimulate the nitrification process (Kumar et al. 2005). The differences between broad- and narrow-spectrum antibiotics may be partially due to (1) selective pressure on the bacteria that do not participate in the nitrification process and (2) stimulation of the bacterial species that are responsible for nitrification.

In the study of Cao et al. (2015), it was implied that the application of OTC decreased the rate of nitrification due to the higher concentration of NH_4^+-N that was found in the soil. Sulfadimethoxine (0, 25, 50, 100, and 200 μg kg^{-1}) inhibited nitrification in an apparent dose-related pattern (Toth et al. 2011) (Fig. 15.4). The effects were statistically significant on days 15, 30, and 50 and at 30 and 200 μg kg^{-1}, respectively, which can have important implications concerning the biogeochemical cycling of nitrogen. Earlier, the nitrate-N concentration was also significantly reduced after the application of streptomycin, thus indicating that nitrifying bacteria are especially susceptible to streptomycin (Ingham and Coleman 1984). Similarly, the inhibition of nitrification by sulfadiazine was also found but at a higher dose of the antibiotic (100 mg kg^{-1}), while the addition of 10 mg kg^{-1} soil to the manure did not significantly change the potential nitrification rates compared to treatments with

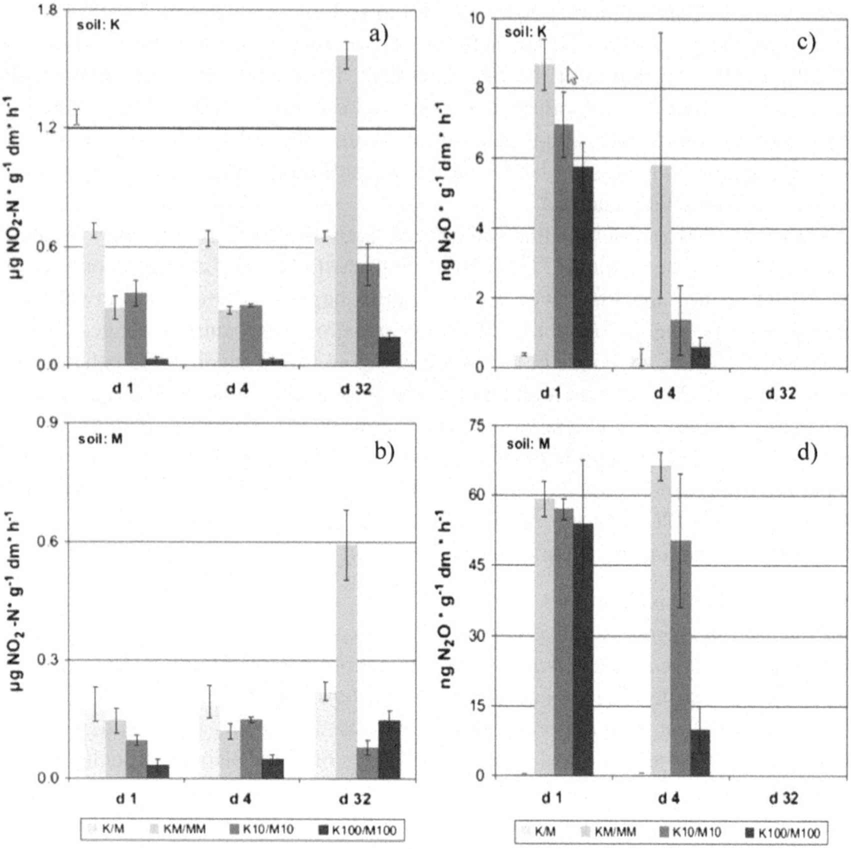

Fig. 15.4 Potential nitrification (**a**, **c**) and denitrification (**b**, **d**) activities of the two different soils treated with different SDZ concentrations and manure at different time points after the application. The data presented here are the mean values of four replicates with the standard deviations. K/M, pure soils K and M; KM/MM, soils K and M amended with manure; K10/M10 and K100/M100, soils K and M amended with manure in combination with SDZ loads of 10 mg SDZ kg^{-1} soil and 100 mg SDZ kg^{-1} soil, respectively; d 1, d 4, d 32—days after spiking (reproduced from Kotzerke et al. 2008)

only the application of manure at the early sampling time points (Kotzerke et al. 2008). Similarly, roxarsone and virginiamycin significantly inhibited this process in soil but only at levels that were several-fold higher than those expected in the poultry litter that was applied soils, which was no higher than 50 and 22 mg kg^{-1}, respectively, while bacitracin-tested soils were not significantly different than the controls at any of the concentrations that were tested (up to 500 mg kg^{-1}) (Banerjee and D'Angelo 2013). Generally, none of the antibiotics that were tested in this study (bacitracin, roxarsone, virginiamycin) influenced denitrification at any of the concentrations tested (0, 1.5, 5, 15, 50, 150, and 500 mg kg^{-1}). These data indicated that both nitrification and denitrification processes would likely be affected by the addition of poultry litter at typical antibiotic levels to these soils. In the study of Kotzerke et al. (2008), two different soils (orthic luvisol and gleyic cambisol) to which antibiotics were added showed reduced denitrification rates, with a larger decrease in those treatments with 100 mg kg^{-1} soil. After 32 days of incubation, no potential denitrification activity could be measured in any treatment.

Iron Reduction

When spiked into manure and mixed with soil at environmentally relevant concentrations in a laboratory experiment, sulfadimethoxine (100 and 200 μg kg^{-1}) blocked soil iron reduction over periods extending from a few days to the end of the study (50 days), while chlortetracycline did not affect Fe reduction (Toth et al. 2011). With monensin, the effect was less clear. The lowest dose of this antibiotic (10 μg kg^{-1}) decreased Fe (II) by 97% on day 1, but the effect was lower by day 8, while the highest concentration (100 μg kg^{-1}) nearly completely blocked Fe reduction. Previously, Thiele-Bruhn (2005) reported ED50 (effective dose with 50% inhibition of iron reduction) at 17.9 mg kg^{-1} of sulfadimethoxine and 25.4 mg kg^{-1} of chlortetracycline. It should be mentioned that these treatment concentrations were several orders of magnitude greater than those of Toth et al. (2011).

15.5 Combined Effects of Antibiotics and Other Disturbing Agents on Soil Microorganisms

To investigate the effect of the addition of a single antibiotic in soil may not reflect the true field conditions because antibiotics and other disturbing factors (such as heavy metals and/or pesticides) often coexist in really polluted agricultural soils because they can accumulate in soils through the application of fertilizer and manure and wastewater irrigation (Liu et al. 2012b). Heavy metals affect antibiotics in soil in various ways. Many antibiotics with different acidic and basic functional groups can be complex with heavy metal ions in a solution, which alters their speciation and as a result their behaviors (Pan et al. 2012). However, there are many antibiotics that have no complex ability or with such an ability only under certain pH conditions (Tolls 2001), and it is unknown how heavy metals affect the adsorption of these antibiotics. The presence of heavy metals may change the properties and structure of soil, e.g., by aggregating and precipitating soil colloid in a solution, by decreasing the negatively charged density on soil surfaces, and by making humic acids and biopolymers more condensed and rigid (Hou et al. 2007; Wang et al. 2007; Pei et al. 2011; Ugochukwu et al. 2011). These changes in soil properties may consequently influence the adsorption of antibiotics to some degree, although the appropriate mechanisms are more widely unknown (Pei et al. 2014). The data presented by Pei et al. (2014) showed that Cu suppressed sulfathiazole (STZ) and tylosin (TYL) adsorption into soil at pH < 5.0 because of the electrostatic competition, while it increased STZ adsorption at pH > 5.0 due to form of

Table 15.4 Results of two-way ANOVA on functional diversity, evenness, AWCD, and substrate utilization in experiment with OTC and Cu singly or in combination

	Shannon diversity	Shannon evenness	AWCD	Amino acid	Carbohydrate	Carboxylic acid
F-value						
Cu (C)	26.89	5.14	273.04	48.31	23.88	31.34
OTC (C)	7.36	2.39	45.76	0.69	13.35	7.42
C × O	14.07	6.44	37.09	0.99	7.55	7.14
P-value						
Cu (C)	0.000	0.043	0.000	0.000	0.000	0.000
OTC (C)	0.008	0.134	0.000	0.527	0.001	0.008
C × O	0.001	0.013	0.000	0.399	0.008	0.009

OTC, oxytetracycline; AWCD, average well color development
Adopted from Kong et al. (2006), modified

STZ-Cu complexes. In contrast, Al only decreased STZ adsorption at $pH < 6.0$ and slightly affected STZ at $pH > 6.0$. Tylosin adsorption was suppressed by both Cu and Al over the entire pH range due to several possible mechanisms: (1) there is competition between cationic Cu and Al as well as TYL for the same adsorption sites in soil; (2) the adsorption of cationic Cu and Al made the surfaces of soil less negatively charged and was unfavorable for the adsorption of TYL through electrostatic attraction; and (3) Cu and Al decreased the pore size of the soil and retarded the diffusion of large-sized TYL to these pores (Pei et al. 2014).

Because of the co-occurrence of OTC and Cu in animal manure, the influence of the combination of both substances on soil microbial communities should be assessed. The combination of OTC and Cu, which had critical values of 11 μM for OTC and 20 μM for Cu, significantly decreased CLPP diversity, evenness, and the utilization of carbohydrates and carboxylic acids compared to only one of the contaminants, thus suggesting a synergistic effect of the two pollutants on soil microbes (Table 15.4, Kong et al. 2006). Although the two pollutants showed specific effects, both of them affected the bacteria-degrading polymers. It was also found that a combination of cypermethrin and heavy metals could result in a great reduction in functional diversity (Xie et al. 2009). Moreover, repeated application of manure containing antibiotics, which is a normal agricultural practice, may lead to intensifying its effect on the soil microbial community structure (Hammesfahr et al. 2008).

15.6 Concluding Remarks and Future Directions

The effects of antibiotics that are applied with manure or other organic materials to soil on the structure and functions of soil microbial communities are not well understood to date and require further studies to investigate the biodiversity of soil microorganisms. Residual concentrations of antibiotics, either unaltered or as metabolites and degradation products, can have strong short-term and long-term impacts on the structure and functions of nontarget bacterial communities and other soil organisms. The major current concerns about these compounds include the potential to induce resistance in bacterial strains (the abundance of resistant bacteria and the transfer of ARGs) in the environment, which could pose a potential ecological and environmental risk to soil quality and human and animal health via the rapid expansion of antibiotic-resistant pathogenic bacteria. As was shown in the reviewed literature, antibiotics in soil differ significantly in their effectiveness on soil microorganisms. These differences can be explained by:

1. The complex interaction of a variety of different natural and anthropogenic factors such as soil texture, pH, availability of nutrient, moisture content, oxidation-reduction status, temperature, and light exposure (Halling-Sørensen et al. 2003; Sarmah et al. 2006).
2. A decreasing mobility and bioavailability of antibiotics by soil adsorption/ complexation, which is why they are more difficult to be transported into microbial cells.
3. The various ranges and directions of antibiotics degradation.
4. Different sources/types of microorganisms and resistances of microorganisms to antibiotics, i.e., soil microbial diversity is huge and the different microbial species in soil could have a substantial overlap in the functional diversity of a microbial community (Chapin et al. 1997). The addition of antibiotics may exert a selective pressure on the metabolic activity of the subgroups of the overall soil microbial community and the suppression could be easily compensated for by other species (Hammesfahr et al. 2008; Thiele-Bruhn and Beck 2005).
5. The dose-response and time-dependent effects of the impact of antibiotics on the soil microbiota.

One or more of these factors might explain why antibiotics, individually applied or applied in mixture, revealed different effects on the structure, diversity, and functions of a soil microbial community.

Additionally, the different data that has been obtained in studies may be the result of the different methods that are used in the determination of microbial communities, e.g., the culturable microbial species in Biolog microplates represented only a small fraction of the total community, and the shifts in the soil microbial community that were caused by an antibiotic disturbance might not be totally described using the Biolog method (de Lipthay et al. 2004; Xie et al. 2009). The Biolog method resulted in a longer period of sulfamethoxazole (SMX) inhibition on the functional diversity of the soil microbial community than SIR (Liu et al. 2012a). The effects of tylosin on the soil microbial community structure measured using DGGE showed that the differences between tylosin and the control diminished after 15 days, while the altered microbial community structure measured using the Biolog method remained over 2 months of incubation (Westergaard et al. 2001). Thus, since no single method can describe the total bacterial community, a combination of different methods is necessary if a more detailed view should be determined.

The lack of the influences of environmentally relative doses of antibiotics on the structure, diversity, and functions of soil microorganisms was found very often, which may be explained by some possible mechanisms of resistance: (1) a low permeability of a cell membrane to antibiotics (particularly in Gram-negative bacteria), (2) the efflux of antibiotics out of the cell (3), a site alteration of antibiotic target site, and (4) the enzymatic inactivation of the antibiotic (Mazel and Davies 1999). Many of these mechanisms are attributed to genes that can be transferred between bacteria in different phylogenetic groups by horizontal gene transfer processes such as transformation, transduction, and conjugation (Davison 1999). One or more of these mechanisms could account for the intrinsic resistance (insensitivity) to antibiotics by the soil microorganisms that have been observed in various studies.

The different effects of antibiotics on the soil microorganisms show that further experiments are required to investigate the biodiversity of how pharmaceuticals affect soil microorganisms. Due to the existence of many factors in soil that can modify the effects of antibiotics on soil microorganisms, future work is needed to evaluate the influence of antibiotics in actual fields, especially in polluted environments under field conditions. Special attention should be paid to the beneficial microorganisms (e.g., arbuscular mycorrhizal fungi, rhizobium, etc.). Moreover, little is known about dose-response relationships and the potential threshold concentrations of antibiotics that are applied to soil with manure, and these parameters should be determined in future studies for different soil types and animal husbandry systems in order to permit the assessment of their short- and long-term risks and effects on human health (Jechalke et al. 2014). To test time- and concentration-related effects, the fact that numerous antibiotics are biostatic and not biotoxic should be considered. Concerning the effect of environmental disturbance on a microbial community, it is important to focus on structural changes and not only on diversity, since the structure seems to be more sensitive to disturbance and the changes last longer. The importance of structural changes to soil functioning is still a challenging question.

References

Akimenko YV, Kazev KS, Kolesnikov SI (2015) Impact assessment of soil contamination with antibiotics (for example, an ordinary chernozem). Am J Appl Sci 12(2):80–88

Andersen R, Grasset L, Thormann MN, Rochefort L, André-Jean Francez AJ (2010) Changes in microbial community structure and function following Sphagnum peatland restoration. Soil Biol Biochem 42:291–301

Aust M, Godlinski F, Travis GR, Hao X, McAllister TA, Leinweber P, Thiele-Bruhn S (2008) Distribution of sulfamethazine, chlortetracycline and tylosin in manure and soil of Canadian feedlots after subtherapeutic use in cattle. Environ Poll 156:1243–1251

Baguer AJ, Jensen J, Krogh PH (2000) Effects of antibiotics oxytetracycline and tylosin on soil fauna. Chemosphere 40:751–757

Banerjee S, D'Angelo E (2013) Livestock antibiotic effects on nitrification, denitrification, and microbial community composition in soils. Open J Soil Sci 3:203–212

Bansal OP (2015) Effects of tetracycline on soil enzyme activities in an alluvial soil. Br J Med Med Res 5(8):1007–1016

Bansal OP, Srivastava V (2014) Impact of tetracycline on population dynamics of soil microorganisms. Int J Pure App Biosci 2(4):112–118

Bastida F, Zsolnay A, Hernández T, García C (2008) Past, present and future of soil quality indices: a biological perspective. Geoderma 147:159–171

Blackwell PA, Kay P, Boxall ABA (2007) The dissipation and transport of veterinary antibiotics in a sandy loam soil. Chemosphere 67:292–299

Bohme L, Langer U, Bohme F (2005) Microbial biomass, enzyme activities and microbial community structure in two European long-term field experiments. Agric Ecosyst Environ 109:141–152

Boleas S, Alonso C, Pro J, Fernández C, Carbonell G, Tarazona JV (2005) Toxicity of the antimicrobial oxytetracycline to soil organisms in a multi-species-soil system (MS·3) and influence of manure co-addition. J Hazard Mater 122:233–241

Boxall A, Kolpin DW, Halling-Sørensen B, Tolls J (2003) Are veterinary medicines causing environmental risks? Environ Sci Technol 37:286A–294A

Brandt KK, Sjøholm OR, Krogh KA, Halling-Sørensen B, Nybroe O (2009) Increased pollution-induced bacterial community tolerance to sulfadiazine in soil hotspots amended with artificial root exudates. Environ Sci Technol 43:2963–2968

Bundy JG, Paton GI, Campbell CD (2004) Combined microbial community level and single species biosensor responses to monitor recovery of oil polluted soil. Soil Biol Biochem 36:1149–1159

Byarugaba DK (2009) Mechanisms of antimicrobial resistance. In: Sosa A de J, Byarugaba DK, Amabile C, Hsueh PR, Kariuki S, Okeke IN (eds) Antimicrobial resistance in developing countries. Springer, New York, pp 15–26

Cao J, Dingge J, Wang C (2015) Interaction between earthworms and arbuscular mycorrhizal fungi on the degradation of oxytetracycline in soils. Soil Biol Biochem 90:283–292

Chapin FS, Walker BH, Hobbs RJ, Hooper DU, Lawton JH, Sala OE, Tilman D (1997) Biotic control over the functioning of ecosystems. Science 277:500–504

Chen W, Liu W, Pan N, Jiao W, Wang M (2013) Oxytetracycline on functions and structure of soil microbial community. J Soil Sci Plant Nutr 13(4):967–975

Chen GX, He WW, Wang Y, Zou YD, Liang JB, Liao XD, Wu YB (2014) Effect of different oxytetracycline addition methods on its degradation behavior in soil. Sci Tot Environ 479:241–246

Chopra I, Roberts M (2001) Tetracycline antibiotics: mode of action, applications, molecular biology, and epidemiology of bacterial resistance. Microb Mol Biol Rev 65(2):232–260

Colinas C, Ingham E, Molina R (1994) Population responses of target and nontarget forest soil organisms to selected biocides. Soil Biol Biochem 26:41–47

Davison J (1999) Genetic exchange between bacteria in the environment. Plasmid 42(2):73–91

De la Torre A, Iglesias I, Carballo M, Ramrez P, Muoz M (2012) An approach for mapping the vulnerability of European Union soils to antibiotic contamination. Sci Tot Environ 414:672–679

De Liguoro M, Cibin V, Capolongo F, Halling-Sørensen B, Montesissa C (2003) Use of oxytetracycline and tylosin in intensive calf farming: evaluation of transfer to manure and soil. Chemosphere 52:203–212

De Lipthay JR, Johnsen K, Albrechtsen HJ, Rosenberg P, Aamand J (2004) Bacterial diversity and community structure of a subsurface aquifer exposed to realistic low herbicide concentrations. FEMS Microbiol Ecol 49:59–69

Derry AM, Staddon WJ, Trevors JT (1998) Functional diversity and community structure of microorganisms in uncontaminated and creosote-contaminated soils as determined by sole-carbon-source-utilization. World J Microbiol Biotechnol 14:571–578

Ding C, He J (2010) Effect of antibiotics in the environment on microbial populations. Appl Microbiol Biotechnol 87:925–941

Ding GC, Radl V, Schloter-Hai B, Jechalke S, Heuer H, Smalla K, Schloter M (2014) Dynamics of soil bacterial communities in response to repeated application of manure containing sulfadiazine. PLoS One 9(3):e92958

Donoho AL (1984) Biochemical studies on monensin. J Anim Sci 58:1528–1539

Du L, Liu W (2012) Occurrence, fate, and ecotoxicity of antibiotics in agro-ecosystems. A review. Agron Sustain Dev 32:309–327

EMEA (2008) Revised guideline on environmental impact assessment for veterinary medicinal products in support of the VICH guidelines GL 6 and GL 38. European Medicines Agency Guidance Document EMEA/CVMP/ERA/418282/2005-Rev.1

Fang H, Han Y, Yin Y, Pan X, Yu Y (2014) Variations in dissipation rate, microbial function and antibiotic resistance due to repeated introductions of manure containing sulfadiazine and chlortetracycline to soil. Chemosphere 96:51–56

Fründ HC, Schoesser A, Westendorp H (2000) Effects of tetracycline on the soil microbiota determined with microtiter plates and respiration measurement. Mitteilungen Dtsch Bodenkundl Gesellsch 93:244–247

Furtula V, Farrell EG, Diarrassouba F, Rempel H, Pritchard J, Diarra MS (2010) Veterinary pharmaceuticals and antibiotic resistance of Escherichia coli isolates in poultry litter from commercial farms and controlled feeding trials. Poult Sci 89(1):180–188

Garland JL (1997) Analysis and interpretation of community-level physiological profiles in microbial ecology. FEMS Microb Ecol 24:289–300

Gavalchin J, Katz SE (1994) The persistence of fecal-borne antibiotics in soil. J AOAC Int 77:481–485

Gomez J, Mendez R, Lema JM (1996) The effect of antibiotics on nitrification processes. Appl Biochem Biotechnol 57/58:869–876

Gomez E, Ferreras L, Toresani S (2006) Soil bacterial functional diversity as influenced by organic amendment application. Bioresour Technol 97:1484–1489

Gutiérrez IR, Watanabe N, Harter T, Glaser B, Radke M (2010) Effect of sulfonamide antibiotics on microbial diversity and activity in a Californian Mollic Haploxeralf. J Soils Sediment 10:537–544

Halling-Sørensen B (2001) Inhibition of aerobic growth and nitrification of bacteria in sewage sludge by antibacterial agents. Arch Environ Contam Toxicol 40:451–460

Halling-Sørensen B, Sengeløv G, Tjørnelund J (2002) Toxicity of tetracyclines and tetracycline degradation products to environmentally relevant bacteria, including selected tetracycline-resistant bacteria. Arch Environ Contam Toxicol 42:263–271

Halling-Sørensen B, Sengeløv G, Ingerslev F, Jensen LB (2003) Reduced antimicrobial potentials of oxytetracycline, tylosin, sulfadiazine, streptomycin, ciprofloxacin, and olaquindox due to environmental processes. Arch Environ Contam Toxicol 44:7–16

Halling-Sørensen B, Jacobsen AM, Jensen J, Sengelov G, Vaclavik E, Ingerslev F (2005) Dissipation and effects of chlortetracycline and tylosin in two agricultural soils: a field-scale study in southern Denmark. Environ Toxicol Chem 24:802–810

Hammesfahr U, Heuer H, Manzke B, Smalla K, Thiele-Bruhn S (2008) Impact of the antibiotic sulfadiazine and pig manure on the microbial community structure in agricultural soils. Soil Biol Biochem 40:1583–1591

Hammesfahr U, Kotzerkeb A, Lamshöftc M, Wilkeb BM, Kandelerd E, Thiele-Bruhn S (2010) Effects of sulfadiazine-contaminated fresh and stored manure on a soil microbial community. Eur J Soil Biol 47:61–68

Hamscher G, Szczęsny S, Höper H, Nau H (2002) Determination of persistent tetracycline residues in soil fertilized with liquid manure by high-performance liquid chromatography with electrospray ionisation tandem mass spectrometry. Anal Chem 74:1509–1518

Hamscher G, Pawelzick TH, Hoeper H, Nau H (2005) Different behaviour of tetracyclines and sulfonamides in sandy soils after repeated fertilization with liquid manure. Environ Toxicol Chem 24:861–868

Heuer H, Smalla K (2007) Manure and sulfadiazine synergistically increased bacterial antibiotic resistance in soil over at least two months. Environ Microbiol 9:657–666

Heuer H, Schmidt H, Smalla K (2011) Antibiotic resistance gene spread due to manure application on agricultural fields. Curr Opin Microbiol 14:236–243

Hou T, Xu RK, Tivari D, Zhao AZ (2007) Interaction between electrical double layers of soil colloids and Fe/Al oxides in suspensions. J Colloid Interface Sci 310:670–674

Hu DF, Coats JR (2007) Aerobic degradation and photolysis of tylosin in water and soil. Environ Toxicol Chem 26:884–889

Hu XG, Zhou QX, Luo Y (2010) Occurrence and source analysis of typical veterinary antibiotics in manure, soil, vegetables and groundwater from organic vegetable bases, northern China. Environ Pollut 158(9):2992–2998

Hund-Rinke K, Simon M, Lukow T (2004) Effects of tetracycline on the soil microflora: function, diversity, resistance. J Soils Sediment 4:11–16

Ingham ER, Coleman DC (1984) Effects of streptomycin, cycloheximide, fungizone, captan, carbofuran, cygon, and PCNB on soil microorganisms. Microbial Ecol 10:345–358

Insam H (1997) A new set of substrates proposed for community characterization in environmental samples. In: Insam H, Rangger A (eds) Microbial communities. Functional versus structural approaches. Springer, Berlin, pp 259–260

Jechalke S, Kopman C, Rosendahl I, Groeneweg J, Weichlet V, Krögerrecklenfort E, Brandes N, Nordwig M, Ding GC, Siemens J, Heuer H, Smalla K (2013) Increased abundance and transferability of resistance genes after field application of manure from sulfadiazine-treated pigs. Appl Environ Microbiol 79:1704–1711

Jechalke S, Heuer H, Siemens J, Amelung W, Smalla K (2014) Fate and effects of veterinary antibiotics in soil. Trends Microbiol 22(9):536–545

Jones A, Bruland G, Agrawal S, Vasudevan D (2005) Factors influencing the sorption of oxytetracycline to soils. Environ Toxicol Chem 24:761–770

Kemper N (2008) Veterinary antibiotics in the aquatic and terrestrial environment. Ecol Indicat 8:1–13

Kong WD, Zhu YG, Fu BJ, Marschner P, He JZ (2006) The veterinary antibiotic oxytetracycline and Cu influence functional diversity of the soil microbial community. Environ Pollut 143:129–137

Kopmann C, Jechalke S, Rosendahl I, Groeneweg J, Krögerrecklenfort E, Zimmerling U, Weichelt V, Siemens J, Amelung W, Heuer H, Smalla K (2013) Abundance and transferability of antibiotic resistance as related to the fate of sulfadiazine in maize rhizosphere and bulk soil. FEMS Microbiol Ecol 83:125–134

Kotzerke A, Sharma S, Schauss K, Heuer H, Thiele-Bruhn S, Smalla K, Wilke BM, Schloter M (2008) Alterations in soil microbial activity and *N*-transformation processes due to sulfadiazine loads in pig-manure. Environ Pollut 153:315–322

Kumar K, Gupta SC, Chander Y, Singh AK (2005) Antibiotic use in agriculture and its impact on the terrestrial environment. Adv Agron 87:1–54

Li QC, Allen HL, Willum A (2004) Microbial biomass and bacterial functional diversity in forest soils: effects of organic matter removal, compaction, and vegetation control. Soil Biol Biochem 36:571–579

Li Z, Chang PH, Jean JS, Jiang WT, Wang CJ (2010) Interaction between tetracycline and smectite in aqueous solution. J Colloid Interface Sci 341:311–319

Liu F, Ying GG, Tao R, Zhao JL, Yang JF, Zhao LF (2009) Effects of six selected antibiotics on plant growth and soil microbial and enzymatic activities. Environ Pollut 157:1636–1642

Liu F, Wu J, Ying GG, Luo Z, Feng H (2012a) Changes in functional diversity of soil microbial community with addition of antibiotics sulfamethoxazole and chlortetracycline. Appl Microbial Biotechnol 95:1615–1623

Liu W, Pan N, Chen W, Jiao W, Wang M (2012b) Effect of veterinary oxytetracycline on functional diversity of soil microbial community. Plant Soil Environ 58(7):295–301

Liu B, Li Y, Zhang X, Wang J, Gao M (2014) Combined effects of chlortetracycline and dissolved organic matter extracted from pig manure on the functional diversity of soil microbial community. Soil Biol Biochem 74:148–155

Lunestad BT, Goksøyr J (1990) Reduction in the antibacterial effect of oxytetracycline in sea water by complex formation with magnesium and calcium. Dis Aquat Org 9:67–72

Ma J, Lin H, Sun W, Fu J (2014) Soil microbial systems respond differentially to tetracycline, sulfamonomethoxine, and ciprofloxacin entering soil under pot experimental conditions alone and in combination. Environ Sci Pollut Res Int 2:7436–7448

Mäder P, Fließbach A, Dubois D, Gunst L, Fried P, Niggli U (2002) Soil fertility and biodiversity in organic farming. Science 296:1694–1697

Maliszewska-Kordybach B, Klimkowicz-Pawlas A, Smreczak B (2007) Ecotoxic effect of phenanthrene on nitrifying bacteria in soils of different properties. J Environ Qual 36:1635–1645

Martinez JL (2009) Environmental pollution by antibiotics and by antibiotic resistance determinants. Environ Pollut 157:2893–2902

Mazel D, Davies J (1999) Antibiotic resistance in microbes. Cell Mol Life Sci 56(9-10):742–754

McGuire JM, Boniece WS, Higgins CE, Hoehn MM, Stark WW, Westhead J, Wolfe RN (1961) Tylosin, a new antibiotic. I. Microbiological studies. Antibiot Chemother 11:320–327

Müller AK, Westergaard K, Christensen S, Sørensen SJ (2002) The diversity and function of soil microbial communities exposed to different disturbances. Microb Ecol 44:49–58

Nannipieri P, Kandeler E, Ruggiero P (2002) Enzyme activities and microbiological and biochemical processes in soil. In: Burns RG, Dick RP (eds) Enzymes in the environment, activity, ecology and application. Marcel Dekker, New York, pp 1–33

Navas M, Benito M, Rodríguez I, Masaguer A (2011) Effect of five forage legume covers on soil quality at the Eastern plains of Venezuela. Appl Soil Ecol 49:242–249

Pan B, Qiu M, Wu M, Zhang D, Peng HB, Wu D (2012) The opposite impacts of Cu and Mg cations on dissolved organic matter-ofloxacin interaction. Environ Pollut 161:76–82

Park H, Choung YK (2007) Degradation of antibiotics (tetracycline, sulfathiazole, ampicillin) using enzymes of glutathione S-transferase. Hum Ecol Risk Assess Int J 13(5):1147–1155

Pei ZG, Shan XQ, Zhang SZ, Kong JJ, Wen B, Zhang J, Zheng LR, Xie YN, Janssens K (2011) Insight to ternary complexes of co-adsorption of norfloxacin and Cu (II) onto montmorillonite at different using EXAFS. J Hazard Mater 186:842–848

Pei Z, Yang S, Li L, Li C, Zhang S, Shan X, Wen B, Guo B (2014) effects of copper and aluminium on the adsorption of sulfathiazole and tylosin on peat and soil. Environ Pollut 184:579–585

Pils JRV, Laird DA (2007) Sorption of tetracycline and chlortetracycline on K- and Ca-saturated soil clays, humic substances, and clay–humic complexes. Environ Sci Technol 41:1928–1933

Pinna MV, Castaldi P, Deiana P, Pusino A, Garau G (2012) Sorption behavior of sulfamethazine on unamended and manure-amended soils and short-term impact on soil microbial community. Ecotoxicol Environ Safe 84:234–242

Preston-Mafham J, Boddy L, Randerson PF (2002) Analysis of microbial community functional diversity using sole-carbon-source utilisation profiles—a critique. FEMS Microbiol Ecol 42:1–14

Rabølle M, Spliid NH (2000) Sorption and mobility of metronidazole, olaquindox, oxytetracycline, and tylosin in soil. Chemosphere 40:715–722

Reichel R, Rosendahl I, Peeters E, Focks A, Groeneweg J, Bierl R, Schlichting A, Amelung W, Thiele-Bruhn S (2013) Effects of manure from sulfadiazine- (SDZ) and difloxacin- (DIF) medicated pigs on the structural diversity of microorganisms in bulk and rhizosphere soil. Soil Biol Biochemi 62:82–91

Salyers AA, Amábile-Cuevas CF (1997) Why are antibiotic resistance genes so resistant to elimination? Antimicrob Agents Chemother 41:2321–2325

Sarmah AK, Meyer MT, Boxall ABA (2006) A global perspective on the use, sales, exposure pathways, occurrence, fate and effects of veterinary antibiotics (VAs) in the environment. Chemosphere 65:725–759

Schloter M, Dilly O, Munch JC (2003) Indicators for evaluating soil quality. Agric Ecosyst Environ 98:255–262

Schmitt H, Haapakangas H, van Beelen P (2005) Effects of antibiotics on soil microorganisms: time and nutrients influence pollution induced community tolerance. Soil Biol Biochem 37:1882–1892

Schulz S, Brankatschk R, Dümig A, Kögel-Knabner I, Schloter M, Zeyer J (2013) The role of microorganisms at different stages of ecosystem development for soil formation. Biogeosciences 10:3983–3996

Smalla K, Wachtendorf U, Heuer H, Liu W, Forney L (1998) Analysis of BIOLOG GN substrate utilization patterns by microbial communities. Appl Environ Microbiol 64:1220–1225

Song WL, Ding YJ, Chiou CT, Li H (2010) Selected veterinary pharmaceuticals in agricultural water and soil from land application of animal manure. J Environ Qual 39:1211–1217

Stoob K, Singer HP, Stettler S, Hartmann N, Mueller SR, Stamm CH (2006) Exhaustive extraction of sulfonamide antibiotics from aged agricultural soils using pressurized liquid extraction. J Chromatogr A 1128:1–9

Swift MJ (1994) Maintaining the biological status of soil: a key to sustainable land management. In: Greenland DJ, Szabolcs I (eds) Soil resilience and sustainable land use. CAB, Wallingford, pp 33–39

Thiele-Bruhn S (2003) Pharmaceutical antibiotic compounds in soils— a review. J Plant Nutr Soil Sci 166:145–167

Thiele-Bruhn S (2005) Microbial inhibition by pharmaceutical antibiotics in different soils-dose-response relations determined with the iron (III) reduction test. Environ Toxicol Chem 24:869–876

Thiele-Bruhn S, Beck IC (2005) Effects of sulfonamide and tetracycline antibiotics on soil microbial activity and microbial biomass. Chemosphere 59:457–465

Thiele-Bruhn S, Seibicke T, Schulten HR, Leinweber P (2004) Sorption of sulfonamide pharmaceutical antibiotics at whole soils and particle-size fractions. J Environ Qual 33:1331–1342

Tolls J (2001) Sorption of veterinary pharmaceuticals in soils: a review. Environ Sci Technol 35:3397–3406

Toth JD, Feng Y, Dou Z (2011) Veterinary antibiotics at environmentally relevant concentrations inhibit soil iron reduction and nitrification. Soil Boil Biochem 43:2470–2472

Ugochukwu N, Ebong E, Uzoma O, Fu QL, Huang L, Hu HQ (2011) Impacts of inorganic ions, and temperature on lead adsorption onto variable charge soils. In: 5th International conference on chemical, biological and environment science, Bangkok

Vaclavik E, Halling-Sørensen B, Ingerslev F (2004) Evaluation of manometric respiration tests to assess the effects of veterinary antibiotics in soil. Chemosphere 56:667–676

Wang XL, Yang K, Tao S, Xing BS (2007) Sorption of aromatic organic contaminants by biopolymers: effects of pH, copper (II) complexation, and cellulose coating. Environ Sci Technol 41:185–191

Wang J, Lin H, Sun W, Xia Y, Ma J, Fu J, Zang Z, Wu H, Qian M (2016) Variations in the fate and biological effects of sulfamethoxazole, norfloxacin and doxycycline in different vegetable–soil systems following manure application. J Hazard Mat 304:49–57

Wei X, Wu SC, Nie XP, Yediler A, Wong MH (2009) The effects of residual tetracycline on soil enzymatic activities and plant growth. J Environ Sci Health B 44(5):461–471

Westergaard K, Müller AK, Christensen S, Bloem J, Sørensen SJ (2001) Effects of tylosin as a disturbance on the soil microbial community. Soil Biol Biochem 33:2061–2071

Wright GD (2005) Bacterial resistance to antibiotics: enzymatic degradation and modification. Adv Drug Deliv Rev 57(10):1451–1470

Xie WJ, Zhou JM, Wang HY, Chen XQ, Lu ZH, Yu JB, Chen XB (2009) Short-term effects of copper, cadmium and cypermethrin on dehydrogenase activity and microbial functional diversity in soils after long-term mineral or organic fertilization. Agric Ecosyst Environ 129:450–456

Xie XM, Liao M, Yang J, Chai JJ, Fang S, Wang RH (2012) Influence of rootexudates concentration on pyrene degradation and soil microbial characteristics in pyrene contaminated soil. Chemosphere 88:1190–1195

Zhang Q, Dick WA (2014) Growth of soil bacteria, on penicillin and neomycin, not previously exposed to these antibiotics. Sci Total Environ 493:445–453

Zhang CB, Wang J, Liu WL, Zhu SX, Ge HL, Chang SX, Chang J, Ge Y (2010) Effects of plant diversity on microbial biomass and community metabolic profiles in a full-scale constructed wetland. Ecol Eng 36:62–68

Zhu YG, Johnson TA, Su JQ, Qiao M, Guo GX, Stedtfeld RD, Hashsham SA, Tiedje JM (2013) Diverse and abundant antibiotic resistance genes in Chinese swine farms. Proc Natl Acad Sci USA 110:3435–3440

Zielezny Y, Groeneweg J, Vereecken H, Tappe W (2006) Impact of sulfadiazine and chlortetracycline on soil bacterial community structure and respiratory activity. Soil Biol Biochem 38:2372–2380

Zuccato E, Bagnati R, Fioretti F, Natangelo M, Calamari D, Fanelli R (2001) Environmental loads and detection of pharmaceuticals in Italy. In: Kummerer K (ed) Pharmaceuticals in the environment. Springer, Berlin, pp 19–27

Chapter 16
Soil Antibiotics and Transfer of Antibiotic Resistance Genes Affecting Wildlife

Vanessa Silva, Isabel Carvalho, Gilberto Igrejas, and Patrícia Poeta

16.1 Soil Antibiotics

The discovery of penicillin marked the beginning of the era of antibiotics. Today, we use over 250 antibiotics, produced mostly by bacteria and fungi. Antibiotics are used both in veterinary and human medicine in order to treat or prevent bacterial infections. Agriculture industries use a large part of the antibiotics used worldwide to increase the growth rate in livestock and poultry to treat sick animals and for crop dusting (Landers et al. 2012). Nowadays, in Europe the use of antibiotics for growth

The original version of this chapter was revised. An erratum to this chapter can be found at DOI 10.1007/978-3-319-66260-2_22.

V. Silva (✉) • I. Carvalho
Veterinary Sciences Department, University of Trás-os-Montes and Alto Douro, Quinta de Prados, 5000-911 Vila Real, Portugal
e-mail: vanessanssilva20@gmail.com; isabelcarvalho93@hotmail.com

G. Igrejas
Department of Genetics and Biotechnology, Functional Genomics and Proteomics' Unit, University of Trás-os-Montes and Alto Douro, Vila Real, Portugal

Functional Genomics and Proteomics Unit, University of Tras-os-Montes and Alto Douro (UTAD), Vila Real, Portugal

Research Unit on Applied Molecular Biosciences (UCIBIO-REQUIMTE), University NOVA of Lisboa, Lisboa, Caparica, Portugal
e-mail: gigrejas@utad.pt

P. Poeta
Veterinary Sciences Department, University of Trás-os-Montes and Alto Douro, Quinta de Prados, 5000-911 Vila Real, Portugal

Research Unit on Applied Molecular Biosciences (UCIBIO-REQUIMTE), University NOVA of Lisboa, Lisboa, Caparica, Portugal
e-mail: ppoeta@utad.pt

M.Z. Hashmi et al. (eds.), *Antibiotics and Antibiotics Resistance Genes in Soils*, Soil Biology 51, DOI 10.1007/978-3-319-66260-2_16

promotion has been prohibited, nevertheless, some countries, such as Canada and the United States, still use antibiotics for this purpose. According to several studies, the total amount of annual use of antibiotics has reached 100,000–200,000 tons worldwide (Wang and Tang 2010). The overuse of antibiotics is a current problem since it led to an antibiotic resistance crisis. When administrated, and depending on the specific antibiotic used, the animal species, and the type and duration of administration, antibiotics are not fully absorbed and metabolized in the organism being partly excreted through urine and feces as parent compound or metabolites, and depending on the compound, the excretion rate can be as high as 80–90% (Bound and Voulvoulis 2004). Thus, a significant percentage of the antibiotics consumed by humans and animals may be introduced into the environment leading to an antibiotic accumulation in the terrestrial and aquatic environments contributing to the development of resistant bacteria (Hu et al. 2010; Zhou et al. 2012). Tetracyclines, sulfonamides, and macrolides are the major groups of antibiotics used by humans and in farm animals to promote animal growth. However, the biggest concern for the entrance of antibiotic in soil are the veterinary usage of antibiotics since wastewater treatment plants are the major source of release of human antibiotics into the environment (Kim and Carlson 2007). A great part of the antibiotics used in farm animals are excreted with the manure. Some antibiotics, such as β-lactams and macrolides, can be degraded in a few days while the manure is still in storage, however, many antibiotics still remain active in manure and are transferred to soil when using manure in agricultural fields (Boxall et al. 2004). Tylosin is commonly used in farm animals, studies reported that the degradation rate of tylosin on pig manure was from 60 to 85% under anaerobic conditions within 24 h, and the degradation of this antibiotic was almost complete in manure within 12 h under aerobic conditions (Kolz et al. 2005). Antibiotic metabolites can transform back into bioactive compounds after excretion (Lanhammer 1989). After antibiotic transformation into their conjugates, the antibiotics become inactive; however, they can become active again in manure and end up in soil (Christian et al. 2003). When manure is applied, the active compounds found in manure reach the upper soil layer where they may accumulate or enter the groundwater leading to a larger dissemination of antibiotics (Boxall et al. 2003).

Tetracyclin is the primarily antibiotic group used for veterinary purposes, as well as for human therapy and agriculture. The most common antibiotics used in veterinary medicine are oxytetracycline, sulfachloropyridazine, and tylosin (Kay et al. 2003). The secretion rates of these antibiotics are very high ranging from 28 to 100% for oxytetracycline, sulfachloropyridazine, and tylosin. The antibiotic degradation in manure is due to the antibiotics composition, the type of livestock manure and composting conditions. The degradation of the antibiotics is due to adsorption and/or degradation rather than abiotic processes (Kim et al. 2011). For instance, tetracycline has the ability to form chelate complexes with organic matter making it strongly adsorb with manure components. Once manure is applied, the antibiotics in manure will be transferred to the soil and their ability to remain in the soil varies with the soil and antibiotic characteristics and the interaction between them.

16.2 Interaction of Antibiotics with Soil

Antibiotics used in farm animals are introduced in soil through pharmaceutical companies, drug manufacturing process, use of manure in agricultural fields and grazing animals. However, the main cause of soil contamination with antibiotics is the direct use of manure (Tasho and Cho 2016). The ability of antibiotics to be adsorbed and fixated to soil particles is determined by their physicochemical properties (the presence of functional groups making them either ionic, amphiphilic, or amphoteric, size, solubility, and hydrophobicity), climatic conditions, and the soil properties (pH, type, and content in organic matter) (Doretto and Rath 2013). Studies have determined the strength of sorption of several antibiotics to soil being the streptomycin the most adsorbed antibiotic followed by erythromycin, tylosin, bacitracin, chlortetracycline, and oxytetracycline in this order (Katz and Katz 1983). Antibiotics can be effective and active during a long period of time which depends on their structure and their persistence in soil. Studies demonstrated that the potential of several antibiotics may increase with time (Dojmi di Delupis et al. 1992; Samuelsen 1994). The degradation and fixation of antibiotics in soil is influenced by the surface compounds and the pores of the soil matrix which may also prevent the degradation of antibiotics. The adsorption of antibiotics to soil organic matter is strong and depends on the composition and the quantity of organic matter in soil. In order to predict the transport behavior of the antibiotics in soil it is common to use the soil-water partition coefficient, K_d. To determinate the K_d values, a computational modeling can be used, or it can be calculated directly or indirectly through the octanol-water partition coefficients (K_{ow}) (Uhrich et al. 2014).

The different classes of antibiotics mostly used in medicine have their own properties which influence their ability to adsorb and fixate to soils. Tetracyclines are amphoteric and water soluble compounds. They have the ability of binding to humic acids, proteins, and clay minerals through the anionic functional groups; therefore, soils with low pH may retain tetracyclines more easily. Moreover, the ionic strength of the soil compounds also influences the adsorption of tetracyclines to soils. Tetracyclines may also be photodegraded since they strongly absorb light (Mitscher 1978). From the class of tetracycline, chlortetracycline is the antibiotic most commonly found in soils and with higher concentrations, which makes this antibiotic very persistent in soil (Daghrir and Drogui 2013). The conformation of compounds found in soil can also contribute to the adsorption strength of antibiotics. The increasing aromaticity of soil compounds increases the adsorption of oxytetracycline. Unlike tetracyclines, sulfonamides are quite insoluble in water. The decrease of soil pH leads to the ionization of the amphoteric sulfonamides and influences their K_d which can increase from 1 to 30. However, the addition of manure to soil increases the pH and consequently the K_d value decreases (Boxall et al. 2002). The composition of soil organic matter is associated with the adsorption of sulfonamides since the concentration of lipids and lignin dimers in soil organic matter is correlated with the adsorption of sulfonamides. However, although the addition of manure to soils leads to an increase in organic matter, the K_d of sulfonamides tends to decrease as the soil become more alkalinic.

Sulfonamides stability is affected by pH since these antibiotics act as weak acids and form salts in acid and basic solutions (Boxall et al. 2002; Thiele-Bruhn 2003). Macrolides are another classes of antibiotics very much used in medicine. They have generally a low water solubility which varies between the several macrolides available. Since macrolides are weak bases, they are unstable in acids, for instance, in acid medium tylosin A is converted to tylosin B, so in manure, which has a high pH, the positively charged macrolides link to the negatively charged ions in manure which is favorable for the adsorption of these antibiotics in manure (Thiele-Bruhn 2003). β-lactams, which comprises the penicillin and cephalosporins, have a ring in their structure which is responsible for their action as antibiotics. However, cephalosporins have a six-membered heterocycle ring whereas penicillin has five-membered heterocycle ring. This ring is also responsible for the degradation rate of β-lactams since it is easily cleaved in basic and acid medium. This instability of the lactam ring is the reason why β-lactams are rarely found in soils since they suffer a fast degradation (Alder et al. 2001; Midtvedt 2001). Antibiotics from the class of quinolones are very stable compounds and are hardly degraded by hydrolysis or high temperatures. Quinolones have a heterocyclic ring and the substituent group at the C-6 position determines the effectiveness of the antibiotic. Theses antibiotics are inactivated through metabolism, however, when in the soil, where it might be activated, it does not suffer degradation under anaerobic conditions. Quinolones have the ability to adsorb to soils, mainly soils with manure since they adsorb strongly to manure rich in organic matter (Halling-Sùrensen et al. 1998; Thiele-Bruhn 2003). Aminoglycosides are basic compounds with two or more amino sugars bound to aminocyclitol in their composition. They are very soluble in water which makes them easily degraded in the environment; for this reason, they are unlikely to be persistent. The sorption of this class of antibiotics is quite weak (Thiele-Bruhn 2003).

Based on the above, it can be concluded that macrolides, β-lactams, and aminoglycosides are antibiotics that are quite degradable in soils whereas tetracyclines, sulfonamides, and quinolones are more resistant to degradation in soils. The degradation of antibiotics is mainly due to biodegradation although other reactions may occur. Although these are the most commonly used antibiotics, other classes of antibiotics are used in agriculture, such as arsenicals, polypeptides, ionophores, and others (Landers et al. 2012). These antibiotics may not contribute so much to antibiotic resistance which makes them a smaller threat to human health.

The presence and persistence of several antibiotics in soil may have an adverse impact on soil microbial communities representing a hazard to human health since it may lead to an increase and spread of antibiotic resistant bacteria.

16.3 Effects of Antibiotics on Soil Bacterial Communities

Bacteria are one of the most important groups of organisms in soil since they are responsible for maintaining the mineral immobilization and for the processes underlying the decomposition that occur in soil, thus, a high number of bacteria exist in soil (Nwosu 2001). Bacteria are indispensable for the nutrient and geochemical cycles and for soil fertility. The soil microbial communities, such as fungi and bacteria, produce antibiotics which are responsible for controlling the dynamics of bacterial populations. When the antibiotics used in medicine and agriculture reach the soil, they will affect the soil microbial communities leading to a disturbance in the bacterial communities affecting the abundance, diversity, and transferability of resistance genes which may cause the bacteria to acquire gene-encoding resistance (Esiobu et al. 2002). These effects on the microbial communities depend on several features such as the microbial groups, the concentration of antibiotic, and on the original soil properties. Antibiotics found on soil can have their selective effects on several groups of microorganisms, from fungi or bacteria to a single genus or species (Table 16.1). Thus, antibiotics may affect relative abundance of microbial species and consequently the interaction between species (Ding and He 2010). The manure applied in soil is the main cause of the antibiotic spread in soil, and it affects the soil bacteria not only because of its content in antibiotics, but also because it has considerable quantities of nutrients which increase the soil content in carbon and nitrogen and also affects the soil pH (Poulsen et al. 2013). Manure containing antibiotics, when applied only once, does not affect the bacterial community in a long-term since it is followed by a quick regeneration of the structure and function of the community. The application of manure repeatedly may have a bigger impact in the soil microbial community since it might have an accumulative effect, however, little is known about this subject (Ding et al. 2014). The repeated application of manure containing antibiotics, in a long-term, and even with concentrations below inhibitory concentrations, may lead to an increase in resistant bacteria in soil stimulating the spread of mobile genetic elements and antibiotics resistant genes through the environment. Antibiotics exert a selective pressure on the soil bacteria which may favor the horizontal gene transfer. Another issue is when the antibiotic administrated to animals is impure it can exert a selective pressure on intestinal bacteria of farm animals leading the bacteria to acquire resistance. Thus, the resistance genes can be excreted in manure and be spread to the environment. Still, studies show that the bacteria present in manure does not have the ability to adapt to the soil which would result in the decrease of resistant bacteria

Table 16.1 Antibiotics spectrum and mode of action

Antibiotic	Gram	Mode of action
Tetracyclines	Positive and negative	Inhibition of protein synthesis
Sulfonamides	Positive and negative	Inhibition of folinic acid and consequently DNA synthesis
Macrolides	Positive	Inhibition of protein synthesis
β-lactams	Positive	Inhibition of cell wall synthesis
Quinolones	Negative	Inhibition of DNA synthesis and replication
Aminoglycosides	Negative	Inhibition of protein synthesis

Kahn and Line (2005); Beers (2006); Ding and He (2010)

(Hammesfahr et al. 2008; Heuer et al. 2008). However, the transference of resistance genes from manure bacteria to soil bacteria may still occur as the resistance genes are located on translocative elements (such as transposons, plasmids, integrons, bacteriophages, and gene cassettes) and can be transferred by conjugative elements (Jechalke et al. 2014).

The use of antibiotics in agriculture and veterinary medicine is the major cause of antibiotic spread in soils, which is known to be responsible for the increase in the incidence of horizontal gene transfer and resistance gene fixation in genomes. However, although the expansion of resistance to antibiotics in the environment is partly due to antibiotic input, it remains a complex process controlled by numerous different aspects.

16.4 Antimicrobial Resistance in the Environment

Antimicrobial resistance is a universal problem in human and veterinary medicine. The presence of multidrug resistant bacteria is probably attended by environment co-contamination foremost to a major clinical and public health concern within the lifetime of most people living today (Guenther et al. 2011). A great part of antibiotic resistant genes obtained by microbial human pathogens have emerged from the conventional environment. Consequently, comprehending causes that stimulate intrinsic levels of antibiotic resistant genes in the environment might be epidemiologically importance.

Drug resistance is widespread in treatment, particularly in the cancer sector, where minor populations of cells that are able to survive anticancer drugs and other therapeutic contributions can restore tumors that originally respond satisfactorily to treatment. Nevertheless, there are single characteristics to antibiotic drug resistance displayed by bacteria. This is since bacteria and other microorganisms such as fungi have progressed genetic and phenotypic competences that allow them to resist antibiotics, which they produce naturally (WHO 2014). In ecosystems, the presence of these distinctive particles permits microorganisms to protect themselves against, or abolish, opponents. Throughout the advance of evolution, which for bacteria is supposed to have begun 3.5 billion years ago, a massive and different pool of genes (the "resistome") has been recognized, and these qualify bacteria to defend themselves against antibiotics which are nowadays used therapeutically by physicians to target them (WHO 2014). Currently, the majority of the antibiotics used are of natural origin or resultant somehow from natural compounds. In this sense, the majority are consequently familiar risks to microorganisms and their capability to survive them is coded into their genomes. Bacteria also have the capacity to develop new modifications in response to artificial antibiotics such as the fluoroquinolones, by producing new resistant strains. In this sense, even when bacteria do not have "evolutionary" familiarity with antibiotics, they show the ability to resist to new molecules of antibiotics that intimidate their viability, producing new variants of themselves and increasing the likelihood that one or more members of the population can restore. According to Darwinian principles, the use of antibiotics in human and veterinary medicine select resistant strains, i.e., those that withstand to the antibiotic and this explains the development of drug resistance nowadays. Perhaps, one of the highest achievements in the history of medicine is the discovery of antibiotics from naturally occurring molecules that are now used to treat a varied range of infections. Nevertheless, antibiotic resistance has now ascended to levels that stance major dangers to the treatment of both common and serious infections. Different pathogens are now multidrug resistant and some essential antibiotics, which were once firstline drugs, are no longer effective. The seriousness of the situation has been increasing and international surveillance data show that resistant pathogens are responsible for enhancing hospital and community acquired infections and mortality. In the developed world, infections caused by drug-resistant pathogens could now be responsible for more than 400,000 deaths per year (WHO 2014).

The growing occurrence of antibiotic resistant bacteria is one of the greatest thoughtful fears to public health in the twenty-first century. One of the ways resistance genes reach the food system is through the application of manure from animals treated with antibiotics, which are defined as reservoirs of these genes, to farmland (Udikovic-Kolica et al. 2014). Management of farm animals with antibiotics and then placing their manure in soil can transform the bacteria in dirt to develop resistant to the drugs. However, some studies advocate that the manure itself could be relevant to resistance, even when it comes from animals that are restricted of antibiotic use suggesting a multifaceted link between antibiotic use in agriculture and resistance in human pathogens. Numerous bacteria naturally transport antibiotic resistance genes, perhaps as protection against the antibiotics produced by some soil fungi and bacteria. Manure itself is recognized to adjust the composition of bacterial communities in soil and in that study, the researchers discover that the soil treated with manure contained expressively greater amounts of bacteria producing β-lactamases than did soil treated with only the nitrogen-based fertilizer (Udikovic-Kolica et al. 2014). By locating genetic markers in the resistant bacteria, it can be verified that these microorganisms came from the soil and not from the manure, suggesting that the manure treatment had aided these natural bacteria from soil to develop resistance (Udikovic-Kolica et al. 2014).

Currently, it is known that the environment is one of the big reservoirs of resistant organisms and genes. Furthermore, this resistome precedes human use of antibiotics (Hughes and Datta 1983; Poirel et al. 2002). Some research has been performed in order to know how the environmental resistome crosses with the nosocomial bacteria resistome (Wright 2010). Resistant organisms in different ecosystems as aquatic, terrestrial and animal can be part of the normal microbial populations and harbor intrinsic resistance. However, in the majority of situations, they can be the consequence of pollution by anthropogenic sources (Osterblad et al. 2001; Xi et al. 2009). The antibiotic agents are released into the environment via wastewater sewage, farming products as fertilizers, and outflow from waste packing services (Sarmah et al. 2006; Le-Minh et al. 2010; Barrett 2012). Therefore, the higher concentration of antibiotics in the environment has as consequences the increase of diversity and amount of resistance genes. The mobile genetic elements such as plasmids, transposons, or integrons are responsible for the regular resistance genes commonly found in bacteria, and they can be acquired or disseminated through horizontal gene transfer (Gillings 2013). Different bacterial species that harbor distinct resistance genes were already found in clinical settings, domestic and wild animals, and in settings apparently not prejudiced by human populations (Poeta et al. 2007; Pallecchi et al. 2008; Stokes and Gillings 2011).

16.5 The Problem of Antibiotic Resistance in Wild Animals

Antibiotic resistance genes present in mobile elements of bacterial pathogens are ubiquitous and can be found in wild animals that supposedly are not in contact with antibiotics. So, we can achieve that antibiotic resistance genes are extensively disseminated and resistance genes can proceed even when a positive selection pressure that doesn't exist (Martinez 2012). The natural environments, principally the wild animals, could be reservoirs of antibiotic resistance genes. Therefore, research studies were performed to analyze the natural ecosystems contribution in the dissemination of resistance (i.e., into human pathogens). These studies investigate different animal reservoirs, which are fundamental to improve and disseminate resistance (Radhouani et al. 2011; Radhouani et al. 2012) or different abiotic sources as wastewater treatment plants (Araújo et al. 2010). Urban and poultry-slaughterhouse wastewater treatment plants are useful targets to evaluate the prevalence of antibiotic resistant isolates in the environment. It appears that these strains could pass through wastewater treatment and be transferred to the surface waters, constituting a public health problem. Possibly, antimicrobial drugs or chemicals released into the sludge and sewage system may sustain resistant strains. The problem may intensify when the co-transference phenomenon is present and the spread of a gene encoding a virulence factor of one strain to another involves the transfer of resistance to one or more antibiotics (Araújo et al. 2010). In situations in which the genes that encode resistance to different antimicrobial agents are located in the same genetic element will be necessary to limit or prohibit the use of both before the effect of resistance could be detected. At the same time, antibiotics and disinfectants are released into the water and may exercise a selective and ecological damage in aquatic communities, resulting in resistance to antibiotics (Araújo et al. 2010).

Approximately since the first reports concerning antibiotic resistance among wild animals, a debate started whether it was related or not with the human use of antibiotics (Gilliver et al. 1999; Osterblad et al. 2001). For example, Gilliver et al. (1999) detected a high prevalence of fecal bacteria with antibiotic resistance from wild rodents living in woodland sites in northwest England where apparently none or at least minimum antibiotic release was applied (Gilliver et al. 1999). On the other hand, Osterblad et al. (2001) described almost an absence of resistance in fecal enterobacteria from moose, deer, and vole living in remote areas of Finland. Based on the results obtained by these studies, an agreement can be performed. The gastrointestinal microbiota of different wildlife animals reveals antibiotic resistance stages that appear closely associated with the ecological settings of the studied wildlife. Wildlife inhabitants existing in earlier contiguity to humans may have advanced levels of antibiotic resistance than those populations with practically no interaction with both humans and anthropogenic antibiotics (Osterblad et al. 2001).

Additionally, bacteria isolated from wildlife animals may express resistance to multiple antibiotic agents not due to the direct contact with antimicrobials, but because the multiresistant organisms have spread to different ecological niches (O'Brien 2002). Therefore, in this type of studies it is imperative that the implication of the collected data must be carefully analyzed and reflect the ecological contextual of both the investigated organisms and the bacterial population (Cole et al. 2005). In the last few years, antibiotic resistance studies performed in indicator bacteria from wildlife became more and more frequent. Reports showing the presence of acquired resistance determinants in the commensal microbiota of wildlife animals are frequent in wild birds as they might serve as reservoirs of antibiotic resistant bacteria with enormous potential for dissemination (Dolejska et al. 2007). Migrant birds can fly amazing distances and inhabit a diversity of environments, which occasionally enclose lots of countries and can theoretically spread resistance genes throughout. Furthermore, during the migratory process the proximity to human activity (urban or rural) could strengthen and intensify the prevalence of antibiotic resistant bacteria (Cole et al. 2005; Dolejska et al. 2007; Radhouani et al. 2009; Silva et al. 2010, 2011; Radhouani et al. 2011). Likewise, also in rodents, the presence of acquired antibiotic resistance genes, the potential as reservoirs, and subsequent dissemination have been analyzed (Cabrita et al. 1992; Gilliver et al. 1999; Osterblad et al., 2001; Mallon et al. 2002; Kozak et al. 2009). Surveys in wildlife have been dedicated on how the exposure of wild animals to antibiotics affect antibiotic resistance in enteric bacteria highlighting to understanding the origins, protagonists, and dissemination of antibiotic resistance genes in gastric microbial communities. However, more reports are required to completely comprehend the impact of antibiotic use and the consequent dissemination of resistance through the environment (Allen et al. 2010). It is imperative to refer that antimicrobial resistance reports among completely nonsynanthropic animal species are uncommon, mainly, studies conducted with endangered species that don not have contact with human activity. Primary, the spread of antimicrobial resistance in wild, and possibly remote, ecosystems could be evaluated. Additional, the presence of resistance genes among commensal bacteria that represent an improved health risk could be measured. Moreover, in conservation programs an increased risk of therapeutic failure due to antimicrobial resistance can consequently represent a serious obstacle for the protection of endangered species.

16.6 Conclusion

Antibiotic resistant bacteria are now considered a global threat and public health problem that can occur naturally in bacterial soil communities. However, one of the major causes of the spread of resistant bacteria in soil is the use of antibiotics in farm animals and the application of their manure in agricultural soils. Antibiotics are not fully metabolized in the organism; therefore, a part of the antibiotics is excreted through manure and urine and when it reaches the soil it will lead to a selective pressure on the soil bacteria resulting in an increase in antibiotic resistance. Thus, the terrestrial and aquatic environments become a reservoir of resistant organisms and genes which may affect both humans and animals. Antibiotic resistance genes have been found in wild animals which may play an important role in the dissemination of these genes. The dispersion of antibiotics into soil and water enhances the risk of breaking natural barriers, and currently it is estimated that the majority of antibiotics used in humans and food-producing animals are excreted and accumulated in the environment, largely metabolized and some molecules such as fluoroquinolones have long half-lives. In a wider context, the presence of multidrug-resistant bacteria is frequently reported in wild birds and mammals with no apparent exposure to antimicrobials indirectly; these may also drive from/into an environmental contamination source. Finally, antibiotic resistance does not respect geographical or biological borders; food animals and foods from animal origin are traded worldwide facilitating the spread of resistance.

References

Alder AC, McArdell CS, Golet EM, Ibric S, Molnar E, Nipales NS, Giger W (2001) Occurrence and fate of fluoroquinolone, macrolide, and sulfonamide antibiotics during wastewater treatment and in ambient waters in Switzerland. In: Daughton CG, Jones-Lepp T (eds) Pharmaceuticals and personal care products in the environment: scientific and regulatory issues. American Chemical Society, Washington, DC, pp 56–69

Allen HK, Donato J, Wang HH, Cloud-Hansen KA, Davies J, Handelsman J (2010) Call of the wild: antibiotic resistance genes in natural environments. Nat Rev Microbiol 8:251–259

Araújo C, Torres C, Silva N, Carneiro C, Gonçalves A, Radhouani H, Correia S, da Costa PM, Paccheco R, Zarazaga M, Ruiz-Larrea F, Poeta P, Igrejas G (2010) Vancomycin-resistant enterococci from Portuguese wastewater treatment plants. J Basic Microbiol 50(6):605–609

Barrett JR (2012) Preventing antibiotic resistance in the wild: a new end point for environmental risk assessment. Environ Health Perspect 120:a321

Beers MH (2006) The Merck manual of diagnosis and therapy, 18th edn. Merck, Whitehouse Station, NJ

Bound JP, Voulvoulis M (2004) Pharmaceuticals in the aquatic environment—a comparison of risk assessment strategies. Chemosphere 56:1143–1155

Boxall ABA, Blackwell P, Cavallo R, Kay P, Tolls J (2002) The sorption and transport of a sulphonamide antibiotic in soil systems. Toxicol Lett 131:19–28

Boxall ABA, Kolpin DW, Halling-Sørensen B, Tolls J (2003) Are veterinary medicines causing environmental risks? Environ Sci Technol 37:286–294

Boxall AB, Fogg LA, Blackwell PA, Kay P, Pemberton EJ, Croxford A (2004) Veterinary medicines in the environment. Rev Environ Contam Toxicol 180:1–91

Cabrita J, Rodrigues J, Braganca F, Morgado C, Pires I, Goncalves AP (1992) Prevalence, biotypes, plasmid profile and antimicrobial resistance of Campylobacter isolated from wild and domestic animals from northeast Portugal. J Appl Bacteriol 73:279–285

Christian T, Schneider R, Farber HA, Skutlarek D, Meyer MT, Goldbach HE (2003) Determination of antibiotic residues in manure, soil, and surface waters. Acta Hydrochim Hydrobiol 31:36–44

Cole D, Drum DJ, Stalknecht DE, White DG, Lee MD, Ayers S, Sobsey M, Maurer JJ (2005) Free-living Canada geese and antimicrobial resistance. Emerg Infect Dis 11:935–938

Daghrir R, Drogui P (2013) Tetracycline antibiotics in the environment: a review. Environ Chem Lett 11:209–227

Ding C, He J (2010) Effect of antibiotics in the environment on microbial populations. Appl Microbiol Biotechnol 87:925–941

Ding GC, Radl V, Schloter-Hai B, Jechalke S, Heuer H, Smalla K, Schloter M (2014) Dynamics of soil bacterial communities in response to repeated application of manure containing sulfadiazine. PLoS One 9(3):e92958

Dojmi di Delupis G, Macri A, Civitareale C, Migliore L (1992) Antibiotics of zootechnical use: effects of high and low dose contamination on Daphnia magna. Aquat Toxicol 22:53–60

Dolejska M, Cizek A, Literak I (2007) High prevalence of antimicrobial-resistant genes and integrons in Escherichia coli isolates from Black-headed Gulls in the Czech Republic. J Appl Microbiol 103:11–19

Doretto KM, Rath S (2013) Sorption of sulfadiazine on Brazilian soils. Chemosphere 90:2027–2034

Esiobu N, Armenta L, Ike J (2002) Antibiotic resistance in soil and water environments. Environ Health Res 12:133–144

Gillings MR (2013) Evolutionary consequences of antibiotic use for the resistome, mobilome and microbial pangenome. Front Microbiol 4:4

Gilliver MA, Bennett M, Begon M, Hazel SM, Hart CA (1999) Antibiotic resistance found in wild rodents. Nature 401:233–234

Guenther S, Ewers C, Wieler LH (2011) Extended-spectrum beta-lactamases producing E. coli in wildlife, yet another form of environmental pollution? Front Microbiol 2:246

Halling-Sùrensen B, Nielsen SN, Lanzky PF, Ingerslev F, HoltenLützhùft HC, Jùrgensen SE (1998) Occurrence, fate and effects of pharmaceutical substances in the environment—a review. Chemosphere 36:357–393

Hammesfahr U, Heuer H, Manzke B, Smalla K, Thiele-Bruhn S (2008) Impact of the antibiotic sulfadiazine and pig manure on the microbial community structure in agricultural soils. Soil Biol Biochem 40:1583–1591

Heuer H, Focks A, Lamshöft M, Smalla K, Matthies M, Spiteller M (2008) Fate of sulfadiazine administered to pigs and its quantitative effect on the dynamics of bacterial resistance genes in manure and manured soil. Soil Biol Biochem 40:1892–1900

Hu X, Zhou Q, Luo Y (2010) Occurrence and source analysis of typical veterinary antibiotics in manure, soil, vegetables and groundwater from organic vegetable bases, northern China. Environ Pollut 158:2992–2998

Hughes VM, Datta N (1983) Conjugative plasmids in bacteria of the 'pre-antibiotic' era. Nature 302:725–726

Jechalke S, Heuer H, Siemens J, Amelung W, Smalla K (2014) Fate and effects of veterinary antibiotics in soil. Trends Microbiol 22:536–545

Kahn CM, Line S (2005) Merck veterinary manual, 9th edn. Merck, Whitehouse Station, NJ

Katz JM, Katz SE (1983) Microbial assay systems for determining antibiotic residues in soils. J Assoc Anal Chem 66:640–645

Kay P, Blackwell PA, Boxall ABA (2003) Fate of veterinary antibiotics in a macroporous tile drained clay soil. Environ Toxicol Chem 23(5):1136–1144

Kim S, Carlson K (2007) Temporal and spatial trends in the occurrence of human and veterinary antibiotics in aqueous and river sediment matrices. Environ Sci Technol 41:50–57

Kim KR, Owens G, Kwon S, So K, Lee D, Ok YS (2011) Occurrence and environmental fate of veterinary antibiotics in the terrestrial environment. Water Air Soil Pollut 214:163–174

Kolz AC, Moorman TB, Ong SK, Scoggin KD, Douglass EA (2005) Degradation and metabolite production of tylosin in anaerobic and aerobic swine-manure lagoons. Water Environ Res 77:49–56

Kozak GK, Boerlin P, Janecko N, Reid-Smith RJ, Jardine C (2009) Antimicrobial resistance in Escherichia coli isolates from swine and wild small mammals in the proximity of swine farms and in natural environments in Ontario, Canada. Appl Environ Microbiol 75:559–566

Landers TF, Cohen B, Wittum TE, Larson EL (2012) A review of antibiotic use in food animals: perspective, policy, and potential. Public Health Rep 127:4–22

Lanhammer JP (1989) Untursuchungen zum Verbleib antimikrobiell wirksamer Aezneistoffe als in Gulle und im landwirtschaftlichen Umfeld. PhD dissertation, Universität Bonn, Germany, p 138

Le-Minh N, Khan SJ, Drewes JE, Stuetz RM (2010) Fate of antibiotics during municipal water recycling treatment processes. Water Res 44:4295–4323

Mallon DJ, Corkill JE, Hazel SM, Wilson JS, French NP, Bennett M, Hart CA (2002) Excretion of vancomycin-resistant enterococci by wild mammals. Emerg Infect Dis 8:636–638

Martinez JL (2012) Natural antibiotic resistance and contamination by antibiotic resistance determinants: the two ages in the evolution of resistance to antimicrobials. Front Microbiol 3:1

Midtvedt T (2001) Antibiotics in the Environment: zinc bacitracin ± environmental toxicity and breakdown. In: Kümmerer K (ed) Pharmaceuticals in the environment: sources, fate, effects and risks. Springer, Berlin, pp 77–79

Mitscher LA (1978) The chemistry of the tetracycline antibiotics. Marcel Dekker, Basel, p 330

Nwosu V (2001) Antibiotic resistance with particular reference to soil microorganisms. Res Microbiol 152:421–430

O'Brien TF (2002) Emergence, spread, and environmental effect of antimicrobial resistance: how use of an antimicrobial anywhere can increase resistance to any antimicrobial anywhere else. Clin Infect Dis 34(3):S78–S84

Osterblad M, Norrdahl K, Korpimaki E, Huovinen P (2001) Antibiotic resistance. How wild are wild mammals? Nature 409:37–38

Pallecchi L, Bartoloni A, Paradisi F, Rossolini GM (2008) Antibiotic resistance in the absence of antimicrobial use: mechanisms and implications. Expert Rev Anti Infect Ther 6:725–732

Poeta P, Costa D, Igrejas G, Rodrigues J, Torres C (2007) Phenotypic and genotypic characterization of antimicrobial resistance in faecal enterococci from wild boars (Susscrofa). Vet Microbiol 125:368–374

Poirel L, Kampfer P, Nordmann P (2002) Chromosome-encoded Ambler class A beta-lactamase of Kluyverageorgiana, a probable progenitor of a subgroup of CTX-M extended-spectrum beta-lactamases. Antimicrob Agents Chemother 46:4038–4040

Poulsen PHB, Al-Soud W, Bergmark K, Magid J, Hansen LH, Sørensen SJ (2013) Effects of fertilization with urban and agricultural organic wastes in a field trial—prokaryotic diversity investigated by pyrosequencing. Soil Biol Biochem 57:784–793

Radhouani H, Poeta P, Igrejas G, Gonçalves A, Vinué L, Torres C (2009) Antimicrobial resistance and phylogenetic groups in isolates of Escherichia coli from seagulls at the Berlengas nature reserve. Vet Rec 165:138–142

Radhouani H, Igrejas G, Pinto L, Gonçalves A, Coelho C, Rodrigues J, Poeta P (2011) Molecular characterization of antibiotic resistance in enterococci recovered from seagulls (Larus cachinnans) representing an environmental health problem. J Environ Monitor 13:2227–2233

Radhouani H, Poeta P, Gonçalves A, Pacheco R, Sargo R, Igrejas G (2012) Wild birds as biological indicators of environmental pollution: antimicrobial resistance patterns of Escherichia coli and enterococci isolated from common buzzards (Buteobuteo). J Med Microbiol 61:837–843

Samuelsen OB (1994) Simultaneous determination of ormethoprim and sulphadimethoxine in plasma and muscle of Atlantic salmon (Salmo salar). J Chromatogr B 660:412–417

Sarmah AK, Meyer MT, Boxall AB (2006) A global perspective on the use, sales, exposure pathways, occurrence, fate and effects of veterinary antibiotics (VAs) in the environment. Chemosphere 65:725–759

Silva N, Felgar A, Goncalves A, Correia S, Pacheco R, Araujo C, Igrejas G, Poeta P (2010) Absence of extended-spectrum-beta-lactamase-producing Escherichia coli isolates in migratory birds: song thrush (Turdus philomelos). J Antimicrob Chemother 65:1306–1307

Silva N, Igrejas G, Rodrigues P, Rodrigues T, Gonçalves A, Felgar AC, Pacheco R, Gonçalves D, Cunha R, Poeta P (2011) Molecular characterization of vancomycin-resistant enterococci and extended-spectrum beta-lactamase-containing Escherichia coli isolates in wild birds from the Azores Archipelago. Avian Pathol 40:473–479

Stokes HW, Gillings MR (2011) Gene flow, mobile genetic elements and the recruitment of antibiotic resistance genes into Gram-negative pathogens. FEMS Microbiol Rev 35:790–819

Tasho RP, Cho JY (2016) Veterinary antibiotics in animal waste, its distribution in soil and uptake by plants: a review. Sci Total Environ 563–564:366–376

Thiele-Bruhn S (2003) Pharmaceutical antibiotic compounds in soils—a review. J Plant Nutr Soil Sci 166:145–167

Udikovic-Kolica N, Wichmanna F, Brodericka NA, Handelsmana J (2014) Bloom of resident antibiotic-resistant bacteria in soil following manure fertilization. PNAS 111:42

Uhrich SRW, Navarro DA, Zimmerman L, Aga DS (2014) Assessing antibiotic sorption in soil: a literature review and new case studies on sulfonamides and macrolides. Chem Cent J 8:1–12

Wang M, Tang JC (2010) Research of antibiotics pollution in soil environments and its ecological toxicity. J Agro Environ Sci 29:261–266

World Health Organization (2014) Antimicrobial resistance: global report on surveillance

Wright GD (2010) Antibiotic resistance in the environment: a link to the clinic? Curr Opin Microbiol 13:589–594

Xi C, Zhang Y, Marrs CF, Ye W, Simon C, Foxman B (2009) Prevalence of antibiotic resistance in drinking water treatment and distribution systems. Appl Environ Microbiol 75:5714–5718

Zhou Q, Zhang MC, Shuang CD, Li ZQ, Li AM (2012) Preparation of a novel magnetic powder resin for the rapid removal of tetracycline in the aquatic environment. Chin Chem Lett 23:745–748

Chapter 17
Genotoxicity and Biochemical Toxicity of Soil Antibiotics to Earthworms

Sarfraz Ahmed, Muhammad Ibrahim, Rashem Waheed, Abu Bakar Hassan Azdee, Muhammad Zaffar Hashmi, and Shabir Ahmed

17.1 Introduction

Antibiotics are the biologically active compounds that are used predominantly to treat microbial diseases. Antibiotics are used in large quantities; every year the consumption of these drugs is exceeding several hundred tons (Daughton and Ternes 1999). Current worldwide consumption of antibiotic compounds is approximately 100,000–200,000 mg/year (Van Boeckel et al. 2014). They are utilized as growth enhancers as well as feed additives in livestock, poultry, and fish farming. Since organic manure is widely used in many countries worldwide, the agricultural soils receiving the organic manure may contain considerable amounts of antibiotics and their metabolites (Cabello 2006). Many antibiotics retain poorly in the animal body during digestion and are excreted directly in feces or urine (Kim et al. 2011). Consequently, there are ample chances for antibiotics to be released into natural ecosystems, where they can affect the structure and activity of inhabitants (Martinez 2009) raising serious concerns and questions as environmental threat. Antibiotics, their residues, and metabolites in soils may destroy the balance of the soil ecosystem and would be hazardous to bacteria and other microorganisms in the soil ecosystem (Kümmerer and Henninger 2003; Pei et al. 2006).

The original version of this chapter was revised. An erratum to this chapter can be found at DOI 10.1007/978-3-319-66260-2_22.

S. Ahmed (✉) • M. Ibrahim • R. Waheed • A.B.H. Azdee • S. Ahmed
Department of Biochemistry, Bahauddin Zakariya University, Multan 60800, Pakistan
e-mail: sarfraz.ahmed@bzu.edu.pk; ibrahhimjam@bzu.edu.pk; rashemwaheed@gmail.com; ab.azdee@yahoo.com; shabbirchemist@gmail.com

M.Z. Hashmi
Department of Meteorology, COMSATS Institute of Information Technology, Islamabad, Pakistan

M.Z. Hashmi et al. (eds.), *Antibiotics and Antibiotics Resistance Genes in Soils*, Soil Biology 51, DOI 10.1007/978-3-319-66260-2_17

During the past few decades, a great focus has been put upon the positive effects of antibiotics, but little attention has been paid to their toxicological effects on living organisms from their excessive use (Laville et al. 2004). Compared with the aquatic ecosystem, soil environment is more complex as soil has high buffering capacity, and the toxic effects of antibiotics can be mitigated in soil (Bao et al. 2008). In recent years, more attention has been focused towards the ecotoxicological effects of the antibiotic residues in the environment and soil ecosystem organisms such as microorganisms and several other inhabitants including earthworms (Chee-Sanford et al. 2001; Kim 2006; Kwon et al. 2011).

Earthworms are one of the most common soil organisms which can be found in various environments and are known to play a decisive role in soil nutrient cycling (Bouché 1977). They are segmented animals with tube-shaped body texture, feeding on live and dead organic matter. They represent a dominant part of the soil biomass as they contain about 80% of the total soil biomass and play a crucial role in health, and fertility of soil ecosystems by acting as soil engineers to regulate important soil processes such as they help greatly in organic matter decomposition, nutrient recycling, soil formation, decomposition, fertilization (Test 1984), and regulation of soil structure and dynamics of organic matter(Edwards and Bohlen 1992; Hunt et al. 1988). Since ancient times, earthworms have been used as oriental medicine as an anti-inflammatory, analgesic, and antipyretic agents (Prakash and Gunasekaran 2010). It has also been reported that earthworms produce certain anti-microbiological substances especially active proteins and enzymes, and their surface excrete exhibits potent antimicrobial activity against certain plant pathogens (Bauer et al. 1966; Li et al. 2011). Earthworms use the sensitive receptors on their body surfaces to sense chemicals in the soil (Bouché 1977). Their ecological importance, high biomass in soil, and highly observed sensitivity to relatively low concentrations of environmental toxins make them one of the most suitable bioindicators for assessment of risk in the soil (Landrum et al. 2006). The abundance of earthworms in soil represents the health of soil ecosystems and the level of environmental safety (Xia et al. 2005). They respond against contaminations through certain reactions (Lukkari et al. 2004). Therefore, earthworms have been used in the standardization of acute and sub-acute highly ecotoxicological assays and are preferable bioindicators for assessing chemical contamination of soils by the Organization for Economic Cooperation and Development (OECD) and International Standards Organization (De Silva and van Gestel 2009; No 1984). Earthworms *Eisenia-fetida* and *Eudrilus eugeniae* have been used mostly as model terrestrial organisms by the American Society for Testing and Materials to collaborate the work done on earthworm contamination by different antibiotics (ISO 1998). *E. fetida* is an epigeic (litter-dwelling) earthworm species which is used in the standardization of chronic and acute ecotoxicological assays for industrial chemicals, as it is thought to be a model organism for environmental surveillance (Aoki et al. 1998). It is used frequently as standard toxicology test organism to investigate its acute and sub-acute toxicity by measuring changes in biochemical markers (i.e., activities of antioxidant enzymes of catalase (CAT) and superoxide dismutase (SOD)) (Li et al. 2015). Genotoxicity and as well as biochemical toxicity through antibiotics have been reported in earthworms presenting their sensitivity to a wide range of toxicants (De Silva and van Gestel

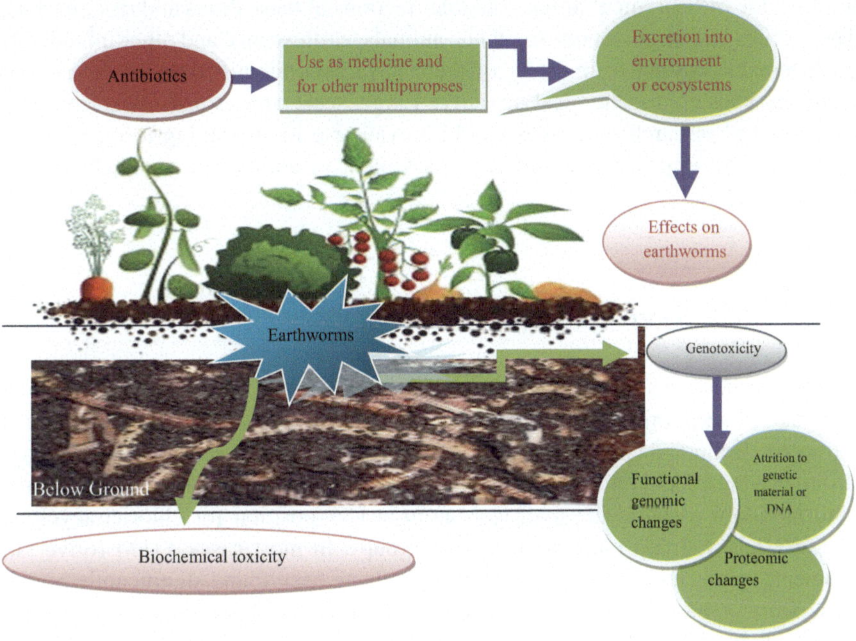

Fig. 17.1 Mode of genotoxicity and biochemical toxicity of soil antibiotics to earthworms

2009) (Fig. 17.1). Functional genomic changes are now being explored owing to molecular "omic" technologies. Similarly, biochemistry, medicinal effects, and results of different antibiotics belonging to different classes on earthworms have been investigated (Lund et al. 2014). Investigation on geno- and biochemical toxic effects of antibiotics on earthworms is an under phase area, so far different types or classes of antibiotics which have been investigated to evaluate their effects on them include tetracycline, enrofloxacin, sulfonamides, cycloheximide, ampicillin, cefotaxime, nystatin, pyridoxal hydrochloride, neomycin, and chloramphenicol.

The main purpose of this chapter is to get a more comprehensive knowledge of the effects of soil antibiotics on earthworms and to provide more information about the potential and ecological risks of antibiotics on soil ecosystems.

17.2 Effects of Different Types or Classes of Antibiotics on Earthworms

Several studies have examined the occurrence of antibiotics in soil, manure, water, and other matrices; however, less investigation has focused on ecological effects, assessing risks, toxicity data, transformation products, contaminant mixtures, and inhabitants in agroecosystems. Various proposed models have been useful in representing toxicity data based on the type and concentration of contaminants. Continuous development of robust and sensitive analytical methods will help to improve the measurement of bioavailable fractions of these drugs and risk analysis. These drugs can be poisonous to plants, animals, earthworms, and other inhabitants of different ecosystems; majority of them have been identified as genotoxic and many as causing biochemical toxicity. Large scale strives involving multiple agencies and research initiatives would be valuable in attempting to unify information and to improve fate and risk assessment of antibiotics on inhabitants in agroecosystems.

17.2.1 Effects of Tetracycline Family on Earthworms

Tetracyclines are one of the most toxic groups of antibiotics in ecosystems. Among this class of antibiotics, tetracycline and chlortetracycline which are derivatives of tetracene (Brain and Grant 2004) can pose a serious threat to ecosystems. The only difference between tetracycline and chlortetracycline in chemical structure is that a chlorine atom replaces the hydrogen atom at C-7 position for chlortetracycline. Halogen has a strong electron attracting groups. In most cases, OTM (olive tail moment) values induced by chlortetracycline have been found significantly higher than tetracycline at the same concentration after 7–14 days exposure effect, increasing the molecular polarity (Zhang et al. 2008). This makes chlortetracycline to integrate with the enzymes system more easily to hinder detoxification, thus increasing toxicity. On the other hand, the presence of chlorine atom increases the fat solubility of chlortetracycline. The outermost layer of the body wall of earthworms is composed of columnar epithelial cells which secrete thin cuticle formation with holes. Thus, chlortetracycline could be more readily taken up by the earth through this membrane and may display acute toxicity (Zhang et al. 2006). This toxicity results in reduced activity of antioxidant enzymes such as SOD (superoxide dismutase) and CAT (chloramphenicol acetyltransferase) and the corresponding increased DNA damage (Honda et al. 2000). The test organism sexcreted SOD or CAT enzymes in response to contaminant exposure to protect against oxidative radicals; however, this mechanism is not triggered at lower doses. The results based on the comet assay showed that two common antibiotics (chlortetracycline and tetracycline) could produce genotoxicity in earthworms by inducing DNA chain breakage with a positive dose–response relationship. Chlortetracycline induces higher DNA damage than tetracycline. Therefore, DNA damage may be used as a more sensitive and effective biomarker for detecting genotoxicity of contaminants in soil (Lin et al. 2012).

Oxytetracycline (OTC) which is another member of tetracycline family is widely used as an antibiotic in veterinary medicines (Robinson et al. 2010). Its applications include growth enhancement, health protection, and disease treatment in livestock industry (Sarmah et al. 2006). Animals cannot completely absorb OTC, so they excrete approximately 25–75% or even 70–90% of OTC in antimicrobial active form in their urine or feces (Hsu et al. 2013). Thus, OTC may have a harmful effect on soil undergoing a series of changes in soil, such as absorption, translocation, and degradation (Robinson et al. 2010). OTC can also cause biochemical toxicity by decreasing the activities of many enzymes like sucrose phosphatase, urease, and hydrogen peroxidase in earthworms (Aad et al. 2010). Research has shown that at lower concentrations, OTC) promotes triticum elongation (Boleas et al. 2005), but induces genotoxicity in earthworms (Wishart et al. 2012).

Irrigation of industrial wastewater, sewage slurry, and precipitation of industrial and traffic waste gases cause production of heavy metals such as Pb, which is considered as a typical heavy metal pollutant in agricultural land (Li et al. 2008). In soil and organic fertilizer samples, antibiotics and heavy metals are often found to coexist (Zhang et al. 2005) and their interactive effect demonstrates potentially hazardous effects to earthworms present in the environment (Chatrchyan et al. 2013). Toxicity of Pb to plants, soil microorganism, and soil fauna are indicated in many studies (Adare et al. 2008). Experiments have shown that the single and combined pollution of OTC and Pb cause coelomocyte apoptosis and damaging of lysosomal membrane stability in earthworms. It suggests that assessing lower toxicity, coelomocyte apoptosis, and membrane stability can be more sensitive and appropriate biomarkers. As compared to single toxicity, combined toxicity can be synergistic and antagonistic (Zhu et al. 2006). It has also been investigated that the interaction of OTC and Pb causes cellular lipid peroxidation because the combined interaction on earthworm lysosomal membrane was a synergistic reaction at treatment (10 mg/kg OTC + 50 mg/kg Pb) (Gao et al. 2014). In this process, oxidative stress-related responses and cellular lipid peroxidation that can damage cell membrane permeability are caused by Pb (Wang et al. 2007). Antagonistic effect of combined toxicity on coelomocyte apoptosis and lysosomal membrane has been observed when OTC concentration is going to be increased (Gao et al. 2014).

17.2.2 Effects of Enrofloxacin on Earthworms

Enrofloxacin (EFLX) is a broad-spectrum antibiotic used against Gram-positive and Gram-negative microorganisms. It is a fluorinated quinolone exhibiting bactericidal activity (Dickens et al. 1997). In its mode of action, it inhibits an enzyme involved in passing genetic information to the daughter cells during cell division, DNA gyrase (Woodward and Rao 2001). EFLX has its applications in veterinary medicine, where it is used for controlling systemic infections. For rapid growth of animals, it is also applied in their feed at subtherapeutic level (Walker et al. 1992; Wu et al. 2005). EFLX is only partially excreted in animal body and is then excreted in feces (Wu et al. 2005). When animal waste is applied as a supplement fertilizer, plenty of EFLX and its metabolites are spread with manure in the environment. So, its continuous release into the environment results its pseudo-persistent occurrence in the environment (Yan et al. 2013). Research on EFLX stability indicates that in dark environment, its half-life exceeds up to 120 days (Wu and Kanamori 2005). This demonstrates that even at lower concentrations, EFLX can be retained in soil for longer periods of time (Pierini et al. 2004). This exposure of EFLX residues may even steer microbial drug resistance (Gao et al. 2008). The potential impact of EFLX release in soils has caused growing public concerns nowadays because the ability of microorganisms to utilize the carbon resource has been reported to be decreased significantly under the pressure of EFLX (Ma et al. 2007).

It has been shown by the acute toxicity test that survival of earthworms is not affected by EFLX residues in soil (No 1984). It has also been probed that lower concentrations of EFLX (below 0.634 mg/cm^2) did not affect survival of earthworms, but higher EFLX concentrations may trigger morphological changes (Gao et al. 2008). Effect of EFLX on the growth of earthworms is dependent on dose and exposure time. Growth can be inhibited by exposure under 2.0 g/kg of EFLX for more than 3 days or 1.0 g/kg of EFLX for more than 2 weeks (Gao et al. 2008). Long-time exposure or high concentrations of EFLX elicit adverse effects on the growth of earthworms (Li et al. 2015). Research has revealed that EFLX induces reproductive toxicity in earthworms; 50 mg/kg of EFLX in soils could stop the reproduction of earthworms. However, EFLX toxicity towards reproduction of earthworms was low as compared to other microbial agents (Li et al. 2015). Several studies indicated that EFLX causes changes in superoxide dismutase (SOD) and catalase (CAT) enzymes which are responsible for the metabolism of earthworms (Dong et al. 2012; Gao et al. 2008). So, it can be presumed that EFLX may induce biochemical stress in earthworms rather than genotoxic effects.

17.2.3 Effects of Sulfamethoxazole and Trimethoprim on Earthworms

Sulfamethoxazole belongs to a major antibiotic class sulfonamide. Whereas, trimethoprim is a pyrimidine inhibitor of dihydrofolate reductase, an antibacterial related to pyrimethamine. Its efficiency is potentiated by sulfonamides as trimethoprim–sulfamethoxazole combinations. It can be used alone as an antimalarial agent as well (Chatrchyan et al. 2014). Trimethoprim when used alone it has bacteriostatic action, but when used in combination with sulamethoxazole, it causes bacteriocidal action and produces marked synergy against Gram-negative bacteria (Reeves et al. 1969). These drugs and/or their degradation products cause soil toxicity or complexity on incorporation into the environment (Pino et al. 2015). They may be accumulated in the soil and can perforate in earthworms. The half-life for the sulfamethoxazole is approximately in the range of 9–60 days (Lin and Gan 2011), while trimethoprim shows variable behavior in two different types of soil. In one type of soil, its half-life was observed 26 days, while in another type of soil it was found hardly degraded (Lin and Gan 2011). These drugs inhibit soil enzymatic activity (Kong et al. 2006; Liu et al. 2009), affecting its fertility. Thus, these substances may change the nature of food source for the earthworms (Kelsey et al. 2005; Klok et al. 2006), which can possibly affect their lives and ecosystem patterns. To how much extent, earthworms can be affected by such patterns, is still a matter of discussion and further investigation .So far, sulfamethoxazole and trimethoprim showed no significant toxic effects on *E. fetida*; thus, there is a need for further studies to determine the toxic effects of sulfamethoxazole and trimethoprim and their degradation products on earthworms (Pino et al. 2015).

17.2.4 *Effects of Cycloheximide, Ampicillin, Cefotaxime, Nystatin, Pyridoxal Hydrochloride, Neomycin, and Chloramphenicolon Earthworms*

Earthworms can often adversely be affected by several types of antibiotics and/or their residues present in soil. In terms of investigation and extended research, this is an under-phase area. Antibiotics can be of microbial origin or purely chemo-synthetic or semi-chemo-synthetic (Murugan et al. 2015) and probably can elicit biochemical as well as genotoxic effects in earthworms. A research conducted toxicity tests of seven different antibiotics such as cycloheximide, ampicillin, cefotaxime, nystatin, pyridoxal hydrochloride, neomycin, and chloramphenicol on earthworm both in nutrient agar medium and in its natural habitat which involved different antibiotics incorporated compost. It was found that the earthworms inoculated into the sterile compost remained alive, while in case of agar plates earthworms died only in nystatin added agar plates; however, they remained alive in all other antibiotics incorporated plates (Murugan et al. 2015). Earthworms incubated in nystatin incorporated agar were found to be dead after 4 h, while earthworms placed in the other antibiotics remained alive even after 24 h. Thus, nystatin has a negative effect on earthworms. Time activity showed that after 270 min, the earthworm's tail was cut off and after 330 min, the earthworms were found to be completely dead. Nystatin caused damage especially in the epidermal and circular muscle layer of earthworms (Murugan et al. 2015). As no biochemical and genotoxic effects were studied, further research is needed to explore the exact reason for the death of earthworms with its tail amputated by nystatin and similarly for other types of antibiotics.

17.3 Possible Mechanisms of Geno- and Biochemical Toxicity in Earthworms

Despite their importance in gene expression, mechanisms are not yet understood in soil invertebrates. Until now, antibiotic-induced changes in genome expression in natural biota are still being studied through structural alteration of DNA (Vasseur and Bonnard 2014). Exposure of terrestrial species to antibiotics that may alter genomic function has become an increasing topic of research in the last decade.

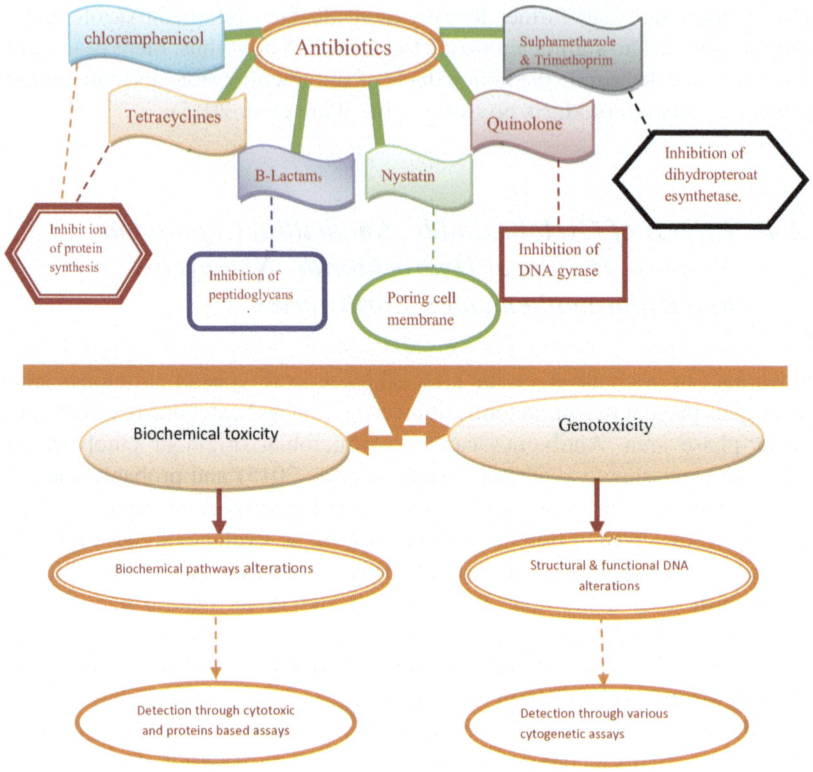

Fig. 17.2 Mode of action of antibiotics and their possible mechanistic effects on earthworms followed by detection

Indeed, genome disturbances due to genetic and epigenetic mechanisms may impair growth as well as reproduction and population dynamics in the long term (Burcham 1999). The major types of antibiotics whose effects on earthworms have been investigated till now include β-lactams, penicillins, cephalosporins, monobactams, carbapenems; tetracyclines (chlortetracyclines, oxytetracyclines, and tetracyclines); nystatin; sulfonamides (sulfamethazole); quinines (enrofloxacin); and diaminopyridine (trimethoprim). In their mechanism of action, nystatin binds with ergosterol in cell membrane; β lactams affect peptidoglycan synthesis by affecting transpeptidase and peptide cross-linking (Sainsbury et al. 2011); tetracyclines, chloramphenicol, and macrolides inhibit protein synthesis (Chopra and Roberts 2001); and quinolones inhibit DNA replication by affecting enzymes, namely DNA gyrase and the eukaryotic topoisomerase II (Barnard and Maxwell 2001) (Fig. 17.2). Different antibiotics based on their mode of action and targets may possibly affect similar or different biochemical pathways and molecular targets in different lower organisms. Likewise to other genotoxic substances, these drugs in their mechanism of action may induce attrition to the genetic material through interactions with the structure and sequence of DNA (Sugden et al. 2001) (Fig. 17.2). They can cause DNA adducts, DNA cross-linking, sister chromatid exchange, DNA breaks in micronuclei, chromosomal aberrations, chromosome mutations, and gene mutations (Chen et al. 2010).

Biochemical toxicity of antibiotics may affect enzyme mechanisms, regulation of drug metabolism, and their related genes and genetic factors that cause differences in responsiveness to antibiotics, and bioactivation of antibiotics to toxic intermediates and drug/drug interactions. Similarly, it can transform enzymes, transporters, and cellular processes in specific organs such as lung, liver, brain, and intestine. Biochemical toxicity as the production of reactive oxygen species (ROS) reflected by changes in antioxidant enzyme activities (superoxide dismutase, catalase) and increased lipid peroxidation (indicated by malondialdehyde) has been measured in parallel to genotoxicity (Nakashima et al. 2008).

17.4 Detection Methods of Geno- and Biochemical Toxicity in Earthworms

Genotoxic studies focused on detection of structural DNA alterations (Dearfield et al. 2002; Müller et al. 2003) and such detections can be through various methods. The measurement of DNA adducts and the comet assays have improved the potential for investigating populations at risk. Indeed, they can be applied in situ and in different cell types without prior knowledge of karyotype and cell turnover (Jha 2008). The single cell gel electrophoresis assay or comet assay are efficient tools to measure DNA damage in individual cells and are widely used in toxicology and ecotoxicology (Frenzili et al. 2009; Jha 2008; Vasseur et al. 2012). It detects single and double strand breaks, alkali labile sites, oxidative DNA damage, and DNA cross-links. Whereas, coelomocytes represent the main cellular type used in earthworms in the evaluation of genotoxicity of contaminants by the comet assay (Lionetto et al. 2012). Using the comet assay, apoptotic cells can be easily visualized and characterized with a disintegrated nucleus. The sprayed genetic material that can be clearly discriminated from damaged nuclei of non-apoptotic cells (Vasseur and Bonnard 2014).The SOS/umu assay test enables to evaluate DNA damage which is based on the alterations in the induction of the SOS response caused by DNA damage (NIB 2013). The micronucleus test for structural and numerical chromosomal aberrations can also be applied for the evaluation of genotoxicity (Furman 2008). A micronucleus is a small structure separate from the nucleus. It contains nuclear DNA arisen from DNA fragments and/or whole chromosomes which are incorporated in the daughter cell during mitotic division. These structure abnormalities can be due to mitotic loss of acentric chromosomal fragments, mechanical problems from chromosomal exchange and breakage, mitotic loss of chromosomes, and apoptosis (Furman 2008). Different types of aberrations detected in cells affected by genotoxic substances are chromosome breaks, chromatid and chromosome gaps, chromatid deletions, translocation, fragmentation, and complex rearrangements which can be detected by various cytogenetic tests (Furman 2008).

Biochemical toxicity as the production of ROS reflected by changes in antioxidant enzymes activities can be measured in parallel to genotoxicity with the neutral red retention time (NRRT) assay. DNA oxidative damage can be quantified by HPLC analysis of the indicator of DNA oxidative lesions (Nakashima et al. 2008). Furthermore, biochemical analysis can be performed in terms of hematology and histopathology. Testing whether a drug inhibits a particular enzyme or binds to a particular receptor or other biomolecule is the most direct way to test a drug for its specific mechanism of action. Enzyme and receptor-binding assays or specific protein assays tend to be reliable. The specific protein, protein complex, receptor, or other biomolecules of interest can also be investigated. Several types of cell-based assays can be employed to predict acute toxicity, such as plasma membrane permeability, cell proliferation, and adenosine triphosphate (ATP) content. Simple cytotoxicity assays have long been used to predict animal toxicity. Methods for measuring cytotoxicity usually involve direct measurement of the fraction of cells that have intact membranes, for example, with neutral red uptake or fluorescent DNA dye uptake; measurement of the metabolism of surviving cells, for instance, with reduction of 3-(4,5-dimethyl-thiazol-2-yl)-2,5-diphenyltetrazolium bromide (MTT), uridine uptake, or reduction of Alomar blue; cell number, measurement of ATP content, total DNA content, cell proliferation, total protein content, and several others (ILO 2011).

17.5 Future Prospects and Challenges

There have been a number of studies on the occurrence of antibiotics in manure, soil, water, and other matrices in the soil; however, the ecological effects of antibiotics and their role in the development of geno- and biochemical toxicity in soil earthworms have not been clearly probed till now. Data available is incomplete and additional research is required to completely clarify the current reservoir of antibiotic related contaminants in soil ecosystem. The challenges in achieving accurate risk assessment of soil antibiotics on earthworms remain formidable till now due to various factors, ranging from the accuracy and variability in analytical techniques measuring bioavailability and toxicity of antibiotic mixtures and their transport. The impact of antibiotics in soil ecosystem and the development of geno- and biochemical toxicity in earthworms as well as the methods that allow the measurement of the bioavailable fraction of antibiotics are extremely valuable. In future, environmental risk assessment of antibiotics should incorporate mixture toxicity and account for toxicity contributions from transports and other forms of contamination. Additional focus on standard research approaches and execution steps in obtaining the reliable data are essential to provide an inclusive assessment of antibiotics and their toxicity in earthworms.

17.6 Conclusion

The degradation of different antibiotics is different and probably slow in soil particles. But, their residues, bulk accumulation, metabolites, and long-term exposure in agroecosystems can cause several adverse effects on growth, reproduction as well as on physiological activities of earthworms. It can be concluded from above discussion that antibiotics may cause acute and sub-acute toxicity to earthworms resulting in genotoxcity as well as biochemical toxicity in earthworms. The long-term exposure of these drugs to earthworms, test conducting, toxicity level, and the mechanisms involved need to be investigated further.

References

Aad G, Abbott B, Abdallah J, Abdelalim A, Abdesselam A, Abdinov O, Abi B, Abolins M, Abramowicz H, Abreu H (2010) The ATLAS simulation infrastructure. Eur Phys J C 70:823–874

Adare A, Afanasiev S, Aidala C, Ajitanand N, Akiba Y, Al-Bataineh H, Alexander J, Al-Jamel A, Aoki K, Aphecetche L (2008) Suppression pattern of neutral pions at high transverse momentum in Au+ Au collisions at s N N= 200 GeV and constraints on medium transport coefficients. Phys Rev Lett 101:232301

Aoki H, Iso S, Kawai H, Kitazawa Y, Tada T (1998) Space-time structures from IIB matrix model. Prog Theor Phys 99:713–745

Bao Y-y, Zhou Q-x, Xie X-j (2008) Influence of tetracycline kind antibiotics on the control of wheat germination and root elongation. China Environ Sci 6:024

Barnard FM, Maxwell A (2001) Interaction between DNA gyrase and quinolones: effects of alanine mutations at GyrA subunit residues Ser(83) and Asp(87). Antimicrob Agents Chemother 45:1994–2000

Bauer AW, Kirby WM, Sherris JC, Turck M (1966) Antibiotic susceptibility testing by a standardized single disk method. Am J Clin Pathol 45:493–496

Boleas S, Alonso C, Pro J, Fernández C, Carbonell G, Tarazona JV (2005) Toxicity of the antimicrobial oxytetracycline to soil organisms in a multi-species-soil system (MS· 3) and influence of manure co-addition. J Hazard Mater 122:233–241

Bouché M (1977) Strategies lombriciennes. Ecol Bull 25:122–132

Brain SD, Grant AD (2004) Vascular actions of calcitonin gene-related peptide and adrenomedullin. Physiol Rev 84:903–934

Burcham PC (1999) Internal hazards: baseline DNA damage by endogenous products of normal metabolism. Mutat Res 443:11–36

Cabello FC (2006) Heavy use of prophylactic antibiotics in aquaculture: a growing problem for human and animal health and for the environment. Environ Microbiol 8:1137–1144

Chatrchyan S, Khachatryan V, Sirunyan A, Tumasyan A, Adam W, Bergauer T, Dragicevic M, Erö J, Fabjan C, Friedl M (2013) Measurement of the B s 0→ μ+ μ− branching fraction and search for B 0→ μ+ μ− with the CMS experiment. Phys Rev Lett 111:101804

Chatrchyan S, Khachatryan V, Sirunyan A, Tumasyan A, Adam W, Bergauer T, Dragicevic M, Erö J, Fabjan C, Friedl M (2014) Search for the standard model Higgs boson produced in association with a W or a Z boson and decaying to bottom quarks. Phys Rev D 89:012003

Chee-Sanford JC, Aminov RI, Krapac I, Garrigues-Jeanjean N, Mackie RI (2001) Occurrence and diversity of tetracycline resistance genes in lagoons and groundwater underlying two swine production facilities. Appl Environ Microbiol 67:1494–1502

Chen T, Mei N, Fu PP (2010) Genotoxicity of pyrrolizidine alkaloids. J Appl Toxicol 30:183–196

Chopra I, Roberts M (2001) Tetracycline antibiotics: mode of action, applications, molecular biology, and epidemiology of bacterial resistance. Microbiol Mol Biol Rev 65:232–260

Daughton CG, Ternes TA (1999) Pharmaceuticals and personal care products in the environment: agents of subtle change? Environ Health Perspect 107:907

Dearfield KL, Cimino MC, McCarroll NE, Mauer I, Valcovic LR (2002) Genotoxicity risk assessment: a proposed classification strategy. Mutat Res 521:121–135

De Silva PM, van Gestel CA (2009) Comparative sensitivity of Eisenia andrei and Perionyx excavatus in earthworm avoidance tests using two soil types in the tropics. Chemosphere 77:1609–1613

Dickens GR, Castillo MM, Walker JC (1997) A blast of gas in the latest Paleocene: simulating first-order effects of massive dissociation of oceanic methane hydrate. Geology 25:259–262

Dong L, Gao J, Xie X, Zhou Q (2012) DNA damage and biochemical toxicity of antibiotics in soil on the earthworm Eisenia fetida. Chemosphere 89:44–51

Edwards CA, Bohlen PJ (1992) The effects of toxic chemicals on earthworms. In: Ware GW (ed) Reviews of environmental contamination and toxicology. Springer, New York, pp 23–99

Frenzili G, Nigro M, Lyons B (2009) A review: use of the comet assay in aquatic environmental impact assessments. Mutat Res 681:80–92

Furman GM (2008) Genotoxicity testing for pharmaceuticals current and emerging practices

Gao Y, Sun X, Sun Z, Zhao N, Li Y (2008) Toxic effects of enrofloxacin on growth rate and catalase activity in Eisenia fetida. Environ Toxicol Pharmacol 26:177–180

Gao M, Zhou Q, Song W, Ma X (2014) Combined effects of oxytetracycline and Pb on earthworm Eisenia fetida. Environ Toxicol Pharmacol 37:689–696

Honda M, Yamada Y, Tomonaga M, Ichinose H, Kamihira S (2000) Correlation of urinary 8-hydroxy-2′-deoxyguanosine (8-OHdG), a biomarker of oxidative DNA damage, and clinical features of hematological disorders: a pilot study. Leukmia Res 24:461–468

Hsu PD, Scott DA, Weinstein JA, Ran FA, Konermann S, Agarwala V, Li Y, Fine EJ, Wu X, Shalem O (2013) DNA targeting specificity of RNA-guided Cas9 nucleases. Nat Biotechnol 31:827–832

Hunt H, Ingham E, Coleman D, Elliott E, Reid C (1988) Nitrogen limitation of production and decomposition in prairie, mountain meadow, and pine forest. Ecology 69:1009–1016

ILO (2011) Toxicology test methods. International labour Organization

ISO (1998) 11268-2: Soil quality—effects of pollutants on earthworms (Eisenia fetida)—Part 2: Determination of effects on reproduction. International Organization for Standardization, Geneva

Jha AN (2008) Ecotoxicological applications and significance of the comet assay. Mutagenesis 23:207–221

Kelsey JW, Colino A, White JC (2005) Effect of species differences, pollutant concentration, and residence time in soil on the bioaccumulation of 2, 2-BIS (p-chlorophenyl)-1, 1-dichloroethylene by three earthworm species. Environ Toxicol Chem 24:703–708

Kim S-C (2006) Occurrence, fate, and transport of human and veterinary antibiotics in the watershed. Colorado State University

Kim K-R, Owens G, Kwon S-I, So K-H, Lee D-B, Ok YS (2011) Occurrence and environmental fate of veterinary antibiotics in the terrestrial environment. Water Air Soil Pollut 214:163–174

Klok C, van der Hout A, Bodt J (2006) Population growth and development of the earthworm Lumbricus rubellus in a polluted field soil: possible consequences for the godwit (Limosa limosa). Environ Toxicol Chem 25:213–219

Kong W-D, Zhu Y-G, Fu B-J, Marschner P, He J-Z (2006) The veterinary antibiotic oxytetracycline and Cu influence functional diversity of the soil microbial community. Environ Pollut 143:129–137

Kümmerer K, Henninger A (2003) Promoting resistance by the emission of antibiotics from hospitals and households into effluent. Clin Microbiol Infect 9:1203–1214

Kwon S, Owens G, Ok Y, Lee D, Jeon W-T, Kim J-G, Kim K-R (2011) Applicability of the Charm II system for monitoring antibiotic residues in manure-based composts. Waste Manag 31:39–44

Landrum M, Cañas JE, Coimbatore G, Cobb GP, Jackson WA, Zhang B, Anderson TA (2006) Effects of perchlorate on earthworm (Eisenia fetida) survival and reproductive success. Sci Total Environ 363:237–244

Laville N, Aıt-Aıssa S, Gomez E, Casellas C, Porcher J (2004) Effects of human pharmaceuticals on cytotoxicity, EROD activity and ROS production in fish hepatocytes. Toxicology 196:41–55

Li Q, Du W, Liu D (2008) Perspectives of microbial oils for biodiesel production. Appl Microbiol Biotechnol 80:749–756

Li W, Wang C, Sun Z (2011) Vermipharmaceuticals and active proteins isolated from earthworms. Pedobiologia 54(suppl):S49–S56

Li Y, Hu Y, Ai X, Qiu J, Wang X (2015) Acute and sub-acute effects of enrofloxacin on the earthworm species Eisenia fetida in an artificial soil substrate. Eur J Soil Biol 66:19–23

Lin K, Gan J (2011) Sorption and degradation of wastewater-associated non-steroidal anti-inflammatory drugs and antibiotics in soils. Chemosphere 83:240–246

Lin J-W, Chen C-W, Peng C-Y (2012) Potential hazard analysis and risk assessment of debris flow by fuzzy modeling. Nat Hazard 64:273–282

Lionetto MG, Calisi A, Schettino T (2012) Earthworm biomarkers as tools for soil pollution assessment. Intech, Rijeka

Liu F, Ying G-G, Tao R, Zhao J-L, Yang J-F, Zhao L-F (2009) Effects of six selected antibiotics on plant growth and soil microbial and enzymatic activities. Environ Pollut 157:1636–1642

Lukkari T, Taavitsainen M, Soimasuo M, Oikari A, Haimi J (2004) Biomarker responses of the earthworm Aporrectodea tuberculata to copper and zinc exposure: differences between populations with and without earlier metal exposure. Environ Pollut 129:377–386

Lund MB, Kjeldsen KU, Schramm A (2014) The earthworm—Verminephrobacter symbiosis: an emerging experimental system to study extracellular symbiosis. Front Microbiol 5:128

Ma Y, Chen Z, Zeng Z (2007) Effects of enrofloxacin on functional diversity of soil microbial communities. Acta Ecol Sin 27:3400–3406

Martinez JL (2009) Environmental pollution by antibiotics and by antibiotic resistance determinants. Environ Pollut 157:2893–2902

Müller L, Blakey D, Dearfield KL, Galloway S, Guzzie P, Hayashi M, Kasper P, Kirkland D, MacGregor JT, Parry JM (2003) Strategy for genotoxicity testing and stratification of genotoxicity test results—report on initial activities of the IWGT Expert Group. Mutat Res 540:177–181

Murugan K, Shiji G, Umamaheswari HSS (2015) Toxicity studies of antibiotics in earthworm, Eudrilus eugeniae. Int J Pure App Biosci 3:241–255

Nakashima T, Okada T, Asahi J, Yamashita A, Kawai K, Kasai H, Matsuno K, Gamou S, Hirano T (2008) 8-Hydroxydeoxyguanosine generated in the earthworm Eisenia fetida grown in metal-containing soil. Mutat Res 654:138–144

NIB (2013) In vitro genotoxicity testing. National Institute of Biology

No T (1984) 207: Earthworm, acute toxicity tests. OECD guidelines for the testing of chemicals, Section 2: Effects on biotic systems

Pei R, Kim S-C, Carlson KH, Pruden A (2006) Effect of river landscape on the sediment concentrations of antibiotics and corresponding antibiotic resistance genes (ARG). Water Res 40:2427–2435

Pierini E, Famiglini G, Mangani F, Cappiello A (2004) Fate of enrofloxacin in swine sewage. J Agric Food Chem 52:3473–3477

Pino MR, Val J, Mainar AM, Zuriaga E, Español C, Langa E (2015) Acute toxicological effects on the earthworm Eisenia fetida of 18 common pharmaceuticals in artificial soil. Sci Total Environ 518:225–237

Prakash M, Gunasekaran G (2010) Gastroprotective effect of earthworm paste (Lampito mauritii, Kinberg) on experimental gastric ulcer in rats. Eur Rev Med Pharmacol Sci 14:171–176
Reeves D, Faiers M, Pursell R, Brumfitt W (1969) Trimethoprim-sulphamethoxazole: comparative study in urinary infection in hospital. Br Med J 1:541–544
Robinson MD, McCarthy DJ, Smyth GK (2010) edgeR: a bioconductor package for differential expression analysis of digital gene expression data. Bioinformatics 26:139–140
Sainsbury S, Bird L, Rao V, Shepherd SM, Stuart DI, Hunter WN, Owens RJ, Ren J (2011) Crystal structures of penicillin-binding protein 3 from Pseudomonas aeruginosa: comparison of native and antibiotic-bound forms. J Mol Biol 405:173–184
Sarmah AK, Meyer MT, Boxall AB (2006) A global perspective on the use, sales, exposure pathways, occurrence, fate and effects of veterinary antibiotics (VAs) in the environment. Chemosphere 65:725–759
Sugden KD, Campo CK, Martin BD (2001) Direct oxidation of guanine and 7,8-dihydro-8-oxoguanine in DNA by a high-valent chromium complex: a possible mechanism for chromate genotoxicity. Chem Res Toxicol 14:1315–1322
Test O (1984) 207: earthworm, acute toxicity tests. Organization for Economic Cooperation and Development (ed). OECD guidelines for testing of chemicals
Van Boeckel TP, Gandra S, Ashok A, Caudron Q, Grenfell BT, Levin SA, Laxminarayan R (2014) Global antibiotic consumption 2000 to 2010: an analysis of national pharmaceutical sales data. Lancet Infect Dis 14:742–750
Vasseur P, Bonnard M (2014) Ecogenotoxicology in earthworms: a review. Curr Zool 60:255–272
Vasseur P, Atienzar F, Cossu-Leguille C, Rodius F, Lemière S (2012) Biomarkers of genotoxicity for in situ studies at individual and population levels. In: Amiard-Triquet C, Amiard J-C, Rainbow PS (eds) Ecological biomarkers: indicators of ecotoxicological effects. CRC, Boca Raton, FL, pp 327–360
Walker R, Stein G, Hauptman J, MacDonald K (1992) Pharmacokinetic evaluation of enrofloxacin administered orally to healthy dogs. Am J Vet Res 53:2315–2319
Wang J, Agrawala M, Cohen MF (2007) Soft scissors: an interactive tool for realtime high quality matting. In: ACM transactions on graphics (TOG). ACM, p 9
Wishart DS, Jewison T, Guo AC, Wilson M, Knox C, Liu Y, Djoumbou Y, Mandal R, Aziat F, Dong E (2012) HMDB 3.0—The human metabolome database in 2013. Nucleic Acids Res 41 (database issue):D801–D807. doi:10.1093/nar/gks1065
Woodward C, Rao S (2001) Ab-initio simulation of isolated screw dislocations in bcc Mo and Ta. Philos Mag A 81:1305–1316
Wu Y-M, Kanamori H (2005) Rapid assessment of damage potential of earthquakes in Taiwan from the beginning of P waves. Bull Seismol Soc Am 95:1181–1185
Wu Y-b, Wang Z-s, Liao X-d, Chen Z (2005) Study on the excretion of enrofloxacin in chicken and its degradation in chicken feces. Acta Veterinaria et Zootechnica Sinica 36:1069
Xia K, Bhandari A, Das K, Pillar G (2005) Occurrence and fate of pharmaceuticals and personal care products (PPCPs) in biosolids. J Environ Qual 34:91–104
Yan W, Zhang J, Jing C (2013) Adsorption of Enrofloxacin on montmorillonite: two-dimensional correlation ATR/FTIR spectroscopy study. J Colloid Interface Sci 390:196–203
Zhang Y, Tan Y-W, Stormer HL, Kim P (2005) Experimental observation of the quantum Hall effect and Berry's phase in graphene. Nature 438:201–204
Zhang J, Liang Y, Dong Y, Huang M (2006) Toxicity of tetracyclines and their epi-isomers on the luminescent bacterium. Chin J Toxicol 20:279–281
Zhang H, Luo Y, Zhou Q (2008) Research advancement of eco-toxicity of tetracycline antibiotics. J Agro Environ Sci 27:407–413
Zhu S, Oberdörster E, Haasch ML (2006) Toxicity of an engineered nanoparticle (fullerene, C 60) in two aquatic species, Daphnia and fathead minnow. Mar Environ Res 62:S5–S9

Chapter 18
Potential Environmental, Ecological and Health Effects of Soil Antibiotics and ARGs

Biljana Balabanova

18.1 Introduction

Since their introduction into medicine in the 1940s, antibiotics have been central to modern healthcare. The generic term "antibiotic" is generally used to denote any class of organic molecule that inhibits/dysfunctions or kills microbes with specific interactions with bacterial targets, without any consideration of the source of the particular compound or class (Davies and Davies 2010). Antibiotics can be declared as the most successful family of drugs so far developed for improving human health. Besides this fundamental application, antibiotics have also been used for preventing and treating animal and plant infections as well as for promoting growth in animal farming (Smith et al. 2002; Singer et al. 2003; Cabello 2006; Martinez 2009). All these applications made antibiotics to be released in large amounts in natural ecosystems. China is the largest producer and user of antibiotics in the world, based on market sales data (Martinez 2009; Pruden et al. 2013). Penicillin was one of the first widely available antibiotics (discovered by Alexander Fleming in 1928) and entered mass production in the early 1940s. It was soon followed by streptomycin, tetracycline and other antibiotics (Bergstrom and Feldgarden 2008; Taubes 2008).

The original version of this chapter was revised. An erratum to this chapter can be found at DOI 10.1007/978-3-319-66260-2_22.

B. Balabanova (✉)
Faculty of Agriculture, University "Goce Delčev", Krste Misirkov bb, 2000 Štip, Republic of Macedonia
e-mail: biljana.balabanova@ugd.edu.mk

M.Z. Hashmi et al. (eds.), *Antibiotics and Antibiotics Resistance Genes in Soils*, Soil Biology 51, DOI 10.1007/978-3-319-66260-2_18

Because of the intensive use of antibiotics for human (domestic and hospital use), veterinary and agriculture purposes, these compounds are continuously released into the environment from anthropogenic introducing sources (Witte 1998). Ferber (2003) and Singer et al. (2003) singled out several "hotspots" of potential evolution and spreading of antibiotic resistance into the environment. Eventually, the increased production capacity of antibiotics encouraged other applications of the drugs outside of medical settings. For example, low levels of antibiotic agents were more frequently being added as a prophylactic to animal feed because they were found to promote growth in livestock (Frost 1991). Although detailed estimates of annual use and production are broadly not available (Sarmah et al. 2006), trade data suggest an exponential increase in antibiotic production prior to 1990 with >50% of that manufacture being for agricultural purposes (Levy 2002; Ungemach 2000).

As stated by the World Health Organization, the increasing emergence of antibiotic resistance in human pathogens is a special concern, not only for treating infectious disease but also for other pathologies in which antibiotic prophylaxis is needed for avoiding associated infections. The spread of "antibiotic-resistant bacteria" means that commonplace medical procedures once previously taken for granted could be consigned to medical limbo. It is important to remark that several antibiotics are produced by environmental microorganisms (Waksman and Woodruff 1940). To understand in full the development of resistance, we will thus need to address the study of antibiotics and their resistance genes, not just in clinics but in natural (non-clinical) environments also (Martinez 2008, 2009; Martinez and Baquero 2000, 2002; Martinez and Perez-Diaz 1990; Martinez et al. 1998, 2007, 2009). The situation concerning antibiotics and their resistances resembles in some aspects to heavy metal contamination. Heavy metals, very similar to antibiotics, are natural compounds present in different ecosystems. However, their utilization by humans has increased their bioavailability, leading to dramatic changes in polluted ecosystems. Differing to heavy metals that challenge all forms of life, antimicrobials mainly alter the microbiosphere, and probably because of this, the consequences of antibiotic pollution on the biodiversity have received less attention (Martinez 2009). Antibiotic resistance is a threat to human and animal health globally, and key measures are required to reduce the risks posed by antibiotic resistance genes that occur in the environment.

The World Health Organization has pointed for the emergence of antimicrobial resistance as a complex occurrence driven by many interconnected factors (WHO 2014, 2015). Over the past two decades, scientists have focused to include the environment as a source for introducing of resistance genes and as a site of antimicrobial resistance evolution (Kümmerer 2009; Pruden et al. 2012; Franklin et al. 2016). In general, dominant trends which contribute to a global scale-up in antibiotic consumption are the following: (a) rising incomes lead to increasing access to antibiotics, resulting with resistance, and (b) increased demand for animal protein and resulting intensification of food animal production are leading to greater use of antibiotics in agriculture, again driving resistance. The greater volume of antibiotics is used; greater is the chance that ARBs will prevail in the contents for surviving at the bacterial level. The US Centers for Disease Control and Prevention

(CDC) estimates that antibiotic resistance is responsible for more than two million infections and 23,000 deaths each year in the United States, estimated 25,000 deaths/per year in Europe and 58,000 neonatal sepsis deaths/per year in India (WHO 2015; Gelband et al. 2015).

18.1.1 Heavy Metal vs. Antimicrobial Resistance in Natural Ecosystems

Heavy metal resistance in natural ecosystems may help to understand antibiotic resistance in the environment. The elements involved in the resistance to heavy metals are encoded in the chromosomes of bacteria like *Ralstonia metallidurans* (Mergeay et al. 2003). However, strong selective pressure due to anthropogenic pollution has made that these chromosomally encoded determinants are now present in gene-transfer units (Silver and Phung 2005; Nies 2003; Berendonk et al. 2015). Antibiotic resistance genes that were naturally present in the chromosomes of environmental bacteria (Okoh et al. 2007; Fajardo and Martínez 2008; Igbinosa and Okoh 2009; Gelband et al. 2015) are present in plasmids that can be transferred to human pathogens. It has been highlighted that the contact of bacteria from human-associated microbiota with environmental microorganisms in sewage plants/in natural ecosystems is an important feature to understand the emergence of mechanisms of resistance in human pathogens (Baquero et al. 2008; Cattoir 2008).

18.2 Linkage of Antibiotics and Antibiotic Resistance to Sources

The terms *antimicrobial* and *antibiotic* are often used interchangeably in various publications. The term *antimicrobial* is defined as a natural, semisynthetic or synthetic chemical that kills or inhibits the growth of microorganisms such as bacteria, fungi, viruses and protozoa. Antimicrobials that kill organisms are called *cidal agents*, while those that inhibit or slow growth are called *static agents*. The term *antibiotic* is used to describe the subset of antimicrobials that target bacteria. Since bacteria and fungi have been producing antibiotics for hundreds of millions of years (D'Costa et al. 2011), ARGs have been a part of the endemic resistome for just as long (Barlow and Hall 2002; Poinar and Wright 2011; Wright and Poinar 2012; D'Costa et al. 2013; Igbinosa and Odjadjare 2015; Gelband et al. 2015).

Selecting what antibiotic resistance input or outcome to analyse is crucial in determining what can be reliably measured, how it can be measured and what statements can be made about antibiotic resistance in that environment (Levy and Marshall 2004). The three antibiotic resistance parameters typically measured are biologically linked in most cases but are not necessarily interchangeable and may not directly correlate to one another (Martínez and Baquero 2002; Davies and Davies 2010; D'Costa et al. 2013; Igbinosa and Odjadjare 2015). If a certain level of drug is present in the environment, then ARB and ARG levels will not necessarily demonstrate the same pattern. Knowledge of the types of genes that confer resistance to particular antibiotics is necessary when analysing an affected environment. For example, when attempting to infer an impact of sulphonamide drugs entering a system, analysing for ARGs known to be associated with sulphonamides rather than ARGs associated with other drugs would produce the most reliable results and findings (Davies and Davies 2010). Additionally, certain ARGs have a limited host range, while others cross many physiological barriers (Kohanski et al. 2010). Furthermore, knowledge of the specific organisms that carry these genes, how transferable these genes are and evidence of epidemiological impacts of these genes need to be considered before drawing any conclusions about their influence on human or animal health. Understanding the susceptibility of bacteria to a particular antibiotic drug is also key information for analyses of ARBs, since certain drugs are not effective against Gram-positive or Gram-negative bacteria because of physiological differences.

18.2.1 Antibiotics and Antibiotic Resistance Elements in Bacterial Ecosystems

Since antibiotics are efficient inhibitors of bacterial growth produced by environmental microorganisms, it has been widely accepted that their role in nature will be to inhibit microbial competitors. Conversely, antibiotic resistance determinants should serve to avoid the activity of antibiotics, in such a way that they would be a good example of the Darwinian struggle for life. Although this can be true in some occasions, an alternative hypothesis stating that antibiotics could be signal molecules that shape the structure of microbial communities has been proposed (Linares et al. 2006; Yim et al. 2007; Fajardo and Martínez 2008). Thus way, the antibiotics can have a hormetic effect, beneficial at low concentrations likely found in most natural ecosystems and harmful at the high concentrations used for therapy (Berendonk et al. 2015). Similarly, it has been stated that some elements that serve to resist high concentrations of antibiotics have disparate functional roles in their original hosts (Martinez 2008, 2009; Davies and Davies 2010; Berendonk et al. 2015). The strong increase of antibiotic concentrations in natural ecosystems as the consequence of human activities (human therapy, farming) shifts the original functions of antimicrobials and resistance elements to the weapon/shield roles they play in hospitals or farms (Martinez 2008). These changes might influence not just the selection of antibiotic-resistant microorganisms but also the structure of the natural microbial populations and may alter the physiology of microorganisms as well. Besides selecting antibiotic-resistant mutants and favouring the acquisition of antibiotic resistance determinants by gene-transfer elements that can spread among the environmental microbiota, antibiotic pollution can enrich the population of intrinsically resistant microorganisms and reduce the population of susceptible microbiota. *Cyanobacteria* are responsible for more than a third of total free O_2 production and fixation of CO_2 and are susceptible to antibiotics (D'Costa et al. 2011). There is no indication that the *Cyanobacteria* population is suffering the impact of antibiotic pollution and the risks for this situation are likely very low. However, the dramatic effect that eliminating *Cyanobacteria* as the consequence of antibiotic pollution might have for the biosphere reinforces the idea that the release of antibiotics in natural environments has relevant consequences not just in terms of resistance but for the maintenance of the global activity of the microbiosphere also (Halling-Sørensen et al. 1998; Hirsch et al. 1999). A very representative study improved that tetracycline has a negative impact on the functional diversity of soil microbial communities (Zak et al. 1994, 2003; Hill et al. 2000). Antibiotics at much higher concentrations that are usually found in natural ecosystems can be found in soils (Franklin et al. 2016). These high concentrations are usually concentrated to areas of human activity, where environments usually have very low content of antibiotics. Risk assessments can be taken into consideration mainly for those areas with higher antibiotic introduction and containing human-associated microorganisms for analysing the effect of antibiotic pollution effect in natural ecosystems (Baquero et al. 2008). Antibiotic resistance genes are found worldwide, due to their origin in bacteria (Davies 1994). Pallecchi et al. (2008) declare that intensive dissemination of genes frequently occurs in human pathogens in places without a high antibiotic load. There is high percent of probability for their maintenance in natural ecosystems. Because of abovementioned facts, antibiotic resistance genes can be considered as pollutants. Antibiotic resistance genes are naturally located in the chromosomes of environmental bacteria (D'Costa et al. 2006; Martinez 2008; Wright 2012). Only those compounds that are present in gene-transfer elements should be considered as pollutants. Contamination by antibiotics is relevant, but not only for their impact on bacteria that can infect humans. The release of antibiotics together with antibiotic resistance bacteria can impact as well the environmental microbiota.

18.2.2 Antibiotic Resistance in Clinically Relevant Bacteria

Rampant use of antibiotics in humans, animals and agriculture settings is the leading cause of increasing trend in microbial resistance. For the first time, the resistance to antibiotics appeared in hospitals with the emergence of sulphonamide-resistant *Streptococcus pyogenes* in the 1930s followed by penicillin-resistant *Staphylococcus aureus*(Levy and Marshall 2004; Liu et al. 2014). The prevalent threat is posed by multiple drug resistance (MDR), and some pathogens such as *Mycobacterium tuberculosis*, methicillin-resistant *Staphylococcus aureus* (MRSA), vancomycin-resistant *Staphylococcus aureus* (VRSA), vancomycin-resistant enterococci (VRE), *Klebsiella pneumonia*, drug-resistant *Escherichia coli*,

Pseudomonas aeruginosa and *Acinetobacter baumannii* are globally notable examples in the hospital community (Levy and Marshall 2004; Berendonk et al. 2015). Penicillin-resistant *Streptococcus pneumoniae* (PRSP), a common pathogen of children, is also mainly community acquired (Brochet et al. 2008; Franklin et al. 2016). This has resulted in an increased frequency and duration of hospitalization, with increased exposure to multidrug-resistant pathogens that are present in healthcare settings (Kumar et al. 2005). Forsberg et al. (2014) refer that most pathogens resistant to multiple antibiotics were isolated from healthcare settings, where antibiotic use was prevalent. Resistance to penicillin, cephalosporins and other *-lactam* antibiotics has increased the use of erythromycin and new macrolides such as azithromycin and clarithromycin (Georgopapadakou 1993). For this reason bacterial infections resistant to antibiotics are becoming increasingly common in clinical settings (Zoutman and Ford 2005). Many pathogenic agents have become resistant to various classes of antibiotics since the 1960s. This fact has become clinically, epidemiologically and socioeconomically important, because infections caused by resistant bacteria can be especially difficult and costly to treat (Zoutman and Ford 2005). Emergence of multidrug resistance in community-acquired pathogens such as *Mycobacterium tuberculosis* and *Streptococcus pneumoniae* demonstrates a worrisome situation in which the effectiveness of the antibiotics that previously successfully treated the infections has deteriorated over time (WHO 2015).

Christian et al. (2003) show that discharge of effluents from wastewater treatment plants provides the introduction pathway for contaminants into environment. Multidrug-resistant *Acinetobacter baumannii* is a rapidly emerging pathogen in healthcare settings where antimicrobial resistance has seriously limited treatment options (Eliopoulos et al. 2008). Environmental pollution can be important reservoir in outbreaks of *A. baumannii*, thereby underscoring the need for increased use of antibiotics with higher potency. Antibiotics as human and animal excretory products enter into soil environment either directly or after passage through wastewater treatment plants.

The introduction of antibiotics into the soil environment causes a selective pressure which results in an increase in the proportion of bacteria that are resistant to antibiotics (Igbinosa and Obuekwe 2014; Igbinosa 2015). Increased resistance to a wide variety of antibiotics has been found in bacteria located in waterbodies receiving run-off from agricultural and abattoir environments (Igbinosa et al. 2012, 2013; Igbinosa 2015). According to Larsson and Fick (2009), antibiotics can be released into environments as inappropriately disposed unused drugs and as part of effluent from drug production facilities. The processes that lead to the development of antibiotic resistance have probably occurred throughout all of microbial evolutionary descent. Resistant bacteria and antibiotic resistance genes that work to inactivate antibiotics and antibiotic molecules are present in the environment at all times, thus distinguishing naturally occurring resistance in organisms from resistance as a result of environmental pollution (Wright and Poinar 2012).

18.3 Dissemination of Antibiotic-Resistant Bacteria in the Environment

The existence of a natural environmental resistance gene pool, which includes all known clinical resistance mechanisms, has been discovered (D'Costa et al. 2006; Aminov 2009). Soil can occurs as recipient of antibiotics from anthropogenic sources that accumulate and persist for long time of period. A large fraction of antibiotics stays in environment in an active form (Baquero et al. 2008; Martinez 2009; Lupo et al. 2012). However, antibiotics are introduced into the environment through various pathways, as given in Fig. 18.1, that include:

- Effluents from disposal of human waste
- Waste from agricultural food animal production and aquaculture
- Direct application to some plants
- Industrial effluents from pharmaceutical facilities
- Agricultural run-off and disposal of ethanol production waste products

The discharge of effluents from wastewater treatment plants (WWTP) represents important point sources of contaminants in the soil environment. WWTP have been described as a significant releasing source of antibiotics and for antimicrobial resistance. Some treatment options appear in order to reduce loading of antibiotic residues that could be distributed in the environment (Michael et al. 2013). Treatment options and removal pathways of antibiotic residues include adsorption, biodegradation, disinfection, membrane separation, hydrolysis, photolysis and volatilization depending on antibiotic properties (Zhang et al. 2009; Zuccato et al. 2010). Tetracyclines can be removed by adsorption onto the

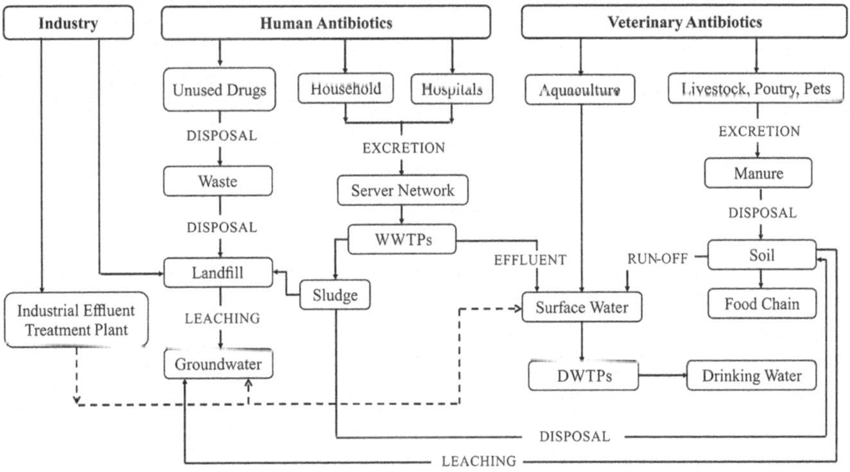

Fig. 18.1 Possible routes of antibiotics into the environment (adapted from Frade et al. 2014)

biomass flocs; beta-lactams are largely degraded by hydrolysis reactions driven by bacteria or physical chemical processes, while erythromycin and ciprofloxacin are recalcitrant toward biodegradation in activated sludge (Li and Zhang 2011). Biological waste treatment processes rely on complex ecological interactions. Novo and Manaia (2010) explains processes to treat wastewater. It has been shown to influence the contribution of antimicrobial resistance elements and resistant strains of bacteria released into the environment. Microbial communities react with drastic changes in ecosystem functioning and species composition and abundance (Fatta-Kassinos et al. 2010, 2011). The connection between aquatic pollution impacts and potentially pathogenic bacteria is of particular relevance for human well-being (Martinez 2009; Nogales et al. 2011; Franklin et al. 2016). High amounts of organic matter deposited in waterbodies lead to nutrient enrichment. These conditions can initiates growth of heterotrophic bacteria. On the other hand, pathogenic bacteria originating from human faeces are released directly into the environment through wastewater discharges, thereby compromising water quality (Figueira et al. 2011; Nogales et al. 2011). Aquatic systems can be highly impacted by human activities, receiving contaminants and pollution from different sources and thereby encouraging the exchange and mixture of genes as genetic platforms.

18.3.1 Resistant Bacteria in Humans

Methicillin-resistant *Staphylococcus aureus* (MRSA) has declined in incidence in Europe, the United States and Canada over the past 8 years, to 18%, 44% and 16%, respectively; in sub-Saharan Africa, India, Latin America and Australia, this percent is still rising, recorded at 47% in India in 2014 and 90% in Latin American hospitals in 2013 (WHO 2015). *Escherichia coli (E. coli)* and related bacteria have become resistant to newer third-generation cephalosporins (WHO 2015).

Escherichia coli (E. coli) and related bacteria have become resistant to newer third-generation cephalosporins, indicating that they are difficult to treat extended spectrum beta-lactamase (ESBL) producers. In 2013, in 17 of 22 European countries, 85–100% of *E. coli* isolates were ESBL positive; in 2009 and 2010, 28% of all *Enterobacteriaceae* from urinary tract infections in 11 countries in Asia were ESBL producers, and resistance to third- and fourth-generation cephalosporins ranged from 26 to 50% (Li et al. 2012). In Latin America in 2014, resistance in *Klebsiella pneumoniae* ranged from 19% in Peru to 87% in Bolivia (WHO 2015).

Carbapenem-resistant *Enterobacteriaceae* (CRE)—for Europe, five countries reported increases in 2013 (WHO 2015). In US hospitals, 11% of *K. pneumoniae* and 2% of *E. coli* were resistant to carbapenems in 2012, while in Latin America in 2013, resistance of *K. pneumoniae* to carbapenems ranged from full susceptibility in the Dominican Republic to 28% resistant in Guatemala. Monitoring conducted in India reveals that 13% of *E. coli* were resistant to carbapenems in 2013 (WHO 2015).

18.4 Behaviour of Antibiotics in the Ecosystems

Antibiotics used for preventing or treating infections in humans or animals as well as for promoting faster growth of livestock are only partially metabolized and are then discharged along the excreta, either to sewage treatment plants or straightforward in waters or soils (Dolliver and Gupta 2008). In addition, antibiotics are topically added to the aerial organs of infected plants, although the amount of antibiotics used in plant agriculture is low compared with human and veterinary medicine and animal production (McManus et al. 2002). To alleviate the effect that the release of antibiotics for nonhuman use may have for the selection of resistance in human pathogens, the European Union banned the feeding of those antibiotics, which are valuable in human medicine, to livestock for growth promotion in 1998. Many countries have restricted the use of antibiotics in aquaculture, including strong restrictions in the use of antibiotic prophylaxis and proscription of the utilization of antibiotics that are still useful in the therapy of human infections.

As stated by the World Health Organization, the amount of antimicrobials used in animals is not known precisely because national statistics on the amount and pattern of use of antimicrobials exist in only a few countries (WHO 2002). This is an important drawback to evaluate the impact of antibiotic utilization in veterinary on the selection of resistance and the release of antibiotics in the environment. Overall, the World Health Organization estimates that about half of the total amount of produced antibiotics is used in food animals (WHO 2002). It is important to remark however that the total amount of antibiotics used in animals is not known with certainty. Cabello reported in 2006 that "in Chile, statistics indicate that annually 10–12 metric tons of quinolones are used in human medicine and approximately 100–110 metric tons of these antibiotics are used veterinary medicine per year, most of them in aquaculture" (Cabello 2006). Nevertheless, taking into consideration that 25–75% of the antibiotics administered to feedlot animals are excreted unaltered in faeces, it is clear that the antimicrobial use in livestock is an important source of antibiotics release into the environment (Dolliver and Gupta 2008). The main impact of antibiotic pollutants is on the environmental microbiota. Utilization on antibiotics can select for antibiotic-resistant bacteria within the treated host. In the case of antibiotics used for farming purposes, selection of resistance can be important for both the treatment of animal infections and for human health. Several evidences support an association between the use of antimicrobial agents in food animals and antimicrobial resistance among bacteria isolated from humans (Angulo et al. 2004). The effect of antibiotics used for farming in human health has mainly focused on food-borne pathogens. These bacteria are present in the animals and can infect humans. Examples of food-borne pathogens are *Campylobacter jejuni*, *E. coli*, *Salmonella* or *Enterococcus faecium*, among others. For those pathogens, both mutation-driven antibiotic resistance and the acquisition of antibiotic resistance genes (Martinez 2009; Franklin et al. 2016) are important concerns for human health, because the same strain can colonize both animals and humans, and antibiotic resistance genes can easily spread among bacterial species that are closely phylogenetically related (Sundsfjord et al. 2001). Livestock are given veterinary medicines to treat disease and protect their health. After application of the drug, the substance may be metabolized, and a mixture of the parent compound and metabolites will then be excreted in the urine and faeces. Releases of veterinary medicines to the environment occur both directly, for example, through the treatment of animals on pasture, and indirectly, via the application of animal manure to land, and once released to land, the veterinary medicines may leach to groundwater or be transported to surface waters in drainage waters and overland flow. Numerous veterinary medicines have been extensively studied: primarily the sheep dip chemicals and anthelmintics. With the exception of a few studies and reviews (Baguer et al. 2000; Loke et al. 2000; Rabolle and Spliid 2000; Tolls 2001), limited information is available on the fate, behaviour and effects of other major classes of veterinary medicines used to treat livestock, and very little data are available on concentrations in the environment. In order to assess the risks posed to the environment by veterinary medicines used to treat livestock, a number of models and guidelines have been developed for predicting concentrations of veterinary medicines in soil (Montforts 1999). The strategy of application to soil and the properties of the veterinary medicines differ from most pesticides and industrial chemicals, so the use of the assumptions may be inappropriate. Besides being a potential vector for veterinary pharmaceutical into soils, manure contains higher levels of ammonia that will increase the pH of the soil solution, thus altering the speciation of the veterinary medicines and thus affecting the sorption of the compound. Transport of manure-associated and dissolved organic carbon-associated veterinary medicines through the soil profile to surface waters and groundwaters may also be important pathways.

18.4.1 Pathways of the Relisted Antibiotics

Increasing overall resistance may impact human health even if the resistance genes are originally selected in microorganisms' lineages that are specific of the animals and cannot infect humans. For instance, it has been recently described that horizontal gene-transfer events link human methicillin-resistant *Staphylococcus aureus* human pathogens to contagious bovine mastitis bacteria (Martinez 2009). Second, antibiotics can remain in the tissues of the animals, so that they can be considered as food pollutants. The effect of these compounds in the human host has not been studied in detail. However, it has been suggested that they might trigger in occasions allergic reactions and contribute to select for antibiotic-resistant bacteria in the human microbiota (Cabello 2006). Third, antibiotics released to soils or waters can modify the local environmental microbiota producing changes in their composition or activity that are not fully understood. The alterations in the bacterial populations include selection of resistant mutants in susceptible species, changes in the distribution of antibiotic resistance genes present in gene-transfer units and selection of resistant species in such a way that the overall composition of the microbiota is modified. For instance, the exposition to ciprofloxacin of salt marsh sediment microbial communities favours selection of sulphate-reducing and Gram-negative bacteria (Cordova-Kreylos and Scow 2007). Selection of resistance and reduction of the complexity of environmental microbiota is not however the unique consequence of antibiotic pollution. Antibiotics can produce transient changes in the activity of microbial populations (Linares et al. 2006; Fajardo and Martínez 2008) that might be relevant for their productivity even at subinhibitory concentrations. It has been described that the pollution of manure with sulphadiazine reduces microbial activity, mainly some processes in nitrogen turnover (Kotzerke et al. 2008), besides increasing resistance in soil (Ghosh and LaPara 2007; Heuer and Smalla 2007). Although the overall utilization of antibiotics for farming purposes might be decreasing as the consequence of the implementation of politics for reducing antibiotic resistance, the removal of antibiotics as growth promoters has been followed by an increase of therapeutic antibiotics used in animals (Singer et al. 2003). On the other hand, although national programmes to control antimicrobial resistance and to improve the rational use of antibiotics in humans are reducing the amount of antibiotics used for human therapy (Muller et al. 2007), this use of antibiotics will obviously remain. It is thus predictable that the amount of antibiotics released in the environment from farms and humans' residues will likely keep at rather high levels in the future. This means that besides control policies in the use of antibiotics, studies for improving their degradation are needed. Treatment of water, sewage and in general residues contaminated with antibiotics should be implemented before their release to natural ecosystem or its transformation in manure to be used as organic fertilizer in agriculture (Dolliver and Gupta 2008), because antibiotics present in manure antibiotics are incorporated in soil (Alanis 2005; Brown et al. 2006). Several techniques (coagulation, activated carbon filtration, ionic treatment or micelle–clay systems) were demonstrated as effective for the removal of different antibiotics (Shellie et al. 2003; Sukul and Spiteller 2007; Gulkowska et al. 2008; Choi et al. 2008). However in all cases a variable percentage of the antibiotics usually remain after wastewater treatment (Brown et al. 2006) and can challenge bacterial populations downstream the wastewater processing plant (Watkinson et al. 2007). It is important to remark that treatment of water and residues from human activity is far to be performed worldwide, so that in most occasions antibiotic-polluted residues are released in the environment without further processing. Antibiotics are naturally degraded by processes that include (a) photodegradation, (b) chemical degradation and (c) biodegradation. These processes are correlated with the influence on moisture, temperature, chemical composition (pH, EC) and the microorganisms. Different habitats will render different paths of antibiotic degradation; binding of antibiotics to clay and sediments delays their degradation, but simultaneously removes antibiotics from water, in such a way that particulate matter present in rivers may reduce antibiotic pollution in waters at long distance of waste drainage, at the cost of increasing locally antibiotic concentration at sediments (Baquero et al. 2008).

18.5 Antibiotic Resistance Gene as a Pollutant

Several works highlight the presence of antibiotic resistance genes in environments that are unlikely contaminated with antibiotics used by humans. These include the deep terrestrial subsurface (Brown and Balkwill 2009). These studies presents good example of the ubiquity of genes that might confer resistance upon expression in a heterologous host, independently on whether or not their primary function is resistance (Wright 2012). Therefore, it is important to underline that the finding of resistant organisms in a given environment should not necessarily be considered as an evidence of pollution by antibiotics or by resistance genes. Contamination by resistance genes would not be necessarily involved. In the case of antibiotic pollution, resistant bacterial species and resistant mutants of the susceptible ones can be selected without involvement of pollutant antibiotic resistance genes. There are two situations in which an antibiotic resistance determinant can be considered as a pollutant: (1) antibiotic selective pressure in natural ecosystems may select the integration and further dissemination of antibiotic resistance genes in gene-transfer units, which can be then considered as contaminants, and (2) residues from hospitals, houses and farms contain bacteria that can carry antibiotic resistance determinants. The finding of specific antibiotic resistance genes, which are already disseminated among human, animal or plant bacterial pathogens, can be considered as indication of a history of contamination (Austin and Anderson 1999). Differing to the situation with antibiotics, this contamination is not necessarily local neither dependant on the constant release of residues, because once those genes are in the environment, they can disseminate among different bacterial species and distinct habitats. It has been demonstrated that antibiotic resistance genes can migrate between connected aquatic systems. It is unclear whether the presence of antibiotic resistance genes is the result of the migration of antibiotic-resistant bacteria or the transmission of resistance genes by HGT (Koike et al. 2007). The fact that remote human populations with minimal antibiotic exposure carry antibiotic-resistant commensal bacteria further support the worldwide dissemination of resistance genes (Grenet 2004; Geiger et al. 2008).

18.5.1 Main Genetic Reactors in Antibiotic Resistance

Genetic reactors are places in which the occasion occurs for genetic evolution, particularly because of high biological connectivity, generation of variation and presence of specific selection (Aminov 2011). Beyond mutational events, significant genetic variation occurs as a consequence of recombinatorial events, frequently resulting from genetic exchanges among organisms inside populations and communities. There are four main genetic reactors in which antibiotic resistance evolves:

1. Human/animal microbiota
2. Hospitals, farms, etc.
3. Wastewater effluents and sewage treatment plants
4. Surface groundwater, soil and sediments

The primary reactor is constituted by the human and animal microbiota, with more than 500 bacterial species involved, in which therapeutic or preventive antibiotics exert their actions. The secondary reactor involves the hospitals, long-term care facilities, farms or any other place in which susceptible individuals are crowded and exposed to bacterial exchange. The tertiary reactor corresponds to the wastewater and any type of biological residues originated in the secondary reactor, including lagoons, sewage treatment plants or compost toilets, in which bacterial organisms from many different individuals have the opportunity to mix and genetically react. The fourth reactor is the soil and the surface or groundwater environments, where the bacterial organisms originated in the previous reactor mix and counteract with environmental organisms. Water is involved as a crucial agent in all four genetic reactors, but particularly in the last ones. The possibility of reducing the resolvability of antibiotic resistance depends on the ability of humans to control the flow of active antimicrobial agents, bacterial clones and genetically based biological information along these genetic reactors.

18.5.2 Industrial Antibiotics in Soil–Water Environments

Water dissolves industrial antibiotics that are bound to environmental matrices. Binding to soil particles (and sediments) delays its biodegradation and explains long-term permanence of the drugs in the environment. Of course, soil particles also remove antibiotics from water, so that a kind of water–soil pharmacokinetics might be considered. Antimicrobial agents are retained in soil by its association with soil chemicals. For instance, Elliot soil humic acids produce complexation of antibiotics (Zhang et al. 2009; Cytryn 2013). Interestingly, heavy metals (as methylmercury) also associate with humic acids, so that in the water film associated with soil organic particles, several antimicrobial effects might be simultaneously present. Indeed it appears that in the presence of humic substances, in both dissolved and mineral-bound forms, environmental mobility of antibiotics might increase (Demanèche et al. 2008; Negreanu et al. 2012). Oxides of some metals, such as Al and Fe, can alter these interactions by changing surface charge. Sorption to such oxides results in different types of ciprofloxacin surface complexes (Knapp et al. 2010). This way, general alterations in soil (as pH changes or ionic strength) might alter these antibiotic–soil interactions, producing different levels of antibiotic release (dissolution) from soil particles (Christian et al. 2003; Schlüsener and Bester 2006; Negreanu et al. 2012).

18.5.3 Industrial Antibiotics in Water–Sludge Environments

Antimicrobial agents as sulphonamides, macrolides, trimethoprim, cephalosporins or fluoroquinolones can be found at potentially active concentrations in activated sludge treatment, and the antibiotic load along the year correlates with the variation in annual consumption data, being higher in the winter (Giger et al. 2003). The wastewater concentration of antimicrobials depends on the sludge–wastewater partition coefficient, but with fluoroquinolones field experiments of sludge application to agricultural land confirmed long persistence of these compounds, but with limited mobility into the subsoil (Schlüsener and Bester 2006; Göbel et al. 2004). In compost toilets, amoxicillin decay is negligible, even in the presence of beta-lactamase-producing bacteria (Zervos et al. 2003). The extensive use of antibiotics in human medicine, animal farming and agriculture leads to antibiotic contamination of manure, which can be used as fertilizer. Leaching tests indicate that in general less than 1% of fluoroquinolones in the sludge reached the aqueous phase, which might indicate a relatively reduced mobility when sludge is used to fertilize soil. Nevertheless, that does not exclude localized biological effects on particulated material. High concentrations of fluoroquinolones were found in secondary sludge. Macrolides were frequently resistant to the processes carried out in sewage treatment plants in South China, and even higher concentrations were found in the final effluents than in the raw sewages (Xu et al. 2007).

18.6 Impact of Antibiotics Resistance Determinant on Human and Environmental Health

Wastewater discharges from domestic sources affect the diversity of resistant bacteria (Czekalski et al. 2012; Thevenon et al. 2012; Vignesh et al. 2012). These impacts also shape the genetic pool of waterbodies by increasing abundance of antibiotic resistance genes within the habitats (Zervos et al. 2003). Thus, hospital effluents have been shown to be rich in resistance genes and resistant bacteria (Zervos et al. 2003). Contamination in aquatic environments contributes to the spread of human pathogens along with the dissemination of antibiotic-resistant bacteria. The overexposure to antibiotics is leading to increasing levels of resistance in the human commensal microbiota (Austin et al. 1999). The resistomes of faecal bacteria, once in environmental locale, contribute antibiotic resistance genes to non-resistant indigenous microorganisms (Aminov 2009, 2011). In some aquatic systems, the cycle can include subsequent transmission of antibiotic resistant to bacteria associated in human (Devirgiliis et al. 2011; Figueira et al. 2011). Enrichment in antibiotic-resistant bacteria is promoted by the presence of antimicrobials in the environment. Bacterial communities in aquatic systems comprise antibiotic producers and bacteria intrinsically resistant to several antibiotics. These two groups are natural carriers of genetic determinants of resistance. Taking these factors together, aquatic environments can be seen as reactor system where drug resistance characteristics spread and recombine. Under such conditions, multidrug resistance features may emerge, and transmission to pathogenic bacteria is facilitated (Lupo et al. 2012). Several studies have shown that in natural systems, pollution promotes antibiotic resistance spread, for which mobile genetic elements have an important contribution (Aminov 2009). Environmental and pathogenic bacteria harbour antibiotic resistance genes; the regulations of these genetic units differ with the source. Pathogens usually carry these genes on mobile genetic elements and express them constitutively. Antibiotic resistance can reach the environment with the potential of adversely affecting aquatic and terrestrial organisms which eventually might reach humans through drinking water and the food chain (Aarestrup et al. 2008). The emergence of resistance is a highly complex process which is not yet fully understood with respect to the significance of the interaction of bacterial populations and antibiotics, even in a medicinal environment (Martinez and Baquero 2000; Alanis 2005). The transfer of resistant bacteria to humans could occur via water or food if plants are watered with surface water or sewage sludge, if manure is used as a fertilizer (Dolliver and Gupta 2008). The transfer pathways of antibiotic resistance from animals to humans are not clearly understood.

18.7 Risk Assessment

Integrated risk assessment of the evolution and emergence of antibiotic resistance in the environment addresses two main issues:

- The potential of subinhibitory concentrations of antibiotics to promote the development of ARB in complex bacterial communities
- The capacity of resistance determinants to transfer from anthropogenic sources (such as treated wastewater, manure or others) to human commensal or pathogenic bacteria

18.7.1 Potential of Subinhibitory Antibiotic Concentrations

The *Guideline on the environmental risk assessment of medicinal products for human use*, produced by the European Medicines Agency, does not recognize that the emergence and proliferation of antibiotic resistance may be the most important risk associated with environmental contamination by antibiotics. Indeed, the endpoints for no-effect concentrations are different from traditional environmental risk assessment, as the effects of antibiotics in promoting antibiotic resistance can go far beyond the toxicological implications. The effects of antibiotics may be potentiated or extended by cofactors (general stress situations and micro-contaminants, such as heavy metals and biocides), which possibly enhance the spread and evolution of antibiotic resistance. Therefore, combined molecular- and culture-based methods are necessary to determine the concentrations at which resistance acquisition and selection are likely to occur in environmental compartments. Although challenging, given the scarcity of knowledge regarding the mechanisms involved at the genetic, cellular and population levels, this addendum is urgently needed. Within this addendum, the gold standard of a reliable risk assessment should determine the range of concentrations at which, under defined conditions, an antibiotic can promote selection and the acquisition of resistance.

18.7.2 Transmission of Resistance Determinants from Anthropogenic Sources

Another important aspect of antibiotic resistance risk assessment refers to the spread and transmission of resistance determinants from hotspots to downstream environments. Mathematical models capable of predicting the influence of potential selective pressures, or the occurrence and the evolutionary success of genetic recombination events, have proven to be promising tools in predicting the spread of antibiotic resistance determinants (Martinez 2009). As they are specifically developed for environmental niches and environment–human interfaces, these mathematical models should rely on parameters such as population size; bacterial population growth rate and survival; occurrence and frequency of horizontal gene transfer and its implications on the population fitness; and the influence of other biotic and abiotic factors (Cabello 2006; Martinez 2009). Such models would allow predictions to be made regarding the dynamics of ARB hosting ARGs and the possible localization of ARGs on MGEs, thus supporting the assessment of their fate from anthropogenic sources to downstream environments. A quantitative risk assessment framework should be developed by coupling data and analyses, such as those outlined above, with a stochastic assessment of exposure to clinically relevant bacteria in the environment. Such a model should then be used to predict the environmental conditions that are associated with the evolution of antibiotic resistance and infer the probability of antibiotic resistance determinants spreading. However, owing to the scarcity of data on the occurrence of antibiotic resistance and horizontal gene transfer in the environment, it is currently difficult to develop validated models that can be applied in the framework of environmental risk assessment guidelines.

18.7.3 ARB and ARGs as Contaminants of Emerging Concern

The European Water Framework Directive establishes the requirements for determining the biological and chemical quality standards of waterbodies in Europe (Berendonk et al. 2015). Annex I of this directive sets obligations for environmental quality standards (EQS) for priority substances and certain other pollutants, and it even identifies priority hazardous substances. The inclusion of ARB and ARGs as priority contaminants would be justified based on the results of numerous scientific studies, which show that the occurrence of antibiotic resistance increases in bodies of water (such as inland surface waters, transitional waters, coastal waters and groundwater) when they are subjected to anthropogenic impacts, such as wastewater effluents, animal manure, agricultural run-off and wildlife living in urban areas (Berendonk et al. 2015). The application of these guidance levels may be especially important for the regulation of specific practices, such as reuse of wastewater or soil fertilization with manure.

The inclusion of ARB and ARGs in the list of contaminants of emerging concern would require clear definitions on the necessary monitoring methods. Although the environmental survival of ARGs primarily depends on the host and the type of MGEs, the estimation of the levels of ARGs seems a reliable and feasible method to monitor antibiotic resistance (Pruden et al. 2012). However, a crucial issue that needs additional investigation is the selection of target ARGs to be monitored as indicators of resistance and the determination of safe concentrations of these genes in water. Such indicator genes should be more abundant in anthropogenic sources and rare in native aquatic and terrestrial ecosystems. However, determination of the maximum acceptable levels of these genes in the environment seems at the current state of knowledge (Pruden et al. 2012). The establishment of a comprehensive database and the use of modelling approaches would be valuable contributions to estimate such limits. Despite these challenges, this knowledge is an essential prerequisite, not only for establishing a strategy of direct action against antibiotic resistance in the environment but also for the application of drugs and interventions directed at preventing the emergence and evolution of ARB and ARGs (Berendonk et al. 2015).

18.7.4 Critical Control Points

Environmental hotspots, where ARBs are abundant or the transfer of ARGs is promoted, are critical points for resistance control. Good examples of such critical points are characterized by a high prevalence of resistance or by the occurrence of resistance determinants of emerging concern. These locations comprise habitats that are influenced by human activities (e.g. wastewater from animal husbandry and intensive food production facilities). Sites subjected to frequent discharge of antibiotic residues have been shown to be potential hotspots for the proliferation and spread of new resistance determinants to human and pathogenic bacteria and should be considered as critical control points.

Although some of the antimicrobials administered to animals are used exclusively in veterinary applications, most belong to the same structural families that are used in human medicine. As they share the same basic chemical molecular structures and mechanisms of action, these antibiotics are assumed to put selective pressures on human commensal and pathogenic bacteria. Large quantities of antibiotics that are administered to animals in intensive production sites are discharged, often unmetabolized with manure and slurry when applied as fertilizer, and thus contaminate soils as well as surface water and groundwater. At present, it is difficult to ascertain whether antibiotics reaching the environment at low concentrations exert a substantial selective pressure on ARGs.

Urban, hospital and pharmaceutical industry wastewater is among the main sources of antibiotic contamination in soil ecosystems. In the environment, these contaminants can reach water resources for drinking water production, enter the food chain or reach clinically relevant niches. These effects can be potentially even more pronounced when irrigation with wastewater effluents (wastewater reuse schemes) is applied. Water reuse is already a common practice in many regions of the world owing to increased water scarcity, mainly in arid and semiarid regions. Most of the wastewater treatment plants worldwide, in particular those using mechanical and biological treatments, are primarily designed to remove organic compounds, nutrients (e.g. nitrogen and phosphorous) and suspended solids. However, the currently available wastewater treatment processes have limited capability to efficiently remove organic micropollutants, including antibiotics and other antimicrobial agents. Similarly, certain ARGs can survive the wastewater treatment processes with a maintenance (or even an increase) of resistance prevalence compared to the pretreatment levels. These features require the immediate implementation of technological solutions capable of mitigating ARB and ARGs in wastewater to safe levels. Although the definition of a "safe level" may be difficult to achieve, it is at least necessary to find an agreement on the threshold values below which the probability of significant proliferation of an ARG is severely impaired.

18.7.5 Management Options

Technologies for the removal of microscopic pollutants, including antibiotics, and microorganisms from wastewater are becoming increasingly available (e.g. membrane filtration, activated carbon, photo-driven technologies and ozonation) (Berendonk et al. 2015). Additional research should be conducted in order to determine the effectiveness of these processes for the elimination of ARB and ARGs and to characterize the associated microbiological risks. Recommendations of effective and economically sustainable interventions at critical points within the wastewater stream are urgently needed. Management options aimed at preventing and controlling antibiotic resistance in the environment comprise several different aspects, including the choice of ARB and ARGs to be listed as contaminants of emerging concern; the determination of differentiated maximum admissible levels of an antibiotic, ARB or ARG; and the identification of critical points of control at which prevention and remediation measures should be implemented.

18.8 Conclusions/Perspectives

The rapidly growing number of antibiotic-resistant bacterial pathogens severely undermines the ability to control infectious diseases, and currently it is one of the most challenging problems in public healthcare. Realization of the circumstances under which the microbial world evolves, i.e. the modus operandi of microorganisms involves much more extensive lateral gene exchange and recombination processes than previously recognized and that there are no isolated compartments in microbial ecosystems, led to the studies integrating the antibiotic resistance research area within the broader evolutionary and ecology contexts. These studies yielded valuable insights into the processes in microbial communities that eventually result in the emergence, dissemination and fixation of antibiotic resistance genes in human and animal bacterial pathogens. In particular, phylogenetic analyses helped to identify the nonantibiotic-producing environmental bacteria, but not the antibiotic producers, as harbouring the readily available, abundant and diverse pool of antibiotic resistance genes, from which the genes can be transferred to bacteria in the human and animal ecological compartments. Further systematic studies of the antibiotic resistome of the environment are necessary, not only for identification of the ancestors and missing links in the evolution of presently well-characterized antibiotic resistance genes, but, more importantly, for identification of the potential threats for newly introduced antibiotics to serve as an early warning system. Besides the soil antibiotic resistome, there are other antibiotic resistance gene reservoirs, the origin and diversity of which is difficult to explain in terms of possible antibiotic exposure (e.g. aquatic environment). These genes might have served some other needs of the bacterial cell rather than conferring antibiotic resistance per se, and identification of these functions would be helpful for understanding the evolution of antibiotic resistance.

Another aspect of ecology of antibiotic resistance that recently emerged with the advent of molecular ecology tools in antibiotic resistance studies in the environment is the realization of the fact that the microbial ecosystems are not isolated and there is extensive gene exchange between different compartments. Antibiotic usage in animals, for example, may lead to dissemination of antibiotic resistance genes into the broader environment. Thus, there is also an urgent need to develop environmental genetic tools to evaluate the gene transfer/flow rates within and between the microbial ecosystems as well as to identify genetic elements and bacteria involved. Although the environmental genomics/metagenomics approach partially addresses this issue, it is still descriptive and specialized genetic tools, adapted for the in situ use, are required.

References

Aarestrup FM, Wegener HC, Collignon P (2008) Resistance in bacteria of the food chain: epidemiology and control strategies. Expert Rev Anti-Infect Ther 6(5):733–750

Alanis AJ (2005) Resistance to antibiotics: are we in the post-antibiotic era? Arch Med Res 36(6): 697–705

Aminov RI (2009) The role of antibiotics and antibiotic resistance in nature. Environ Microbiol 11(12):2970–2988

Aminov RI (2011) Horizontal gene exchange in environmental microbiota. Front Microbiol 2:158

Angulo FJ, Nargund VN, Chiller TC (2004) Evidence of an association between use of anti-microbial agents in food animals and anti-microbial resistance among bacteria isolated from humans and the human health consequences of such resistance. J Veterinary Med Ser B 51(8–9):374–379

Austin DJ, Anderson RM (1999) Studies of antibiotic resistance within the patient, hospitals and the community using simple mathematical models. Philos Trans R Soc B 354(1384):721–738

Austin DJ, Kristinsson KG, Anderson RM (1999) The relationship between the volume of anti-microbial consumption in human communities and the frequency of resistance. Proc Natl Acad Sci USA 96(3):1152–1156

Baguer AJ, Jensen J, Henning Krogh P (2000) Effects of the antibiotics oxytetracycline and tylosin on soil fauna. Chemosphere 40:751–757

Baquero F, Martínez JL, Cantón R (2008) Antibiotics and antibiotic resistance in water environments. Curr Opin Biotechnol 19(3):260–265

Barlow M, Hall BG (2002) Phylogenetic analysis shows that the OXA beta-lactamase genes have been on plasmids for millions of years. J Mol Evol 5:314–321

Berendonk TU, Manaia CM, Merlin C, Fatta-Kassinos D, Cytryn E, Walsh F, Bürgmann H, Sørum H, Norström M, Pons MN, Kreuzinger N (2015) Tackling antibiotic resistance: the environmental framework. Nat Rev Microbiol 13(5):310–317

Bergstrom CT, Feldgarden M (2008) The ecology and evolution of antibiotic-resistant bacteria in evolution in health and disease. In: Stearns S, Koella J (eds) Evolution in health and disease, 2nd edn. Oxford University Press, New York

Brochet M, Couvè E, Zouine M, Poyart C, Glaser P (2008) A naturally occurring gene amplification leading to sulfonamide and trimethoprim resistance in Streptococcus agalactiae. J Bacteriol 190:672–680

Brown MG, Balkwill DL (2009) Antibiotic resistance in bacteria isolated from the deep terrestrial subsurface. Microb Ecol 57(3):484

Brown KD, Kulis J, Thomson B, Chapman TH, Mawhinney DB (2006) Occurrence of antibiotics in hospital, residential, and dairy effluent, municipal wastewater, and the Rio Grande in New Mexico. Sci Total Environ 366(2):772–783

Cabello FC (2006) Heavy use of prophylactic antibiotics in aquaculture: a growing problem for human and animal health and for the environment. Environ Microbiol 8(7):1137–1144

Cattoir V (2008) Unexpected occurrence of plasmid-mediated quinolone resistance determinants in environmental Aeromonas spp. Emerg Infect Dis 14(2):231–237

Choi KJ, Kim SG, Kim SH (2008) Removal of antibiotics by coagulation and granular activated carbon filtration. J Hazard Mater 151:38–43

Christian T, Schneider RJ, Färber HA, Skutlarek D, Meyer MT, Goldbach HE (2003) Determination of antibiotics residues in manure, soil, and surface waters. Acta Hydrochim Hydrobiol 31:34–44

Cordova-Kreylos AL, Scow KM (2007) Effects of ciprofloxacin on salt marsh sediment microbial communities. ISME J 1:585–595

Cytryn E (2013) The soil resistome: the anthropogenic, the native, and the unknown. Soil Biol Biochem 63:18–23

Czekalski N, Berthold T, Caucci S, Egli A, Bürgmann H (2012) Increased levels of multiresistant bacteria and resistance genes after wastewater treatment and their dissemination into Lake Geneva, Switzerland. Front Microbiol 3:106

Davies J (1994) Inactivation of antibiotics and the dissemination of resistance genes. Science 264(5157):375–383

Davies J, Davies D (2010) Origins and evolution of antibiotic resistance. Microbiol Mol Biol Rev 74(3):417–433

D'Costa VM, McGrann KM, Hughes DW, Wright GD (2006) Sampling the antibiotic resistome. Science 311(5759):374–377

D'Costa VM, King CE, Kalan L, Morar M, Sung WW, Schwarz C, Froese D, Zazula G, Calmels F, Debruyne R, Golding GB (2011) Antibiotic resistance is ancient. Nature 477(7365):457–461

D'Costa VM, King CE, Kalan L, Morar M, Sung WW, Schwarz C, Froese D, Zazula G, Calmels F, Debruyne R, Golding GB (2013) Diverse and abundant antibiotic resistance genes in Chinese swine farms. Proc Natl Acad Sci USA 110:3435–3440

Demanèche S, Sanguin H, Potè J, Navarro E, Bernillon D, Mavingui P, Wildi W, Vogel TM, Simonet P (2008) Antibiotic-resistant soil bacteria in transgenic plant fields. Proc Natl Acad Sci USA 105:3957–3962

Devirgiliis C, Barile S, Perozzi G (2011) Antibiotic resistance determinants in the interplay between food and gut microbiota. Genes Nutr 6(3):275

Dolliver H, Gupta S (2008) Antibiotic losses in leaching and surface runoff from manure-amended agricultural land. J Environ Qual 37(3):1227–1237

Eliopoulos GM, Maragakis LL, Perl TM (2008) Acinetobacter baumannii: epidemiology, antimicrobial resistance, and treatment options. Clin Infect Dis 46(8):1254–1263

Fajardo A, Martínez JL (2008) Antibiotics as signals that trigger specific bacterial responses. Curr Opin Microbiol 11(2):161–167

Fatta-Kassinos D, Hapeshi AA, Meric S, Gros M, Petrovic M, Barcelo D (2010) Existence of pharmaceutical compounds in tertiary treated urban wastewater that is utilized for reuse applications. Water Resour Manag 25:1183–1193

Fatta-Kassinos D, Meric S, Nikolaou A (2011) Pharmaceutical residues in environmental waters and wastewater: current state of knowledge and future research. Anal Bioanal Chem 399:251–275

Ferber D (2003) Antibiotic resistance. WHO advises kicking the livestock antibiotic habit. Science 301:1027

Figueira V, Vaz-Moreira I, Silva M, Manaia CM (2011) Diversity and antibiotic resistance of Aeromonas spp. in drinking and waste water treatment plants. Water Res 45(17):5599–5611

Frade VMF, Dias M, Teixeira ACSC, Palma MSA (2014) Environmental contamination by fluoroquinolones. Braz J Pharm Sci 50(1):41–54

Franklin AM, Aga DS, Cytryn E, Durso LM, McLain JE, Pruden A, Roberts MC, Rothrock MJ, Snow DD, Watson JE, Dungan RS (2016) Antibiotics in agroecosystems: introduction to the special section. J Environ Qual 45(2):377–393

Forsberg KJ, Patel S, Gibson MK, Lauber CL, Knight R, Fierer N, Dantas G (2014) Bacterial phylogeny structures soil resistomes across habitats. Nature 509(7502):612

Frost AJ (1991) Antibiotics and animal production. In: Woolcock JB (ed) Microbiology of animals and animal products. Elsevier, Amsterdam

Geiger T, Goerke C, Mainiero M, Kraus D, Wolz C (2008) The virulence regulator Sae of Staphylococcus aureus: promoter activities and response to phagocytosis-related signals. J Bacteriol 190(10):3419–3428

Gelband H, Molly Miller P, Pant S, Gandra S, Levinson J, Barter D et al (2015) The state of the world's antibiotics 2015. Wound Healing South Afr 8(2):30–34

Georgopapadakou NH (1993) Penicillin-binding proteins and bacterial resistance to beta-lactams. Antimicrob Agents Chemother 37(10):2045

Ghosh S, LaPara TM (2007) The effects of subtherapeutic antibiotic use in farm animals on the proliferation and persistence of antibiotic resistance among soil bacteria. ISME J 1:191–203

Giger W, Alder AC, Golet EM, Kohler H-PE, McArdell CS, Molnar E, Siegrist H, Suter MJ-F (2003) Occurrence and fate of antibiotics as trace contaminants in wastewaters sewage sludge and surface waters. Chimia 57:485–491

Göbel A, McArdell CS, Suter MJF, Giger W (2004) Trace determination of macrolide and sulfonamide antimicrobials a human sulfonamide metabolite and trimethoprim in wastewater using liquid chromatography coupled to electrospray tandem mass spectrometry. Anal Chem 76:4756–4764

Grenet K (2004) Antibacterial resistance, Wayampis Amerindians, French Guyana. Emerg Infect Dis 10(6):1150–1153

Gulkowska A, Leung HW, So MK, Taniyasu S, Yamashita N, Yeung LW, Richardson BJ, Lei AP, Giesy JP, Lam PK (2008) Removal of antibiotics from wastewater by sewage treatment facilities in Hong Kong and Shenzhen, China. Water Res 42:395–403

Halling-Sørensen B, Nielsen SN, Lanzky PF, Ingerslev F, Lützhøft HH, Jørgensen SE (1998) Occurrence, fate and effects of pharmaceutical substances in the environment-A review. Chemosphere 36(2):357–393

Heuer H, Smalla K (2007) Manure and sulfadiazine synergistically increased bacterial antibiotic resistance in soil over at least two months. Environ Microbiol 9(3):657–666

Hill GT, Mitkowski NA, Aldrich-Wolfe L, Emele LR, Jurkonie DD, Ficke A, Maldonado-Ramirez S, Lynch ST, Nelson EB (2000) Methods for assessing the composition and diversity of soil microbial communities. Appl Soil Ecol 15(1):25–36

Hirsch R, Ternes T, Haberer K, Kratz KL (1999) Occurrence of antibiotics in the aquatic environment. Sci Total Environ 225(1):109–118

Igbinosa IH (2015) Prevalence and detection of antibiotic-resistant determinant in Salmonella isolated from food-producing animals. Trop Anim Health Prod 47(1):37–43

Igbinosa EO, Obuekwe IS (2014) Evaluation of Antibiotic Resistant Gene in Abattoir Environment. J Appl Sci Environ Manag 18(2):165–170

Igbinosa EO, Odjadjare EE (2015) Antibiotics and antibiotic resistance determinants: an undesired element in the environment. In: Méndez-Vilas A (ed) The battle against microbial pathogens: basic science, technological advances and educational programs, vol 2. Formatex Research Center, Badajoz, pp 858–866

Igbinosa EO, Okoh AI (2009) Impact of discharge wastewater effluents on the physico-chemical qualities of a receiving watershed in a typical rural community. Int J Environ Sci Technol 6(2): 175–182

Igbinosa IH, Igumbor EU, Aghdasi F, Tom M, Okoh AI (2012) Emerging Aeromonas species infections and their significance in public health. Sci World J 2012:625023

Igbinosa EO, Odjadjare EE, Chigor VN, Igbinosa IH, Emoghene AO, Ekhaise FO, Igiehon NO, Idemudia OG (2013) Toxicological profile of chlorophenols and their derivatives in the environment: the public health perspective. Sci World J 2013:460215

Knapp CW, Dolfing J, Ehlert PAI, Graham DW (2010) Evidence of increasing antibiotic resistance gene abundances in archived soils since 1940. Environ Sci Technol 44:580–587

Kohanski MA, Dwyer DJ, Collins JJ (2010) How antibiotics kill bacteria: from targets to networks. Nat Rev Microbiol 8(6):423–435

Koike S, Krapac IG, Oliver HD, Yannarell AC, Chee-Sanford JC, Aminov RI, Mackie RI (2007) Monitoring and source tracking of tetracycline resistance genes in lagoons and groundwater adjacent to swine production facilities over a 3-year period. Appl Environ Microbiol 73(15): 4813–4823

Kotzerke A, Sharma S, Schauss K, Heuer H, Thiele-Bruhn S, Smalla K, Wilke BM, Schloter M (2008) Alterations in soil microbial activity and N transformation processes due to sulfadiazine loads in pig-manure. Environ Pollut 153:315–322

Kumar K, Gupta SC, Changer Y, Singh AK (2005) Antibiotic use in agriculture and their impact on the terrestrial environment. Adv Agron 87:1–54

Kümmerer K (2009) Antibiotics in the aquatic environment–a review–Part I. Chemosphere 75(4): 417–434

Larsson DJ, Fick J (2009) Transparency throughout the production chain—a way to reduce pollution from the manufacturing of pharmaceuticals? Regul Toxicol Pharmacol 53(3):161–163

Levy SB (2002) Factors impacting on the problem of antibiotic resistance. J Antimicrob Chemother 49(1):25–30
Levy SB, Marshall B (2004) Antibacterial resistance worldwide: causes, challenges and responses. Nat Med 10(12 Suppl):S122–S129
Li B, Zhang T (2011) Mass flows and removal of antibiotics in two municipal wastewater treatment plants. Chemosphere 83:1284–1289
Li W, Shi Y, Gao L, Liu J, Cai Y (2012) Occurrence of antibiotics in water, sediments, aquatic plants, and animals from Baiyangdian Lake in North China. Chemosphere 89(11): 1307–1315
Linares JF, Gustafsson I, Baquero F, Martinez JL (2006) Antibiotics as intermicrobial signaling agents instead of weapons. Proc Natl Acad Sci 103(51):19484–19489
Liu Y, Zhang J, Gao B, Feng S (2014) Combined effects of two antibiotic contaminants on *Microcystis aeruginosa*. J Hazard Mater 279:148–155
Loke ML, Ingerslev F, Halling Sørensen B, Tjornelund J (2000) Stability of tylosin A in manure containing test systems determined by high performance liquid chromatography. Chemosphere 40:759–765
Lupo A, Coyne S, Berendonk TU (2012) Origin and evolution of antibiotic resistance: the common mechanisms of emergence and spread in water bodies. Front Microbiol 3:18
Martinez JL (2008) Antibiotics and antibiotic resistance genes in natural environments. Science 321:365–367
Martinez JL (2009) Environmental pollution by antibiotics and by antibiotic resistance determinants. Environ Pollut 157:2893–2902
Martinez JL, Baquero F (2000) Mutation frequencies and antibiotic resistance. Antimicrob Agents Chemother 44:1771–1777
Martínez JL, Baquero F (2002) Interactions among strategies associated with bacterial infection: pathogenicity, epidemicity, and antibiotic resistance. Clin Microbiol Rev 15(4):647–679
Martinez JL, Perez-Diaz JC (1990) Cloning of the determinants for microcin D93 production and analysis of three different D-type microcin plasmids. Plasmid 23:216–225
Martinez JL, Alonso A, Gomez-Gomez JM, Baquero F (1998) Quinolone resistance by mutations in chromosomal gyrase genes. Just the tip of the iceberg? J Antimicrob Chemother 42:683–688
Martinez JL, Baquero F, Andersson DI (2007) Predicting antibiotic resistance. Nat Rev Microbiol 5:958–965
Martinez JL, Fajardo A, Garmendia L, Hernandez A, Linares JF, Martínez-Solano L, Sanchez MB (2009) A global view of antibiotic resistance. FEMS Microbiol Rev 34:44–65
McManus PS, Stockwell VO, Sundin GW, Jones AL (2002) Antibiotic use in plant agriculture. Annu Rev Phytopathol 40(1):443–465
Mergeay M, Monchy S, Vallaeys T, Auquier V, Benotmane A, Bertin P, Taghavi S, Dunn J, van der Lelie D, Wattiez R (2003) Ralstonia metallidurans, a bacterium specifically adapted to toxic metals: towards a catalogue of metal-responsive genes. FEMS Microbiol Rev 27(2–3):385–410
Michael I, Rizzo L, McArdell CS, Manaia CM, Merlin C, Schwartz T, Dagot C, Fatta-Kassinos D (2013) Urban wastewater treatment plants as hotspots for the release of antibiotics in the environment: a review. Water Res 47(3):957–995
Montforts MHHM (1999) Environmental risk assessment for veterinary medicinal products Part 1. Other than GMO-containing and immunological products. RIVM report 601300 001, April 1999. National Institute for Public Health and Environment, Bilthoven, The Netherlands
Muller A, Coenen S, Monnet DL, Goossens H (2007) European Surveillance of antimicrobial consumption (ESAC): outpatient antibiotic use in Europe, 1998–2005. Euro Surveill 12: E071011–E071011
Negreanu Y, Pasternak Z, Jurkevitch E, Cytryn E (2012) Impact of treated wastewater irrigation on antibiotic resistance in agricultural soils. Environ Sci Technol 46(9):4800–4808

Nies DH (2003) Efflux-mediated heavy metal resistance in prokaryotes. FEMS Microbiol Rev 27(2–3):313–339

Nogales B, Lanfranconi MP, Piña-Villalonga JM, Bosch R (2011) Anthropogenic perturbations in marine microbial communities. FEMS Microbiol Rev 35(2):275–298

Novo A, Manaia CM (2010) Factors influencing antibiotic resistance burden in municipal wastewater treatment plants. Appl Microbiol Biotechnol 87(3):1157–1166

Okoh AI, Odjadjare EE, Igbinosa EO, Osode AN (2007) Wastewater treatment plants as a source of microbial pathogens in receiving watersheds. Afr J Biotechnol 6(25)

Pallecchi L, Bartoloni A, Paradisi F, Rossolini GM (2008) Antibiotic resistance in the absence of antimicrobial use: mechanisms and implications. Expert Rev Anti-Infect Ther 6(5):725–732

Poinar HN, Wright GD (2011) Antibiotic resistance is ancient. Nature 477:457–461

Pruden A, Arabi M, Storteboom HN (2012) Correlation between upstream human activities and riverine antibiotic resistance genes. Environ Sci Technol 46:11541–11549

Pruden A, Larsson DGJ, Amňzquita A, Collignon P, Brandt KK, Graham DW, Lazorchak JM, Suzuki S, Silley P, Snape JR, Topp E, Zhang T, Zhu Y-G (2013) Management options for reducing the release of antibiotics and antibiotic resistance genes to the environment. Environ Health Perspect 121:878–885

Rabolle M, Spliid NH (2000) Sorption and mobility of metronidazole, olaquindox, oxytetracycline and tylosin in soil. Chemosphere 40:715–722

Sarmah AK, Meyer MT, Boxall ABA (2006) A global perspective on the use, sales, exposure pathways, occurrence, fate and effects of veterinary antibiotics (VAs) in the environment. Chemosphere 65:725–759

Schlüsener MP, Bester K (2006) Persistence of antibiotics such as macrolides, tiamulin and salinomycin in soil. Environ Pollut 143(3):565–571

Shellie R, Marriott P, Leus M, Dufour JP, Mondello L, Dugo G, Sund K, Winniford B, Griffith J, Luong J (2003) Retention time reproducibility in comprehensive two-dimensional gas chromatography using cryogenic modulation. II. An interlaboratory study. J Chromatogr A 1019:273–278

Silver S, Phung LT (2005) Genes and enzymes involved in bacterial oxidation and reduction of inorganic arsenic. Appl Environ Microbiol 71(2):599–608

Singer RS, Finch R, Wegener HC, Bywater R, Walters J, Lipsitch M (2003) antibiotic resistance—the interplay between antibiotic use in animals and human beings. Lancet Infect Dis 3:47–51

Smith DL, Harris AD, Johnson JA, Silbergeld EK, Morris JG Jr (2002) Animal antibiotic use has an early but important impact on the emergence of antibiotic resistance in human commensal bacteria. Proc Natl Acad Sci USA 99:6434–6439

Sukul P, Spiteller M (2007) Fluoroquinolone antibiotics in the environment. Rev Environ Contam Toxicol 191:131–162

Sundsfjord A, Skov Simonsen G, Courvalin P (2001) Human infections caused by glycopeptide-resistant Enterococcus spp: are they a zoonosis? Clin Microbiol Infect 7(4):16–33

Taubes G (2008) The bacteria fight back. Science 321:356–361

Thevenon F, Adatte T, Wildi W, Poté J (2012) Antibiotic resistant bacteria/genes dissemination in lacustrine sediments highly increased following cultural eutrophication of Lake Geneva (Switzerland). Chemosphere 86(5):468–476

Tolls J (2001) Sorption of veterinary pharmaceuticals—a review. Environ Sci Technol 35:3397–3406

Ungemach FR (2000) Figures on quantities of antibacterials used for different purposes in the EU countries and interpretation. Acta Vet Scand Suppl 93:89–98

Vignesh S, Muthukumar K, James RA (2012) Antibiotic resistant pathogens versus human impacts: a study from three eco-regions of the Chennai coast, southern India. Mar Pollut Bull 64(4):790–800

Waksman SA, Woodruff HB (1940) The soil as a source of microorganisms antagonistic to disease-producing bacteria. J Bacteriol 40:581–600

Watkinson AJ, Murby EJ, Costanzo SD (2007) Removal of antibiotics in conventional and advanced wastewater treatment: implications for environmental discharge and wastewater recycling. Water Res 41:4164–4176

WHO (2002) Use of antimicrobials outside human medicine and resultant antimicrobial resistance in humans. Fact Sheet N° 268, World Health Organization, Geneva
WHO (2014) Antibiotic resistance: global report on surveillance 2014. World Health Organization, Geneva
WHO (2015) Antimicrobial resistance: draft global action plan on antimicrobial resistance. World Health Organization, Geneva
Witte W (1998) Medical consequences of antibiotic use in agriculture. Science 279:996–997
Wright GD (2012) Antibiotics: a new hope. Chem Biol 19(1):3–10
Wright GD, Poinar H (2012) Antibiotic resistance is ancient: implications for drug discovery. Trends Microbiol 20(4):157–159
Xu W, Zhang G, Li X, Zou S, Li P, Hu Z, Li J (2007) Occurrence and elimination of antibiotics at four sewage treatment plants in the Pearl River Delta (PRD), South China. Water Res 41(19): 4526–4534
Yim M, Shen WM, Salemi B, Rus D, Moll M, Lipson H, Klavins E, Chirikjian GS (2007) Modular self-reconfigurable robot systems [grand challenges of robotics]. IEEE Robot Autom Mag 14(1):43–52
Zak JC, Willig MR, Moorhead DL, Wildman HG (1994) Functional diversity of microbial communities: a quantitative approach. Soil Biol Biochem 26(9):1101–1108
Zak DR, Holmes WE, White DC, Peacock AD, Tilman D (2003) Plant diversity, soil microbial communities, and ecosystem function: are there any links? Ecology 84(8):2042–2050
Zervos MJ, Hershberger E, Nicolau DP, Ritchie DJ, Blackner LK, Coyle EA, Donnelly AJ, Eckel SF, Eng RH, Hiltz A, Kuyumjian AG (2003) Relationship between fluoroquinolone use and changes in susceptibility to fluoroquinolones of selected pathogens in 10 United States teaching hospitals, 1991–2000. Clin Infect Dis 37(12):1643–1648
Zhang Y, Marrs CF, Simon C, Xi C (2009) Wastewater treatment contributes to selective increase of antibiotic resistance among Acinetobacter spp. Sci Total Environ 407:3702–3706
Zoutman DE, Ford BD (2005) Canadian Hospital Epidemiology Committee and Canadian Nosocomial Infection Surveillance Program, The relationship between hospital infection surveillance and control activities and antibiotic-resistant pathogen rates. Am J Infect Control 33(1):1–5
Zuccato E, Castiglioni S, Bagnati R, Melis M, Fanelli R (2010) Source, occurrence and fate of antibiotics in the Italian aquatic environment. J Hazard Mater 179:1042–1048

Chapter 19
Risk Assessment of Antibiotics and Antibiotic-Resistant Genes in Soil

Khushbu Salian and Vladimir Strezov

19.1 Introduction

In order to treat bacterial infections in both animals and humans, use of antibiotics has increased over the past few decades resulting in contamination of ground and surface waters of various aquatic and soil environments (Allen et al. 2010; Thiele-Bruhn 2003). The environmental risks of pharmaceuticals, in general, were first identified in the 1990s followed by a series of monitoring and effect studies (Agerstrand et al. 2015). However, the environmental risks of antibiotics were more rigorously studied in the later years. Similar to many pharmaceutical drugs, antibiotics were designed to act effectively at low doses with short time of residence and to be flushed out of the body through excretion (Thiele-Bruhn 2003).

The excrements have potential to enter into the environment through a number of pathways (Chap. 4) and have resulted in residual concentrations of antibiotics and increased abundance of antibiotic-resistant microbes in the environment. Large number of studies have been performed in order to understand the environmental and ecological processes involved in the acquisition of resistance; however, the complexity of the processes have made it difficult to saturate the knowledge gap (Berglund 2015). According to Winckler and Grafe (2001), antibiotic residues found in soils are most likely due to the application of contaminated excrements on agricultural lands as fertilizers. A large number of veterinary antibiotics are given to livestock animals as feed supplements and growth promoters, which in many cases are excreted in the form of hazardous unmetabolised veterinary pharmaceutical (Ho et al. 2014). Continuous

The original version of this chapter was revised. An erratum to this chapter can be found at DOI 10.1007/978-3-319-66260-2_22.

K. Salian (✉) • V. Strezov
Department of Environmental Sciences, Faculty of Science and Engineering, Macquarie University, North Ryde, NSW 2109, Australia
e-mail: khushbusalian45@gmail.com; vladimir.strezov@mq.edu.au

M.Z. Hashmi et al. (eds.), *Antibiotics and Antibiotics Resistance Genes in Soils*, Soil Biology 51, DOI 10.1007/978-3-319-66260-2_19

application of such excrements, as estimated by various studies, can add up to kilograms per hectare of antibiotics into the agricultural soils (Tang et al. 2015; Thiele-Bruhn 2003; Winckler and Grafe 2001).

Antibiotics in soil either inhibit some microbial growth or cause acquisition of microbial resistance, which then promote growth of microbes, thus changing the microbial community of the soil (Thiele-Bruhn 2003). Higher concentration of antibiotics or resistant bacteria have seen to affect other organisms present in the soil, often causing bioaccumulation of antibiotics in the plants, contamination of nearby surface or groundwater and many other effects which are yet to be accessed (Chap. 17) (Kemper 2008). The risks posed by different antibiotic agents must be identified to reduce the impacts on the surrounding environment.

The aim of this chapter is to discuss risk assessment processes developed to identify the potential risks due to the antibiotics or ARGs present in soils. The chapter highlights a risk assessment process that can be used for any antibiotics, either those already released into the environment or new antibiotics which are not yet registered. The chapter also discusses various risk assessments performed in literature providing insight into the available data required for the assessment. For ARGs in soil, the chapter highlights some of the standard assessment processes used in various literatures in order to assess the abundance of ARGs present in soil. Also, the risks of using antibiotic-resistant marker genes in genetically modified organisms (GMOs) that pose various environmental risks as well as affect human and animal health is also discussed.

19.2 Risk Assessment

Risk assessment is an evolving process (for any given agent) that can be altered when any additional data is gathered (Durham and Swenberg 2013). Risk, according to Durham and Swenberg (2013), is the probability of hazard and exposure being expressed. A conclusive risk assessment is obtained when a broad scientific knowledge is combined with biological base and mechanistic information. Whether a particular hazard will develop or not can be determined from mechanistic information, further providing quantitative data that highlights the probability of a risk occurring.

Major government bodies prefer risk assessment to be carried out before a particular antibiotic is released into the market (Kümmerer 2008), for which the applicant must provide various information that the drug would be safe for the environment. However, there are various antibiotics that have already reached the environment and must be assessed for any future risks. Therefore, risk assessment is a process that can be used before as well as after the release of the antibiotic in to the environment.

The risk assessment of human and veterinary pharmaceuticals established by the European Union focuses on the potential environmental risks generated by the use of various pharmaceuticals (Kümmerer 2008; EMA 2016). The legal document focuses on all pharmaceuticals in general and thus can also be used for the antibiotics. Most of the antibiotics reaching the soil come from manure fertilization and sewage sludge irrigation. Antibiotics can reach the soil either directly or indirectly—application of manure collected from animals being treated, direct excretion of urine and dung of grazing animals, spreading of contaminated sludge or slurry (RIVM 1999). For assessment of the potential environmental risks, a new guidance document has been released, which is not legally binding but is followed, in general, by various parties, including regulatory and industrial bodies (VICH 2000; EMA 2016; Kümmerer 2008). A replacement of the existing guidance document for veterinary medicines occurred in 2000 and 2006. This is when the International Cooperation on Harmonization of Technical Requirements for Registration of Veterinary Products (VICH) was established between Europe, the USA and Japan and two main guidance documents, GL8 (2000) and GL38 (2006), were produced, which describe the steps to be followed while carrying out the environmental risk assessment (ERA) for any target pharmaceutical. In addition to the establishment of VICH, New Zealand and Australia joined in as observers (Koschorreck et al. 2002).

19.3 Environmental Risk Assessment in Accordance to the European Medicines Agency Guidelines

According to the European Medicines Agency (EMA) and Committee for Medicinal Products for Veterinary Use (CVMP), risk assessment of a product is evaluation of the exposure, fate and effects of the product (EMA 2016). The ERA is designed to assess the environmental risks caused by the use of various medicinal products (Kümmerer 2008). The entire risk assessment process, as described in the VICH guidelines, has been structured around the risk quotient (RQ), that indicates the probability of an adverse effect occurring. RQ is defined as the ratio between the predicted environmental concentration (PEC) and the predicted no-effect concentration (PNEC) (VICH 2006; Tihulca 2013; EMA 2016). The entire ERA process is divided into two phases, where Phase I assesses the extent of exposure of the product to the environment, whereas Phase II assesses the fate, behaviour and effects of the product on the environment. Phase II is further divided into two stages, where Tier A assesses the data on fate and effects of the veterinary medicine, which is used for risk characterization. If the risk quotient obtained cannot be excluded, then the assessment moves forward to Tier B, wherein the evaluation is subject to expert judgement (RIVM 1999, EMA 2016).

In order to evaluate the medicinal product all relevant information, must be included in the application to carry out the ERA (EMA 2016). This would include all favourable as well as unfavourable information, i.e. any abandoned or incomplete trial or test concerning the veterinary product. Published data must be included along with the proprietary data, which must be discussed along with the open literature data. Further information on the different types of data essential for ERA as well as the way in which it must be presented can be gathered from the EU Directive 2001/82/EC.

19.3.1 Phase I and Assessment of Environmental Concentration of Antibiotics in Soil

Phase I is where the investigator assesses the product on the basis of a number of questions on the chemical and physical properties, administration routes, use, frequency of dose, routes of excretion into the environment and animal husbandry (Koschorreck et al. 2002). For the risk assessment of antibiotics in soil, the question of prime importance is if the predicted environmental concentration (PEC) of the target drug in the soil (PECsoil) is less than 100 μg/kg (VICH 2000; EMA 2016).

The amount of antibiotic expected in the environment, which is called predicted environmental concentration (PEC) or measured environmental concentration (MEC), can be estimated by measuring the concentration levels present in the manure (or in some cases sludge or compost) (Martin et al. 2012). There are a number of exposure screening models discussed in literature to determine the PEC of veterinary medicines in soil (Montforts 2006; Montforts et al. 1999; Spaepen et al. 1997; RIVM 1999; Montforts 2006). Spaepen et al. (1997), proposed a single conservative method that can be used across international borders providing default values collected from different databases to simplify the calculations. The rate at which the manure is incorporated into the soils along with the data of drug residue levels present in the manure, the amount of antibiotic residues in the soil can be calculated. Following are the steps to calculate PEC_{soil}:

Step I: Calculating the total quantity of active ingredient (Q) from Eq. (19.1).

$$Q = \mathrm{ID} \times \mathrm{BW} \times T \times N \tag{19.1}$$

where ID is individual dose rate (mg per kg body weight); BW is the body weight of the animal type (kg); T is number of individual treatment per animal (per animal) and N is the number of animals raised per year on each place in the animal housing location, (per year per place).

Step II: Calculating the concentration of residues of the active ingredient in the combined excreta (C_{E}, mg per kg excreta—faeces and urine together) from Eq. (19.2).

$$C_{\mathrm{E}} = Q/P_{\mathrm{E}} \tag{19.2}$$

where Q is the total quantity of active ingredients and P_{E} is the yearly output of excreta (kg per place per year).

Step III: Calculating maximum quantities of manure applied to land (M, kg excreta per hectare per year).

There are two ways, direct and indirect ways to calculate the maximum quantities of manure applied. Direct way would be to collect the data on the rates of manuring from the practice followed. However, an indirect method is preferred as it is a more conservative approach. According to EU regulations, a limit exists for the inputs of fertilizing substances in order to reduce the release of nutrients and in turn minimize pollution rates. The maximum amount of nitrogen (A_N, kg nitrogen per hectare per year) and/or phosphorus (A_p, kg P_2O_5 per hectare per year) are stated in these regulations.

Maximum quantities of manure applied to land can be calculated using Eqs. (19.3) and (19.4) from

- The yearly production of nitrogen (P_N, kg N per place per year)
- The yearly production of phosphorus (P_P, kg P_2O_5 per place per year)
- The yearly manure output (P_E) known for the target animal

$$M_N = \frac{A_N}{P_N} \times P_E \tag{19.3}$$

$$M_P = \frac{A_P}{P_P} \times P_E \tag{19.4}$$

Depending on the agricultural use of the land, allowed quantities of N and P may vary. The most restrictive rate (mainly nitrogen) obtained from either of the above equation must be used as M (kg excreta per hectare per year) (Spaepen et al. 1997).

Step IV: Calculating the amount of active substance-related residues applied per unit area of land (C_{SA}, mg per hectare per year)

Combining steps II and III, Eq. (19.5) is determined.

$$C_{SA} = M \times C_E \tag{19.5}$$

The variable P_E is cancelled out in this assessment. C_{SA} can also be obtained directly by combining the most relevant ratio from A_N/P_N and A_P/P_P and Q. However, the above-mentioned procedure involving P_E is preferred for the possible expansion of the calculation routine and clarity.

Step V: Calculating PEC

Initially, the concentration of C_{SA} is transformed into concentrations in ploughed soil, i.e. C_{SV} (mg per kg soil and per year).

C_{SV} is calculated from:

- The weight of the ploughed layer (W, kg per hectare)
- The soil bulk density (ρ, kg/m^3)
- The volume of the ploughed layer (V, m^3/ha)

Whereas, V depends on the depth of the furrow (D, in cm). After converting hectare into m^2 and cm into m:

$$V = \frac{D}{100} \times 100 \times 100 = D \times 100 \tag{19.6}$$

$$W = V \times \rho \tag{19.7}$$

If the land is ploughed, a furrow depth of 25 cm can be used as D that depends on the regional agricultural practice. However, in case of no soil disturbance, the substance distribution over the top 5 cm of the soil can be assumed (Spaepen et al. 1997). Different soil textures have varying soil bulk densities. According to the EU Directive 81/852/EEC, an average value for the soil density of 1500 kg/m^3 can be used.

Predicted environmental concentration (PEC) in soil (C_{sv}) can then be calculated according to Eq. (19.8):

$$C_{SV} = \frac{C_{SA}}{W + M} \tag{19.8}$$

The expanded/modified calculations can be used to modify the above-mentioned basic equations in order to refine the PEC in soil. Thus, if the calculated C_{sv} is below the threshold trigger value of 100 μg/kg then no further assessment for the antibiotic is required, whereas, if the C_{sv} exceeds 100 μg/kg then the assessment moves on to Phase II of the ERA process.

A study performed by Dahshan et al. (2015), analysed the potential ecological risks generated by applying poultry litter as fertilizers on agricultural lands in Egypt. The method proposed by Spaepen et al. (1997) was used by the authors in order to calculate the PEC of antibiotics in soil. The analysis showed that the predicted concentration of tetracycline in broiler chicken litter was 443.34 μg/kg, which is above the trigger value (threshold) of 100 μg/kg and thus the risk assessment must proceed to the next stage.

However, the method proposed by Spaepen et al. (1997) can only be applied in situations, wherein the excreta is collected from housed animals (animals housed completely throughout their production cycle) and stored all together. The direct excretion of dung or urine from grazing animals (not housed) is not considered. The model for calculating the PEC in soil proposed by Spaepen et al. (1997) is still considered as the baseline model, which underwent a number of modifications to improve the assessment.

In the recent guidelines developed by the European Medical Agency (2016), two sets of formulas have been discussed depending on the direct and indirect addition of manure containing antibiotic residues into the soil; one for intensively reared animals and the other for pasture animals+. A total residue approach is adopted, wherein it is assumed that the total amount of dose given to the animal is excreted and degradation/metabolism is not taken into consideration. For intensively reared animals, the initial $PEC_{soil\ initial}$ depends on the amount of manure that contains the active residue, which can be applied onto the land (Tihulca 2013). The assumption under which the formula has been developed is that the intensively reared animals are housed indoors throughout the production cycle, veterinary medicinal products treatment is provided in housing and the excreta contains the active ingredient, which is then collected from the stable and applied onto lands as manure. For animals that are on pasture throughout their production cycle, which means that the treatment is carried out in the field and the faeces containing residues of veterinary medicine are directly excreted onto the land, the number of animals kept on a particular area of land is the key factor for calculating the $PEC_{soil\ initial}$ for pasture animals (Tihulca 2013; EMA 2016). Thus, if the calculated $PEC_{soil\ initial}$ is below the threshold trigger value of 100 μg/kg no further assessment for the antibiotic is required, whereas, if the $PEC_{soil\ initial}$ exceeds 100 μg/kg then the assessment moves on to Phase II of the ERA process (EMA 2016).

19.3.2 Phase II and Predicted No-Effect Concentration

In Phase II, a more detailed evaluation of exposure of the antibiotic (active ingredient) to the environment is performed, which is then followed by the fate and effect assessment.

19.3.2.1 Tier A and Predicted Environmental Concentration Refinement

Risk assessment performed using the initial Predicted Environmental Concentration (PEC) shows that when RQ is ≥ 1 refinement of the PEC must be considered. The exposure assessment in Phase II is meant to refine the initial PEC calculated previously by considering data on degradation and metabolism of the antibiotics in the animal body and soil (EMA 2016). For instance, the excreted metabolite can be more water soluble and persistent and/or mobile as compared to the parent compound. Thus, it is essential to assess the relevant metabolites of the parent compounds. In case of biodegradation data, persistent compounds, i.e. if DT90 (time to degrade 90% of the original concentration of the compound in the tested soils) is >1 year in soil (annual application), which means that the compound has a possibility to accumulate in the soil; thus, the initial PEC must be recalculated. There are a number of options available to refine PEC, depending on the characteristics of the active ingredient and other scenarios, such as:

- Refinement based on excretion pattern
- Refinement based on metabolism
- Refinement based on degradation in soil and
- Refinement based on degradation in slurry/manure

The detailed refinement procedure for each of the above-mentioned options have been discussed in the 2016 EMA guideline. Data on metabolism and degradation of various antibiotics can be obtained from literature (Chen et al. 2016; Pan and Chu 2016; Wu et al. 2014). A study performed by Slana and Dolenc (2013) evaluated the environmental risk of enrofloxacin (EF) in soil, when cattle manure was applied onto the land. The initial PEC concentration without subtracting metabolism and degradation was 145.7 μg/kg, whereas, when the values for metabolism were included, the refined PEC value was 21.9 μg/kg. As the value of PEC was reduced when accounting for the metabolism and was below the trigger value of 100 μg/kg, it meant that no risks existed due to EF in the environment. However, if the values are above 100 μg/kg after subtracting the mitigation values, the assessment would continue onto the next stage. Table 19.1 shows the concentration of various antibiotics in soils worldwide.

19.3.2.2 Tier B and Effects Assessment

There are two steps in the effects assessment procedure (European Commission 2003):

1. Hazard identification, wherein the effects of concern are identified
2. Dose–response assessment, where the predicted no-effect concentration (PNEC) is determined

Evaluation of the data for both steps is of prime importance for their completeness and adequacy, thus it has been suggested that the effects assessment must start with the evaluation of available ecotoxicological data (European Commission 2003). Ecotoxicological tests are performed in order to assess the adverse effects of antibiotics on the ecosystem services, which also includes all non-target organisms (Brandt et al. 2015; Koschorreck et al. 2002). Ecosystem functioning in soil is a collaborative action of all organisms present in the soil; thus, it is essential to protect the entire soil community, which plays an important role in maintaining the structure and function of the ecosystem (European Commission 2003). Therefore, it is central to include a suite of soil tests that provides data relevant to primary producers (plants), consumers (invertebrates, e.g. earthworms) and decomposers (microbes).

The predicted no-effect concentration is the concentration below which an effect most likely would not occur and is calculated by dividing the lowest long-term no-observed effect concentration (NOEC) or short-term L(E)C50 (lethal concentration required to kill 50% of the population) value by the assessment factor (AF). Extrapolation from various laboratory toxicity test data (for a limited number of species) contains a level of uncertainty, which is reflected using the assessment factor (AF) (European Commission 2003). AF value is considered to be between 1000 and 10, i.e. 1000 depicts conservative and protective (applied when limited data is available or lowest effect value of acute tests is available) and can be reduced to 10 as more evidence on toxicity is obtained (for instance, more number of sensitive species present, compound is not degraded easily thus chance of chronic exposure exists and long-term chronic effects test results are available) (VICH 2006). Dahshan et al. (2015) calculated PNEC by substituting toxicity of tetracycline (TOX) with the lowest median effective concentration (EC50), which was adopted from studies performed by Thiele-Bruhn and Beck (2005) who studied the effects of tetracycline on soil microbial activity. The authors also used

Table 19.1 Antibiotic concentration worldwide

Antibiotic class	Compound	Sites	Concentration in soil (μg/kg)
Tetracyclines (TETs)	Oxytetracycline (OTC)	Beijing, China	80
		Austria	nd
		Guangdong, China	9.6
		Kenya, Africa	nd–29.38
	Doxycycline (DC)	Kenya, Africa	nd–3.85
		Malaysia	63–728
	Tetracycline (TC)	Beijing, China	5.2
		Austria	nd
		Guangdong, China	44.1
		Kenya, Africa	nd–16.02
	Chlortetracycline (CTC)	Guangdong, China	31.1
		Beijing, China	17
		Kenya, Africa	nd–38.79
		Denmark	10–15
Fluoroquinolones (FQs)	Ciprofloxacin (CIP)	Guangdong, China	26.9
		Beijing, China	23
		Shandong, China	104.4
		Kenya, Africa	9.88
		Turkey	nd
	Norfloxacin (NOR)	Guangdong, China	61.9
		Beijing, China	13
		Kenya, Africa	nd–10.34
		Shandong, China	55.7
		Malaysia	nd–96
	Enrofloxacin (ENR)	Guangdong, China	99.4
		Beijing, China	47
		Kenya, Africa	nd–16.91
		Shandong, China	18.6
		Turkey	50
		Austria	50
		Malaysia	36–378
Sulfonamides (SAs)	Sulfamethazine (SMZ)	Guangdong, China	5.5
		Beijing, China	0.37
		Kenya, Africa	nd–24.23
	Sulfadiazine (SD)	Guangdong, China	13.4
		Beijing, China	0.11
		Kenya, Africa	nd–3.85
		Austria	nd
		Malaysia	nd

(continued)

Table 19.1 (continued)

Antibiotic class	Compound	Sites	Concentration in soil (μg/kg)
	Sulfameter (SME)	Kenya, Africa	nd–5.79
		Guangdong, China	51.4
	Sulfamethoxazole (SMX)	Guangdong, China	23.5
		Beijing, China	0.06
		Kenya, Africa	nd–14.47
	Sulfonamides (SAs)	Turkey	400

Note:
nd not detected
Ho et al. (2014); Karcı and Balcıoğlu (2009); Martínez-Carballo et al. (2007); Yang et al. (2016)

assessment factor (AF) of 1000 as the risks were estimated by acute toxicity tests and obtained a PNEC value of 270 μg/kg tetracycline in broiler chicken farms in Egypt (Dahshan et al. 2015). This means when the toxicity data is available for a producer, a consumer or/and a decomposer, PNEC in soil can be calculated using assessment factors (AF).

The availability of toxicity data for soil compartments is limited when compared with data availability of water compartments. Where soil compartment data is available, they will mostly represent test results obtained from short-term studies. In cases where the toxicity data is limited, equilibrium partitioning method is adopted, wherein the $PNEC_{soil}$ value is derived from $PNEC_{water}$ by using the partition coefficient (K_d) (European Commission 2003; EMA 2016). A study performed by Yang et al. (2016), investigated the occurrence and environmental risks of 12 antibiotics in soils collected from four sampling sites in Kenya. The $PNEC_{soil}$ values for 8 of the antibiotics were obtained from literature, whereas the $PNEC_{soil}$value for sulfamethoxazole (SMZ) was derived from De Liguoro et al. (2009) who studied the toxicity of SMZ to *Daphnia magna* using the solid-water equilibrium partition coefficient (Yang et al. 2016). Similar study performed by Wu et al. (2014), assessed the impacts of quinolone antibiotic in soil samples collected from five vegetable farms in Southern China. The $PNEC_{soil}$ values were calculated from $PNEC_{water}$ values by using the equilibrium partition approach, wherein the $PNEC_{water}$ values were calculated using the lowest acute toxicity data (which is available for water compartments) using the AF of 1000. Another study performed by Li et al. (2015), analysed the occurrence of 15 antibiotics in soils and manures collected from 11 greenhouse vegetable production (GVP) bases in Beijing, China, which used the $PNEC_{soil}$ values calculated by Wu et al. (2014). This means that there are a number of scientific papers that can be used to derive the toxicity data of antibiotics in soil to calculate the $PNEC_{soil}$ values or can be derived by using data on toxicity of antibiotics assessed for the aquatic environment.

19.3.3 Risk Quotient

When the ratio of predicted environmental concentration (in this case, PEC_{soil}) and predicted no-effect concentration is lower than 1 [Risk Quotient (RQ) < 1], then no further assessment is required. However, if the RQ is >1, then a more detailed study on the effects and fate of the antibiotic is performed to refine the PNEC and PEC values. PEC can decline as a result of biodegradation (biotransformation and mineralization), photolysis (chemical transformation) and hydrolysis of the antibiotic in the environment (Koschorreck et al. 2002; VICH 2006; EMA 2016). The results obtained from the short-term ecotoxicity tests can be further refined by including long-term field or semi-field data. After the refinement, if the risk quotient of the antibiotic is still greater than 1, potential risk to the environment due to the antibiotic is assumed.

The RQ value for tetracycline obtained by Dahshan et al. (2015) was 1.64, meaning that the residues of tetracycline found in poultry litter in Egypt displayed environmental risks. It was also noted that the toxicity of a single pollutant can be altered due to the presence of other veterinary medicines and pollutants in soils. The authors also highlighted some of the limitation of the RQ equation. The type of effects or mechanism adopted by a number of compounds may be different than the once considered in the application of RQ equation, which can lead to the overestimation of the potential effects. Although, if only one or two substances are causing the RQ to increase above 1 unit the above possibility will not be of concern (Dahshan et al. 2015). In spite of such limitations, the risk quotient method is being accepted internationally and is being used to assess the environmental risks of antibiotics in soil. Table 19.2 shows various studies performed worldwide using the risk quotient method. However, if the RQ obtained after all the possible refinements is still ≥1, then various risk mitigation strategies must be adopted for reducing the impacts.

19.4 Risk Assessment of Antibiotic-Resistant Genes

The release of antibiotics into the environment can affect pathogens as well as commensal bacteria. Resistance is provoked by the continuous sublethal dosage of antibiotics, which can be due to the repeated spreading of contaminated manure onto agricultural soils. Resistance in many soil organisms is not only due to the input of antibiotics into the soil but also due to the application of manure containing resistance genes to the soil (Tang et al. 2015; Thiele-Bruhn 2003). A study performed by (Fründ et al. 2000), concluded that the application of manure containing tetracycline-induced antibiotic resistance, which lasted for weeks in soil microorganisms. Once entered into the soil, antibiotic-resistant genes (ARGs) can alter the soil community (direct effect) or can bioaccumulate in the plants growing on that soil. This causes the ARGs to enter into the food chain, inducing

Table 19.2 Environmental risks of various antibiotics in soil worldwide

Antibiotic	Area of study	Ecotoxicity	$PNEC_{soil}$ (μg/kg)	Risk quotient > 1-high; 0.1 < RQ < 1-medium; <0.1-low	References
Norfloxacin (NOR)	Guangdong, China	0.022 mg/L	29.68	Medium	Backhaus et al. (2000), Wu et al. (2014)
Ciprofloxacin (CIP)	Guangdong, China	0.005 mg/L	25.64	High	Wu et al. (2014)
Lomefloxacin (LOM)	Guangdong, China	0.022 mg/L	93.85	Medium	Robinson et al. (2005), Wu et al. (2014)
Enrofloxacin (ENR)	Guangdong, China	0.049 mg/L	24.00	High	Backhaus et al. (2000), Wu et al. (2014)
Norfloxazin (NFX)	Beijing, China	0.022 mg/L	29.68	High	Li et al. (2015)
Tetracycline (TC)	Kenya, Africa	30 mg/kg	30	High	Thiele-Bruhn and Beck (2005), Yang et al. (2016)
Oxytetracycline (OTC)	Kenya, Africa	50 mg/kg	50	Medium	Yang et al. (2016)
Chlortetracycline (CTC)	Kenya, Africa	270 mg/kg	270	High	Thiele-Bruhn and Beck (2005), Yang et al. (2016)

resistance in humans (indirect effect). Thus, lowering the success rate of various pharmacotherapies used to cure humans as well as animals. Therefore, it is essential to monitor the occurrence and the level of resistance already existing in organisms present in a particular area of soil to reduce the development of risk to the ecosystem.

19.4.1 Identification of Antibiotic-Resistant Bacteria

A recent study performed by Kim et al. (2016), examined the impact of applying livestock (pig and cattle) manure compost on South Korean agricultural soils. The main aim of the study was to determine the occurrence, diversity and abundance of oxytetracycline-, tetracycline- and chlortetracycline-resistant bacteria and genes in three agricultural soils. Various soil samples were collected from the agricultural lands, from which tetracycline-resistant bacteria were isolated using the viable cell count method (Kim et al. 2016). The rate of antibiotic resistance of the bacterial colonies that grew on the plates were assessed using the following formula:

$$\text{Antibiotic resistance rate} = \frac{\text{Total number of bacteria that grew on plates containing antibiotics}}{\text{Total number of bacteria that grew on plates with no antibiotics}} \times 100$$

The analysis showed that the agricultural soils treated with manure showed 2–11 times higher amount of resistant bacteria as compared to the natural soil (Kim et al. 2016). Heuer et al. (2011) also reported that manure from livestock animals carry a considerable amount of antibiotic-resistant bacteria; the repeated application of such manure on agricultural soil can increase antibiotic resistance. The review also highlighted that microbial communities in natural soils (non-agricultural soils) can also possess a vast diversity of antimicrobial resistance (Heuer et al. 2011). Thus, also the analysis performed by Kim et al. (2016) found small amount of viable antibiotic-resistant bacteria in the control (natural) soil as well. The rationale behind this was given by Allen et al. (2010) that resistance is introduced by a natural process into the ecosystem, which includes migratory wild birds and wind.

19.4.2 Abundance of ARGs in Soil

Once the presence of resistant bacteria is determined, the prevalence of ARGs in the genomic DNA of the isolated bacteria is examined using polymerase chain reaction (PCR). This method has been adopted by a number of studies to amplify the ARGs using gene-specific primers (Kim et al. 2016; Ng et al. 2001; Tang et al. 2015; Wang et al. 2014). Wang et al. (2014), studied the distribution pattern of sulfonamide-resistant (*sul1*, *sul2* and *sul3*) genes in four chicken farms, four pig farms, one mountain forest and one non-arable agricultural area in the Jiangsu province, Southeastern China. The PCR results showed that the DNA from the pig-manured soils had higher copy numbers of *sul1* gene, compared to *sul2* gene. However, in the chicken-manured soils, the abundance of *sul2* gene was higher than the *sul1* gene. Also, *sul3* gene was detected at a relatively lower quantity in the eight manured soils, but was not detected in the forest soil (Wang et al. 2014). The study concluded that repeated application of manure obtained from chicken and pigs treated with sulfadiazine (SDZ) elevates the abundance and transfer of ARGs in soil.

Another study performed by Tang et al. (2015) investigated the occurrence of ARGs due to long-term application of manure in paddy soils in China. The manure applied to the target sites varied, i.e. JX site—12,800 kg/ha/year dw, CS site—3480 kg/ha/year dw, YT site—4500 kg/ha/year dw and NC site—4200 kg/ha/year dw. The abundance of ARGs in the soil were analysed using real-time PCR. This method was used to quantify six tetracycline genes and two sulfonamide genes from the soil samples. The abundance of ARGs were higher in JX, NC and YT sites, whereas only one gene was influenced by the manure application in the CS site (lowest amount of manure applied). A study performed by Knapp et al. (2009) in Netherlands found that a site (Heino), which received ten times more manure, failed to show proportionate ARGs levels, when compared to the other four sites. The authors claimed that it might be because of local differences in cropping patterns, irrigation water sources, soil type and other factors. This means that the abundance and dissemination of ARGs from resistant bacteria to indigenous microbes can be affected due to various spatial and temporal changes. However, there are no specific risk assessment procedures developed for assessing the risks of ARGs in soil as different 'ARG-environment' combinations have to be evaluated on a case-by-case basis in order to identify a specific ARG as environmental pollutant (Woegerbauer et al. 2015).

19.4.3 Genetically Modified Organisms in Soil

Large number of genetically modified bacterial strains are being introduced into the soil in order to promote plant growth, degrade polluting compounds (xenobiotics) or for pest control (against pathogens)(Van Elsas et al. 1998). Although naturally occurring bacteria can perform these functions, the rate of success is limited; thus, the use of bacteria isolated from the same environment, which can adapt to the in situ ecological stresses, is genetically modified to enhance their performance rate in the environment (Doyle et al. 1995; Van Elsas et al. 1998). Introduction of genetically modified organisms (GMOs) can cause undesirable effects in the surrounding environment as well as affect the microbial community present in the soil. Amarger (2002) provides an in-depth information on the various applications of modified bacteria in agriculture.

Gene markers are introduced into the inoculant bacteria that can be detected against the natural background of indigenous organisms. This includes insertion of ARGs into the genome of the inoculant bacteria (Amarger 2002; Van Elsas et al. 1998). Once introduced into the soil, marker genes are known to inactivate antibiotics, which are used in clinical as well as veterinary medicine. Dissemination of ARGs from the parent bacteria (original resistant gene donor) to the following generation or recipient bacteria can also occur, thus increasing the abundance of ARGs in soil (EFSA 2007). The use of transgenic plants result in the accumulation of transgenic DNA in soil. DNA from transgenic plants enter into the soil either during decomposition of dying cells, when microorganisms perform active secretion, by dispersal of pollen, during growth or decay of plant litter. Number of studies have indicated that a minor proportion of transgenic DNA, in this case ARGs, can remain intact for a sufficiently long period and can be taken up by any competent bacteria in the environment (de Vries et al. 2003; Meier and Wackernagel 2003; EFSA 2007).

The European Food and Safety Authority (EFSA) (2007) prepared a report, which can be used as supplement information in order to tackle antibiotic-resistant marker (ARM) genes used in GMOs by competent authorities, in accordance with the EU Directive 2001/18/EC. The report highlights various aspects of ARM, such as methods used for construction of ARM, frequency of horizontal gene transfer and DNA stability in soil. All these information, along with the biochemical characteristics and function of the ARM gene and its products are required for the determination of possible risks to the environment and human health. The process of risk assessment starts by determining a risk question 'What is the impact of an ARM gene in a transgenic plant?'

The potential hazards are identified by assessing two types of risks:

1. Direct risks. Induction of toxic reactions upon consumption or contact of plant tissue in the respective consumer
   - Transformation of plant cell DNA into a toxic DNA fragment
   - Production of toxic proteins or toxic RNA molecules due to coding of sequences on the bacterial plasmid vector
   - Production of toxic substances, when the plant metabolism is intervened by the vector encoded bacterial proteins
2. Indirect risks. Adverse effects, which are unintended, on the environment, human and animal health from a direct toxic impact
   - Escape and spread of ARM genes from transgenic to conventional plants.
   - Dissemination rate of ARM genes in bacterial population increases and thus pathogenic bacterial population affects animals and humans reducing the options of therapy.

After potential hazard identification or characterization, exposure assessment is performed, i.e. frequency of ARM gene transfer leading to resistance in bacterial strains. A study performed by (Kim et al. 2004) investigated the possibility of gene transfer from GMOs to soil microorganisms over a period of 6 months. The study concluded that no gene transfer was observed between the introduced GMO soybeans to any soil-borne bacteria, including *Rhizobium*, the root nodule bacteria. Although no gene transfer was observed, the risk evaluation step in the case of ARGs is usually challenged since an extremely rare single event can have a dramatic impact on the environment as well as human and animal health (EFSA 2007). Both, hazard characterization and exposure assessment results are then used for assessing the risks. The EFSA report also provides information on various ARM genes, which are used frequently for the production and development of transgenic plants. The report also highlights the biochemical functions of various ARM genes and existing background level of resistance in the natural habitats that can be used for the risk assessment of target ARM gene.

19.5 Conclusion

Use of antibiotics and ARGs is increasing at a global scale. Risk assessment is a complex process, which can be used for identification or estimation of potential environmental and/or health hazards arising from the extensive use of antibiotics and ARGs. Risk assessment of antibiotic using the risk quotient (RQ) method is the most extensively used procedure for estimating the risks posed by any antibiotic entering the soil. However, not all data required by the European Medical Association process is readily available. For the predicted environmental concentration (PEC) refinement, the absence of reliable, good and consistent fate data makes the refinement process difficult (Grung et al. 2008). Also, one of the major impediments for application of the risk quotient methods is the estimation of predicted no-effect concentration (PNEC) in soil. Most of the ecotoxicity analysis have been performed for the water compartment, which complicates the estimation of PNEC in soil. ARGs/GMOs were introduced to increase agricultural yield; however, the use of these genes/organisms are affecting the recipient environment as well as the consumers. For ARGs, the uncertainty of dissemination of resistance, changing climatic conditions, soil type, geographical changes, bacterial genome and other factors precipitate a number of challenges to assess the risks of ARGs in soil.

References

Agerstrand M, Berg C, Bjorlenius B, Breitholtz M, Brunstrom B, Fick J, Gunnarsson L, Larsson DG, Sumpter JP, Tysklind M, Ruden C (2015) Improving environmental risk assessment of human pharmaceuticals. Environ Sci Technol 49:5336–5345

Allen HK, Donato J, Wang HH, Cloud-Hansen KA, Davies J, Handelsman J (2010) Call of the wild: antibiotic resistance genes in natural environments. Nat Rev Microbiol 8:251–259

Amarger N (2002) Genetically modified bacteria in agriculture. Biochimie 84:1061–1072

Backhaus T, Scholze M, Grimme LH (2000) The single substance and mixture toxicity of quinolones to the bioluminescent bacterium Vibrio fischeri. Aquat Toxicol 49:49–61

Berglund B (2015) Environmental dissemination of antibiotic resistance genes and correlation to anthropogenic contamination with antibiotics. Infect Ecol Epidemiol 5:28564

Brandt KK, Amezquita A, Backhaus T, Boxall A, Coors A, Heberer T, Lawrence JR, Lazorchak J, Schonfeld J, Snape JR, Zhu YG, Topp E (2015) Ecotoxicological assessment of antibiotics: a call for improved consideration of microorganisms. Environ Int 85:189–205

Chen G, Liu X, Tartakevosky D, Li M (2016) Risk assessment of three fluoroquinolone antibiotics in the groundwater recharge system. Ecotoxicol Environ Saf 133:18–24

Dahshan H, Abd-Elall AM, Megahed AM, Abd-El-Kader MA, Nabawy EE (2015) Veterinary antibiotic resistance, residues, and ecological risks in environmental samples obtained from poultry farms, Egypt. Environ Monit Assess 187:2

De Liguoro M, Fioretto B, Poltronieri C, Gallina G (2009) The toxicity of sulfamethazine to Daphnia magna and its additivity to other veterinary sulfonamides and trimethoprim. Chemosphere 75:1519–1524

de Vries J, Heine M, Harms K, Wackernagel W (2003) Spread of recombinant DNA by roots and pollen of transgenic potato plants, identified by highly specific biomonitoring using natural transformation of an Acinetobacter sp. Appl Environ Microbiol 69:4455–4462

Doyle JD, Stotzky G, McClung G, Hendricks CW (1995) Effects of genetically engineered microorganisms on microbial populations and processes in natural habitats. Adv Appl Microbiol 40:237–287

Durham SK, Swenberg JA (2013) Risk assessment. In: Haschek and Rousseaux's handbook of toxicologic pathology. Elsevier, Amsterdam, pp 989–997

EFSA (European Food and Safety Authority) (2007) Risk assessment of antibiotic resistance marker genes in genetically modified organisms. Available at: https://www.bmgf.gv.at/cms/home/attachments/7/4/2/CH1052/CMS1198058130229/woegerbauer_2007.pdf. Accessed 18 May 2017

European Commission (2003) Technical guidance document on risk assessment. Available at: https://echa.europa.eu/documents/10162/16960216/tgdpart2_2ed_en.pdf. Accessed 26 April 2017

European Medicines Agency (EMA) (2016) Guideline on environmental impact assessment for veterinary medicinal products in support of the VICH guidelines GL6 and GL38. Available at: http://www.ema.europa.eu/docs/en GB/document_library/Scientific_guideline/2009/10/WC500004386.pdf. Accessed 25 April 2017

Fründ H-C, Schlösser A, Westendarp H (2000) Effects of tetracycline on the soil microflora determined with microtiter plates and respiration measurement. Mitteilgn Dtsch Bodenkundl Gesellsch 93:244–247

Grung M, Kallqvist T, Sakshaug S, Skurtveit S, Thomas KV (2008) Environmental assessment of Norwegian priority pharmaceuticals based on the EMEA guideline. Ecotoxicol Environ Saf 71:328–340

Heuer H, Schmitt H, Smalla K (2011) Antibiotic resistance gene spread due to manure application on agricultural fields. Curr Opin Microbiol 14:236–243

Ho YB, Zakaria MP, Latif PA, Saari N (2014) Environmental risk assessment for veterinary antibiotics and hormone in Malaysian agricultural soil. Iranian J Public Health 43:67

Karcı A, Balcıoğlu IA (2009) Investigation of the tetracycline, sulfonamide, and fluoroquinolone antimicrobial compounds in animal manure and agricultural soils in Turkey. Sci Total Environ 407:4652–4664

Kemper N (2008) Veterinary antibiotics in the aquatic and terrestrial environment. Ecol Indic 8:1–13

Kim YT, Byoung K, Hwang N, Kang TH, Sung U (2004) Investigation of possible gene transfer to soil microorganisms for environmental risk assessment of genetically modified organisms. J Microbiol Biotechnol 14(3):498–502

Kim SY, Kuppusamy S, Kim JH, Yoon YE, Kim KR, Lee YB (2016) Occurrence and diversity of tetracycline resistance genes in the agricultural soils of South Korea. Environ Sci Pollut Res Int 23:22190–22196

Knapp CW, Dolfing J, Ehlert PA, Graham DW (2009) Evidence of increasing antibiotic resistance gene abundances in archived soils since 1940. Environ Sci Technol 44:580e587

Koschorreck J, Koch C, Rönnefahrt I (2002) Environmental risk assessment of veterinary medicinal products in the EU—a regulatory perspective. Toxicol Lett 131:117–124

Kümmerer K (2008) Pharmaceuticals in the environment: sources, fate, effects and risks. Springer, Berlin

Li C, Chen J, Wang J, Ma Z, Han P, Luan Y, Lu A (2015) Occurrence of antibiotics in soils and manures from greenhouse vegetable production bases of Beijing, China and an associated risk assessment. Sci Total Environ 521–522:101–107

Martin J, Camacho-Munoz MA, Santos JL, Aparicio I, Alonso E (2012) Distribution and temporal evolution of pharmaceutically active compounds alongside sewage sludge treatment. Risk assessment of sludge application onto soils. J Environ Manag 102:18–25

Martínez-Carballo E, González-Barreiro C, Scharf S, Gans O (2007) Environmental monitoring study of selected veterinary antibiotics in animal manure and soils in Austria. Environ Pollut 148:570–579

Meier P, Wackernagel W (2003) Monitoring the spread of recombinant DNA from field plots with transgenic sugar beet plants by PCR and natural transformation of Pseudomonas stutzeri. Transgenic Res 12:293–304

Montforts MH (2006) Validation of the exposure assessment for veterinary medicinal products. Sci Total Environ 358:121–136
Montforts MH, Kalf DF, van Vlaardingen PL, Linders JB (1999) The exposure assessment for veterinary medicinal products. Sci Total Environ 225:119–133
Ng LK, Martin I, Alfa M, Mulvey M (2001) Multiplex PCR for the detection of tetracycline resistant genes. Mol Cell Probes 15:209–215
Pan M, Chu LM (2016) Adsorption and degradation of five selected antibiotics in agricultural soil. Sci Total Environ 545–546:48–56
RIVM Dutch National Institute for Public Health and the Environment (1999) Environmental risk assessment for veterinary medicinal products. Part 1. Other than GMO-containing and immunological products. First update. Available at: http://www.rivm.nl/en/Documents_and_publications/Scientific/Reports/1999/april/Environmental_risk_assessment_for_veterinary_medicinal_products_Part_1_Other_than_GMO_containing_and_immunological_products_First_update?sp=cml2bXE9ZmFsc2U7c2VhcmNoYmFzZT0yMzkzMDtyaXZtcT1mYWxzZTs=&pagenr=2394. Accessed 26 April 2017
Robinson AA, Belden JB, Lydy MJ (2005) Toxicity of fluoroquinolone antibiotics to aquatic organisms. Environ Toxicol Chem 24:423–430
Slana M, Dolenc MS (2013) Environmental risk assessment of antimicrobials applied in veterinary medicine—a field study and laboratory approach. Environ Toxicol Pharmacol 35:131–141
Spaepen KR, Van Leemput LJ, Wislocki PG, Verschueren C (1997) A uniform procedure to estimate the predicted environmental concentration of the residues of veterinary medicines in soil. Environ Toxicol Chem 16:1977–1982
Tang X, Lou C, Wang S, Lu Y, Liu M, Hashmi MZ, Liang X, Li Z, Liao Y, Qin W, Fan F, Xu J, Brookes PC (2015) Effects of long-term manure applications on the occurrence of antibiotics and antibiotic resistance genes (ARGs) in paddy soils: evidence from four field experiments in south of China. Soil Biol Biochem 90:179–187
Thiele-Bruhn S (2003) Pharmaceutical antibiotic compounds in soils—a review. J Plant Nutr Soil Sci 166:145–167
Thiele-Bruhn S, Beck IC (2005) Effects of sulfonamide and tetracycline antibiotics on soil microbial activity and microbial biomass. Chemosphere 59:457–465
Tihulca IV (2013) Methodology for environmental risk assessment associated with the use of veterinary medicinal products. Med Vet/Vet Drug 1:74–80
Van Elsas J, Duarte G, Rosado A, Smalla K (1998) Microbiological and molecular biological methods for monitoring microbial inoculants and their effects in the soil environment. J Microbiol Methods 32:133–154
VICH (Harmonisation of Technical Requirements for Registration of Veterinary Medicinal Products) (2000) Environmental impact assessment (EIAS) for Veterinary medicinal products (VMPS)—Phase I. Available at: http://www.vichsec.org/guidelines/pharmaceuticals/pharma-safety/environmental-safety.html. Accessed 25 April 2017
VICH (Harmonisation of Technical Requirements for Registration of Veterinary Medicinal Products) (2006) Environmental impact assessment (EIAS) for Veterinary medicinal products (VMPS)—Phase II. Available at: http://www.vichsec.org/guidelines/pharmaceuticals/pharma-safety/environmental-safety.html. Accessed 25 April 2017
Wang N, Yang X, Jiao S, Zhang J, Ye B, Gao S (2014) Sulfonamide-resistant bacteria and their resistance genes in soils fertilized with manures from Jiangsu Province, Southeastern China. PLoS One 9:e112626
Winckler C, Grafe A (2001) Transfer of veterinary drugs and pharmacologically-active feed additives into soil with special consideration of tetracyclines—executive summary. J Soils Sediments 1:58–62
Woegerbauer M, Zeinzinger J, Gottsberger RA, Pascher K, Hufnagl P, Indra A, Fuchs R, Hofrichter J, Kopacka I, Korschineck I, Schleicher C, Schwarz M, Steinwider J, Springer B, Allerberger F, Nielsen KM, Fuchs K (2015) Antibiotic resistance marker genes as environmental pollutants in GMO-pristine agricultural soils in Austria. Environ Pollut 206:342–351

Wu XL, Xiang L, Yan QY, Jiang YN, Li YW, Huang XP, Li H, Cai QY, Mo CH (2014) Distribution and risk assessment of quinolone antibiotics in the soils from organic vegetable farms of a subtropical city, Southern China. Sci Total Environ 487:399–406

Yang Y, Owino AA, Gao Y, Yan X, Xu C, Wang J (2016) Occurrence, composition and risk assessment of antibiotics in soils from Kenya, Africa. Ecotoxicology 25:1194–1201

Chapter 20
Antibiotics in the Soil: Sources, Environmental Issues, and Bioremediation

Umesh B. Jagtap

20.1 What Is Antibiotic?

The Merriam-Webster dictionary defines an antibiotic as "*a substance produced by or a semi-synthetic substance derived from a microorganism and able in dilute solution to inhibit or kill another microorganism.*" Antibiotics are of particular interest because of their designated function and specific biological activity. A large number of antibiotics are used in medical treatment, veterinary, and agriculture farms to cure or prevent bacterial infections in humans, to increase feed efficiency as well as growth performance in animals and plants, respectively (Jechalke et al. 2014; Tasho and Cho 2016). No doubt the use of antibiotics as a medicine increases the quality and expectancy of life.

The overuse and misuse of antibiotics and its recalcitrant nature to biological degradation make them persistent or pseudo-persistent in the environment and toxic to non-target flora and fauna including human beings (Gothwal and Shashidhar 2015).

20.2 From Where They Came to Soil?

Enormous quantities of antibiotics have been released into the environment from different sources, pathways, and anthropogenic activities. The major sources are wastewater or waste released from pharmaceutical industry, hospitals, animal

The original version of this chapter was revised. An erratum to this chapter can be found at DOI 10.1007/978-3-319-66260-2_22.

U.B. Jagtap (✉)
Department of Botany, Government Vidarbha Institute of Science and Humanities, Amravati 444604, Maharashtra, India
e-mail: umeshjagtap93@gmail.com

M.Z. Hashmi et al. (eds.), *Antibiotics and Antibiotics Resistance Genes in Soils*,
Soil Biology 51, DOI 10.1007/978-3-319-66260-2_20

Table 20.1 The list of antibiotics detected in environment (Roig and D'aco 2016)

List of antibiotics
Amoxicylin, ampicillin, azithromycin, carbadox, cefaclor, cefotaxime, cefoperazone, cefradine, ceftazidime, ceftriaxone, cephalexcin, chloramphenicol, chlortertracycline, ciprofloxacin, clarithromycin, clinafloxacin, clindamycin, cloxacillin, danofloxacine, domeclocycline, dimetridazole, doxicycline, enoxacin, enrofloxacin, erythromycin, fleroxacin, flucloxacillin, flumequine, furaltadone, gatifloxacin, josamycin, levofloxacin, lincomycin, metronidazole, mezlocillin, minocycline, moxifloxacin, nalidixic acid, nifuroxazide, norfloxacin, norfluoxetine, ofloxacin, oleandomycin, oxacillin, oxolinic acid, oxytetracycline, penicillin, pipemidic acid, rifampin, ronidazole, roxithromycin, sarafloxacin, spiramycin, sparfloxacin, sulfachloropyridazine, sulfadiazine, sulfadimethoxine, sulfadimidine, sulfaguanidine, sulfamerazine, sulfamethazine, sulfamethiazole, sulfamethoxazole, sulfamethoxypiridazine, sulfamoxole, sulfanilamide, sulfanitran, sulfapyridine, sulfaquinoxaline, sulfasalazine, sulfasoxazole, sulfathiazole, sulfidomidin, sulfisoxazole, tetracycline, thiamphenicol, tilmicosin, triclosan, tylosin, trimethoprim, vancomycin

husbandries, aquaculture, and agriculture practices including manure fertilization (Jechalke et al. 2014; Pruden et al. 2013; Tasho and Cho 2016). The list of antibiotics detected in environment is given in Table 20.1.

20.3 What Happens to Them in the Soil?

Appearance, accumulation, and spread of antibiotic in the soil has become a global health threat and is frequently associated with overuse and misuse of antibiotics in clinic, veterinary, and agriculture pharm as chemotherapeutic agents. Modern industrial-scale animal feeding operations rely extensively on veterinary pharmaceuticals, including antibiotics, to augment animal growth and other purposes. An estimated 63,151 tons of antimicrobials was consumed by livestock across the globe in 2010. By 2030, an alarming rise of 67%, from 63,151 $\pm$ 1560 tons to 105,596 $\pm$ 3605 tons is to be projected (Van Boeckel et al. 2015). Subsequently, following their use, high amount of antibiotics, and its metabolites released into the environment through animal excretion via dung and urine (McEachran et al. 2015). The tons of animal excreta generated from the livestock pharming is used as manure fertilizers as it is rich in nutrients and organic matters. Animal manures from industrial livestock farms in comparison to those from the farmer's households contains a higher concentration of antibiotic residues (Tasho and Cho 2016). The repeated application of such animal manures in agricultural fields leads to high accumulation of antibiotics in soil. Recently, it was found that the amount of florfenicol antibiotic used on swine farms and the spreading of soils with swine waste could promote the prevalence and abundance of florfenicol resistance genes including the linezolid resistance genes in adjacent soils, and agricultural application of swine manure with florfenicol may have caused a residual level of florfenicol in the soils (Zhao et al. 2016). However, most of the people discard the remaining, unused, expired pills and liquid pharmaceuticals by pouring them into the toilet or sink (Kümmerer 2010). Similarly, the use of wastewater effluent from animal farms, pharmaceutical industries, and hospitals for irrigation of agriculture fields is also responsible for transmission and accumulation of antibiotics in soil (Sallach et al. 2015).

Once, reached in the soil, antibiotics interact with soil and microorganism present in the soil. A lot of antibiotics remain persistently in soil or various environmental compartments (plants, manure, soil, sediment, and water) as they are recalcitrant to degradation. Furthermore, several antibiotics are taken up and accumulated in the roots of the crops grown on agricultural fields supplied with manures from antibiotic treated animals or irrigated with contaminated wastewater. For example, the presence of antibiotics ciprofloxacin and narasin in barley roots was reported by Eggen et al. (2011). Whereas, bioconcentration of zwitterionic antibiotics viz., oxytetracycline, chlortetracycline, and norfloxacin were reported in rice roots (Hawker et al. 2013). Another report shows bioconcentration of oxytetracycline in the edible parts of aquatic plants when fertilized with contaminated swine manure (Boonsaner and Hawker 2015). Consumption of such contaminated plant parts causes a significant threat to human health. In addition, they can inhibit crop growth, leading to a decrease in crop production. Challenges in analysis and measurement of antibiotics and their impacts in agroecosystems have been recently reviewed by Aga et al. (2016). For additional comprehensive information and discoveries, the reader is counseled to refer several books and specific review articles that have been already published elsewhere (Du and Liu 2012; Jechalke et al. 2014; Tasho and Cho 2016).

20.4 Why They Matter?

20.4.1 The Rise of Antibiotic Resistance

Resistance genes exist naturally in the environment owing to a range of selective pressures in nature. Humans have applied additional selective pressure for antibiotic resistance genes because of the large quantities of antibiotics that we produce, consume, and apply in medicine and agriculture. Physical and biological forces also cause widespread dissemination of resistance genes throughout many environments (Allen et al. 2010).

The dispersal and accumulation of antibiotics in the soil and environment results into the problem of antibiotic resistance, i.e., reducing an effectiveness of antibiotics in treating common infections in human and animals (Laxminarayan et al. 2013). So far, a number of studies have been carried out on the occurrence of antibiotic and antibiotic resistance genes (ARG) in the manure, soil, water, and agriculture systems. Recently, existence of ARGs has been detected on harvested vegetables grown in manure-amended soil, including root endophytes, leaf endophytes, and phyllosphere microorganisms (Wang et al. 2015). In another report, prevalence and abundance of florfenicol and linezolid resistance genes were identified in soils adjacent to swine feedlots in China (Zhao et al. 2016). The myriad reports published showed that application of antibiotic containing manure to the soil increase abundance and transferability of antibiotic resistance genes that might be contributed to the development of multi-resistant human pathogens. Since, emergence and dissemination of antibiotic resistance genes has become a serious and growing human health threat globally.

20.4.2 Toxic Effects on Flora and Fauna

The antibiotics in soil adversely affect not only soil flora and fauna but also ultimately affect the human beings. Comprehensive analytical, biological, and ecotoxicological methods are developed for measurement of antibiotic residues in soil and assessing their impacts on microorganisms, plants (Table 20.2), and animals. Consequently, Dong et al. (2012) with the help of comet assay showed that chlortetracycline and tetracycline antibiotics exposure could induce DNA damage and alter enzymatic activities in earthworm (*Eisenia fetida*). Reviews on the effects of antibiotics on agroecosystem, the structure, and function of soil microbial communities in bulk soils and the rhizosphere are available (Jechalke et al. 2014; Du and Liu 2012).

20.5 How Can We Deal with Antibiotic Pollution?

The physical (e.g., sorption) and chemical (e.g., photodegradation, oxidation) methods were employed to remediate contaminated soil. These methods are expensive due to high maintenance cost. Since, alternative biological methods such as bioremediation were developed for the removal of antibiotics from contaminated soil.

20.5.1 Bioremediation: Use of Living Organisms to Clean Up Antibiotic Pollution

Bioremediation involves the use of living organisms to remove or detoxify pollutants within a given environment. Although bacteria are the most common group of organisms used for bioremediation, the use of plants (phytoremediation), algae (phycoremediation), and fungi (mycoremediation) is increasing.

Several bacteria are reported to accelerate the biodegradation of antibiotics by enzymatic transformation under aerobic and anaerobic conditions. Alexy et al. (2004) evaluated the biodegradability of 18 clinically important antibiotics and their effects on environmental bacteria in the closed bottle test according to the test guidelines of the Organization for Economic Cooperation and Development (OECD). The present study revealed that all the antibiotics examined had degraded

Table 20.2 Toxic effect of some antibiotics on plants

Antibiotic	Target plant	Effect	References
Enrofloxacin	*Cucumis sativus* L., *Lactuca sativa* L., *Phaseolus vulgaris* L., *Raphanus sativus* L.	Modifying the length of primary root, hypocotyl, cotyledons, and the number/ length of leaves	Luciana et al. (2003)
Oxytetracycline	Alfalfa	Inhibits growth of stem and root	Kong et al. (2007)
Oxytetracycline, tetracycline	Wheat	Inhibits root and shoot elongation	Bao et al. (2008)
Oxytetracycline	Lettuce	Inhibition of growth	Cui et al. (2008)
Chlortetracycline, tetracycline, tylosin, sulfamethazine, trimethoprim	Sweet oat, rice, and cucumber	Inhibits germination	Liu et al. (2009)
Sulfadiazine, sulfadimidine, enrofloxacin	Chinese cabbage and tomato	Inhibits root and shoot elongation	Jin et al. (2009)
Tetracycline	Ryegrass	Plant biomass, especially the roots reduced; Plant P assimilation decreased	Wei et al. (2009)
Oxytetracycline	63 wheat species	Decreased biomass and chlorophyll in leaves	Xie et al. (2009)
Chlortetracycline, levofloxacin, and sulfamethoxazole	Lettuce, alfalfa, carrot	Inhibits root elongation	Hillis et al. (2011)
Enrofloxacin	Narrow-leaved lupin	Inhibits root growth	Adomas et al. (2013)
Tetracycline, sulfamethazine, norfloxacin, erythromycin, and chloramphenicol	Lettuce, tomato, carrot, and cucumber	Inhibits root elongation	Pan and Chu (2016)

Modified from Du and Liu (2012)

by less than 60% during the test period of 28 days. Thus, they must all be classified as not readily biodegradable. Only benzylpenicillin had biodegraded by 27% within 28 days. In another report, the inherent biodegradability of 17 antibiotics was determined in a combined test design based on the Zahn–Wellens test and the CO_2-evolution test performed according to the OECD guidelines. The results showed that only benzylpenicillin sodium salt (Penicillin G) proved to be ultimately biodegradable, reaching $ThCO_2$ degradation extents of 78–87%. Among the others, only amoxicillin, imipenem, and nystatin showed certain ultimate biodegradation in few of the parallel flasks and can be regarded as partially biodegradable with formation of stable metabolites (Gartiser et al. 2007).

Furthermore, Rodríguez-Rodríguez et al. (2012) were assessed the degradation of the sulfonamides sulfapyridine (SPY) and sulfathiazole (STZ) by the white-rot fungus *Trametes versicolor*. Complete degradation was accomplished in fungal cultures at initial pollutant concentrations of approximately 10 mg L^{-1} although a longer period of time was needed to completely remove STZ in comparison to SPY. When cytochrome P450 inhibitors were added to the fungal cultures, STZ degradation was partially suppressed, while no additional effect was observed for SPY. Experiments with purified laccase and laccase mediators caused the removal of greater than 75% of each antibiotic. A fluidized bed reactor with *T. versicolor* pellets degraded a mixture of sulfonamides (SPY, STZ, and sulfamethazine) by greater than 94% each at a hydraulic residence time of 72 h. In another study, (Singh et al. 2017) demonstrated the degradation of ciprofloxacin (CIP) by an edible white-rot fungus *Pleurotus ostreatus*. It was found that CIP has a stimulatory effect on growth and enzyme activity of *P. ostreatus*. Maximum enzyme (glucanase, ligninases, laccase) production was observed at the highest concentration of CIP (500 ppm). Antibiotic degradation of about 68.8, 94.25, and 91.34% was estimated after 14 days of incubation at 500 ppm CIP using Titrimetric, Indigo carmine, and Methyl orange assay, respectively. High performance liquid chromatography revealed 95.07% degradation while microbiological test also exhibited a decreased antimicrobial activity of degraded products against *Escherichia coli*, *Staphylococcus aureus*, and *Streptococcus pyogenes*.

20.5.2 Phytoremediation: Plants to Clean Up Antibiotic Pollution

Phytoremediation is one of the biological approaches, which involves the use of natural or genetically engineered plants and their associated microorganisms for the removal/detoxification of environmental pollutants to remediate contaminated soils. Phytoremediation has arisen as an inexpensive, environment-friendly, and publically acceptable strategy as compared to the other physical and chemical remediation technologies. However, phytoremediation can be used efficiently for the remediation of heavy metals, organic compounds including trinitrotoluene, tetrachloroethylene, polycyclic aromatic hydrocarbons, herbicides/pesticides, and explosives. For an overview of the phytoremediation and developments in the phytoremediation technologies, see the following articles and references therein: Arthur et al. (2005); Pilon-Smits (2005).

Besides, the phytoremediation technologies are less utilized for the removal of antibiotics from soil. Phytoremediation potential of *Myriophyllum aquaticum* and *Pistia stratiotes* to modify antibiotic growth promoters, tetracycline, and oxytetracycline in aqueous wastewater systems have been reported by Gujarathi et al. (2005). In another report, Gahlawat and Gauba (2016) demonstrated phytoremediation potential of *Brassica juncea* to remediate tetracycline. In addition to this, *Chrysopogon zizanioides*

L. Nash (vetiver) plants treated with tetracycline capable of transformation/detoxification of tetracycline (Sengupta et al. 2016). Several pharmaceuticals may also be taken up by plants, but their concentrations in plant tissues are commonly so small that plant uptake might not represent a major pathway for the removal of antibiotics from soil (Jechalke et al. 2014).

20.6 Conclusion

Antibiotics in soil possess a serious threat to human and soil healthcare. The proverb that "an ounce of prevention is worth a pound of cure" is applicable to both human and soil healthcare. Clearly, the best prevention for antibiotic pollution in soil is not to release antibiotic contaminants into the soil and environment (Gómez-Sagasti et al. 2016). Therefore, ideal management practices such as correct use of antibiotics, proper treatment/disposal of animal effluent/manure, treatment of wastewater from domestic, hospital, and pharmaceutical industries, development of biological remediation technology, and mass media-assisted awareness programs is needed to decrease the release of antibiotics in the environment (Pruden et al. 2013). Phytoremediation in the broad sense offers a powerful technology for the removal of various contaminants form environment. Therefore, there is rapid need of exploration and exploitation of newer plants facilitating efficient removal/degradation of antibiotics from soil.

References

Adomas B, Antczak-Marecka J, Nalecz-Jawecki G, Piotrowicz-Cieslak AI (2013) Phytotoxicity of enrofloxacin soil pollutant to narrow-leaved lupin plant. Pol J Environ Stud 22:71–76

Aga DS, Lenczewski M, Snow D, Muurinen J, Sallach JB, Wallace JS (2016) Challenges in the measurement of antibiotics and in evaluating their impacts in agroecosystems: A critical review. J Environ Qual 45:407–419

Alexy R, Kumpel T, Kümmerer K (2004) Assessment of degradation of 18 antibiotics in the closed bottle test. Chemosphere 57:505–512

Allen HK, Donato J, Wang HH, Cloud-Hansen KA, Davies J, Handelsman J (2010) Call of the wild: antibiotic resistance genes in natural environments. Nat Rev Microbiol 8:251–259

Arthur EL, Rice PJ, Rice PJ, Anderson TA, Baladi SM, Henderson KL, Coats JR (2005) Phytoremediation—an overview. Crit Rev Plant Sci 24:109–122

Bao YY, Zhou QX, Xie XJ (2008) Influence of tetracycline kind antibiotics on the control of wheat germination and root elongation. China Environ Sci 28:566–570

Boonsaner M, Hawker DW (2015) Transfer of oxytetracycline from swine manure to three different aquatic plants: implications for human exposure. Chemosphere 122:176–182

Cui X, Qiao XL, Han CW, Wang Z (2008) Uptake of oxytetracycline and its phytotoxicity to lettuce. J Agro Environ Sci 27:1038–1042

Dong L, Gao J, Xie X, Zhou Q (2012) DNA damage and biochemical toxicity of antibiotics in soil on the earthworm *Eisenia fetida*. Chemosphere 89:44–51

Du L, Liu W (2012) Occurrence, fate, and ecotoxicity of antibiotics in agro-ecosystems. A review. Agron Sustain Dev 32:309–327
Eggen T, Asp TN, Grave K, Hormazabal V (2011) Uptake and translocation of metformin, ciprofloxacin and narasin in forage-and crop plants. Chemosphere 85:26–33
Gahlawat S, Gauba P (2016) Phytoremediation of aspirin and tetracycline by *Brassica juncea*. Int J Phytoremediation 18:929–935
Gartiser S, Urich E, Alexy R, Kümmerer K (2007) Ultimate biodegradation and elimination of antibiotics in the inherent test. Chemosphere 67:604–613
Gómez-Sagasti MT, Epelde L, Alkorta I, Garbisu C (2016) Reflections on soil contamination research from a biologist's point of view. Appl Soil Ecol 105:207–210
Gothwal R, Shashidhar T (2015) Antibiotic pollution in the environment: a review. Clean Soil Air Water 43:479–489
Gujarathi NP, Haney BJ, Linden JC (2005) Phytoremediation potential of *Myriophyllum aquaticum* and *Pistia stratiotes* to modify antibiotic growth promoters, tetracycline, and oxytetracycline, in aqueous wastewater systems. Int J Phytoremediation 7:99–112
Hawker DW, Cropp R, Boonsaner M (2013) Uptake of zwitterionic antibiotics by rice (*Oryza sativa* L.) in contaminated soil. J Hazard Mater 263:458–466
Hillis DG, Fletcher J, Solomon KR, Sibley PK (2011) Effects of ten antibiotics on seed germination and root elongation in three plant species. Arch Environ Contam Toxicol 60:220–232
Jechalke S, Heuer H, Siemens J, Amelung W, Smalla K (2014) Fate and effects of veterinary antibiotics in soil. Trends Microbiol 22:536–545
Jin CX, Chen QY, Sun RL, Zhou QX, Liu JJ (2009) Eco-toxic effects of sulfadiazine sodium, sulfamonomethoxine sodium and enrofloxacin on wheat, Chinese cabbage and tomato. Ecotoxicology 18:878–885
Kong WD, Zhu YG, Liang YC, Zhang J, Smith FA, Yang M (2007) Uptake of oxytetracycline and its phytotoxicity to alfalfa (*Medicago sativa* L.) Environ Pollut 147:187–193
Kümmerer K (2010) Pharmaceuticals in the environment. Annu Rev Environ Resour 35:57–75
Laxminarayan R, Duse A, Wattal C, Zaidi AK, Wertheim HF, Sumpradit N, Vlieghe E, Hara GL, Gould IM, Goossens H, Greko C (2013) Antibiotic resistance—the need for global solutions. Lancet Infect Dis 13:1057–1098
Liu F, Ying GG, Tao R, Zhao LL, Yang JF, Zhao LF (2009) Effects of six selected antibiotics on plant growth and soil microbial and enzymatic activities. Environ Pollut 157:1636–1642
Luciana M, Salvatore C, Maurizio F (2003) Phytotoxicity to and uptake of enrofloxacin in crop plants. Chemosphere 52:1233–1244
McEachran AD, Blackwell BR, Hanson JD, Wooten KJ, Mayer GD, Cox SB, Smith PN (2015) Antibiotics, bacteria, and antibiotic resistance genes: aerial transport from cattle feed yards via particulate matter. Environ Health Perspect 123:337
Pan M, Chu LM (2016) Phytotoxicity of veterinary antibiotics to seed germination and root elongation of crops. Ecotoxicol Environ Saf 126:228–237
Pilon-Smits E (2005) Phytoremediation. Annu Rev Plant Biol 56:15–39
Pruden A, Larsson DJ, Amézquita A, Collignon P, Brandt KK, Graham DW, Lazorchak JM, Suzuki S, Silley P, Snape JR, Topp E (2013) Management options for reducing the release of antibiotics and antibiotic resistance genes to the environment. Environ Health Perspect 121:878–885
Rodríguez-Rodríguez CE, Garcia-Galan MJ, Blanquez P, Diaz-Cruz MS, Barcelo D, Caminal G, Vicent T (2012) Continuous degradation of a mixture of sulfonamides by *Trametes versicolor* and identification of metabolites from sulfapyridine and sulfathiazole. J Hazard Mater 213–214:347–354
Roig B, D'Aco V (2016) Distribution of pharmaceutical residues in the environment. In: Hester RE, Harrison RM (eds) Pharmaceuticals in the environment. Royal Society of Chemistry, Cambridge, pp 34–69

Sallach JB, Zhang Y, Hodges L, Snow D, Li X, Bartelt-Hunt S (2015) Concomitant uptake of antimicrobials and Salmonella in soil and into lettuce following wastewater irrigation. Environ Pollut 197:269–277

Sengupta A, Sarkar D, Das P, Panja S, Parikh C, Ramanathan D, Bagley S, Datta R (2016) Tetracycline uptake and metabolism by vetiver grass (*Chrysopogon zizanioides* L. Nash). Environ Sci Pollut Res 23:24880–24889

Singh SK, Khajuria R, Kaur L (2017) Biodegradation of ciprofloxacin by white rot fungus *Pleurotus ostreatus*. 3 Biotech 7:69

Tasho RP, Cho JY (2016) Veterinary antibiotics in animal waste, its distribution in soil and uptake by plants: a review. Sci Total Environ 563:366–376

Van Boeckel TP, Brower C, Gilbert M, Grenfell BT, Levin SA, Robinson TP, Teillant A, Laxminarayan R (2015) Global trends in antimicrobial use in food animals. Proc Natl Acad Sci USA 112:5649–5654

Wang FH, Qiao M, Chen Z, Su JQ, Zhu YG (2015) Antibiotic resistance genes in manure-amended soil and vegetables at harvest. J Hazard Mater 299:215–221

Wei X, Wu SC, Nie XP, Yediler A, Wong MH (2009) The effects of residual tetracycline on soil enzymatic activities and plant growth. J Environ Sci Health Part B 44:461–471

Xie XY, Zhang YQ, Li ZJ, Liang YC, Yao JH, Zhang SQ (2009) Cultivar differences in toxic effects of oxytetracycline on wheat (*Triticum durum*). Asian J Ecotoxicol 4:577–583

Zhao Q, Wang Y, Wang S, Wang Z, Du XD, Jiang H, Xia X, Shen Z, Ding S, Wu C, Zhou B (2016) Prevalence and abundance of florfenicol and linezolid resistance genes in soils adjacent to swine feedlots. Sci Rep 6:32192

Chapter 21
Management and Regulation of Antibiotics and Antibiotics Resistance Genes in Soils

Muhammad Zaffar Hashmi, Wajid Nasim, Hussani Mubarak, Nosheen Mirza, Sohaib Khan, Xiaomei Su, Sarfraz Ahmed, and Ahmad Hasnain

21.1 Introduction

Antibiotic drugs have become essential for many medical interventions due to their role in reduction of common infectious diseases (Amini and Tavazoie 2011; Kingston 2000). Antibiotics occur as naturally and synthetic chemical compounds

---

The original version of this chapter was revised. An erratum to this chapter can be found at DOI 10.1007/978-3-319-66260-2_22.

M.Z. Hashmi (✉)
Department of Meteorology, COMSATS Institute of Information Technology, Islamabad, Pakistan
e-mail: hashmi_qau@yahoo.com

W. Nasim
Department of Environmental Sciences, COMSATS Institute of Information Technology, Vehari, Pakistan

H. Mubarak
Department of Soil and Environmental Sciences, Ghazi University, Dera Ghazi Khan, Pakistan

N. Mirza
Department of Environmental Sciences, COMSATS Institute of Information Technology, Abbottabad 22060, Pakistan

S. Khan
Department of Anthropology, Bahauddin Zakariya University, Multan, Pakistan

X. Su
College of Geography and Environmental Science, Zhejiang Normal University, Yingbin Road 688#, Jinhua 321004, China

S. Ahmed
Department of Biochemistry, Bahauddin Zakariya University, Multan 60800, Pakistan

A. Hasnain
Department of Geography, Bahauddin Zakariya University, Multan, Pakistan

M.Z. Hashmi et al. (eds.), *Antibiotics and Antibiotics Resistance Genes in Soils*,
Soil Biology 51, DOI 10.1007/978-3-319-66260-2_21

with antimicrobial activity that are widely used in veterinary and human medicine (Laxminarayan et al. 2013). Antibiotics have application in the prevention and treatment of bacterial infection which may either inhibit or kill the growth of bacteria. Revolution of antibiotics in medicine industry start in the twentieth century and together with vaccination led to the near eradication of diseases, such as tuberculosis in the developed world. Antibiotics global datasets suggest that from 2010 to 2030 antibiotic use in food-animal production will increase by (67%) (Gelband et al. 2015).

The application of antibiotics has led to the production of antibiotic resistance which represents a growing and serious human health threat worldwide. Recent research highlighted that the main sources, reservoirs, and recipients of ARGs are water, air, and soil (Davies and Davies 2010; Martinez 2009, Tang et al. 2015). Likewise, air, soil, and water environments receive inputs of antibiotics and antimicrobials, which can serve to amplify ARGs (Heuer et al. 2011; Zhu et al. 2013). Nonpathogenic bacteria around us are also one of the factors for resistance genes in soil (Bonomo and Szabo 2006; Magiorakos et al. 2012).

Antibiotics and ARGs are abundant in human and animal fecal material; thus, active and from environmental resistance reservoirs. Both soil and water can be directly affected by the wastewater, agricultural, and industrial input of antibiotics, which impose selection pressure and enable the spread, amplification, and maintenance of ARGs (Hashmi et al. 2017; Heuer et al. 2011; Rizzo et al. 2013). The alternative use of animal growth promoters such as metals or biocides will not necessarily aid in limiting the spread of antibiotic resistance because they can also select for antibiotic resistance through co-resistance or cross-resistance (Baker-Austin et al. 2006; Knapp et al. 2009). In addition to end-of-pipe options, source control is key. Hence, this chapter highlights the rationale for use of antimicrobial compounds in animals and humans, and potential advantages of limiting or managing antimicrobial use that impact the feasibility of management approaches. Here, we identify and provide an overview of potential mitigation options for minimizing the spread of antibiotics and antibiotic resistance along these pathways. Limiting impacts to soil environments is of special interest because these environments serve as a source of exposure to humans via recreational use, food, ingestion, and aerosol inhalation.

This chapter identifies potential management options such as simple management practices that work synergistically with existing goals and policies, such as bioremediation for soil management, runoff control, nutrient management, or infrastructure upgrades, that may be put into effect immediately.

21.2 Management Options

21.2.1 Antibiotics Contaminated Soil Remediation

Antibiotics are introduced into the fields in the form of natural fertilizers (utilized in animal husbandry) or by watering the plants with wastewaters. After entering in the soil environment antibiotics persist in the soil, sediment, or groundwater. The rate of degradation of antimicrobials in the environment varies and is dependent on a range of environmental conditions, for example: antibiotic concentration, chemical structure of the compound, composition and structure of soil/sediment, humic acids content, humidity, pH, temperature, sorption capacity, chemical composition of the environment, presence of other sources of carbon, presence of inorganic matter, availability of oxygen, and microorganisms that support biodegradation (Alexy et al. 2004; Vasconcelos et al. 2009). When the antibiotics are added to the soil solid phase, they are liable to microbial transformation. However, bio-transformation results in the re-transformation of metabolites into the parent compound similarly as in fertilizers (Giang et al. 2015; Jechalke et al. 2014). Less than 2% if the added compounds the mineralization accounts for many of the antibiotics (Forster et al. 2009; Junge et al. 2012). Because of the poor light penetration the photodegenration in the soils as an alternative pathway of pharmaceutical degradation is limited (Ozaki et al. 2011). Final pathways to be considered are potential transfers of the antibiotics from soil into the atmosphere, hydrosphere, and biosphere. Low vapor pressure is responsible for the fate of the pharmaceuticals in soil and here volatilization is not relevant. For the removal of antibiotics from the soil, the plant uptake by several pharmaceuticals is done (Felizeter et al. 2012; Sabourin et al. 2012) but there concentrations are commonly so small in the plant tissues that there plant uptake might not represent a major pathway (Engelhardt et al. 2015; Fang et al. 2014; Rosendahl et al. 2012).

21.2.2 Reduce/Optimizing Antibiotic Use

Limiting in the usage of antibiotics could provide the solution in optimizing antibiotics in the environment. A large proportion of the overall consumption of antibiotics worldwide is from the agricultural usage of antibiotics although the specific antibiotics used vary extensively among countries (Kemper 2008; Sarmah et al. 2006). Annual production of antibiotics in China is about 210 million kg and 46% of these are estimated to be used in livestock (Wang and Ma 2008; Yun-peng and Yue 2008). In general, uncontrolled use of antibiotics and metals is increasing in Chinese agriculture and industry, corresponding to enrichment of ARGs in the manure (Zhu et al. 2013) and affected environment, particularly in soils (Wu et al. 2010). In the United States, recent reports indicated that $\geq$70% of total antibiotics are administered to livestock (Food and Drug Administration 2011) and Australia (Joint Expert Advisory Committee on Antibiotic Resistance 1999).

The most direct route of controlling agricultural antibiotic release into the environment, and likely also antibiotic resistance is the reduction in the types and use of antibiotics in animal production. To reduce the high risk of antibiotic resistance transfer from animals to humans, some countries have adapted that regulations on antibiotics dosing based on clinical efficacy, and banned the usage of growth promoters (Angulo et al. 2006; Smith et al. 2005). Antibiotics were phased out as growth promoters in 1986 in Sweden, followed by Denmark in the late 1990s, and subsequently the European Union. Adaptations of organic feed practices in poultry industry of the United States reduce the multidrug resistance rates of *Enterococcus faecium* from 84 to 17% (Sapkota et al. 2011). In Denmark, decline in the total use of veterinary antibiotics was achieved from ≥200 metric tons in 1994 to around 70 metric tons in 1999 (Monitoring 2003). Reports suggest that banning subtherapeutic use of antibiotics in Denmark led to marked reductions of fecal *enterococci*, antibiotic resistance, in the animal populations (Aarestrup et al. 2001), demonstrating that it is indeed possible to reverse the occurrence of antibiotic resistance among a national population of food animals through regulations restricting antibiotic use. Further, monitoring for response of resistance carriage in humans significant confounds are related to consumption of imported meat and international travel that may carry higher loads of resistant bacteria could provide the strategy in reduction of antibiotics and ARGs (Hammerum et al. 2007; Lewis et al. 2008).

21.2.3 Alternatives to Antibiotics

Alternative to antibiotics several other growth promoters such as metals (Arsenic, copper, and zinc) are commonly used in animal feeds (Bolan et al. 2004; Poulsen 1998). However, alternative to antibiotics could not provide the effective solution in the reduction of ARGs because antibiotic resistance can be co-selected by metals (Berg et al. 2010; Seiler and Berendonk 2012; Song et al. 2017). It is apparent that antibiotic resistance becomes worse if replace with metals. Further, metals (notably Cu) can accumulate in agricultural soils (Bolan et al. 2004; Gräber et al. 2005), and thus serve as even stronger long-term selective agents for antibiotic resistance in manure-amended soils than do antibiotic residues, which are more prone to degradation and/or sequestration (Knapp et al. 2011; Song et al. 2017). Herbal materials could be worth pursuing alternative (Hanczakowska and Szewczyk 2007; Orozco Hernández et al. 2009); however, resistance should be monitored in it.

Ideal management practices should focus to control the flow of genetic elements from animal manure to soil ecosystems. Although antibiotic resistance may decline after relaxation of selection pressures, low yet detectable levels of resistance determinants are likely to persist for decades because of the low fitness costs associated with many antibiotic resistance mechanisms (Andersson and Hughes 2010; Johnsen et al. 2011). This indicates that even under minimal antibiotic use conditions (organic), there is a potential for release of ARGs.

21.2.4 Maintaining the Good Health of Animals

Antibiotics usage could be reduce by keeping animals healthy. To control infectious diseases on farms best management practices, such as improved nutritional programs and low animal density can be adopted and developed. For example, high level of nutrition to the calves with antibiotic amendment in dairy calf milk limit the additional health benefit of subtherapeutic antibiotics (Thames et al. 2012) suggesting knowledgeable and healthy animal husbandry is the most important factor in reducing antibiotic use.

21.2.5 Management of Antibiotics and ARGs in Manure

Control of animal wastes provides a practical strategy with other advantages of nutrient management and protection of soil. Limiting sediment erosion and prevention of manure transport from animal farms, control of surface runoff, lagoon spills, and seepage are the best strategies for the containment of manure. Improved manure collection and increased storage capacity could limit surface runoff. The application of manure to land only when crop demands for water and nutrients. Long-term manure storage reduced prevalence of tetracycline-resistant bacteria and tetracycline residues (Chee-Sanford et al. 2009). Manure separation technologies act to concentrate solids from manure slurries through processes such as screening, filtration, or sedimentation and may also provide an avenue to mitigate the release of antibiotic residues and ARGs. Benefits of manure separation include reduced nutrient content, prolonged storage potential, improved biological treatment, and minimization of odors.

Composting and digestion of livestock waste can treat antibiotic residues. Antibiotics on average 50–70% could be eliminated by composting (Sharma et al. 2009; Wang et al. 2012). Aeration, watering, and turning of compost offered some advantage to accelerating antibiotic decay of monensin, chlortetracycline, and tylosin, but even simple storage of manure stockpiles resulted in significant antibiotic degradation (Storteboom et al. 2007). Digestion of livestock waste can eliminate antibiotic residues; 5-week fermentation effectively removed most sulfonamides and trimethoprim (Mohring et al. 2009), whereas sulfamethoxazole and oxytetracycline were reduced more effectively under aerobic than anaerobic incubation of dairy lagoon water (Pei et al. 2007).

21.2.6 Biological Treatment and Energy Recovery

Biological treatments such as composting and lagoons have complex microbial ecology involved, and its impact varies on ARGs. A 100-fold reduction of tetracycline ARGs was recorded after the storage of composting and manure, but *tet*(O) increased when horse manure was composted, even in the absence of measurable antibiotics (Storteboom et al. 2007). Similarly, a tenfold reduction of *tet* ARGs across six anaerobic livestock lagoons were monitored, but *sul* ARGs tended to increase with treatment time (McKinney et al. 2010). Rysz et al. (2013)

revealed that an agricultural *E. coli* strain under anaerobic treatment may be a promising way to impose a high metabolic burden on bacteria and thus limit their capability to engage in horizontal gene transfer. However, some studies reported that ARGs (erythromycin resistance methylase) persist in environment after composting (Sharma et al. 2009).

On-farm methanogenic biogas facilities may provide added incentive for improved waste treatment (Mohring et al. 2009). The increased intensification and geographical concentration of livestock production facilities further solidifies incentives to consider novel manure management technologies (Steinfeld et al. 2006). At a policy level, standards on concentrations of antibiotics in animal manures for land application should be established and monitored. Using animal manures as organic fertilizer also reduces the runoff from animal farms and the risk of lagoon spills and seepages while allowing nutrient recovery. Enacting controls on manure management is challenging because it requires agreement, cooperation, and enforcement among a large number of stakeholders.

21.2.7 Wastewater Treatment from Domestic, Hospital, and Industrial

Wastewater from domestic, hospital, and industrial sectors is the main source of antibiotics and antimicrobial resistance spread and development in soil ecosystem (Rizzo et al. 2013). So there is need for treatment of wastewater to minimize the spread of resistance. Sanitation and sewage in the developing world is serious concern. Globally, 2.6 billion people lack access to basic sanitation WHO (2012), which likely results in direct releases of resistance bacteria and pathogens into the environment. Thus, basic hygiene is likely a critical step to mitigating the spread of resistance. Walsh et al. (2011) detected *NDM-1* gene in chlorinated tap water and polluted surface waters in India. *NDM-1* provides bacteria with resistance to a large number of antibiotics; it is highly mobile and is found in multiple waterborne pathogens, including *Vibrio cholera* (Walsh et al. 2011) and *E. coli* (Kumarasamy et al. 2010).

WWTPs may represent a critical node for control of the global spread of antibiotic resistance. Thermophilic anaerobic sludge digestion appears particularly promising and may achieve superior ARG removal relative to mesophilic digestion, potentially because of the much narrower host ecology of the microorganisms (Diehl and LaPara 2010; Ma et al. 2011). More advanced treatment technologies (e.g., membrane separation) could be applied to retain bacterial cells, including their genetic material (Riquelme Breazeal et al. 2013). In addition, ozone has been proposed to disinfect ARBs and destroy ARGs (Dodd 2012). Because costs of advanced treatments will be significant, an ideal place to start may be to consider ARGs alongside other issues of concern if upgrades are already planned.

At least 56 antibiotics belonging to six different classes have been widely detected at nanogram-per-liter to microgram-per-liter levels in sewage of East Asia, North America, Europe, and Australia (Zhang and Li 2011). Antibiotic residues from different sources (household, pharmaceutical industry, and hospital) enter into municipal sewage along with other co-selecting factors, such as metals and surfactants. Removal pathways include adsorption, biodegradation, disinfection, and membrane separation (Pruden et al. 2013). Other pathways, such as hydrolysis, photolysis, and volatilization, also contribute to removal (Zhang and Li 2011), depending on antibiotic properties. For example, tetracyclines are removed mainly by adsorption onto the biomass flocs; beta-lactams are largely degraded by hydrolysis reactions driven by bacteria or physical chemical processes; and erythromycin and ciprofloxacin are recalcitrant toward biodegradation in activated sludge (Li and Zhang 2010).

The most direct route of removal of ARGs is via solids separation, such as sedimentation. Researchers have observed ARGs from industrial and municipal WWTP sources to persist in river sediment (Kristiansson et al. 2011; Storteboom et al. 2010). However, land application of sludge/biosolids from WWTPs, another means of resource recovery, could also enhance ARGs in soil (Brooks et al. 2007; Munir et al. 2011). For example, in a recent study comparing land application of manure versus biosolids, Munir and Xagoraraki (2011) found elevated levels of tetracycline and sulfonamide ARGs in soils amended with biosolids during the 4-month monitoring period.

Incineration is a zero-risk solution with regard to reduction of antibiotics, and ARGs, although there are trade-offs with air quality and cost of alternative fertilizers. If used appropriately, incineration may provide a source of alternative energy. Landfills still pose some risks because leachates may pollute groundwater and surface water, and they are commonly redirected to a municipal WWTP (Renou et al. 2008).

Hospital and industrial waste considered as hot spots for antibiotics and antibiotic resistance spread in soil. Managing "hot spots," such as drug manufacturers units and hospitals is of high concern because these provide the base to resistant microbes to rapidly spread from one corner of the world across the entire planet (Kovalova et al. 2012; Walsh et al. 2011). Membrane bioreactors are used as targeted pretreatment systems to treat hospitals waste, and they can partially remove antibiotics and other drugs, as well as antibacterial resistance, before discharging into public sewer systems (Kovalova et al. 2012). Other potential hot spots for antibiotic-resistance development are pharmaceutical drug manufacturing sites (Larsson et al. 2007; Fick et al. 2009; Kristiansson et al. 2011). Li et al. (2009, 2010) observed that in China release of therapeutic levels of penicillin and oxytetracycline downstream from a factory increased the resistance rates.

Some industries treat their own wastes from its generation through to discharge, while others discharge to a third party wastewater treatment plants (WWTPs) with or without pretreatment (e.g., pH adjustment, chelation, precipitation). Thus, WWTPs that receive wastes from drug manufacturers will benefit from requiring pretreatment or establishing limits to antibiotic discharge. Industrial WWTPs maintain the high antibiotic concentrations thus inevitably will exert strong selection for antimicrobial resistance. Due to this reason, activated sludge is not recommended for highly antibiotic-contaminated waste streams because of the high density of microbial populations. It is discouraged to seed biological treatment systems with microbes originating from human feces, as well as land application of residual biosolids from hot spot sources.

However, several policy measures could be effective in curtailing the spread of antimicrobial resistance from hot spots.

- First, the industry itself could take a leading role in developing voluntary standards for pharmaceutical wastes (Murray-Smith et al. 2012).
- Second, greater transparency through the supply chain is urgently needed in order to indicate where human drugs are coming from and where they are going (Larsson 2010).
- Third, national purchasers of medicines could aim to take greater responsibility of the issue [Swedish Environmental Management Council (SEMC) 2011].
- Finally, extension of good manufacturing practices to include environmental considerations could be of benefit [Medical Products Agency (MPA) 2011].

21.2.8 Strategic Implementation and Monitoring Needs

To define safe exposure levels is not possible in a strict sense. The scientific community should put effort to develop standards to provide regulators with a basis for defining and implementing standards. Various mitigation strategies would be easy and possible once standards are defined. However, we must acknowledge that the uncertainty is still high regarding ultimate benefits for individual measures. At present, efficacy of mitigation efforts can best be evaluated on the basis of surrogate measures, such as the abundance of antibiotics, and ARGs in the environment. Routine monitoring programs are required to provide baseline data on which to contrast measurements before and after mitigation activities. Establishing and/or maintaining existing biobanks of soil will allow retrospective analyses.

21.3 Conclusion

Antibiotics and ARGs are global issue; this chapter identified several management options across soil bioremediation, wastewater treatment, antibiotics and ARGs in agriculture, and pharmaceutical manufacturing that could aid in mitigating risks of antibiotics and antibiotic resistance in the soil environment. Many of these are practical strategies that are economically feasible and that can be synergistically implemented with other benefits. Outreach, education, communication, monitoring, and transparency are vital for the success of management schemes for limiting the spread of antibiotic resistance via environmental pathways.

References

Aarestrup FM, Seyfarth AM, Emborg H-D, Pedersen K, Hendriksen RS, Bager F (2001) Effect of abolishment of the use of antimicrobial agents for growth promotion on occurrence of antimicrobial resistance in fecal enterococci from food animals in Denmark. Antimicrob Agents Chemother 45:2054–2059

Alexy R, Kümpel T, Kümmerer K (2004) Assessment of degradation of 18 antibiotics in the closed bottle test. Chemosphere 57:505–512

Amini S, Tavazoie S (2011) Antibiotics and the post-genome revolution. Curr Opin Microbiol 14:513–518

Andersson DI, Hughes D (2010) Antibiotic resistance and its cost: is it possible to reverse resistance? Nat Rev Microbiol 8:260–271

Angulo FJ, Heuer OE, Hammerum AM, Collignon P, Wegener HC (2006) Human health hazard from antimicrobial-resistant enterococci in animals and food. Clin Infect Dis 43:911–916

Baker-Austin C, Wright MS, Stepanauskas R, McArthur J (2006) Co-selection of antibiotic and metal resistance. Trends Microbiol 14:176–182

Berg J, Thorsen MK, Holm PE, Jensen J, Nybroe O, Brandt KK (2010) Cu exposure under field conditions coselects for antibiotic resistance as determined by a novel cultivation-independent bacterial community tolerance assay. Environ Sci Technol 44:8724–8728

Bolan N, Adriano D, Mahimairaja S (2004) Distribution and bioavailability of trace elements in livestock and poultry manure by-products. Crit Rev Environ Sci Technol 34:291–338

Bonomo RA, Szabo D (2006) Mechanisms of multidrug resistance in Acinetobacter species and Pseudomonas aeruginosa. Clin Infect Dis 43:S49–S56

Brooks JP, Maxwell SL, Rensing C, Gerba CP, Pepper IL (2007) Occurrence of antibiotic-resistant bacteria and endotoxin associated with the land application of biosolids. Can J Microbiol 53:616–622

Chee-Sanford JC, Mackie RI, Koike S, Krapac IG, Lin Y-F, Yannarell AC, Maxwell S, Aminov RI (2009) Fate and transport of antibiotic residues and antibiotic resistance genes following land application of manure waste. J Environ Qual 38:1086–1108

Danish Integrated Antimicrobial Resistance Monitoring and Research Programme (2003) Research Programme (DANMAP). 2010. Use of antimicrobial agents and occurrence of antimicrobial resistance in bacteria from food animals, food, and humans in Denmark

Davies J, Davies D (2010) Origins and evolution of antibiotic resistance. Microbiol Mol Biol Rev 74:417–433

Diehl DL, LaPara TM (2010) Effect of temperature on the fate of genes encoding tetracycline resistance and the integrase of class 1 integrons within anaerobic and aerobic digesters treating municipal wastewater solids. Environ Sci Technol 44:9128–9133

Dodd MC (2012) Potential impacts of disinfection processes on elimination and deactivation of antibiotic resistance genes during water and wastewater treatment. J Environ Monit 14:1754–1771

Engelhardt I, Sittig S, Šimůnek J, Groeneweg J, Pütz T, Vereecken H (2015) Fate of the antibiotic sulfadiazine in natural soils: experimental and numerical investigations. J Contam Hydrol 177:30–42

Fang H, Han Y, Yin Y, Pan X, Yu Y (2014) Variations in dissipation rate, microbial function and antibiotic resistance due to repeated introductions of manure containing sulfadiazine and chlortetracycline to soil. Chemosphere 96:51–56

Felizeter S, McLachlan MS, de Voogt P (2012) Uptake of perfluorinated alkyl acids by hydroponically grown lettuce (Lactuca sativa). Environ Sci Technol 46:11735–11743

Fick J, Söderström H, Lindberg RH, Phan C, Tysklind M, Larsson DGJ (2009) Contamination of surface, ground, and drinking water from pharmaceutical production. Environ Toxicol Chem 28:2522–2527

Food and Drug Administration (2011) CVM reports on antimicrobials sold or distributed for food producing animals for 2010

Forster M, Laabs V, Lamshoft M, Groeneweg J, Zuhlke S, Spiteller M, Krauss M, Kaupenjohann M, Amelung W (2009) Sequestration of manure-applied sulfadiazine residues in soils. Environ Sci Technol 43:1824–1830

Gelband H, Molly Miller P, Pant S, Gandra S, Levinson J, Barter D, White A, Laxminarayan R (2015) The state of the world's antibiotics 2015. Wound Healing Southern Africa 8:30–34

Giang CND, Sebesvari Z, Renaud F, Rosendahl I, Minh QH, Amelung W (2015) Occurrence and dissipation of the antibiotics sulfamethoxazole, sulfadiazine, trimethoprim, and enrofloxacin in the Mekong Delta, Vietnam. PLoS One 10:e0131855

Gräber I, Hansen JF, Olesen SE, Petersen J, Østergaard HS, Krogh L (2005) Accumulation of copper and zinc in Danish agricultural soils in intensive pig production areas. Geografisk Tidsskrift-Danish J Geogr 105:15–22

Hammerum AM, Heuer OE, Emborg H-D, Bagger-Skjøt L, Jensen VF, Rogues A-M, Skov RL, Agersø Y, Brandt CT, Seyfarth AM (2007) Danish integrated antimicrobial resistance monitoring and research program. Emerg Infect Dis 13:1633

Hanczakowska E, Szewczyk A (2007) Application of herbs and herbal preparations in pig feeding. Ann Anim Sci 7:3–11

Hashmi MZ, Mahmood A, Kattel DB, Khan S, Hasnain A, Ahmed Z (2017) Antibiotics and antibiotic resistance genes (ARGs) in soil: occurrence, fate, and effects. In: Xenobiotics in the soil environment. Springer, Cham, pp 41–54

Heuer H, Schmitt H, Smalla K (2011) Antibiotic resistance gene spread due to manure application on agricultural fields. Curr Opin Microbiol 14:236–243

Jechalke S, Heuer H, Siemens J, Amelung W, Smalla K (2014) Fate and effects of veterinary antibiotics in soil. Trends Microbiol 22:536–545

Johnsen P, Townsend J, Bøhn T, Simonsen G, Sundsfjord A, Nielsen K (2011) Retrospective evidence for a biological cost of vancomycin resistance determinants in the absence of glycopeptide selective pressures. J Antimicrob Chemother 66:608–610

Joint Expert Advisory Committee on Antibiotic Resistance (1999) The use of antibiotics in food-producing animals: antibiotic-resistant bacteria in animals and humans. Available at: http://www.health.gov.au/internet/main/publishing.nsf/content/2A8435C711929352CA256F180057901E/$File/jetacar.pdf

Junge T, Claßen N, Schäffer A, Schmidt B (2012) Fate of the veterinary antibiotic 14C-difloxacin in soil including simultaneous amendment of pig manure with the focus on non-extractable residues. J Environ Sci Health B 47:858–868

Kemper N (2008) Veterinary antibiotics in the aquatic and terrestrial environment. Ecol Indic 8:1–13

Kingston W (2000) Antibiotics, invention and innovation. Res Policy 29:679–710

Knapp CW, Dolfing J, Ehlert PA, Graham DW (2009) Evidence of increasing antibiotic resistance gene abundances in archived soils since 1940. Environ Sci Technol 44:580–587

Knapp C, McCluskey S, Singh B, Campbell C, Hudson G, Gilbert JA (2011) Antibiotic resistance gene abundances correlate with metal and geochemical conditions in archived Scottish soils. PLoS One 6(11):e27300

Kovalova L, Siegrist H, Singer H, Wittmer A, McArdell CS (2012) Hospital wastewater treatment by membrane bioreactor: performance and efficiency for organic micropollutant elimination. Environ Sci Technol 46:1536–1545

Kristiansson E, Fick J, Janzon A, Grabic R, Rutgersson C, Weijdegård B et al (2011) Pyrosequencing of antibioticcontaminated river sediments reveals high levels of resistance

and gene transfer elements. PLoS One 6(2):e17038. https://doi.org/10.1371/journal.pone.0017038
Kumarasamy KK, Toleman MA, Walsh TR, Bagaria J, Butt F, Balakrishnan R, Chaudhary U, Doumith M, Giske CG, Irfan S (2010) Emergence of a new antibiotic resistance mechanism in India, Pakistan, and the UK: a molecular, biological, and epidemiological study. Lancet Infect Dis 10:597–602
Larsson DGJ, de Pedro C, Paxeus N (2007) Effluent from drug manufactures contains extremely high levels of pharma ceuticals. J Haz Materials 148:751–755
Larsson D (2010) Release of active pharmaceutical ingredients from manufacturing sites—need for new management strategies. Integr Environ Assess Manag 6:184–186
Laxminarayan R, Duse A, Wattal C, Zaidi AK, Wertheim HF, Sumpradit N, Vlieghe E, Hara GL, Gould IM, Goossens H (2013) Antibiotic resistance—the need for global solutions. Lancet Infect Dis 13:1057–1098
Lewis HC, Mølbak K, Reese C, Aarestrup FM, Selchau M, Sørum M, Skov RL (2008) Pigs as source of methicillin-resistant Staphylococcus aureus CC398 infections in humans, Denmark. Emerg Infect Dis 14:1383
Li B, Zhang T (2010) Biodegradation and adsorption of antibiotics in the activated sludge process. Environ Sci Technol 44:3468–3473
Li D, Yang M, Hu J, Zhang J, Liu R, Gu X et al (2009) Antibiotic-resistance profile in environmental bacteria isolated from penicillin production wastewater treatment plant and the receiving river. Environ Microbiol 11:1506–1517
Li D, Yu T, Zhang Y, Yang M, Li Z, Liu M et al (2010) Antibiotic resistance characteristics of environmental bacteria from an oxytetracycline production wastewater treatment plant and the receiving river. Appl Environ Microbiol 76:3444–3451
Ma Y, Wilson CA, Novak JT, Riffat R, Aynur S, Murthy S, Pruden A (2011) Effect of various sludge digestion conditions on sulfonamide, macrolide, and tetracycline resistance genes and class I integrons. Environ Sci Technol 45:7855–7861
Magiorakos AP, Srinivasan A, Carey R, Carmeli Y, Falagas M, Giske C, Harbarth S, Hindler J, Kahlmeter G, Olsson-Liljequist B (2012) Multidrug-resistant, extensively drug-resistant and pandrug-resistant bacteria: an international expert proposal for interim standard definitions for acquired resistance. Clin Microbiol Infect 18:268–281
Martinez JL (2009) The role of natural environments in the evolution of resistance traits in pathogenic bacteria. Proc R Soc Lond B Biol Sci 276:2521–2530
McKinney CW, Loftin KA, Meyer MT, Davis JG, Pruden A (2010) Tet and sul antibiotic resistance genes in livestock lagoons of various operation type, configuration, and antibiotic occurrence. Environ Sci Technol 44:6102–6109
Mohring SA, Strzysch I, Fernandes MR, Kiffmeyer TK, Tuerk J, Hamscher G (2009) Degradation and elimination of various sulfonamides during anaerobic fermentation: a promising step on the way to sustainable pharmacy? Environ Sci Technol 43:2569–2574
MPA (Swedish Medical Products Agency) (2011) Underlag för att möjliggöra initieringen av en revidering av EUlagstiftningen om god tillverkningssed, GMP, med syfte att lagstiftningen även ska omfatta miljöhänsyn [in Swedish]. Available at: http://www.lakemedelsverket.se/upload/nyheter/2011/2011-06-16_regeringsuppdrag-miljo-GMP.pdf
Munir M, Xagoraraki I (2011) Levels of antibiotic resistance genes in manure, biosolids, and fertilized soil. J Environ Qual 40:248–255
Munir M, Wong K, Xagoraraki I (2011) Release of antibiotic resis tant bacteria and genes in the effluent and biosolids of five wastewater utilities in Michigan. Water Res 45:681–693
Murray-Smith RJ, Coombe VT, Grönlund MH, Waern F, Baird JA (2012) Managing emissions of active pharmaceutical ingredients from manufacturing facilities: an environmental quality standard approach. Integr Environ Assess Manag 8:320–330
Orozco Hernández JR, Olmedo Sánchez JA, Curiel Flores A (2009) The effect of a herbal grow promoter feed additive on shrimp performance. Res J Biol Sci 4(9):1022–1024

Ozaki N, Bester K, Moldrup P, Henriksen K, Komatsu T (2011) Photodegradation of the synthetic fragrance OTNE and the bactericide triclosan adsorbed on dried loamy sand—results from models and experiments. Chemosphere 83:1475–1479

Pei R, Cha J, Carlson KH, Pruden A (2007) Response of antibiotic resistance genes (ARG) to biological treatment in dairy lagoon water. Environ Sci Technol 41:5108–5113

Poulsen HD (1998) Zinc and copper as feed additives, growth factors or unwanted environmental factors. J Anim Feed Sci 7:135–142

Pruden A, Larsson DJ, Amézquita A, Collignon P, Brandt KK, Graham DW, Lazorchak JM, Suzuki S, Silley P, Snape JR (2013) Management options for reducing the release of antibiotics and antibiotic resistance genes to the environment. Environ Health Perspect (Online) 121:878

Renou S, Givaudan JG, Poulain S, Dirassouyan F, Moulin P (2008) Landfill leachate treatment: review and opportunity. J Hazard Mater 150:468–493

Rizzo L, Manaia C, Merlin C, Schwartz T, Dagot C, Ploy M, Michael I, Fatta-Kassinos D (2013) Urban wastewater treatment plants as hotspots for antibiotic resistant bacteria and genes spread into the environment: a review. Sci Total Environ 447:345–360

Riquelme Breazeal MV, Vikesland PJ, Novak JT, Pruden A (2013) Effect of wastewater colloids on membrane removal of microconstituent antibiotic resistance genes. Water Res 47:130–140

Rosendahl I, Siemens J, Kindler R, Groeneweg J, Zimmermann J, Czerwinski S, Lamshöft M, Laabs V, Wilke B-M, Vereecken H (2012) Persistence of the fluoroquinolone antibiotic difloxacin in soil and lacking effects on nitrogen turnover. J Environ Qual 41:1275–1283

Rysz M, Mansfield WR, Fortner JD, Alvarez PJJ (2013) Tetracycline resistance gene maintenance under varying bacterial growth rate, substrate and oxygen availability, and tetracycline concentration. Environ Sci Technol 47(13):6995–7001. https://doi.org/10.1021/es3035329

Sabourin L, Duenk P, Bonte-Gelok S, Payne M, Lapen DR, Topp E (2012) Uptake of pharmaceuticals, hormones and parabens into vegetables grown in soil fertilized with municipal biosolids. Sci Total Environ 431:233–236

Sapkota AA, Hulet RM, Zhang G, McDermott P, Kinney EL, Schwab KJ, Joseph SW (2011) Lower prevalence of antibiotic-resistant enterococci on US conventional poultry farms that transitioned to organic practices. Environ Health Perspect 119:1622

Sarmah AK, Meyer MT, Boxall AB (2006) A global perspective on the use, sales, exposure pathways, occurrence, fate and effects of veterinary antibiotics (VAs) in the environment. Chemosphere 65:725–759

Seiler C, Berendonk TU (2012) Heavy metal driven co-selection of antibiotic resistance in soil and water bodies impacted by agriculture and aquaculture. Front Microbiol 3:399

SEMC (Swedish Environmental Management Council) (2011) Requirement MSR 10057. Environmental Procedures in the Supplier Chain-Advanced. Available at: http://www.msr.se/en/Upphandling/SEMCOProcurement-criteria/Nursing-and-care/Pharmaceuticals/Pharmaceutical-products/Environmental-procedures-inthe-supplier-chain—advanced/

Sharma R, Larney FJ, Chen J, Yanke LJ, Morrison M, Topp E, McAllister TA, Yu Z (2009) Selected antimicrobial resistance during composting of manure from cattle administered sub-therapeutic antimicrobials. J Environ Qual 38:567–575

Smith DL, Dushoff J, Morris JG Jr (2005) Agricultural antibiotics and human health. PLoS Med 2:e232

Song J, Rensing C, Holm PE, Virta M, Brandt KK (2017) Comparison of Metals and Tetracycline as Selective Agents for Development of Tetracycline Resistant Bacterial Communities in Agricultural Soil. Environ Sci Technol 51:3040–3047

Steinfeld H, Gerber P, Wassenaar TD, Castel V, De Haan C (2006) Livestock's long shadow: environmental issues and options. rome: food and agriculture organization of the United Nations. Available at: ftp://ftp.fao.org/docrep/fao/010/a0701e/a0701e00.pdf

Storteboom H, Arabi M, Davis JG, Crimi B, Pruden A (2010) Tracking antibiotic resistance genes in the South Platte River basin using molecular signatures of urban, agri cultural, and pristine sources. Environ Sci Technol 44:7397–7404

Storteboom HN, Kim S-C, Doesken KC, Carlson KH, Davis JG, Pruden A (2007) Response of antibiotics and resistance genes to high-intensity and low-intensity manure management. J Environ Qual 36:1695–1703

Tang X, Lou C, Wang S, Lu Y, Liu M, Hashmi MZ, Liang X, Li Z, Liao Y, Qin W (2015) Effects of long-term manure applications on the occurrence of antibiotics and antibiotic resistance genes (ARGs) in paddy soils: evidence from four field experiments in south of China. Soil Biol Biochem 90:179–187

Thames CH, Pruden A, James RE, Ray PP, Knowlton KF (2012) Excretion of antibiotic resistance genes by dairy calves fed milk replacers with varying doses of antibiotics. Front Microbiol 3:139

Vasconcelos TG, Henriques DM, König A, Martins AF, Kümmerer K (2009) Photo-degradation of the antimicrobial ciprofloxacin at high pH: identification and biodegradability assessment of the primary by-products. Chemosphere 76:487–493

Walsh TR, Weeks J, Livermore DM, Toleman MA (2011) Dissemination of NDM-1 positive bacteria in the New Delhi environment and its implications for human health: an environmental point prevalence study. Lancet Infect Dis 11:355–362

Wang Y, Ma Y (2008) Potential public hazard of using antibiotics in livestock industry. Chin J Antibiot 33:519–523

Wang L, Oda Y, Grewal S, Morrison M, Michel FC, Yu Z (2012) Persistence of resistance to erythromycin and tetracycline in swine manure during simulated composting and lagoon treatments. Microb Ecol 63:32–40

WHO (World Health Organization). 2012. Antimicrobial resistance. Available at: http://www.who.int/mediacentre/factsheets/fs194/en/

Wu N, Qiao M, Zhang B, Cheng WD, Zhu YG (2010) Abundance and diversity of tetracycline resistance genes in soils adja cent to representative swine feedlots in China. Environ Sci Technol 44:6933–6939

Yun-peng W, Yue M (2008) Potential public hazard of using antibiotics in livestock industry. Chin J Antibiot 9:001

Zhang T, Li B (2011) Occurrence, transformation, and fate of antibiotics in municipal wastewater treatment plants. Crit Rev Environ Sci Technol 41:951–998

Zhu Y-G, Johnson TA, Su J-Q, Qiao M, Guo G-X, Stedtfeld RD, Hashsham SA, Tiedje JM (2013) Diverse and abundant antibiotic resistance genes in Chinese swine farms. Proc Natl Acad Sci USA 110:3435–3440

Erratum to: Antibiotics and Antibiotics Resistance Genes in Soils

Monitoring, Toxicity, Risk Assessment and Management

Muhammad Zaffar Hashmi, Vladimir Strezov, and Ajit Varma

Erratum to:
M.Z. Hashmi et al. (eds.), *Antibiotics and Antibiotics Resistance Genes in Soils,* Soil Biology 51,
DOI: 10.1007/978-3-319-66260-2

The updated online versions of these chapters can be found at
DOI: 10.1007/978-3-319-66260-2_3
DOI: 10.1007/978-3-319-66260-2_4
DOI: 10.1007/978-3-319-66260-2_5
DOI: 10.1007/978-3-319-66260-2_6
DOI: 10.1007/978-3-319-66260-2_7
DOI: 10.1007/978-3-319-66260-2_8
DOI: 10.1007/978-3-319-66260-2_9
DOI: 10.1007/978-3-319-66260-2_10
DOI: 10.1007/978-3-319-66260-2_11
DOI: 10.1007/978-3-319-66260-2_12
DOI: 10.1007/978-3-319-66260-2_13
DOI: 10.1007/978-3-319-66260-2_14
DOI: 10.1007/978-3-319-66260-2_15
DOI: 10.1007/978-3-319-66260-2_16
DOI: 10.1007/978-3-319-66260-2_17
DOI: 10.1007/978-3-319-66260-2_18
DOI: 10.1007/978-3-319-66260-2_19
DOI: 10.1007/978-3-319-66260-2_20
DOI: 10.1007/978-3-319-66260-2_21
DOI: 10.1007/978-3-319-66260-2

M.Z. Hashmi et al. (eds.), *Antibiotics and Antibiotics Resistance Genes in Soils*,
Soil Biology 51, DOI 10.1007/978-3-319-66260-2_22

The original version of this book contains errors which have been corrected. The corrections are given below:

Chapter 3: A Review on Antibiotics Consumption, Physico-Chemical Properties and Their Sources in Asian Soil

The contents of the following sub-headings have been updated:

1) Abstract
2) 3.1 Introduction
3) 3.2 Antibiotics: How Does It Work?
4) 3.3.2 Livestock and Agricultural Usage
5) 3.4.1 Tetracycline and Sulfonamides
6) 3.4.2 Fluoroquinolones
7) 3.4.3 Macrolides
8) 3.4.4 Aminoglycoside
9) 3.5 Sources of Antibiotics in Asian Soil
10) 3.6 Occurrence of antibiotics in Asian soil

The following subheadings and their contents have been added to the chapter:

3.6 Occurrence of Antibiotics in Asian Soil

From the sources discussed previously in this chapter, it is evident that antibiotics can end up in the soil, and finally via plant uptake, such chemicals may enter the food chain. Despite such existing sources, limited data is available on the level of antibiotics in Asian soil. Antibiotic levels in soil in China and Korea among Asian countries have been given in Table 3.5. All the tetracycline compounds showed highest range in Shenyang region, China. Agriculture, animal husbandry, and agricultural product processing units were dominant in north-eastern part of Shenyang city. Sulfonamide compounds also showed very high concentration in Shenyang city. Sulfonamides and tetracycline are the two antibiotics most commonly used to promote growth in livestock production. So it is evident that the high concentration of these two compounds in Shenyang city is due to the use of these two antibiotics in livestock production. As compared to China, the antibiotic concentration in Korean soil is less. This may be because, since 2005, South Korea is gradually decreasing the use of antibiotic growth promoters in livestock production.

3.7 Conclusion

Like several other organic pollutants, anitibiotics also reach soil mostly after partially treated or untreated antibiotics present in wastewater stream. Studies across the globe have reported the uptake of antibiotics by plants. The pathway of these pollutants generally depends on its physio-chemical properties. Studies have shown the development of antibiotic resistance genes in soil, plants, and humans. Further studies are required to find the fate of these pollutants in the environment and the associated risk due to the occurrence of these antibiotics in the environment.

The following tables have been added to the chapter:

Table 3.2 Physico-chemical properties and structure of antibiotics

Class	Molar mass (g/mol)	Water solubility (mg/L)	Log Kow	pKa	Henry's constant (Pal/mol)	Structure
Tetracycline	444.5–527.6	230–52,000	−1.3 to 0.05	3.3/7.7/9.3	1.7×10^{-23}–4.8×10^{-22}	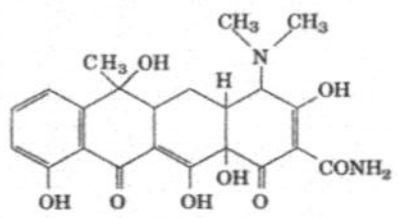
Sulfonamides	172.2–300.3	7.5–1500	−0.1 to 1.7	2–3/4.5–10.6	1.3×10^{-12}–1.8×10^{-8}	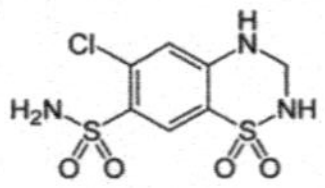
Aminoglycosides	332.4–615.6	10–500	−8.1 to −0.8	6.9–8.5	8.5×10^{-12}–4.1×10^{-18}	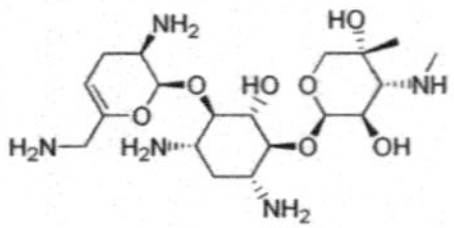
ß-Lactam	334.4–470.3	22–10,100	0.9–2.9	2.7	2.5×10^{-19}–1.2×10^{-12}	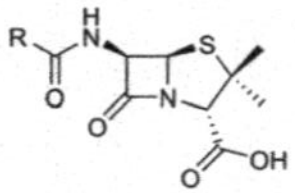
Macrolides	687.9–916.1	0.45–15	1.6–3.1	7.7–8.9	7.8×10^{-36}–2.0×10^{-26}	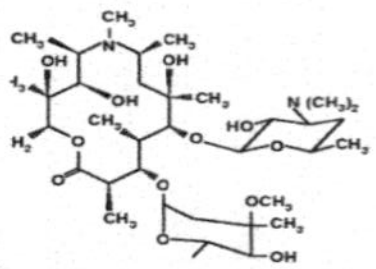
Fluoroquinolones	229.5–417.6	3.2–17,790	−1.0 to 1.6	8.6	5.2×10^{-17}–3.2×10^{-8}	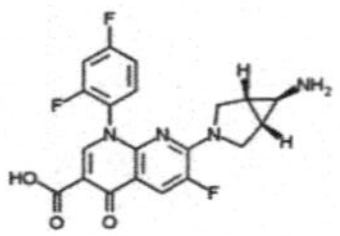

Table 3.3 Antibiotics in the sludge from wastewater treatment plant used as manure

Compounds	Sludge (μg/g)	Country	Reference
Tetracycline			
Tetracycline	2174.46	Sheyang, China(NE)	An et al. (2015)
Oxytetracycline	7369.67	Sheyang, China(NE)	An et al. (2015)
Chlortetracycline	3843.79	Sheyang, China(NE)	An et al. (2015)
Doxycycline	2104.27	Sheyang, China(NE)	An et al. (2015)
Sulfonamides			
Sulfamethoxazole	665	Sheyang, China(NE)	An et al. (2015)
Sulfadiazine	50.32	Sheyang, China(NE)	An et al. (2015)
Sulfamerazine	37.21	Sheyang, China(NE)	An et al. (2015)
Sulfadimidine	27.14	Sheyang, China(NE)	An et al. (2015)

Table 3.4 Antibiotics from livestock excreta used as manure

Compounds	Livestock manure (mg/g)	Country	Reference
Tetracycline			
Tetracycline	56.95	Sheyang, China(NE)	An et al. (2015)
	43.5	North China	Hu et al. (2010)
	3.5	Beijing	Li et al. (2015)
Oxytetracycline	47.25	Sheyang, China(NE)	An et al. (2015)
	183.5	North China	Hu et al. (2010)
	23.271	Beijing	Li et al. (2015)
Chlortetracycline	143.97	Sheyang, China(NE)	An et al. (2015)
	26.8	North China	Hu et al. (2010)
	26.218	Beijing	Li et al. (2015)
Doxycycline	6.5	Sheyang, China(NE)	An et al. (2015)
Fluoroquinolones			
Enrofloxacin	8.684	Beijing	Li et al. (2015)
Ciprofloxacin	4.3	North China	Hu et al. (2010)
	9.342	Beijing	Li et al. (2015)
Norfloxacin	4.187	Beijing	Li et al. (2015)
Ofloxacin	15.7	North China	Hu et al. (2010)
Perfloxacin	24.7	North China	Hu et al. (2010)
Lomefloxacin	0.038	Beijing	Li et al. (2015)
Macrolide			
Roxithromycin	0.067	Beijing	Li et al. (2015)
Lincomycin	3.8	North China	Hu et al. (2010)
Sulfonamides			
Sulfamethoxazole	18.5	Sheyang, China(NE)	An et al. (2015)
	5.7	North China	Hu et al. (2010)
	0.102	Beijing	Li et al. (2015)

(continued)

Table 3.4 (continued)

Compounds	Livestock manure (mg/g)	Country	Reference
Sulfadiazine	4.98	Sheyang, China(NE)	An et al. (2015)
	0.022	Beijing	Li et al. (2015)
Sulfamerazine	4.59	Sheyang, China(NE)	An et al. (2015)
Sulfadimidine	1.95	Sheyang, China(NE)	An et al. (2015)
Sulfamethazine	0.061	Beijing	Li et al. (2015)
Sulfadoxin	32.7	North China	Hu et al. (2010)
Sulfachloropyridazine	2.76	North China	Hu et al. (2010)

Table 3.5 Wastewater containing antibiotics used for irrigation

Compounds	Wastewater (ng/l)	Country	Reference
Tetracycline			
Tetracycline	560	New York	Batt et al. (2006)
	48,000	Wisconsin, USA	Karthikeyan et al. (2006)
Oxytetracycline	47,000	Wisconsin, USA	Karthikeyan et al. (2006)
Fluoroquinolones			
Enrofloxacin	250	Wisconsin, USA	Karthikeyan et al. (2006)
Ciprofloxacin	10	USA	He et al. (2015)
	970	New York	Batt et al. (2006)
	310	Wisconsin, USA	Karthikeyan et al. (2006)
Norfloxacin	250	Wisconsin, USA	Karthikeyan et al. (2006)
Ofloxacin	9	USA	He et al. (2015)
Macrolide			
Erithromycin	3900	Wisconsin, USA	Karthikeyan et al. (2006)
Roxithromycin	1500	Wisconsin, USA	Karthikeyan et al. (2006)
Salfonamides			
Sulfamethoxazol	156	South Africa	Rahzia et al. (2012)
	1340	New York	Batt et al. (2006)
	310	Wisconsin, USA	Karthikeyan et al. (2006)
Sulfamethazine	50	China	Sun et al. (2014)
	300	Wisconsin, USA	Karthikeyan et al. (2006)

The following Figures have been updated:

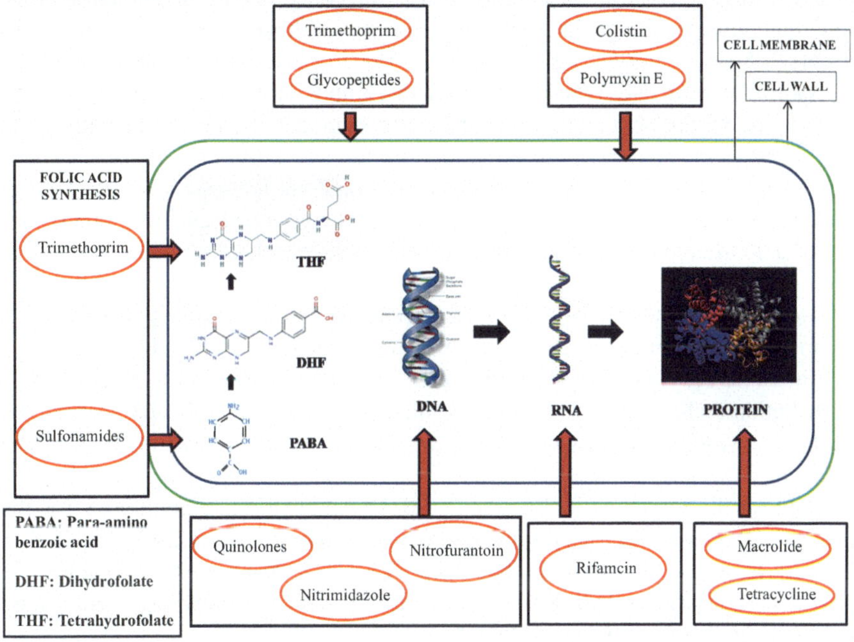

Fig. 3.1 Hypothetical schematic representation showing the mode of action of different antibiotics against bacteria

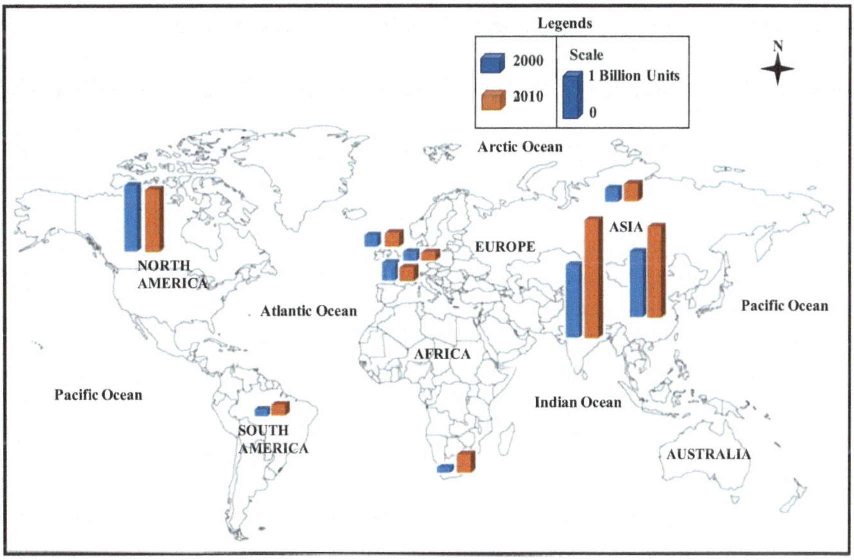

Fig. 3.2 Consumption of antibiotics by different countries across the globe during 2000 and 2010. (Data courtesy-Van Boeckel et al. 2015)

The following References have been added to the chapter:

An J, Chen H, Wei S, Gu J (2015) Antibiotic contamination in animal manure, soil, and sewage sludge in Shenyang, northeast China. Environ Earth Sci 74 (6):5077–5086

Awad YM, Kim S-C, Abd El-Azeem SAM, Kim K-H, Kim K-R, Kim K, Jeon C, Lee SS, Ok YS (2014) Veterinary antibiotics contamination in water, sediment, and soil near a swine manure composting facility. Environ Earth Sci 71(3):1433–1440

Bbosa GS, Mwebaza N (2013) Global irrational antibiotics/antibacterial drugs use: a current and futurehealth and environmental consequences. Microbial pathogens and strategies for combating them: science,technology and education. Formatex, Badajoz

Bound JP, Voulvoulis P (2004) Pharmaceuticals in the aquatic environment – a comparison of risk assessment strategies. Chemosphere 56:1143–1155.

Chen F, Ying G-G, Kong L-X, Wang L, Zhao J-L, Zhou L-J, Zhang L-J (2011) Distribution and accumulation of endocrine-disrupting chemicals and pharmaceuticals in wastewater irrigated soils in Hebei, China. Environ Pollut 159 (6):1490–1498

Defoirdt T, Sorgeloos P, Bossier P (2011) Alternatives to antibiotics for the control of bacterial disease in aquaculture. Curr Opin Microbiol 14(3):251–258

Fatta-Kassinos D, Meric S, Nikolaou A (2011) Pharmaceutical residues in environmental waters and wastewater: current state of knowledge and future research. Anal Bioanal Chem 1:251–275

Gao P, Mao D, Luo Y, Wang L, Xu B, Xu L (2012) Occurrence of sulfonamide and tetracycline-resistant bacteria and resistance genes in aquaculture environment. Water Res 46(7):2355–2364

Ghosh K, Mandal S (2010) Antibiotic resistant bacteria in consumable fishes from Digha coast, West Bengal, India. Proc Zool Soc 63(1):13–20

Heberer T (2002) Occurrence, fate, and removal of pharmaceutical residues in the aquatic environment: a review of recent research data. Toxicol Lett 131 (1–2):5–17

Hu X, Luo Y, Zhou Q (2010) Simultaneous analysis of selected typical antibiotics in manure by microwave assisted extraction and LC–MS. Chromatographia 71:217–223

Hu X, Zhou Q, Luo Y (2010) Occurrence and source analysis of typical veterinary antibiotics in manure, soil, vegetables and groundwater from organic vegetable bases, northern China. Environ Pollut 158(9):2992–2998

Kim S-C, Carlson K (2007) Temporal and spatial trends in the occurrence of human and veterinary antibiotics in aqueous and river sediment matrices. Environ Sci Technol 41:50–57

Kim S-C, Carlson K (2007) Temporal and spatial trends in the occurrence of human and veterinary antibiotics in aqueous and river sediment matrices. Environ Sci Technol 41:50–57

Kinney CA, Furlong ET, Werner SL, Cahill JD (2006) Presence and distribution of wastewater-derived pharmaceuticals in soil irrigated with reclaimed water. Environ Chem 25:317–325

Kinney CA, Furlong ET, Werner SL, Cahill JD (2006) Presence and distribution of wastewater-derived pharmaceuticals in soil irrigated with reclaimed water. Environ Toxicol Chem 25(2):317

Kummerer K, Henniger A (2003) Promoting resistance by the emission of antibiotics from hospitals and households into effluent. Clin Microbiol Infect 9:1203–14

Kwon SI, Owens G, Ok YS, Lee DB, Jeon W-T, Kim JG, Kim K-R (2011) Applicability of the Charm II system for monitoring antibiotic residues in manure-based composts. Waste Manag 31(1):39–44

Li B, Zhang T (2011) Mass flow and removal of antibiotics in two municipal wastewater treatment plants. Chemosphere 831284–1289

Li C, Chen J, Wang J, Ma Z, Han P, Luan Y, Lu A (2015) Occurrence of antibiotics in soils and manures from greenhouse vegetable production bases of Beijing, China and an associated risk assessment. Sci Total Environ 521–522:101–107

Nowara A, Burhenne J, Spiteller M (1997) Binding of fluoroquinolone carboxylic acid derivatives to clay minerals. J Agric Food Chem 45(4):1459–1463

Rabùlle M, Spliid NH (2000) Sorption and mobility of metronidazole, olaquindox, oxytetracycline and tylosin in soil. Chemosphere 40:715–722.

Shi Y, Gao L, Li W, Liu J, Cai Y (2012). Investigation of fluoroquinolones, sulfonamides and macrolides in long-term wastewater irrigation soil in Tianjin, China. Bull Environ Contam Toxicol 89:857–861

Sim W-J, Lee J-W, Lee E-S, Shin S-K, Hwang S-R, Oh J-E (2011) Occurrence and distribution of pharmaceuticals in wastewater from households, livestock farms, hospitals and pharmaceutical manufactures. Chemosphere 82(2):179–186

Tokura Y, Nishijima T, Yagi H, Furukawa F, Takigawa M (1996) Photohaptenic properties of fluoroquinolones. Photochem Photobiol 64(5):838–844

United Nations, Department of Economic and Social Affairs, Population Division (2015) World population prospects: the 2015 revision, key findings and advance tables. Working Paper No. ESA/P/WP.241

Van Boeckel TP, Gandra S, Ashok A, Caudron Q, Grenfell BT, Levin SA, Laxminarayan R (2015) Global antibiotic consumption 2000 to 2010: an analysis of national pharmaceutical sales data. Lancet Infect Dis 14(8):742–750

Watkinson AJ, Murby EJ, Costanzo SD (2007) Removal of antibiotics in conventional and advanced wastewater treatment: Implications for environmental discharge and wastewater recycling. Water Res 41(18):4164–4176

The page numbers of the chapter has been updated to 39–54

Chapter 4: Entry Routes of Veterinary Antibiotics in the Environment
The page numbers of the chapter has been updated to 55–70

Chapter 5: Monitoring of Antibiotics and Antibiotic Resistance Genes in Agroecosystems
The page numbers of the chapter has been updated to 71–96

Chapter 6: Role of Antibiotics in Climate Change
The page numbers of the chapter has been updated to 97–104

Chapter 7: Potential Dissemination of ARB and ARGs into Soil Through the Use of Treated Wastewater for Agricultural Irrigation: Is It a True Cause for Concern?
The page numbers of the chapter has been updated to 105–140

Chapter 8: Antibiotic Resistance Gene Due to Manure Application
The page numbers of the chapter has been updated to 141–150

Chapter 9: Antibiotics and Antibiotics Resistance Genes Dissemination in Soils
The page numbers of the chapter has been updated to 151–190

Chapter 10: Dissemination Mechanism of Antibiotic Resistance Genes in Environment
The page numbers of the chapter has been updated to 191–206

Chapter 11: Fate of Antibiotics in Soil
The page numbers of the chapter has been updated to 207–220

Chapter 12: Uptake of Antibiotics by Plants
The page numbers of the chapter has been updated to 221–238

Chapter 13: Recent Advances in Methods for the Detection of Antibiotics and Antibiotics Resistance Genes in Soil
The page numbers of the chapter has been updated to 239–258

Chapter 14: Elucidation of Emerging Nanomaterials Impacts on Antibiotic Resistance Against Soil and Aquatic Microflora
The page numbers of the chapter has been updated to 259–282

Chapter 15: The Effects of Antibiotics on the Structure, Diversity, and Function of a Soil Microbial Community
The page numbers of the chapter has been updated to 283–312

Chapter 16: Soil Antibiotics and Transfer of Antibiotic Resistance Genes Affecting Wildlife

The affiliations of the following authors have been corrected and the correct affiliations are:

Gilberto Igrejas
Department of Genetics and Biotechnology, Functional Genomics and Proteomics' Unit, University of Trás-os-Montes and Alto Douro, Vila Real, Portugal
Functional Genomics and Proteomics Unit, University of Tras-os-Montes and Alto Douro (UTAD), Vila Real, Portugal
Research Unit on Applied Molecular Biosciences (UCIBIO-REQUIMTE), University NOVA of Lisboa, Lisboa, Caparica, Portugal

Patrícia Poeta
Veterinary Sciences Department, University of Trás-os-Montes and Alto Douro, Quinta de Prados, 5000-911 Vila Real, Portugal
Research Unit on Applied Molecular Biosciences (UCIBIO-REQUIMTE), University NOVA of Lisboa, Lisboa, Caparica, Portugal

The contents of the following sub-headings have been updated:

1) **16.2 Interaction of Antibiotics with Soil**
2) **16.4 Antimicrobial Resistance in the Environment**
3) **16.5 The Problem of Antibiotic Resistance in Wild Animals**

The page numbers of the chapter has been updated to 313–326

Chapter 17: Genotoxicity and Biochemical Toxicity of Soil Antibiotics to Earthworms
The page numbers of the chapter has been updated to 327–340

Chapter 18: Potential Environmental, Ecological and Health Effects of Soil Antibiotics and ARGs
The page numbers of the chapter has been updated to 341–366

Chapter 19: Risk Assessment of Antibiotics and Antibiotic-Resistant Genes in Soil
The page numbers of the chapter has been updated to 367–386

Chapter 20: Antibiotics in the Soil: Sources, Environmental Issues, and Bioremediation
The page numbers of the chapter has been updated to 387–396

Chapter 21: Management and Regulation of Antibiotics and Antibiotics Resistance Genes in Soils
The page numbers of the chapter has been updated to 397–410

INDEX of the book has been updated.
The page numbers of the INDEX has been updated to 411–420

Index[1]

A
Agriculture
  GHG emissions, 98
  PCR, 84
  streptomycin and oxytetracycline, 163
Agroecosystems
  agro-environmental quality, 89
  ARB (*see* Antibiotic-resistant bacteria (ARB))
  ARGs (*see* Antibiotic resistance genes (ARGs))
  culture-based methods, 78, 79
  LC/MS/MS, 91
  microbes, 72
  quantitative analysis, 91
  sensitive analytical methods, 90
  toxicological effects, 90
2-Alkyl-4-quinolones (AQs), 156
Aminoglycoside, 45
Animals health, 400, 401
Antibiotic resistance (AR)
  agroecosystems
    DNA extraction and purification, 83
    extraction, 81–82
    heterotrophic plate counts, 82
    HPLC-MS/MS spectroscopy, 82
    PCR, 83
    primer design, 83
    quantification, 82
    sampling, 80, 81
  ARB, 72
  ARGs, 74
  HGT, 72
  molecular aspects, 79, 80
  polymerase chain reaction-based assays, 83, 84
Antibiotic resistance genes (ARGs), 380
  acquired resistance, 22
  actinomycetes species, 191
  agriculture, 240
  AMR, 191
  bacteria, 141
  baseline abundance, 108, 109
  community and hospital associated infections, 34, 35
  culture-based assays, 86, 87
  determination, soil bound antibiotics, 143
  development, 240
  dispersal and accumulation, 389
  dissemination of antibiotics
    abundance and diversity, 180
    anthropogenic activities, 178–180
    Burkholderia, 175
    Enterobacteriales, 175
    Escherichia coli, 173
    mechanism, 175–177
    metagenomic analysis, 172
    microbiomes, 181
    sporadic indications, 174
    β-lactamase, 173
  DNA transfer, 24
  drug-resistant bacteria, 23

[1]The original version of this index was revised. An erratum to this index can be found at DOI 10.1007/978-3-319-66260-2_22.

M.Z. Hashmi et al. (eds.), *Antibiotics and Antibiotics Resistance Genes in Soils*, Soil Biology 51, DOI 10.1007/978-3-319-66260-2

Antibiotic resistance genes (ARGs) (*cont.*)
Enterobacteriaceae, 20
environmental fungi and bacteria, 192, 193
florfenicol and linezolid, 389
functional metagenomics, 144, 145
genetic change, 22–24
global health care management, 146–147
harvested vegetables, 389
implications, human health, 142, 143
industry and policy makers, 147
infections, 146
intrinsic resistance, 22, 23
manure-applied soils, 109
mechanisms, 22, 147
metagenomic analysis, 108
molecular methods, 87–89
non-resistant bacteria, 23
physical and biological forces, 389
pristine environments, 106–109
public databases and global surveillance projects, 146
rates and trends, 33, 34
risk assessment (*see* Risk assessment process)
16S rRNA gene analysis, 110
screening, 147
soil microbiota, 141
spread and development, 141
spreading manure, 141
tests, 20
tetracycline resistance genes, 22, 25–32
third-generation cephalosporins, 146
transfer of antibiotics, 74, 76, 77
treatment strategies, 145, 146
VA, manure and soil, 142
vegetables, 141
waste disposal processes, 141
Antibiotic-resistant bacteria (ARB)
agroecosystems (*see* Agroecosystems)
ARGs (*see* Antibacterial resistance genes (ARGs))
dissemination of antibiotics
16S rDNA, 171
acquired resistance, 166
CM, 170
MBC, 167
metagenomes, 167
MIC, 167
microbiota, 168
PBP, 170
PCR, 171
protein synthesis, 169
Antibiotics
antimicrobial contamination, 240
applications, 6
assessment, 251
cell wall synthesis inhibitors, 6
classification, 3
cultural method, 249
cytoplasmic membrane function, 6
definition, 2
detection in soil, 251
detection, pharmaceutical compounds, 241
development, 240
disposal of drugs, 251
drug innovation and implementation, 239
environment, 240
excessive, 240
functions and resistance, 193–194
fungal, 1
level, 240
livestock industry, 239
mechanism of action, 5
microbes, 2
microorganisms, 4
misuse, 240
molecular methods
Gene-Z, 250–251
next-generation DNA sequencing, 247, 248
PCR, 246
POC, 250
qPCR, 247
Wi-Fi devices, 251
monitoring, 73, 74
protein synthesis inhibitors, 5–6
replication, 4
sampling and extraction, 241
separation (*see* Chromatography)
soil (*see* Soil antibiotics)
source and sinks, 251
Streptomyces, 13, 14
sulfonamides and β-lactam, 3
teixobactin, 15
transcription inhibitors, 4
uses, 1
Antimicrobial resistance
environment
artificial, 318
bacteria, 318, 321
cancer sector, 318
Darwinian principles, 319
farm animals, 319
food system, 319
genes, 318, 319
human and veterinary medicine, 318
infections, 319
mobile genetic elements, 320
nosocomial bacteria resistome, 319
organisms, ecosystems, 319
risks, 318
wild animals (*see* Wild animals)

Aquaculture, 62–63
Asian soil
  aquaculture, 49
  consumption
    agricultural usage, 42–43
    combating microbial infections, 41, 42
    livestock production, 39, 42, 43
    penicillin β-lactam, 40
    physico-chemical property, 40
  irrigation, 48, 49
  manure, 46
  physico-chemical characteristics
    aminoglycoside, 45
    fluoroquinolones, 43
    macrolides, 45
    sulfonamides, 43
    tetracycline, 43
    β-lactam, 45
  solid waste dumpsite, 49

B
Bacteria
  antimicrobial resistance, 318
  soil antibiotics effects
    agriculture and veterinary medicine, 318
    maintaining mineral immobilization, 316
    manure, 317
    microbial communities, 317
    nutrient and geochemical cycles, 317
    spectrum and mode of action, 317
Bacteriophages
  application, 129
  biocontrol, 128
  CRISPR-Cas system, 128
  potential strategy, 127
  synthetic phages, 128
  treatment, 128
Biodegradation of antibiotics, 216, 217
Biological treatment and energy recovery, 401, 402
Bioremediation
  algae, 390
  bacteria, 390
  benzylpenicillin, 391
  fungi, 390
  liquid chromatography, 392
  OECD guidelines, 391
  plants, 390
  Pleurotus ostreatus, 392
  remove/detoxify pollutants, 390
  Trametes versicolor, 392
Broad-spectrum antibiotics, 222

C
Carbapenem-resistant *Enterobacteriaceae* (CRE), 348
Chlorofluorocarbons (CFCs)
Chloramphenicol acetyltransferase (CAT), 330
Chromatography
  gas, 242, 243
  liquid, 243, 244
  mass spectrometry, 243–245
Climate change
  GHGs, 98
  livestock production system, 99, 101
  methane (CH_4) production, 99–101
  microbes, 102
  ozone depletion, 98
  stratosphere, 99
  temperature and precipitation/rainfall, 97
Community-level physiological profile (CLPP), 297, 298
Concentrated animal feeding operation (CAFOs), 64
Conjugation process, 194
Consumption, antibiotics
  combating microbial infections, 42
  penicillin a β-lactam, 41
Contaminated soil remediation, 398–399
CRISPR-Cas system, 128

D
Degradation rate, VA, 57
Denaturing gradient gel electrophoresis (DGGE), 293
Dialkylresorcinols (DARs), 155
Dissemination of antibiotics, 167, 172–181
  agriculture, 163
  animal husbandry, 163–165
  antimicrobials, 154
  ARGs (*see* Antibacterial resistance genes (ARGs))
  bacteria and fungi, 158
  beta-lactams, 155
  DARs, 155
  degradation rates, 165, 166
  degradation rates, pharmaceuticals, 166
  feed supplemented, 159, 161
  HGT, 153
  HPR, 155
  metagenomic and functional analyses, 154
  MGEs, 154
  MIC, 167
  nonlethal antibiotic, 162
  nutrient modification, 151
  plant pathogens, 152
  pollutant decomposition, 151

Dissemination of antibiotics (*cont.*)
serine and metallo-β-lactamases, 157
veterinary medicine, 160, 161
wastewater purification plants, 156
WWTP, 160

E
Ecosystems
ARB and ARGs, 357
bacterial, 344, 345
behaviour and effects, 349–351
degrading, 351
environmental resistance, 347
food pollutants, 350
human and environmental health, 354, 355
pollutant, 342, 352–354
potential of subinhibitory, 355–356
preventing and control, 358, 359
Ralstonia metallidurans, 343
rampant, 345
resistance control, 357, 358
risk assessment, 355
transmission of resistance determinants, 356
Enrofloxacin (EFLX), 331
Environment
VA
academic and research institutes, 66–67
agriculture sources, 62, 63, 67
different entry routes, 61
direct/indirect methods, 60
grazing animals, 62
intensive livestock and agricultural farming practices, 62
land, 64
leaching and runoff, 66
livestock waste, 62
manure application, 63, 64
pre-treatment method, 68
stormwater runoff, 67
waste/manure storage, 65
wastewater/sewage treatment plants, 65–66
Enzymatic activity
carbohydrates and carboxylic acids, 304
dehydrogenase and urease activities, 299, 300
heavy metals/pesticides, 303
iron reduction, 303
N-transformation processes, 301, 303
oxytetracyclines, 299, 300
strategy, 287
tetracyclines, 300
Enzyme-linked immunosorbent assay (ELISA), 91
European Food and Safety Authority (EFSA), 380
European Medicines Agency (EMA), 370
CVMP, 369
PEC (*see* Predicted environmental concentration (PEC))
Extended spectrum beta-lactamase (ESBL), 348

F
Fate and persistence, 119
Fate, veterinary antibiotics
active and inactive metabolites, 227
degradation of veterinary pharmaceuticals, 228
distribution coefficient (Kd), 228
oxytetracycline and tylosin, 228
pH, 228
Fluoroquinolones, 43–45
Fungal antibiotics, 1

G
Gas chromatography, 242
Gene transfer agents (GTAs)
DNA defense and protection, 201
features, 200
RcGTA, 200
Genetically modified organisms (GMOs)
EFSA, 380
inoculant bacteria, 380
types of risks, 381
Gene-Z, 250–251
Genotoxicity and biochemical toxicity
detection methods, 335, 336
mechanisms, 333–335
Greenhouse gasses (GHGs)
CH_4, 102
direct and indirect sources, 98
livestock production system, 99
microbial control, 103

H
Heavy metal *vs.* antimicrobial resistance, 343
2-Hexyl-5-propyl-alkylresorcinol (HPR), 155
Horizontal gene transfer (HGT), 72
abiotic factors, 121
bacterial lineages, 119

conjugation process, 194
DNA degradation, 120
extracellular DNA, 120
integrons process, 196
mechanism, 119
pathogens, 123, 124
plant surfaces, 122, 123
plasmids, 121
transduction process, 195
transformation process, 194, 195
Horizontal gene transfer assays, 85
Human and environmental health, 354, 355
Human health
growth-promoting and prophylactic, 234
manure amendment, 233
MRSA, 233
resistant bacteria, 232

I
Integrons process, 196
Intensive farming operations, VA
animal manure, 58
aquaculture, 60
consumption, 59, 60
ESVAC, 60
FAOSTAT, 58
fish, 59
food security, farm animals, 58
guestimate, 59
human medicine, 59
livestock farming and aquaculture, 58
metaphylactic, 59
over-reliance pressurizes industrial, 59
penicillin and oxytetracycline (OTC), 59
prophylactic, 59
salmon cultivation, 59
tetracycline, 60
iPod Touch application, 251

L
β-Lactam, 45
Liquid chromatography, 243, 244

M
Macrolides, 45, 199
Management and regulation, antibiotics and ARGs, 402
air, soil and water environments, 398
animal growth promoters, 398
biological treatment and energy recovery, 401, 402
compounds, 397
contaminated soil remediation, 398–399
herbal materials, 400
human and animal fecal material, 398
implementation and monitoring, 404
maintaining animals health, 400–401
manure, 401
metals, 400
production, 398
reduce/optimizing, 399–400
reduction, infectious diseases, 397
revolution, 398
risks, 404
schemes, 405
wastewater (*see* Wastewater treatment)
Manure
analytical techniques, 226
destiny and persistence, 225
lipid soluble antibiotics, 225
physical and chemical properties, 226
veterinary antibiotics, 227
Mass spectrometry
GC–MS, 245
GC–MS/MS, 245
identification/quantification, 243
LC–MS/MS, 245
and liquid chromatography, 244
TBA, 244
Measured environmental concentration (MEC), 370
Metagenomics analysis, 85, 86, 247, 248
16S rRNA, 144
agricultural manure, 144
agricultural soils, 144
bacterial groups, 144
DNA microarrays, 145
Inc18-type plasmids, 145
IncN and *IncW* plasmids, 145
IncP-1 and *IncQ*, 145
intI1, 144
microbial communities, 144
molecular techniques, 144
Metallic nanoparticles (MNPs)
Ag_2ONPs, 266
AgNPs, 265, 267
doxycycline, 267
Methane (CH_4) production
GHG, 99
livestock emissions, 101
ruminants, 100

Methicillin-resistant *Staphylococcus aureus* (MRSA), 20, 33, 233, 345, 348
Microbial community
biological activity, 284
CLPP, 297, 298
efficacy, 287
enzymatic activity, 298, 300, 301
human and animal digestive system, 283
PLFA, 293
property, 290–292
respiratory activity, 294–297
SDZ, 284, 285
SIR, 288
sorption-desorption, 285, 286
structure and diversity, 288–293
tylosin, 305, 306
Microorganism
actinomycetes, 9
algae, 9
antimicrobial compounds, 12, 13
bacteria, 8
biotic and abiotic conditions, 1
components, 7
fungi, 9
isolation and screening, 11
protozoa, 9
soil microbes, 9, 10
Streptomyces, 12
Minimal bactericidal concentration (MBC), 167
Minimal inhibitory concentration (MIC), 167
Mobile genetic elements (MGEs), 72, 153
Molecular methods, AGRGs analysis
Gene-Z, 250–251
iPod Touch application, 251
metagenomics, 247, 248
PCR, 246
qPCR, 247
smartphone, 249, 250

N
Nanomaterials, 260
Nanoparticles (NPs)
AgNPs, 261, 262
antibiotic *vs*. multidrug resistance, 262–265
CuNPs, 261
enzyme activity, 269
microflora, 269
MNPs, 260
nanomaterials, 259
plant growth, 269, 271
Pseudomonas stutzeri, 261
toxicity and health impacts, 273, 274
VAM, 271
Neutral red retention time (NRRT), 336
New Delhi metallo-beta-lactamase (NDM)
carbapenem, 112
Enterobacteriaceae, 112–114, 117
MBL activity, 118
wastewater, 118, 119
Next-generation DNA sequencing, 247, 248
N-transformation processes, 301, 303
Nystatin, 333

O
Oxytetracycline (OTC), 288, 315, 330, 331

P
Penicillin-binding protein (PBP), 170
Penicillin-resistant *Streptococcus pneumoniae* (PRSP), 346
Phospholipid fatty acids (PLFAs), 289
Physico-chemical characteristics
aminoglycoside, 45
fluoroquinolones, 43, 45
β-Lactam, 45
macrolides, 45
sulfonamides, 43
tetracycline, 43
Phytoremediation, 392–393
Plants
disease-causing bacteria, 230
growth, 232
uptake and accumulation, 230, 231
Point of care (POC) device, 250
Pollution
contamination, 352
genetic reactors, 352–353
industrial antibiotics, 353
Polymerase chain reaction (PCR), 83, 84, 171, 246
Predicted environmental concentration (PEC)
antibiotic residues, 370–372
compounds, 375, 376
effects assessment, 374–376
MEC, 370
refinement procedure, 373
Propagation, ARGs
macrolide, 199
quinolone, 197
sulphonamide, 197
tetracycline, 198–199
trimethoprim, 197–198

vancomycin, 198

Q
Quantitative PCR (qPCR), 88, 247
Quinolones, 197, 316

R
Reversed-phase liquid chromatography (RPLC), 243
Rhodobacter capsulatus (RcGTA), 200
Risk assessment process
ARGs
abundance, 379, 380
GMOs, 380, 381
identification, 378–379
EMA (*see* European Medicines Agency (EMA))
environmental risks, 367
hazard and exposure, 368
human and veterinary pharmaceuticals, 368
risk quotien (RQ), 377, 378
16S rRNA gene analysis, 110

S
Smartphone, 249, 250
Soil antibiotics
absorption and fixation, 208
accumulation, 389
agriculture industries, 313
agroecosystems, 389
animal feed, 207
animal manures, 388
antibiotic resistance crisis, 314
ARGs, 389, 390
bacteria (*see* Bacteria)
biodegradation, 216, 217
biological activity, 387
bioremediation, 390–392
chemicals, 208, 209
ciprofloxacin and narasin, 389
and climate, 208
consumption, 314, 389
detection, environment, 388
dispersion, 322
farmyard manure (FYM), 208
fluoroquinolones, 322
industrial-scale animal feeding operations, 388
interaction
aminoglycosides, 316
cephalosporins, 316
degradation and fixation, 315
K_d value, 315
β-lactams, 316
macrolides, 316
oxytetracycline, 315
particles, 315
pH, 315
quinolones, 316
streptomycin, 315
sulfonamides, 315
tetracyclines, 315
Kd value, 208
livestock pharming, 388
macrolides, 314
medical and veterinary, 208
metabolites, 388
microorganism, 389
organic, 208
overuse and misuse, 388
oxytetracycline, 389
penicillin, 313
phytoremediation, 392
quantities, 387
runoff and leaching, 211
sorption (*see* Sorption of antibiotics in soil)
sources, 387
sulfonamides, 314
terrestrial and aquatic environments, 322
tetracyclin, 314
toxic effect, plants, 391
toxic effects, flora and fauna, 390
transformation, 314
transport, 210
uptake by plants
concentrations and toxicity, 214
dryand fresh weight, 216
growth, 215
loamy sand and sandy loam, 215
phytoremediation, 214
rates, 214
vegetables, 214
Soil biomass, 328
Soil bound antibiotics, 143
Soil mobility, VA, 58
Solar inactivation
biocidal effect, 124
disinfection, 125
efficacy, 126
inefficacy, 126
irradiance, 127
solar photoinactivation, 125
Sorption of antibiotics, soil

Sorption of antibiotics (*cont.*)
amines, carboxyls and hydroxyls, 212
aromaticity and electropolarity, 211
cation linking ability, 212
hydrophobic and hydrophilic moieties, 212
interaction, 211
particles, clay minerals, 212
rate, 212
sulfonamides, 212
surfaces, 212
toxicity, 213
Substrate-induced respiration (SIR), 288
Sulfamethoxazole effects, 332, 333
Sulfonamides, 32, 43, 197, 316
Superoxide dismutase (SOD), 328, 330

T
Tetracyclines
CAT, 330
efflux, 25–29
enzymatic, 30
GTP-binding sequences, 32
integrative and conjugative element, 198
mode of action, 317
mosaic, 30
OTC, 330, 331
p*K*a values, 43
ribosomal protection, 29, 30
SOD, 330
Toxic effects of soil antibiotics, earthworms, 333–335
acute and sub-acute, 337
ampicillin, 333
anti-microbiological substances, 328
aquatic ecosystem, 328
biochemical markers, 328
cefotaxime, 333
chemical contamination, 328
chloramphenicol, 333
consumption, 327
cycloheximide, 333
degradation, 337
digestion, animal body, 327
EFLX, 331
environmental threat, 327
functional genomic changes, 329
geno-and biochemical (*see* Genotoxicity and biochemical toxicity, earthworms)
irrigation, 331
long-term exposure, 337
metabolites, 327
microbial diseases, 327
neomycin, 333
nystatin, 333
oriental medicine, 328
OTC, 330, 331
positive effects, 328
prospects and challenges, 336
pyridoxal hydrochloride, 333
risk, 328
sensitive receptors, 328
soil biomass, 328
soil nutrient cycling, 328
standard toxicology test, 328
sulfamethoxazole, 332, 333
tetracyclines, 330
trimethoprim, 332, 333
types/classes, 329–333
Transduction process, 195
Transformation process, 194, 195
Tributylamine (TBA), 244
Trimethoprim, 197, 198, 332, 333

V
Vancomycin, 198
Vancomycin-resistant enterococci (VRE), 345
Vancomycin-resistant *Staphylococcus aureus* (VRSA), 345
Vesicular-arbuscular mycorrhizae (VAM), 271
Veterinary antibiotics (VA's)
animal production, 223, 224
antibiotic-mediated growth promotion, 68
antimicrobial agents, 55, 56
bacteria and fungi, 56
characteristics in soil, 57
control, 56
degradation rate, 57
direct and indirect method, 56
disease-fighting, 222
environment (*see* Environment)
fate, 227–229
fertilization, 67
growth-promoting capabilities, 222
herd/flock antibiotics, 56
intensive farming (*see* Intensive farming operations, VA)
livestock farming, 56, 68
physical and chemical properties, 57
pre-treatment methods, 68
quantity, 56
residual, 67
sciences, 56
semisynthetic and synthetic, 56

soil mobility, 58
surveillance, legal regulation and public awareness, 68
uses, 223, 224
water solubility, 58

W
Waste/manure storage, 65
Wastewater treatment plant (WWTP)
activated sludge process, 111
agricultural irrigation, 111
ARGs (*see* Antibacterial resistance genes (ARGs))
bacteriophages, 127–129
costs, 402
detection, *NDM-1* gene, 402
hospital and industrial, 403
incineration, 403
minimize spread of resistance, 402
NDM, 112–119
policy measures, 404
removal, 403
residues, 403
sanitation, 402
solar inactivation (*see* Solar inactivation)
thermophilic anaerobic sludge digestion, 402
Wastewater/sewage treatment plants, 65–66
Water
physiochemical properties, 229
wastewater and sewage treatment plants, 229
Water Solubility, VA, 58
Wi-Fi Interface, 251
Wild animals
antibiotic resistance
bacteria, 321
conservation programs, 321
disinfectants, 320
environments, 320
gastrointestinal microbiota, 321
microbiota, 321
migrant birds, 321
multiresistant organism, 321
natural ecosystems, 320
prevalence, fecal bacteria, 320
sludge and sewage system, 320
spread, 321
surface waters, 320
surveys, 321
wastewater treatment plants, 320

The manufacturer's authorised representative in the EU is Springer Nature Customer Service Centre GmbH, Europaplatz 3, 69115 Heidelberg, Germany. If you have any concerns regarding our products, please contact ProductSafety@springernature.com

Printed and bound by CPI Group (UK) Ltd, Croydon, CR0 4YY
01/10/2026
02225895-0004